AF612579

UNDER THE SUPERINTENDENCE OF THE SOCIETY FOR THE DIFFUSION OF USEFUL KNOWLEDGE.

THE

Companion to the Almanac;

OR

YEAR-BOOK

OF

GENERAL INFORMATION

FOR

1833.

CONTAINING,

INFORMATION CONNECTED WITH THE CALENDAR,

AND EXPLANATIONS OF

The Celestial Changes,

AND THE

NATURAL PHENOMENA OF THE YEAR, &c.;

GENERAL INFORMATION ON SUBJECTS OF

CHRONOLOGY, GEOGRAPHY, STATISTICS, &c.

THE LEGISLATION, STATISTICS, AND PUBLIC IMPROVEMENTS OF GREAT BRITAIN;

A CHRONICLE OF EVENTS FOR 1832;

TOGETHER WITH A

LIST OF THE NEW HOUSE OF COMMONS.

LONDON:
CHARLES KNIGHT, PALL-MALL EAST

PRICE TWO SHILLINGS AND SIXPENCE.

Figure 1 Title page from the 1833 Companion

WRITINGS

FOR

THE COMPANION TO THE BRITISH ALMANAC

1831 - 1857

BY

AUGUSTUS DE MORGAN

Compiled by Gary Menchen

2022
pageofpages.com
Waterville, ME

ISBN: 979-8-9855566-0-5 (paper)

For more information contact: editor@pagesofpages.com

From 1831 to 1857 Augustus De Morgan contributed essays to the *Companion to the Almanac,* a companion volume to the British Almanac published each year. The Companion was published under the imprint of Charles Knight through 1854, and subsequently by Knight and Company. They were produced "under the supervision" of the Society for the Diffusion of Useful Knowledge through 1846, when the Society was dissolved.

De Morgan was Professor of Mathematics at University College in London, an author, bibliographer, sometime secretary of the Royal Astronomical Society, first president of the London Mathematical Society, contributor to the publications of the Society for the Diffusion of Useful Knowledge, author of nearly one-sixth of the *Penny Cyclopaedia* (27 volumes, 1833-1843, and supplements), consulting actuary, formulator of De Morgan's Laws, campaigner for a decimal coinage, antiquarian, and a prolific contributor of book reviews for the *Athenaeum.*[1] De Morgan is best known for his contributions to logic, but he had wide interests, and is notable for his contributions to the history of mathematics and astronomy, and as a bibliographer and antiquarian. His contributions to the *Companion* reflect those interest.

Some essays are signed. The complete list is given in Sophia De Morgan's *Memoir of Augustus De Morgan* (London, 1882), pages 405-406. There exist cases of the individual essays being extracted from the Companion volumes and bound as a collection. Several are catalogued in American libraries[2].

A.N.L. Munby, in "The History of Bibliography of Science in England", a lecture given in 1968 as the Eighth Annual Zeitlin & Ver Brugge Lecture in Bibliography, makes mention of De Morgan's writings for the *Companion.*

> Between 1831 and 1857 he wrote twenty-five papers, a great many of them bibliographical, for the Companion to the [British] Almanac, an annual publication. The total obscurity of this series of essays deserves illumination and I think that a group of them would repay the trouble of collecting them together in a reprint. Even in his twenties De Morgan displays a profound knowledge of the by-ways of his subject. In 1836, for example, he published an article on "Old Arguments against the Motion of the Earth." This was mainly a translation of an anti-Copernican dialogue of Thomas Fienus in his Disputatio an coelum moveatur et terra quiescat, first printed in 1619, but reprinted in London as late as

[1] pagesofpages.com expects to publish Volume One of complete text of De Morgan's Athenaeum book reviews in 2022.

[2] See WorldCat OCLC no. 13477682.

> 1670, and thereby providing a basis for De Morgan's discussion of seventeenth century scientists who still clung to a pre-Copernican conception of the universe. His 1837 article was on a larger scale, "Notices of English Mathematical and Astronomical Writers between the Norman Conquest and the year 1600," much concerned, of course, with writers such as Recorde, Dee, Leonard and Thomas Digges, and the demonstration that the first of these was a Copernican. De Morgan had complete mastery of the earlier writers in his field, best seen in his article contributed to the Companion of 1843, "References for the History of the Mathematical Sciences," twenty-five pages, in minute type, of packed bibliographical information. I would like to draw attention to two other articles in this series, "On the Earliest Printed Almanacs" (1846), a long account of a subject never treated on this scale before, and, particularly, "On the Difficulty of Correct Description of Books" (1852), an essay which grapples with bibliographical problems hitherto unrecognised and indeed not seriously tackled by anyone else until another three decades had elapsed.[3]

No one appears to have taken the "trouble of collecting them together in a reprint." prior to now. Anyone who wonders at "Between 1831 and 1857 he wrote twenty-five papers", particularly after noticing that this volume's table of contents includes entries for both 1831 and 1857 and has no missing years, should read the article for 1850, "On ancient and modern use in reckoning". The quantity of 25 perhaps originated in Sophia De Morgan's *Memoir*, p. 25.

The essays do not all fall neatly into separate categories, but they can be roughly grouped as follows. There are entries not explicitly under bibliography that have a bibliographical component.

Bibliography

- 1837 Notices of English mathematical and astronomical writers between the Norman conquest and 1600
- 1843 References for the history of the mathematical sciences
- 1846 On the earliest printed almanacs
- 1853 On the difficulty of correct description of books

[3] Quoted from "Jeremy Norman's HistoryofScience.com" at https://www.historyofscience.com/articles/munby-history-biblio-science-england.php

Historical

- 1845 On the ecclesiastical calendar
- 1848 Short supplementary remarks on the first six books of Euclid's geometry
- 1850 On ancient and modern usage in reckoning[4]
- 1851 On some points in the history of arithmetic
- 1852 A short account of some recent discoveries in England and Germany relative to the controversy on the invention of fluxions

The actuarial basis for life assurance or annuities

- 1831 Life assurances
- 1840 On the calculation of single life contingencies
- 1842 On life contingencies, no. 2.

Astronomy

- 1832 Eclipses
- 1833 On comets
- 1834 On the moon's orbit
- 1835 Halley's comet
- 1836 Old arguments against the motion of the earth
- 1838 On Cavendish's experiment
- 1847 Recurrence of eclipses and full moons
- 1855 The progress of the doctrine of the earth's motion between the times of Copernicus and Galileo; being notes on the antegalilean Copernicans.

Computation

- 1839 Notices of the progress of the problem of evolution[5]
- 1841 On the use of small tables of logarithms in commercial calculations, and on the practicability of a decimal coinage
- 1844 On arithmetical computation

---

[4] On reckoning *time*. De Morgan's mastery of sources in this essay is staggering.

[5] Evolution was the name given to a means of extracting the roots of numbers.

Coinage

- 1841 (Second part: on the practicability of a decimal coinage
- 1848 On decimal coinage
- 1854 On a decimal coinage
- 1856 Notes on the history of the English coinage
- 1857 Notes on the state of the decimal coinage system

The original volumes that these texts are copied from are all readily available on-line at sites such as The Internet Archive (https://archive.org/details/pub_british-almanac-companion). The purpose of this volume is to collect De Morgan's essays in one convenient location, clean up the reproductions, and make them more accessible.

On each page of reproduced text the original pagination is at the top; refer to the table of contents to identify the year. The pagination for this volume is printed at the bottom center of each page. The table of contents contains the complete chronological list.

.

[6] The volume pagination is at the bottom of each page.

XIV. LIFE ASSURANCE.

Life Assurance companies may be placed amongst the most useful institutions of modern times, and their increase shews that their value is becoming duly appreciated. Still the nature of them is very imperfectly understood by the generality of that class whose peculiar interest it is to have a correct knowledge on the subject. It is of consequence to those of moderate income, that every facility should be afforded them for acquiring information, by which they may readily ascertain how the advantages which this system offers may be obtained in the best and cheapest manner.

A Table, shewing the respective premiums of the Assurance Offices in London, with a short account of the relative advantages of each, will be found useful as a reference to those who wish to assure their lives. The subjoined Table shews immediately the lowest rate of premium at each age; which, with a summary of the different plans pursued at each office, will enable any one, without difficulty, to select that which will best suit his wishes.

There are three classes of assurance companies. Those of the first class are joint stock companies, which assure to a person paying a fixed annual premium the payment of a fixed sum at his death. Those of the second class are companies also, but, instead of dividing the whole of their profits among their proprietary, they allot a certain proportion of these to the assured. Those of the third class divide the whole of their profits among the assured, after deducting the expenses of management, and reserving an accumulating fund. No observation is necessary on the first class; the companies of this differ only in their rates of premiums, which may be ascertained by a reference to the Table. The whole of the profits are divided among the proprietors, without the assurers receiving the slightest benefit from them.

The companies of the second class divide, or propose to divide among the parties assured, a certain proportion of profits as under:—

Established.	Names.		
1714	Union,	Afford no statement of the division of profits which they divide.	Divide every seven years.
1806	Provident,		
1808	Atlas,		
1819	European,		
1820	B. Commercial,		
1821	Guardian,		
1806	Rock,	Divide two-thirds of their profits.	
1807	Hope,		
1807	Crown,		
1824	Palladium,	Divide four-fifths of their profits.	
1807	Eagle,		
1823	Law,		
	Alliance . .	Affords no statement of division of profits to be divided.	
	Clerical and Medical,	a proportional part, after paying 3 or 5 per cent. to the proprietors.	Every five years.
	University, . .	Four-fifths, after 5 per cent. has been paid to the proprietors.	
1824	Economic, . .	Three-fourths profit divided.	
	Imperial, . .	Two-thirds profit divided, time of division not specified.	
1830	National, . .	Two fifteenths profit divided annually, which goes to the reduction of premiums.	

It is evident, however, that the profits proposed, or promised by companies having a proprietary, are very uncertain. In the first place, the manner of estimating profits differs in different societies; and in the second place, the expense of management, and the advantages reserved for proprietors, vary in all. In this uncertainty, the only criterion that can be given, is a statement of the proportion of profits which has been actually assigned to the assurers by each of the companies who have published the account.

The Rock has added to policies effected on or before 1806; first in 1819, 24*l.* per cent. decreasing two per cent. yearly on all policies effected after 1806 till 1817; and in 1826, a still further bonus of 19*l.* per cent. on the amount of their policies in a decreasing ratio of one per cent. to 1824. So that one assured previously to 1807, has an addition of 24*l.* + 19 = 43. One assured in 1807 22*l.* + 18 = 40, in 1808, 20*l.* + 17 = 37, &c. added to the original 100*l.* assured.

The first division of the profits of the Provident was 13*l.* 8*s.* per cent. on the premiums paid; the second 26*l.* 12*s.* per cent.; and the last in 1827, gave 30*l.* 10*s.* So that if a man aged forty paying a premium of 3*l.* 7*s.* 11*d.*, assured his life in 1806, the sums added at the three different periods to his policy, will be found by calculation to have amounted, in 1827, to 37*l.* 11*s.* 5*d.* per cent.

The Atlas, in 1823, added to 1000*l.* policies, effected at the respective ages of

	£.		£.		£.
20	129	35	139	50	178
25	132	40	149	55	196
30	135	45	160	60	219

These advantages may be applied at the option of the assurer in equivalent reductions of the future payment of premiums, instead of being added to their policies.

The Guardian made its first division of profits in December, 1828, when additions were allotted to the different policies, amounting to rather more than one per cent. per annum on the sums assured by such policies. So that a person having assured 100*l.* for seven years would have an addition of 7*l.* to his policy at the end of that time.

The prospectuses of the other companies in this class furnish no information of the sums which they have respectively allotted to policies effected at their offices. The assurance companies of the third class have no proprietary; but the manner and time of division, the mode of calculating profits, and the proportion reserved for accumulation, appear to differ in all.

Established.	Names.	Periods of division.
1706	Amicable, seven-eighths divided among assured	Yearly.
1762	Equitable	Every ten years.
	Norwich Union	Seven years.
1806	London Life Association, one-fifth profit divided	Every year.

The Amicable office apportions a sum equal to the average annual payment received by the society during the last five years, to be divided among those who die in every year. The executors of an assurer is *secure* of receiving the sum assured, and to this there is always some addition. In 1828, the claims for 200*l.* were paid an additional sum of 25*l.* 19*s.* per cent. Here the young and old, those who have assured for a length of years, and those who have just become members, are all put on an equal footing.

The Equitable Society has of late added three per cent. per annum on the sum assured from the date of their policies in decennial periods, only among the 5000* members who have been longest assured. Those beyond that number must wait an indefinite period, before they are admitted into this favoured class, and allowed to participate in profits.

The Norwich Union adds the whole of the surplus premiums at stated periods to the policies of the members, in proportion to the sums they have respectively contributed; on which principle a bonus was declared in 1816 of 20 per cent. and in 1823 of 24 per cent. upon such contributions. So that a man aged forty assuring his life at this office in 1809, paying a premium annually of 3*l.* 2*s.*, would, at the end of 1823, have an addition to his policy of 14*l.* 15*s.* 1*d.*

The London Life Association now diminishes the premiums of the assured after they have been assured seven years†. An account is taken annually of the state of the concern, and one-fifth of the apparent profits, or of that portion of the assets, which is over and above the present value of the engagements of the office, is divided among the members; the remaining four-fifths are reserved for contingencies and full security. This fifth goes to the reduction of premiums, which, it is said by the office, will always be fifty-five per cent. or *more;* so that a person paying for seven years 100*l.* per cent. would after that period only have to pay 45*l.* per annum for the same policy.

The National, which belongs to the second class, is formed on the same plan, with some modifications. After being assured five years, the first thousand of the assured are to participate in the profits, in the same manner as the members of the London Life Association, but in a different proportion, one-third of the fifth being appropriated to the proprietors, the assurers getting the remaining $\frac{2}{3}$ of $\frac{1}{5} = \frac{2}{15}$ of the profits.

The Economic and University are at present joint stock companies; but they propose to pay off the proprietary, and thenceforward they will be on the plan of mutual assurance societies. The first, when 200,000*l.* profit shall be realized, and the second, when the proprietors can withdraw their capital with 100*l.* per cent. profit; one-tenth of the profits of the company is set apart as a fund to be applied to this purpose.

The following Table will shew the comparative merits of the London Life Association with those offices which take the lowest premiums, and make no returns—assuming that the former reduces the premiums of one assuring for seven years, 55 per cent.

* This limitation was made in 1816.

† This is a new regulation, very recently passed; five years having previously been the term.

after that period; but this assumption is probably very much below the average rate of reduction.

1st. Age.	2d. Premiums paid per annum for 7 years.	3d. Premiums paid after that time.	4th. Equalized Premium.	5th. Lowest Prem. charged at any of the offices.	6th. Periods when the Premiums of the London Life Association will be equalized to the same.
	£. s. d.	£. s. d.	£. s. d.	£. s. d.	
30	2 19 0	1 6 6¼	1 17 6	2 0 5	In nearly 20 years.
40	3 15 0	1 13 9	2 9 5	2 15 10	Rather more than 14years.
50	5 4 0	2 6 9¼	3 13 3	4 0 8	Rather more than 13years.
60	7 8 0	3 6 7	5 14 10	5 14 9	In nearly 13 years.

Here the first column is the age of the assurer when he begins assuring his life; the second, the annual premium he has to pay for seven years; the third, the annual premium when 55 per cent. is deducted; the fourth the high premium for seven years, and the low premium for the remainder of life equalized; that is to say, 1*l.* 17*s.* 6*d.* paid for the whole life annually by a man aged thirty, is equivalent to paying 2*l.* 19*s.* for seven years, and then 1*l.* 6*s.* 6½*d.* for the remainder of life; reckoning, interest at 3½ per cent., and taking the probabilities of life from the government Annuity Tables. The sixth column is the number of years a person must assure at the respective ages, before he has any advantage over those offices whose premiums are quoted in the fifth column.

If a man forty years of age assured at the Rock Office in 1806, at the end of twenty-four years, the probability of life at that age, his policy would be increased 43*l.* per cent.; an annuity at 3½ per cent. interest, which would amount to 43*l.* in twenty-four years, would be 1*l.* 3*s.* 5*d.*; which deducted from 3*l.* 7*s.* 11*d.*, the premium at the Rock for the age of forty, gives 2*l.* 4*s.* 6*d.*; therefore, if this bonus of 43*l.* per cent. had gone to the reduction of premiums, it would have been equal to paying 2*l.* 4*s.* 6*d.* annually, instead of 3*l.* 7*s.* 11*d.* The equalized premium of the London Life Association is at that age 2*l.* 9*s.* 5*d.*; but here the most favourable circumstances are assumed in the former, while in the latter office, the calculations are made, supposing the premiums are reduced only 55*l.* per cent. after seven years.

The experience of assurers at this society has, till now, been much more favourable; after five years they participated in the profits, and their premiums have been reduced in some years, as much as 70*l.* per cent.; there is every reason to suppose, that they may be lessened as much in future, and then this office will certainly continue to be the cheapest under examination, while this manner of applying the profits is the most equitable, and the most advantageous to the parties assured. It is an inexpressible relief, after using every exertion to make the first heavy payments, to be at length freed so very materially from the burthen*.

* It may here be observed, however, that different assurers have different objects. For those who wish to begin at an early period of life, and whose prospects are that every year will make the payments more easy, those offices are best whose premiums are *at first* the lowest. For those whose incomes are likely to decrease, the offices which divide profits are best. To those who have a fixed income for life, those are best which, *on the whole*, are the cheapest, on whatever plan they are framed.

The duty of life assurance is becoming so universally acknowledged, and the advantages accruing from it so well known, that it is not in the plan of this article to dilate on them at large. In the middle ranks of society, there are now but few whose means of supporting their families depend on their own personal exertions, or are derived from income which ceases at their death, who do not gladly seize upon the power afforded them, of leaving some provision for the future maintenance of their surviving dependents.

Previously to these institutions being so general, the man who could save a very small portion of a very limited income, despaired of ever accumulating sufficient for any pittance for his children, and he became reckless of attaining so apparently hopeless an object; the trouble and difficulty of investing very small sums yearly, and of obtaining accumulated interest, were almost sufficient preventives to deter even a man anxiously desirous for the future welfare of his family, from the endeavour of acquiring, by small savings, anything which he could reasonably hope would, though fostered through length of years, swell into a patrimony for his children. But now the more than probability, the certainty is his. The system of life assurance incites all to the moral obligation of exercising foresight and prudence; since through its means these virtues may be successfully practised, and their ultimate reward secured. If a man's income be sufficient to allow of his setting apart somewhat considerable for an accumulating fund, it is still no slight advantage, if we consider the constitution of human nature, that he can put it out of his power to encroach on his accumulations. By laying out an annual sum in assuring his life, he cannot, without difficulty and considerable loss, apply to his present uses his past savings, while there is little fear that he will be so improvident as to forfeit his testamentary claim to them, by neglecting to make his periodic payments.

These benefits accrue to the individual; but assurance companies are likewise highly beneficial to society at large, inasmuch as they are the means of gradually accumulating productive capital. While the annual premiums are considered as a part of expenditure, they and the continually growing amount of interest on them are so much added to the productive capital of the community.

It may not, perhaps, be useless, briefly to exhibit the comparative effects of putting by annual savings, and allowing them to accumulate, or of expending them in a life assurance.

A man with a moderate income may feel that, in a few years, he shall have saved sufficient to leave a competence for his family; but he feels likewise, that at any moment of time he may be snatched from them, ere the proposed sum is amassed; he therefore has recourse to an assurance on his life, and "renders that certain which nature has made uncertain." If at the age of twenty-five, he could save from his income 24*l.* per annum, it would be twenty-six years before his savings, laid out at 3½ per cent. interest, would amount to 1000*l.* But if he employed this annual saving in assuring his life, at any even of those offices which demand the highest premiums, the *hoard* is already his to leave in reversion, and the twenty-six years of the best period of his life are not on this account wasted in anxiety and care. The anticipation of future evil no longer robs him of present enjoyment; by an annual fixed payment he is secure of leaving a fixed sum at his death; and he does not feel it his harassing and comfort-destroying duty to save to the utmost from his present income. All that is required of him is, carefully and punctually to supply the annual tribute which secures to him so invaluable a blessing. Should he live beyond the period at which his savings would have accumulated to the sum assured, he will not be disposed to repine at a bargain, the improvidence of which is caused by his continuation of life, and if he take into the account the exemption from corroding solicitude through so many years, he will think it is cheaply purchased.

It is most desirable, however, that so important a benefit should be obtained at something approximating to its real, not its moral value; which would then still farther extend its beneficial effects, either by enabling the assurer to increase the future provision for his family, or by allowing him, when his means are very limited, to avail himself of it without too much curtailing his present comforts. Experience and calculation shew, that the rates of premiums for assuring life might be very much reduced with perfect security to the assured, and with reasonable profit to the proprietary.

In all tables which have been constructed to exhibit the probabilities of human life, the calculations have been made on a certain number of beings taken indiscriminately from all classes of the community. It is evident that, in such tables, the average duration of life must be shortened, by including the working poor; some prematurely worn out by labour, others wasting life in unhealthy occupations, many dying from neglected disease and scanty nourishment. These of necessity form no part of those whose deaths affect the assurance offices. Those who assure their lives, are generally the healthiest of the most healthy class, the greater proportion of whom are under the most favourable circumstances for longevity. If any labour under disease, they are rejected, or must seek by a much higher payment to obtain equal advantages*. Those who *have not* had the small-pox, nor have been vaccinated, and those who *have* had the gout, are alike obliged to purchase, by an enhanced premium, immunity from this negative and this positive evil†.

It appears, then, that if the rates of premiums are calculated from the probabilities of life of the whole mass of the people, they would leave enormous profits. Some of the offices, however, not

* It is proper to state, as a qualification of this principle, that the Carlisle Tables of Mortality and those founded on the experience of the Equitable Office, do not materially vary; and, to account for this, it ought to be borne in mind that while, on the one hand, that class of the community amongst whom the average duration of life is shortened by habitual privations, are not amongst insurers; on the other hand, many persons apply to insurance offices with fraudulent intentions, which the vigilance of a medical man is often unable to detect.

† At the Equitable Society, the premium is increased 11 per cent.; at the same office, persons who have been vaccinated are admitted on the usual terms.

content with this, add a per centage profit on the rates thus deduced, and all of them in their calculations reckon the interest of money at 3 per cent., much lower than the rate of interest at which they actually employ their capital. From these causes, those who assure their lives are obliged to pay much more for the benefit to be obtained than they ought. The premiums demanded by some offices are disproportionably higher than others, and one is at a loss to understand why any should be willing to pay 7*l.* 14*s.* 11*d.* per annum for an advantage which he can equally well acquire by paying 5*l.* 14*s.* 9*d.* per annum. The Assurance Offices make large profits, of which the Equitable Society is an example; in fifty-seven years it accumulated six millions, and it is understood that its capital is at the present time about ten millions and a half, valuing the 3 per cents. at 90; while the very large bonuses assigned to policies prove that no inconsiderable part of this capital is profit.

This office, up to the year 1820, added to a policy of

20 years date 77 per cent.
30 161
40 280
50 401

and one member, aged 90 years, has 497*l.* per cent. added to his policy.

The probabilities of life, taken from the Government Annuity Tables lately published, are at the respective ages as under; whence annuities at each period are calculated, interest being reckoned at 3½ per cent. To these are subjoined, in the following Table, the highest and lowest premiums, together with the premiums from Mr. Babbage's Table at the respective ages; thus a comparative view may be clearly taken of the whole.

Age.	Expectation of life.	Annuity which would amount to 100*l.* in that time.	Highest Premium demanded by the offices for 100*l.* Policy.	Lowest Premium demanded by the Offices for 100*l.* Policy.	Premiums calculated just to be equal to the rise.
		£ s. d.	£ s. d.	£ s. d.	£ s. d.
20	32	1 14 11	2 3 7	1 10 7	1 9 6
30	29	2 0 11	2 13 5	2 0 5	1 18 6
40	24	2 14 6½	3 8 0	2 15 10	2 10 9
50	18	4 1 7¾	4 14 2	4 0 8	3 11 0
60	13	6 4 1½	7 14 11	5 14 9	5 1 3

The highest premiums taken at the respective

Ages of 20, 30, 40, 50, 60 amount to 100*l.* in 27 years and a little more, 24, 20½, 16, 10½.

The lowest premium at 60 years of age will amount to 100*l.* in fourteen years; the lowest premiums at the other ages are so similar to annuity, that it is unnecessary to go into any calculations concerning them.

It will clearly appear, from what has been said, that the rate of probability assumed in these calculations is much too high, and yet the premiums required are, with the exception of three offices, much higher. Mr. Morgan, in his address at the meeting of the Equitable Society in 1816, stated as the result of an experience of forty-five years, that not one in sixty of those assured at that office died annually. If twelve lives, at each of the respective ages of 20, 30, 40, 50, and 60, be taken, and it is assumed that one dies every year, till at the end of sixty years the whole are dead, an assumption greater than the fact of the Equitable experience, then

$$(\text{£}2\ 3s.\ 7d. + \text{£}2\ 13s.\ 5d. + \text{£}3\ 7s.\ 11d. + \text{£}4\ 10s.\ 8d. + \text{£}6\ 7s.\ 8d.) \times 12 = \text{£}229\ 15s.$$

will be equal to the amount of the first annual premiums paid by these sixty persons; to which, if 8*l.* be added for interest on the sum paid, and 100*l.* be deducted for the one who dies, a profit is left at the end of the year of 137*l.* 15*s.* It is not necessary to go into a rigorous calculation on an hypothesis, but by approximating a series, and taking 229*l.* 15*s.* as the first term of a decreasing arithmetical ratio, whose common difference is four, for the premiums paid in each year, it is found that, at the end of sixty years, after paying annually 100*l.* for the annual death, there will be left much more than 8000*l.* clear profit on sixty lives, each insuring only 100*l.**

In the address at the meeting of the same society in 1801, Mr. Morgan furnishes data, by which it is shewn how very much the probabilities of life are under-rated by the offices. The deaths which actually took place, compared to those in the Table of mortality, by which the premiums were calculated, have been from the ages of

10 to 20 as 1 to 2	40 to 50 as 3 to 5
20 to 30 as 1 to 2	60 to 60 as 5 to 7
30 to 40 as 3 to 5	60 to 80 as 4 to 5

Mr. Babbage, in his excellent and valuable work on assurance of lives, has by means of this Equitable experience, formed the following Table of the rates of premiums, which would provide for the payment of the sum assured, without leaving any profit to the office, computing interest of money at 3 per cent. per annum:—

Age.	Premium.			Age.	Premium.			Age.	Premium.			Age.	Premium.			Age.	Premium.		
	£	s.	d.		£	s.	d.		£	s.	d.		£	s.	d.		£	s.	d.
8 to 14	1	5	1	25	1	13	9	36	2	5	3	47	3	3	11	58	4	14	1
15	1	5	9	26	1	14	8	37	2	6	6	48	3	6	3	59	4	17	7
16	1	6	6	27	1	15	7	38	2	7	10	49	3	8	7	60	5	1	3
17	1	7	2	28	1	16	7	39	2	9	3	50	3	11	0	61	5	5	2
18	1	7	11	29	1	17	6	40	2	10	9	51	3	13	6	62	5	9	5
19	1	8	9	30	1	18	6	41	2	12	4	52	3	16	1	63	5	13	11
20	1	9	6	31	1	19	7	42	2	14	0	53	3	18	10	64	5	18	9
21	1	10	4	32	2	0	8	43	2	15	9	54	4	1	7	65	6	3	11
22	1	11	2	33	2	1	9	44	2	17	8	55	4	4	7	66	6	9	6
23	1	12	0	34	2	2	11	45	2	19	8	56	4	7	7	67	6	15	6
24	1	12	11	35	2	4	1	46	3	1	9	57	4	10	9				

* It is right, however, to state that this is somewhat hypothetical reasoning; and that Mr. Morgan does not agree with the conclusions drawn by different authors with regard to the profits of the Equitable. He says that the profits derived from

Whence Mr. Babbage thinks it may be fairly stated, that those offices which calculate their premiums according to the Northampton Tables of mortality, make a gross profit of 30 per cent., without including the very large additional profit that arises from the average rate of interest being above 3 per cent., and from those premiums which are paid only for a short period, and then are discontinued, leaving them as entire profit to the society.

These remarks on the excessive premiums for assuring life must not be understood as dissuading from the performance of that duty. On the contrary, it is certainly much better to pay something more than its value for this benefit than not to acquire it at all. It should be considered as a moral necessary of life, which, where the possibility exists, *must* be obtained at any price.

However large the list of companies appears, there is still so vast a field for action, that no doubt the wants of the community will progressively call for the establishment of new societies; it may, therefore, perhaps, be useful to inquire into the principles upon which the formation of these valuable institutions should be founded.

A company of mutual assurers must certainly be the cheapest way of assuring life, but there are two difficulties attending this plan. The want of capital to answer the demands of the society, before sufficient funds are formed from the profits, and the most equitable manner of apportioning the surplus money to be divided. The projectors of the London Life Association, with a disinterested generosity very rarely to be found, subscribed a guarantee capital among themselves, until the funds of the society were sufficiently large to enable them to withdraw it, which they did without deriving any profit or advantage from the loan or responsibility. Such conduct is not to be expected often in our money-making mercantile country. At that time, this advance of capital was supposed to be attended with positive risk; and although twenty-four years longer experience of all the assurance companies has now proved, that the capital required is merely nominal—that the annual premiums have been found, from the very beginning, quite sufficient to answer all the demands made on the funds of the society,—yet, even now, few would venture to assure their lives at a new office, however respectably and ably conducted, without some commencing capital beyond the collective premiums sub-

calculating the premiums by the Northampton rate of mortality are not so great as they have represented them; and when we look at the profits that have been shared by the members of the Equitable, it must not be forgotten that a great portion of them have been obtained from a source which no prudent office would ever think of calculating upon, viz., investing their premiums at a low price of stocks, thus obtaining good interest for their capital, and these stocks afterwards rising, by which the value of their capital was proportionably increased. In addition to this advantage, the Equitable originally charged much higher premiums than at present; a great number of policies have also fallen in from parties neglecting to pay their premiums, or surrendering them on terms advantageous to the office. It may be expected that a rate of mortality so favourable as that which has taken place among the members of the Equitable will not be experienced by all Insurance Offices, many of them, for the convenience of persons residing in the country, who are desirous of insuring their lives, having appointed agents, on whose good judgment and integrity they must depend, whose interest it is not to be *too particular* in the selection of lives, a circumstance which is not unlikely to have a material influence on the rate of mortality.

scribed, or responsible guarantee, that the sum assured would be paid in the event of their death. This guarantee must, therefore, be obtained in the formation of a new society; and it is but reasonable to suppose, that those who give it will expect some equivalent advantage.

It is found, from observation extended through many years, that the decrement of life varies very little. However uncertain the duration of individual life, this uncertainty does not extend to an aggregate multitude of individuals. The uniformity in the number of deaths in a community is remarkable; the excess or diminution, in any one year, rarely exceeds above or below the average number a small fractional part of the whole—not more than one-thirteenth, or one-fifteenth part. If from this community are excluded the aged, the infants, and that portion of a population which is most exposed to the casual effects of disease and want, the variations from the mean number of deaths will be still less. And it is always found that the variation is on the side of longevity, it being a fact, that for the last fifty years, the rate of mortality has been very gradually, but progressively lessening. The guarantee required will, therefore, be merely wanted to satisfy the fears of those who have not inquired into the subject, and who cannot understand how the aggregate number of small premiums will quickly increase into a fund amply sufficient to meet all the engagements of the society.

If the premiums were as low as they should be, the profits would not be very great; yet, for the proper stability of the company, there must always be a surplus capital. How, then, and what portion of this is to be divided?

The inconvenience of inadequate plans has, it appears, been felt by those offices which have been long enough established to accumulate large profits. The Equitable, not choosing to allow the newly assured to participate in its immense accumulations, suddenly shut the door on new claimants. Happy was he who had just passed the threshold, while he who lingered on the way, found himself excluded from those great advantages, which the more fortunate obtained by a few days priority. A person assuring at the latter end of 1816 participated in the division made of the profits in 1820 and 1830; while he who began to assure his life at the beginning of 1817 did not receive any benefit from these divisions of profits, and may by possibility find himself excluded from the division of 1840.*

The London Life Association was likewise unwilling, as its profits increased, to share them with new subscribers; a few years

* The Asylum office proposes to secure the certainty of this profit, at stated premiums, to those who assure at the Equitable Society. There are very ingenious calculations on the subject to be found in the prospectus of this office. Its conductors appear, likewise, to have taken much trouble in ascertaining the rates of premiums on those lives which are beyond the limits of ordinary assurance. They have fixed rates for those about to reside in various countries, and for those who, from ill health or any other circumstances, are rejected by other offices. This office is here put in the first class, since in its prospectus there is no mention made of division of profits, but in a list of assurance companies given by an unquestionable authority, it is said to divide its profits every five years.

back its premiums were, in consequence, raised 10 per cent. to those about assuring their lives; this the old members still found was not adequate to the advantages of being admitted into their society; and recently they have added two years to the probationary time, making seven years before those who assure are allowed to participate in the profits already accumulated. Seven years are a weary time to pay very heavy premiums, before any advantages can be obtained from these enhanced rates. Yet, certainly, there is no justice in allowing new assurers the profit arising from the premiums paid by old contributors.

The National Association is desirous of guarding against this difficulty, by limiting the number of participators to a thousand, with power vested in the company to extend that number according to its discretion; how far this provision will remedy the evil remains to be proved, but it has the objection of rendering the advantages of assuring at this society uncertain. Now it is of the greatest importance that these advantages should be as determinate as the nature of the subject will allow.

The principal points to be considered in the formation of an assurance company are,—1st. Capital or responsible guarantee;—2ndly, The rates of premiums to be demanded;—3dly, The proportion of profits to be assigned to the assured;—4th and 5thly, The manner and periods of dividing such profits.

1st. *Capital.*—It is not, of course, necessary to dwell for a moment on the high consequence to the public, as well as to their own success, that these companies should be established by none but persons of the first respectability, who should possess efficient ability to conduct the business of the society judiciously, and recognised responsibility to inspire confidence in those assuring their lives. Twelve such, uniting to effect this purpose, might each guarantee a certain sum—say 2000*l.*, rendering them collectively responsible for 24,000*l.*; for this, and for the trouble of forming the infant society, they might be sufficiently remunerated by receiving fees as directors; it remaining for future consideration, whether these fees should be lessened when the increasing funds of the society should render a guarantee no longer necessary.

2d. *Rate of Premiums.*—The premiums might be those given in Mr. Babbage's Table, enhanced in a certain proportion. An additional fifth of the premiums, it is presumed, would be found amply sufficient, not only to raise an adequate capital, and to pay the expenses of management, but also, in a few years, to allow of dividing some surplus among the subscribers. To give some idea of the operation of forming capital and profit, let there be taken 200 lives, at each of the respective ages of 20, 30, 40, 50, 60, making in all 1000, and assuming that each of these assures his life for 100*l.*, and that sixteen out of these die annually, which is, perhaps, taking it beyond the usual rate of mortality in these societies. Then calculating by the rates of premiums given in Mr. Babbage's Tables, taking for granted that the annual sums paid will be just equal to pay off the policies of those who die in the respective years, and to form an accumulating fund to meet future demands; so that when the last of the thousand die, there will be just 100*l.* left to pay this claim—the additional fifth on these premiums will form an accumulating fund, and the whole will stand thus:—

£	s.	d.			
1	9	6	Premium at	20	years old.
1	18	6	"	30	"
2	10	9	"	40	"
3	11	0	"	50	"
5	1	5	"	60	"
14	11	2	× 200 = 2911*l.* 0*s.* 4*d.*		

the average rate being nearly 3*l.* Then the premiums at the end of the first year, increased by the interest on them, would amount to 2998*l.*; and the payments to be made on account of sixteen deaths being deducted, would leave 1398*l.* At the commencement of the second year, the payments on the 984 remaining lives would be 2863*l.*, which added to the 1398*l.*, would produce 4261*l.*, which at the end of the year would amount to 4388*l.*, from which 1600*l.* being deducted on account of the sixteen deaths, it would leave 2788*l.* Pursuing this calculation in the same manner, the capital of the society, which would just be sufficient for its liabilities, would be as follows:—

	£		£	
At the end of the first year,...	1398,	after paying	1600	in claims.
———— second year,	2788,	"	1600	
———— third year,..	4171,	"	1600	
———— fourth year,	5546,	"	1600	
———— fifth year, ..	6912,	"	1600	

The fifth of the premiums paid over and above those given in the table, with the accumulations on it, would be clear profit. This fifth will be, on 2911*l.*, nearly 600*l.*, which would thus increase:—At the end of the first year it would amount to 618*l.*, the interest being added; this being added to 590*l.*, amount of profits paid by the 984 survivors, with interest on both sums, would be the amount of profits at the end of the second year. This would stand thus at the end of each year:—

		£
First year,	amount of profits,	618
Second year,	do.	1244
Third year,	do.	1880
Fourth year,	do.	2524
Fifth year,	do.	3178

There would, therefore, be at the end of five years 3178*l.* profit over and above the 6912*l.*, which is equal to meet the liabilities of the company.

It is found by the experience of the offices that the average sum assured by each individual is more than 1000*l.* If, then, the value of the policies of the thousand persons assumed above were increased tenfold, the profits would be 31,780*l.* on a rental of

F

33,110*l.*: this would be ample to meet all the engagements, to pay the expenses of management, and would still leave a surplus to be divided among the members. Besides this manner of acquiring profits, a very large source of profit is obtained by improving the capital at a higher rate of interest than 3 per cent. per ann., the rate which has been assumed in the foregoing calculations. There is also to be considered that portion of premiums which is paid for any particular purpose by individuals for a short time, and then discontinued: this the Equitable Society has found to be very considerable indeed. As a company increased its rental, the proportion of profits taken for management and incidental expenses would of course be relatively much less. It may, therefore, be reasonably expected that in the course of a few years there would be sufficient surplus profit, at even these reduced premiums, to sanction their still further material reduction.

3d. *Proportion of Profits to be assigned to the Assured.*—This appears to be at present taken in a very unscientific manner. Why is so large a proportion put aside?—for contingencies, it is said, which may occur to the liabilities of the society. Then surely this part of profits should bear some relative proportion to those liabilities, instead of being allowed to accumulate without rule or limit. It has been shewn how unlikely it is, that a surplus capital will ever be required, provided the estimate of liabilities be taken accurately. It would be very unwise, however, not to make some extra provision for this purpose. A per centage on the estimated capital required, would be, perhaps, the most rational way of retaining profits. Prudence and experience could regulate the rate of per centage; 10 per cent. might, perhaps, be deemed sufficient: so that if the liabilities of a society, in any given year, be estimated at 69,120*l.*; then 69,120*l.* + 6912*l.* = 76,032*l.* might be considered as efficient capital for that year, and the rest of the funds, after the necessary reductions, might with safety be divided among the assured. Then there could by no possibility be overgrown and disproportionate accumulations—these would always bear some relation to the demands on the society, and the early members would reap the benefit in an equitable proportion with those assuring at a later period.

4th. *Manner of Division of Profits.*—The first subscribers to an infant society are usually placed on a much more unfavourable footing than those who become members after it is well established and flourishing. There is, therefore, both justice and policy in holding out some countervailing advantages to those first subscribing, which they shall always retain over members admitted afterwards. It might, perhaps, be eligible to allow those who subscribe in the first, second, third, fourth and fifth years, a greater proportion of profits than those subscribing after that period. This could be so arranged: at the end of the fifth year, when the profits are estimated and first divided, the surplus of those accruing from the first four years might be allotted to the early members: persons having assured in the first year should have the largest proportion relative to their numbers, and so on through the four succeeding years in a decreasing ratio. The portion of this sum distributed each year among the early members might be allotted in such sort, that the fund should decrease as the members decreased by death, so that when the last participator died this money should be wholly expended, having been entirely divided among the original members. The proportion of profits generally distributed among the whole body of members might be allotted in a certain per centage on the amount of premiums paid by each individual.

Time of Division.—The system of taking an annual estimate of profits, and of allowing a certain proportion of these premiums to go to the annual reduction of premiums, as practised by the London Life and National Associations, appears to be highly judicious and most worthy of adoption. It is quite prudent and necessary that the funds of a society should be suffered to accumulate for a certain period of years—say five—before any estimate of profits is declared, or any portion divided. Now, as the first members did not allow themselves to appropriate any of the profits till the end of five years, it is not just that future members should participate in the advantages at a shorter period, unless they purchase the privilege by an enhanced premium. For example, if one assuring his life wishes to participate in the profits at the end of three years, instead of paying 1*l.* 15*s.* 6*d.* per annum for five years, let him pay 2*l.* 7*s.* 4*d.* for three years, and then be admitted as a member; unless the profits of the company enable it to reduce the premiums more than one-half, it will be the gainer by this arrangement, while the option may materially convenience the assurer, who, of course, must be allowed whatever diminution there may be on 1*l.* 15*s.* 6*d.*, and not on 2*l.* 7*s.* 4*d.* The minutiæ of arrangement cannot be investigated in a mere general sketch: at the establishment of such societies there must of necessity be many difficulties in detail, but there are none which may not be readily overcome by intelligent, practical men of business.

Some of the offices grant policies on lives on receiving a certain number of annual payments, and tables are constructed for this purpose; while some receive premiums in an increasing and decreasing ratio. Any of these plans may be convenient, according to the circumstances of those wishing to assure. It would exceed the limits of this article to give any comparative statement of their respective merits.

 A Comparative Table of the Rates demanded at the various

	8 to 14	15	16	17	18	19	20	21
FIRST CLASS. Companies in which the assured do not participate in the profits.	£ s. d.	£ s. d.	£ s. d.	£ s. d.	£ s. d.	£ s. d.	£ s. d.	£ s. d.
Royal Exchange N	1 17 6	1 18 6	1 19 9	2 0 9	2 1 9	2 2 9	2 3 6	2 4 6
London Assurance	1 15 8	1 16 8	1 17 9	1 18 9	1 19 8	2 0 7	2 1 4	2 2 2
Sun	....	1 12 8	1 13 6	1 14 3	1 15 1	1 16 0	1 16 11	1 17 11
Pelican	....	1 11 11	1 12 9	1 13 6	1 14 4	1 15 3	1 16 1	1 16 10
Westminster N	1 17 7	1 18 7	1 19 8	2 0 8	2 1 8	2 2 8	2 3 7	2 4 6
Globe N	1 17 7	1 18 7	1 19 8	2 0 8	2 1 8	2 2 8	2 3 7	2 4 6
Albion N	1 17 7	1 18 7	1 19 8	2 0 8	2 1 8	2 2 8	2 3 7	2 4 6
West of England	1 13 10	1 14 9	1 15 9	1 16 8	1 17 6	1 18 6	1 19 3	2 0 0
† British Commercial	* 1 9 0	1 10 0	1 11 0	1 12 0	1 13 0	1 14 0	1 15 0	1 16 0
Asylum	....	* 1 6 10	* 1 7 6	* 1 8 3	* 1 9 0	* 1 9 9	* 1 10 7	* 1 11 5
‡ London Life Associ. N	1 17 7	1 18 7	1 19 8	2 0 8	2 1 8	2 2 8	2 3 7	2 4 6
Promoter	....	1 7 11	1 8 8	1 9 5	1 10 1	1 10 11	1 11 8	1 12 6
SECOND CLASS. Companies in which the assured and the proprietary participate in profits.								
Union N	1 17 7	1 18 7	1 19 8	2 0 8	2 1 8	2 2 8	2 3 7	2 4 6
Rock N	1 17 7	1 18 7	1 19 8	2 0 8	2 1 8	2 2 8	2 3 7	2 4 6
Hope N	1 17 7	1 18 7	1 19 8	2 0 8	2 1 8	2 2 8	2 3 7	2 4 6
Economic	....	1 10 8	1 11 5	1 12 3	1 13 0	1 13 10	1 14 7	1 15 5
Provident N	1 17 7	1 18 7	1 19 8	2 0 8	2 1 8	2 2 8	2 3 7	2 4 6
Eagle	1 16 5	1 17 6	1 18 8	1 19 9	2 0 9	2 1 8	2 2 6	2 3 4
Atlas N	1 17 7	1 18 7	1 19 8	2 0 8	2 1 8	2 2 8	2 3 7	2 4 6
European	1 13 7	1 14 5	1 15 4	1 16 2	1 17 1	1 18 1	1 19 0	1 19 11
Imperial N	1 17 7	1 18 7	1 19 8	2 0 8	2 1 8	2 2 8	2 3 7	2 4 6
† Brit. Commercial N.	1 17 7	1 18 7	1 19 8	2 0 8	2 1 8	2 2 8	2 3 7	2 4 6
Guardian	1 15 8	1 16 2	1 17 2	1 18 2	1 19 2	2 0 1	2 1 0	2 1 10
Alliance	....	1 12 8	1 13 6	1 14 3	1 15 1	1 16 0	1 16 11	1 17 11
Palladium N	1 17 6	1 18 7	1 19 8	2 0 8	2 1 8	2 2 8	2 3 7	2 4 6
Law Life Assurance N	1 17 7	1 18 7	1 19 8	2 0 8	2 1 8	2 2 8	2 3 7	2 4 6
Clerical, Medical, &c.	1 15 0	1 15 11	1 16 10	1 17 9	1 18 8	1 19 7	2 0 6	2 1 5
Crown	1 14 10	1 15 9	1 16 7	1 17 5	1 18 3	1 19 1	1 19 11	2 0 10
National	1 17 6	1 18 6	1 19 6	2 0 6	2 1 6	2 2 6	2 3 6	2 4 6
University	1 15 9	1 16 8	1 17 9	1 18 8	1 19 7	2 0 7	2 1 5	2 2 4
THIRD CLASS. Companies in which there is no proprietary, and where the contributors are consequently mutual assurers.								
Equitable N	1 17 7	1 18 7	1 19 8	2 0 8	2 1 8	2 2 8	2 3 7	2 4 6
Amicable	1 14 6	1 15 6	1 16 6	1 17 6	1 18 6	1 19 6	2 0 6	2 1 6
Norwich Union	1 13 10	1 14 9	1 15 9	1 16 9	1 17 8	1 18 6	1 19 6	2 0 6
‡ London Life Associ.	....	....	....	....	....	....	....	2 12 0

* Lowest rate of premium at this age.

† Belongs to the first and second class at different premiums.

22	23	24	25	26	27	28	29	30	31	32
£. s. d.	£. s. d.	£. s. d.	£. s. d.	£. s. d.	£. s. d.	£. s. d.	£. s. d.	£. s. d.	£. s. d.	£. s. d.
2 5 3	2 6 3	2 7 0	2 8 0	2 9 0	2 10 0	2 11 0	2 12 3	2 13 3	2 14 6	2 15 9
2 3 0	2 3 11	2 4 8	2 5 7	2 6 6	2 7 6	2 8 6	2 9 8	2 10 7	2 11 9	2 13 0
1 18 11	2 0 1	2 1 3	2 2 6	2 3 9	2 5 2	2 6 7	2 7 11	2 9 2	2 10 6	2 11 10
1 17 7	1 18 4	1 19 2	2 0 1	2 1 3	2 2 7	2 3 11	2 5 2	2 6 4	2 7 7	2 8 10
2 5 4	2 6 3	2 7 1	2 8 1	2 9 0	2 10 0	2 11 1	2 12 3	2 13 4	2 14 6	2 15 9
2 5 4	2 6 3	2 7 1	2 8 1	2 9 0	2 10 1	2 11 1	2 12 3	2 13 5	2 14 7	2 15 9
2 5 4	2 6 3	2 7 1	2 8 1	2 9 0	2 10 1	2 11 1	2 12 3	2 13 5	2 14 6	2 15 9
2 0 10	2 1 3	2 2 6	2 3 3	2 4 0	2 5 0	2 6 0	2 7 0	2 8 0	2 9 0	2 10 3
1 17 0	1 18 0	1 19 0	2 0 0	2 1 0	2 2 0	2 3 0	2 4 0	2 5 0	2 6 0	2 7 0
•1 12 3	•1 13 2	•1 14 1	•1 15 1	•1 16 1	•1 17 1	•1 18 2	•1 19 3	•2 0 5	•2 1 7	•2 2 10
2 5 4	2 6 3	2 7 1	2 8 1	2 9 1	2 10 1	2 11 1	2 12 3	2 13 5	2 14 7	2 15 9
1 13 5	1 14 4	1 15 5	1 16 5	1 17 6	1 18 8	1 19 11	2 1 1	2 2 2	2 3 3	2 4 5
2 5 4	2 6 3	2 7 1	2 8 1	2 9 1	2 10 1	2 11 1	2 12 3	2 13 5	2 14 7	2 15 9
2 5 4	2 6 3	2 7 1	2 8 1	2 9 1	2 10 1	2 11 1	2 12 3	2 13 5	2 14 7	2 15 9
2 5 4	2 6 3	2 7 1	2 8 1	2 9 1	2 10 1	2 11 1	2 12 3	2 13 5	2 14 7	2 15 9
1 16 3	1 17 2	1 18 1	1 19 0	2 0 0	2 1 0	2 2 0	2 3 1	2 4 3	2 5 5	2 6 8
2 5 4	2 6 3	2 7 1	2 8 1	2 9 1	2 10 1	2 11 1	2 12 3	2 13 5	2 14 7	2 15 9
2 3 10	2 4 4	2 4 10	2 5 6	2 6 2	2 7 0	2 7 10	2 8 10	2 9 10	2 10 11	2 12 0
2 5 4	2 6 3	2 7 1	2 8 1	2 9 1	2 10 1	2 11 1	2 12 3	2 13 5	2 14 7	2 15 9
2 0 10	2 1 10	2 2 9	2 3 9	2 4 10	2 5 10	2 6 11	2 8 1	2 9 3	2 10 6	2 11 10
2 5 4	2 6 3	2 7 1	2 8 1	2 9 1	2 10 1	2 11 1	2 12 3	2 13 5	2 14 7	2 15 9
2 5 4	2 6 3	2 7 1	2 8 1	2 9 1	2 10 1	2 11 1	2 12 3	2 13 5	2 14 7	2 15 9
2 2 8	2 3 6	2 4 5	2 5 4	2 6 4	2 7 4	2 8 4	2 9 6	2 10 7	2 11 10	2 13 0
1 18 11	2 0 1	2 1 3	2 2 6	2 3 9	2 5 2	2 6 7	2 7 11	2 9 2	2 10 6	2 11 10
2 5 4	2 6 3	2 7 1	2 8 1	2 9 1	2 10 1	2 11 1	2 12 3	2 13 5	2 14 7	2 15 9
2 5 4	2 6 3	2 7 1	2 8 1	2 9 1	2 10 1	2 11 1	2 12 3	2 13 5	2 14 7	2 15 9
2 2 4	2 3 3	2 4 2	2 5 2	2 6 2	2 7 2	2 8 2	2 9 3	2 10 4	2 11 6	2 12 8
2 1 9	2 2 9	2 3 9	2 4 10	2 5 10	2 6 11	2 8 1	2 9 2	2 10 4	2 11 6	2 12 9
2 5 0	2 6 0	2 7 0	2 8 0	2 9 0	2 10 0	2 11 0	2 12 0	2 13 0	2 14 6	2 15 6
2 3 1	2 4 0	2 4 9	2 5 9	2 6 3	2 7 7	2 8 7	2 9 9	2 10 9	2 11 11	2 13 0
2 5 4	2 6 3	2 7 1	2 8 1	2 9 1	2 10 1	2 11 1	2 12 3	2 13 5	2 14 7	2 15 9
2 2 6	2 3 6	2 4 6	2 5 6	2 6 6	2 7 6	2 8 6	2 9 6	2 10 6	2 11 6	2 12 6
2 1 3	2 2 0	2 2 9	2 3 8	2 4 8	2 5 8	2 6 8	2 7 9	2 8 10	2 10 0	2 11 1
2 13 0	2 13 6	2 14 0	2 14 6	2 15 0	2 16 0	2 17 0	2 18 0	2 19 0	3 0 0	3 1 6

‡ Belongs to the first and third class at different premiums.
N. Rates of premium calculated by the Northampton Table of Mortality.

F 3

	33	34	35	36	37	38	39	40
FIRST CLASS. Companies in which the assured do not participate in the profits.	£ s. d.	£ s. d.	£ s. d.	£ s. d.	£ s. d.	£ s. d.	£ s. d.	£ s. d.
Royal Exchange N ...	2 17 0	2 18 6	2 19 9	3 1 3	3 2 9	3 4 6	3 6 3	3 8 0
London Assurance....	2 14 2	2 15 7	2 16 9	2 18 2	2 19 8	3 1 3	3 0 3	3 4 7
Sun	2 13 4	2 14 11	2 16 8	2 18 5	3 0 4	3 2 4	3 4 5	3 6 6
Pelican............	2 10 3	2 11 9	2 13 5	2 15 1	2 16 10	2 18 9	3 0 8	3 2 8
Westminster N.......	2 17 1	2 18 5	2 19 10	3 1 4	3 2 10	3 4 6	3 6 2	3 7 11
Globe N	2 17 1	2 18 5	2 19 10	3 1 4	3 2 10	3 4 6	3 6 2	3 7 11
Albion N	2 17 1	2 18 5	2 19 10	3 1 4	3 2 10	3 4 6	3 6 2	3 7 11
West of England	2 11 6	2 12 9	2 13 10	2 15 3	2 16 8	2 18 0	2 19 6	3 1 3
† British Commercial .	2 8 0	2 9 6	2 11 0	2 13 6	2 15 0	2 16 6	2 18 0	3 0 0
Asylum..............	* 2 4 2	* 2 5 7	* 2 7 1	* 2 8 8	* 2 10 4	* 2 12 1	* 2 13 11	* 2 15 10
‡ London Life Associ. N	2 17 1	2 18 5	2 19 10	3 1 4	3 2 10	3 4 6	3 6 2	3 7 11
Promoter	2 5 8	2 7 1	2 8 7	2 10 1	2 11 8	2 13 6	2 15 2	2 17 0
SECOND CLASS. Companies in which the assured and the proprietary participate in profits.								
Union N	2 17 1	2 18 5	2 19 10	3 1 4	3 2 10	3 4 6	3 6 2	3 7 11
Rock N..............	2 17 1	2 18 5	2 19 10	3 1 4	3 2 10	3 4 6	3 6 2	3 7 11
Hope N..............	2 17 1	2 18 5	2 19 10	3 1 4	3 2 10	3 4 6	3 6 2	3 7 11
Economic	2 8 0	2 9 5	2 10 11	2 12 6	2 14 2	2 15 11	2 17 9	2 19 9
Provident N	2 17 1	2 18 5	2 19 10	3 1 4	3 2 10	3 4 6	3 6 2	3 7 11
Eagle	2 13 3	2 14 7	2 16 0	2 17 6	2 19 1	3 0 9	3 2 6	3 4 4
Atlas N..............	2 17 1	2 18 5	2 19 10	3 1 4	3 2 10	3 4 6	3 6 2	3 7 11
European	2 13 2	2 14 7	2 16 0	2 17 6	2 19 1	3 0 9	3 2 6	3 4 3
Imperial N	2 17 1	2 18 5	2 19 10	3 1 4	3 2 10	3 4 6	3 6 2	3 7 11
† Brit. Commercial N.	2 17 1	2 18 5	2 19 10	3 1 4	3 2 10	3 4 6	3 6 2	3 7 11
Guardian............	2 14 4	2 15 8	2 17 0	2 18 6	3 0 0	3 1 7	3 3 3	3 5 0
Alliance	2 13 4	2 14 11	2 16 8	2 18 5	3 0 4	3 2 4	3 4 5	3 6 6
Palladium N..........	2 17 1	2 18 5	2 19 10	3 1 4	3 2 10	3 4 6	3 6 2	3 7 11
Law Life Assurance N	2 17 1	2 18 5	2 19 10	3 1 4	3 2 10	3 4 6	3 6 2	3 7 11
Clerical, Medical, &c.	2 13 11	2 15 2	2 16 6	2 17 10	2 19 3	3 0 8	3 2 2	3 3 8
Crown...............	2 14 0	2 15 4	2 16 9	2 18 2	2 19 19	3 1 2	3 2 10	3 4 7
National............	2 17 0	2 18 6	2 19 6	3 1 0	3 2 6	3 4 6	3 6 0	3 8 0
University..........	2 14 3	2 15 6	2 16 11	2 18 4	2 19 9	3 1 4	3 2 11	3 4 7
THIRD CLASS. Companies in which there is no proprietary, and where the contributors are consequently mutual assurers.								
Equitable N..........	2 17 1	2 18 5	2 19 10	3 1 4	3 2 10	3 4 6	3 6 2	3 7 11
Amicable	2 14 0	2 15 6	2 17 0	2 18 6	3 0 0	3 1 6	3 3 0	3 5 0
Norwich Union.......	2 12 3	2 13 6	2 14 10	2 16 2	2 17 6	2 19 0	3 0 6	3 2 0
† London Life Associ. .	3 3 0	3 4 6	3 6 0	3 7 6	3 9 0	3 11 0	3 13 0	3 15 0

* Lowest rate of premium at this age.

† Belongs to the first and second class at different premiums.

Periods of Life by the different Life Assurance [illegible]ces.

41	42	43	44	45	46	47	48	49	50	51
£ s. d.	£ s. d.	£ s. d.	£ s. d.	£ s. d.	£ s. d.	£ s. d.	£ s. d.	£ s. d.	£ s. d.	£ s. d.
3 9 9	3 11 9	3 13 9	3 15 9	3 18 0	4 0 3	4 2 6	4 5 0	4 7 9	4 10 9	4 13 6
3 8 0	3 10 0	3 12 0	3 13 11	3 16 1	3 18 3	4 0 5	4 2 10	4 5 6	4 8 6	4 13 6
3 8 7	3 10 9	3 12 11	3 15 3	3 17 8	4 0 5	4 3 3	4 6 6	4 10 2	4 14 2	4 18 9
3 5 2	3 7 8	3 10 4	3 13 1	3 16 1	3 18 8	4 1 6	4 4 8	4 8 2	4 12 2	4 16 7
3 9 9	3 11 8	3 13 8	3 15	3 17 11	4 0 2	4 2 7	4 5 1	4 7 10	4 10 10	4 13 6
3 9 10	3 11 8	3 13 8	3 15 9	3 17 11	4 0 2	4 2 7	4 5 1	4 7 10	4 10 7	4 13 6
3 9 9	3 11 8	3 13 8	3 15 9	3 17 11	4 0 2	4 2 7	4 5 1	4 7 10	4 10 8	4 13 6
3 2 10	3 4 6	3 6 4	3 8 3	3 10 3	3 12 2	3 14 6	3 16 9	3 19 3	4 1 8	*4 4 3
3 2 0	3 4 0	3 6 0	3 8 0	3 10 0	3 13 0	3 16 0	3 18 10	4 2 3	4 6 6	4 10 10
*2 17 10	*2 19 11	*3 2 1	*3 4 4	3 6 10	3 9 7	3 12 7	3 16 1	3 19 10	4 3 10	4 8 0
3 9 9	3 1 8	3 13 8	3 15 9	3 17 11	4 0 2	4 2 7	4 5 1	4 7 10	4 10 8	4 13 6
2 18 10	3 0 8	3 2 6	3 4 6	*3 6 7	*3 8 11	*3 11 5	*3 14 2	*3 17 3	*4 0 8	4 4 7
3 9 9	3 11 8	3 13 8	3 15 9	3 17 11	4 0 2	4 2 7	4 5 1	4 7 10	4 10 8	4 13 6
3 9 9	3 11 8	3 13 8	3 15 9	3 17 11	4 0 2	4 2 7	4 5 1	4 7 10	4 10 8	4 13 6
3 9 9	3 11 8	3 13 8	3 15 9	3 17 11	4 0 2	4 2 7	4 5 1	4 7 10	4 10 8	4 13 6
3 1 10	3 4 1	3 6 6	3 9 0	3 11 0	3 14 7	3 17 8	4 0 11	4 4 4	4 8 0	4 11 11
3 9 9	3 11 8	3 13 8	3 15 9	3 17 11	4 0 2	4 2 7	4 5 1	4 7 10	4 10 8	4 13 6
3 6 3	3 8 4	3 10 8	3 13 0	3 15 8	3 18 6	4 1 7	4 5 0	4 8 7	4 12 4	4 16 4
3 9 9	3 11 8	3 13 8	3 15 9	3 17 11	4 0 2	4 2 7	4 5 1	4 7 10	4 10 8	4 13 6
3 6 3	3 8 3	2 10 5	3 12 7	3 15 0	3 17 5	4 0 0	4 2 8	4 5 6	4 8 6	4 11 7
3 9 9	3 11 8	3 13 8	3 15 9	3 17 11	4 0 2	4 2 7	4 5 1	4 7 10	4 10 8	4 13 6
3 9 9	3 11 8	3 13 8	3 15 9	3 17 11	4 2 2	4 4 10	4 7 3	4 10 0	4 12 11	4 17 2
3 6 9	3 8 8	3 10 8	3 12 8	3 14 11	3 17 3	3 19 8	4 2 4	4 5 1	4 8 0	4 11 0
3 8 7	3 10 9	3 12 11	3 15 3	3 17 8	4 0 5	4 3 3	4 6 6	4 10 2	4 14 2	4 18 9
3 9 9	3 11 8	3 13 8	3 15 9	3 17 11	4 0 2	4 2 7	4 5 1	4 7 10	4 10 8	4 13 6
3 9 9	3 11 8	3 13 8	3 15 9	3 17 11	4 0 2	4 2 7	4 5 1	4 7 10	4 10 8	4 13 6
3 5 4	3 7 0	3 9 0	3 11 0	3 13 0	3 15 6	3 18 0	4 1 0	4 4 0	4 7 3	4 10 6
3 6 5	3 8 4	3 10 6	3 12 8	3 15 0	3 17 6	4 0 1	4 2 11	4 5 10	4 8 11	4 12 1
3 9 6	3 11 6	3 13 6	3 15 6	3 18 0	4 0 0	4 2 6	4 5 0	4 7 6	4 10 6	4 13 6
3 6 4	3 8 1	3 0 0	3 12 0	3 14 1	3 16 2	3 18 6	4 1 4	4 4 4	4 7 6	4 10 9
3 9 9	3 11 8	3 13 8	3 15 9	3 17 11	4 0 2	4 2 7	4 5 1	4 7 10	4 10 8	4 13 6
3 7 0	3 9 0	3 11 0	3 13 0	3 15 0	3 17 6	4 0 0	4 2 6	4 5 0	4 8 0	4 11 0
3 3 6	3 5 2	3 7 0	3 9 0	3 11 0	3 13 8	3 16 3	3 19 6	4 2 9	4 6 0	4 9 8
3 17 0	3 19 0	4 1 6	4 0 4	4 6 6	4 9 6	4 13 0	4 16 6	5 0 0	5 4 0	5 8 0

‡ Belongs to the first and third class at different premiums.
N. Rates of premium calculated by the Northampton Table of Mortality.

F 4

A Comparative Table of the Rates demanded at the various

	52	53	54	55	56	57	58	59
	£. s. d.	£. s. d.	£. s. d.	£. s. d.	£. s. d.	£. s. d.	£. s. d.	£. s. d.
FIRST CLASS. Companies in which the assured do not participate in the profits.								
Royal Exchange N	4 16 6	4 19 6	5 2 9	5 6 3	5 10 0	5 14 0	5 18 3	6 2 9
London Assurance	4 16 6	4 19 6	5 2 9	5 6 3	5 10 0	5 14 0	5 18 3	6 2 9
Sun	5 3 6	5 8 7	5 14 1	5 19 11	6 6 4	6 13 2	7 0 5	7 7 9
Pelican	5 1 3	5 6 4	5 11 7	5 17 4	6 3 7	6 10 4	6 17 5	7 4 6
Westminster N	4 16 5	4 19 7	5 2 10	5 6 4	5 10 1	5 14 0	5 18 2	6 2 8
Globe N	4 16 5	4 19 7	5 2 10	5 6 4	5 10 1	5 14 0	5 18 2	6 2 7
Albion N	4 16 5	4 19 7	5 2 10	5 6 4	5 10 1	5 14 0	5 18 2	6 2 8
West of England	*4 6 9	*4 9 9	*4 12 9	*4 15 9	*4 19 0	*5 2 6	*5 6 6	*5 10 6
† British Commercial	4 15 6	4 19 8	5 3 0	5 6 6	5 11 0	4 14 6	5 19 0	6 4 0
Asylum	4 12 4	4 16 10	5 1 7	5 6 5	5 11 5	5 16 7	6 1 11	6 7 5
‡ London Life Associ. N	4 16 5	4 19 7	5 2 10	5 6 4	5 10 1	5 14 0	5 18 2	6 2 8
Promoter	4 8 8	4 13 1	4 17 10	5 2 9	5 8 4	5 14 2	6 0 4	6 6 7
SECOND CLASS. Companies in which the assured and the proprietary participate in profits.								
Union N	4 16 5	4 19 7	5 2 10	5 6 4	5 10 1	5 14 0	5 18 2	6 2 8
Rock N	4 16 5	4 19 7	5 2 10	5 6 4	5 10 1	5 14 0	5 18 2	6 2 8
Hope N	4 16 5	4 19 7	5 2 10	5 6 4	5 10 1	5 14 0	5 18 2	6 2 8
Economic	4 16 1	5 0 6	5 5 3	5 10 3	5 15 7	6 1 3	6 7 4	6 13 9
Provident N	4 16 5	4 19 7	5 2 10	5 6 4	5 10 1	5 14 0	5 18 2	6 2 8
Eagle	5 0 5	5 4 8	5 8 11	5 13 0	5 17 10	6 2 10	6 7 10	6 12 7
Atlas N	4 16 5	4 19 7	5 2 10	5 6 4	5 10 1	5 14 0	5 18 2	6 2 8
European	4 15 0	4 18 7	5 2 6	5 6 8	5 11 2	5 15 8	6 0 7	6 5 8
Imperial N	4 16 5	4 19 7	5 2 10	5 6 4	5 10 1	5 14 0	5 18 2	6 2 8
† Brit. Commercial N	5 0 3	5 3 6	5 7 0	5 10 9	5 16 8	6 0 10	6 5 2	6 10 0
Guardian	4 14 2	4 17 5	5 0 11	5 4 8	5 8 7	5 12 10	5 17 4	6 2 2
Alliance	5 3 6	5 8 7	5 14 1	5 19 11	6 6 4	6 13 2	7 0 5	7 7 9
Palladium N	4 16 5	4 19 7	5 2 10	5 6 4	5 10 1	5 14 0	5 18 2	6 2 8
Law Life Assurance N	4 16 5	4 19 7	5 2 10	5 6 4	5 10 1	5 14 0	5 18 2	6 2 8
Clerical, Medical, &c.	4 14 0	4 17 6	5 1 3	5 5 0	5 9 0	5 13 0	5 17 6	6 2 0
Crown	4 15 3	4 18 6	5 1 11	5 5 7	5 9 6	5 13 6	5 18 0	6 2 4
National	4 16 6	4 19 6	5 2 6	5 6 0	5 10 0	5 14 0	5 18 0	6 2 6
University	4 14 1	4 17 8	5 1 4	5 5 4	5 9 7	5 14 0	5 18 2	6 2 8
THIRD CLASS: Companies in which there is no proprietary, and where the contributors are consequently mutual assurers.								
Equitable N	4 16 5	4 19 7	5 2 10	5 6 4	5 10 1	5 14 0	5 18 2	6 2 8
Amicable	4 14 0	4 17 0	5 0 0	5 3 6	5 7 6	5 11 6	5 15 6	6 0 0
Norwich Union	4 13 3	4 17 0	5 1 0	5 5 3	5 9 6	5 13 6	5 17 6	6 2 6
‡ London Life Associ.	5 12 0	5 16 0	6 0 0	6 4 6	6 9 0	6 13 6	6 18 0	7 3 0

* Lowest rate of premium at this age.

† Belongs to the first and second class at different premiums.

60	61	62	63	64	65	66	67	68	69	70
£ s. d.	£ s. d.	£ s. d.	£ s. d.	£ s. d.	£ s. d.	£ s. d.	£ s. d.	£ s. d.	£ s. d.	£ s. d.
6 7 3	6 12 2	6 17 9	7 3 6	7 9 9	7 16 9	8 4 0	8 12 0			
6 7 3										
7 14 11										
7 11 7										
6 7 4	6 12 4	6 17 9	7 3 7	7 9 10	7 16 9	8 4 1	8 12 1			
6 7 4	6 12 4	6 17 9	7 3 7	7 9 10	7 16 9					
6 7 4	6 12 4	6 17 9	7 3 7	7 9 10	7 16 9	8 4 1	8 12 1	9 0 10	9 10 2	10 0 4
*5 14 9	*5 19 2	*6 4 0	*6 9 3	*6 15 0	*7 1 3	*7 7 8	*7 14 0	*8 2 10	*8 11 2	*9 0 6
6 9 6	6 15 0	7 2 6	7 10 0	7 19 0	8 10 0					
6 13 0	6 18 10	7 4 11	7 11 6	7 18 9	8 8 0	8 18 0	9 8 6	10 0 0	10 13 2	11 6 10
6 7 4	6 12 4	6 17 9	7 3 7	7 9 10	7 16 9	8 4 1	8 12 1			
6 12 10										
6 7 4										
6 7 4	6 12 4	6 17 9	7 3 7	7 9 10	7 16 9	8 4 1	8 12 1			
6 7 4										
7 0 7										
6 7 4	6 12 4	6 17 9	7 3 7	7 9 10	7 16 9	8 4 1	8 12 1			
6 18 2										
6 7 4	6 12 4	6 17 9	7 3 7	7 9 10	7 16 9	8 4 1	8 12 1			
6 11 1										
6 7 4										
6 14 11	7 5 4	7 11 6	7 17 11	8 4 10	8 16 4					
6 7 2										
7 14 11										
6 7 4										
6 7 4	6 12 4	6 17 9	7 3 7	7 9 10	7 16 9	8 4 1	8 12 1			
6 7 2	6 12 4	6 17 9	7 3 7	7 9 10	7 16 9	8 4 1	8 12 1			
6 7 2										
6 7 6	6 12 6	6 18 0	7 4 0	7 10 0	7 17 0					
6 7 4	6 12 4	6 17 9	7 3 7	7 9 10	7 16 9	8 4 1	8 12 1			
6 7 4	6 12 4	6 17 9	7 3 7	7 9 10	7 16 9	8 4 1	8 12 1			
6 5 0	6 10 0	6 15 6	7 1 0	7 7 6	7 14 6	8 2 0	8 10 0	8 19 6	9 9 0	9 19 6
6 7 3										
7 8 0	7 13 6	7 19 6	8 6 0	8 13 0	9 0 6	9 8 0	9 16 0			

‡ Belongs to the first and third class at different premiums.

N. Rates of premium calculated by the Northampton Table of Mortality.

F 5

COMPANION TO THE ALMANAC,

FOR

1832.

PART I.

INFORMATION CONNECTED WITH THE CALENDAR AND THE NATURAL PHENOMENA OF THE YEAR; AND WITH NATURAL HISTORY AND PUBLIC HEALTH.

I. ECLIPSES.

In times of ignorance men are alarmed at all celestial phenomena, the recurrence of which takes place at periods too remote to be readily calculated, and which accordingly appear to be guided by no fixed laws; whilst more important and obvious appearances excite no surprise because they are frequent, and because they occur at well known periods: thus the changes of the moon are observed without alarm, while the less obvious occurrence of an eclipse has scattered dismay over an army or a nation. An incidental and considerable assistance has been derived to chronology from this superstitious feeling. Eclipses were thought to be connected, in some secret manner, with the destinies of nations, and their occurrence has been carefully recorded, when near the time of some great battle, or other political event, of whose epoch we should otherwise remain in ignorance. Astronomical science shows how to determine with the greatest exactness the hour of any given eclipse; and we are thus enabled to fix with precision the date of any event which may have been thus accompanied, to confirm the statement of an historian, or to correct his errors. But as the computation of eclipses is attended with considerable difficulty, a few only of the readers of history are able to carry on these researches for themselves. On this account catalogues of eclipses have been calculated by astronomers for many thousand years, by

B 3

a reference to which any chronological point connected with these phenomena may be at once determined.

It may be useful to point out shortly the order in which eclipses follow each other, omitting all scientific details, and taking it for granted that all our readers are aware that eclipses of the sun are caused by the interposition of the body of the moon between our eyes and the sun, and eclipses of the moon by a real deprivation of light from the moon, occasioned by the shadow of the earth.

Eclipses take place every half year, and at each period there may be one, two, or three eclipses; if one only, it must be an eclipse of the sun; if two, there will be one of each luminary; and if three, there will be two of the sun, with one of the moon between them. We will take a view of the eclipses of the last six years to show this more clearly. In 1826 there were two eclipses in the first half year; one of the moon on the 21st of May, and one of the sun, June 5. In the latter half year there were three; two of the sun, October 31 and November 29, with one of the moon between them on the 14th November. Half a year after, April 26, 1827, there was another of the sun, accompanied by one of the moon at a fortnight's interval on May 11; in the next half year two more, one of the sun, October 30, and one of the moon on the 3d of November. In 1828 one of the sun, April 14, and in the last half year another on the 9th of October. In 1829 there were two eclipses in March and April, and two in September. In 1830 three in February and March, and three in August and September. In 1831, two in February, and two in August. This succession will constantly be found to take place, and at the end of eighteen years and ten or eleven days*, the same order will recommence, and eclipses will succeed each other at the same intervals as eighteen years before. Accordingly any one provided with a list of all the eclipses which have occurred during eighteen years, may easily find the epochs of other eclipses taking place at intervals preceding or following that period. To determine, for example, the eclipses of 1832, take those of 1814, and add to each the ecliptic period; thus there was an eclipse of the sun on the 21st January at a quarter after two in the afternoon.

1814	Jan.	21	2h.	15m.
18		11†	7h.	43m.
1832	Feb.	1	9	58

And another on the 17th July at half-past six in the morning.

1814	July	17	6	30
18		10	7	43
1832		27	14	13

There will be then in 1832 two eclipses of the sun; the first on

* The exact period is 6585 days, 7 hours, 43¼ minutes, and the number of days above the 18 years must be taken at 10 or 11, as there may have been 4 or 5 leap years included in the 18 years. In some cases 12 days must be added, as from 1797 to 1815, in which interval there were only 3 leap years.

† Eleven days are added, because the additional day, or 29th February, occurred four times only between January, 1814, and February, 1832. In the next example 10 days are added, because the 29th February had occurred 5 times.

the 1st February, about ten o'clock at night, and the second on the 27th of July, about a quarter-past two in the afternoon.

In pursuance of what has been said, it may be imagined that a correct list of eclipses for eighteen years would be sufficient for all purposes; as by adding the ecliptic period as many times as required, the period of an eclipse might be known at any distance of time. This would be correct if every eclipse appeared under precisely the same circumstance as its corresponding eclipse in the preceding or following period; but this is not the case. An eclipse of the moon, which, in the year 565, for example, was of 6 digits, was in the year 583 of 7 digits, and in 601 nearly 8. In 908 the eclipse became total, and it remained so for about twelve periods, or until the year 1088; this eclipse continued to diminish until the commencement of the fifteenth century, when it totally disappeared in the year 1413. In like manner an eclipse of the sun, which first appeared at the north pole in June 1295, proceeded more southerly at each period. On the 27th August, 1367, it made its first appearance in the north of Europe; in 1439 it was visible all over Europe; in 1601, which was its nineteenth appearance, it was central in London; in 1818, on the 5th May, it was visible in London, and will be again nearly central in the same place on the 15th May, 1836. At its thirty-ninth appearance, August 10th, 1980, the moon's shadow will have passed the equator, and as the eclipse will take place nearly at midnight, it will be invisible in Europe, Africa, and Asia. At every subsequent period the eclipse will go more and more towards the south, until, finally, on the 30th September, 2665, which will be its seventy-eighth appearance, it will go off at the south pole of the earth, and disappear altogether.

The subjoined catalogue contains all the eclipses which will take place before the end of the present century, with the exception of such solar eclipses as will be hardly visible to any inhabited part of the earth; we also omit the eclipses of the moon under one digit.

To enable the reader to comprehend fully our catalogue, it will be proper to make some preliminary observations. Eclipses of the moon taking place under totally different circumstances from those of the sun, and being much more readily understood, we shall begin with them.

The time of the middle of each eclipse of the moon is given to the nearest half hour, reduced to the meridian of London; and the interval between the given time and midnight, reduced to degrees, will show the meridian on which the eclipse will be central. For example, on the 31st March, 1847, there will be an eclipse of the moon at 9½h. a. (half-past nine, afternoon): this eclipse will accordingly be central at 37½° *east* of London, because it occurs* 2 hours *before* midnight. That of July 13, 1851, will be central at 112½° *west* of London, because it will happen at 7½h.m., or *after* midnight. The name of some country has been given where the eclipse will be central on the meridian; by adding about 100 degrees of latitude, both east and west of the given place, the

* Hours are reduced to degrees by multiplying by 15.

limits of the visible eclipse will be known, observing always that these limits will be more extensive in winter than in summer.

The duration of a lunar eclipse depends chiefly on its magnitude; in a total eclipse, the darkness may last nearly four hours, though usually the whole eclipse will be scarcely so long. An eclipse from six to twelve digits* will continue from 2½ to 3½ hours. An eclipse of three to six digits will be of two or three hours duration, and a smaller eclipse of one or two hours only.

With eclipses of the sun the case is very different, and far more difficult than with lunar eclipses. In the latter the obscuration is the same in degree at all parts of the earth where the moon is visible; but an eclipse of the sun, which may be total perhaps in Africa, may be of six digits only at Madrid, and of two in London. The part of the earth's surface to which an eclipse of the sun can be total, is a line of the breadth of 170 miles at the greatest. The central line is, however, rarely so broad, and is often a mere nominal line, without total darkness, the moon being then at her apogee, or greatest distance from the earth, and too small to cover the disc of the sun; in this case the eclipse is said to be annular, (from the Latin word *annulus*, a ring,) as the sun appears to surround the dark body of the moon with a ring of light.

This line extends itself in March from S. W. to N. E., and in September from N. W. to S. E. In June the central line is a curve, going, first, to the N. E., and then to the S. E.; in December, on the contrary, first to the S. E., and then to the N. E. To all places within two thousand miles at least of the central line, the eclipse will be visible; and the nearer the place of observation is situate to the line, so much larger will the eclipse be. If the central trace be but a little northward of the equator in winter, or 25° N. lat. in summer, the eclipse will be visible all over the northern hemisphere. As a general rule, though liable to many modifications, we may observe, that places from 200 to 250 miles from the central line, will have an eclipse of 11 digits; from thence to 500 miles, 10 digits, and so on, diminishing one digit in about 250 miles.

In our table the trace of this central line is indicated by naming some of the principal places through which it passes. As an example, may be adduced the eclipse of the 8th July, 1842, which is marked in the catalogue (c.) Madeira, Caspian, China. Draw on a map a curved line from Madeira to China, through the Caspian Sea; and this line will pass through the south of Italy, Turkey, Georgia, and the Tartarian plains; to all these places the eclipse will be central. Places from 200 to 400 miles from the central line will have the eclipse of 10 or 11 digits, as in Spain, France, Germany, &c. In England the eclipse will be from 8 to 10 digits, and it will decrease as it approaches the pole. On the south of the central line the eclipse will decrease in magnitude,

* A digit is the twelfth part of the surface of the moon or sun, and of course an eclipse of 6 digits will be understood to signify one, in which a half of the body of the luminary is hidden. When, in eclipses of the moon, the magnitude is stated to be above twelve digits, it is understood that the shadow of the earth extends itself so many digits beyond the surface of the moon.

as it recedes from the north, and will cease to be visible about the middle of Africa.

The letters S.E. or N.E., occurring after the name of a place, are intended to show the direction taken by the eclipse after leaving the inhabited part of the earth.

In those cases where the central line passes very near the north or south pole, the line is not traced; but those places only are mentioned where the eclipse was visible; this is indicated by the letter (v) placed before the names of those parts of the world where the eclipse was visible; (t) indicates total.

The foregoing observations will enable a person tolerably acquainted with geography to trace on a map the progress of any given eclipse, and consequently to ascertain to what part of the globe its visible limits will be extended.

1832—⊙ Feb. 1, 10 h. a., cent., N.E. of New Holland, Otaheite, West Indies.
⊙ July 27, 2 h. a., c., Vera Cruz, Cape Verde, Congo.
☾ Jan. 6, 8 h. m., 5$\frac{3}{4}$ dig., New Albion.
1833—⊙ „ 20, 9 h. a., c., New Zealand, South Pacific, Buenos Ayres.
☾ July 2, 1 h. m., 10$\frac{1}{4}$ dig., Canaries.
⊙ „ 17, 7 h. m., v., all Europe, small at south.
☾ Dec. 26. 10 h. a., t., 17 dig., Poland.
⊙ June 7, 10 h. m.. v., South of Africa, Van Diemen's Land.
1834—☾ „ 21, 8$\frac{1}{2}$ h. m., t., 16 dig., Nootka.
⊙ Nov. 30, 7 h. a., c., Russian America, United States, Newfoundland.
☾ Dec. 16, 5h. m., 8 dig., Jamaica.
1835—⊙ May 27, 1$\frac{1}{2}$ h. a., c., Chili, Guinea, Zanguebar.
⊙ Nov. 20, 11 h. m., c., Cape Verd Islands, Congo, Madagascar.
1836—☾ May 1, 8$\frac{1}{4}$ h. m., 4$\frac{1}{4}$ dig., Nootka.
⊙ „ 15, 2$\frac{1}{2}$ h. a., c., California, Labrador, Caspian Sea.
☾ Oct. 24, 1$\frac{1}{2}$ h. a., 1$\frac{1}{2}$ dig., Kamchatka.
⊙ Nov. 9, 2 h. m., c., Van Diemen's Land, New Zealand, N.E.
1837—☾ April 20, 9 h. a., t., 20$\frac{1}{4}$ dig., Armenia.
⊙ May 4, 7$\frac{1}{2}$ h. a., v., North America, small.
☾ Oct. 13, 11$\frac{1}{2}$ h. a., t., 19$\frac{1}{2}$ dig., France.
1838—⊙ March 25, 9 h. a., c., Cape Horn, N.E.
☾ April 10, 2 h. m., 7 dig., Terceira.
⊙ Sept. 18, 9$\frac{1}{2}$ h. a., c., Hudson's Bay, Newfoundland.
☾ Oct. 3, 3 h. a., 10$\frac{3}{4}$ dig., Japan.
1839—⊙ March 15, 2$\frac{1}{2}$ h. a., c., Easter Island, Brazil, Egypt.
⊙ Sept. 7, 10$\frac{1}{2}$ h. a., c., Japan, Owhyhee, west of Peru.
1840—☾ Feb. 17, 2 h. a., 4$\frac{1}{4}$ dig., Sidney.
⊙ March 4, 4 h. m., c., Nubia, China, Japan, N.E.
☾ Aug. 13, 7$\frac{1}{2}$ h. m., 7$\frac{1}{4}$ dig., California.
⊙ „ 27, 7 h. m., c., south-west of Africa, S.E.
1841—☾ Feb. 6, 2$\frac{1}{2}$ h. m., t., 19$\frac{1}{2}$ dig., Brazil.
⊙ „ 21, 11 h. m., v., Greenland, Europe, small at south.
⊙ July 18, 2 h. a., v., Europe and west of Asia, small at south.
☾ Aug. 2, 10 h. m., t., 18 dig., Otaheite.
1842—☾ Jan. 26, 6 h. a., 9 dig., Bengal.
⊙ July 8, 7 h. m., c., Madeira, Caspian, China.
☾ „ 22, 11 h. m., 3 dig., Behring's Straits.

B 5

1842—⊙ Dec. 31, 7½ h. a., c., New Hebrides, South Pacific, South America.
1843—⊙ June 27, 7 h. a., c., Otaheite, Gallipagos, Paraguay.
☾ Dec. 6, midnight, 2¼ dig., London.
⊙ „ 21, 5½ h. m., c., Arabia, Malacca, the Ladrones.
1844—⊙ June 14, midnight, v., New Holland, New Zealand.
☾ May 31, 11 h. a., t., 15½ dig., Germany.
☾ Nov. 24, midnight, t., 18¼ dig., London.
⊙ Dec. 9, 8 h. a., v., North America, small at south.
1845—⊙ May 6, 10½ h. m., v., Canada, all Europe, except S.E.
☾ „ 21, 4½ h. a., t., 12¾ dig., China.
⊙ Oct. 30, 10½ h. a., v., New Holland, New Zealand.
☾ Nov. 14, 1 h. m., Canaries.
1846—⊙ April 25, 5 h. a., c., the Marquesas, Cuba, Senegal.
⊙ Oct. 20, 8½ h. m., c., Guinea, Madagascar, Swan River.
1847—☾ March 31, 9½ h. a., 2¾ dig., Russia.
⊙ April 15, 6½ h. m., v., Cape of Good Hope, c., N. of New Holland.
☾ Sept. 24, 3 h. a., 4½ dig., Japan.
⊙ Oct. 9, 9½ h. m., c., Scotland, Persia, Cochin China.
1848—⊙ March 5, 1½ h. a., v., Canada, small.
☾ „ 19, 9½ h. a., t., 17 dig., Russia.
☾ Sept. 13, 6½ h. m., t., 20 dig., Mexico.
⊙ „ 27, 10 h. m., v., Russia, Siberia, small.
1849—⊙ Feb. 23, 1½ h. m., c., China, North of Japan, N.W. America.
☾ March 9, 1 h. m., 8½ dig., Canaries.
⊙ Aug. 18, 5½ h. m., c., Mozambique, S.E.
☾ Sept. 2, 5½ h. a., 7 dig., Burmese Empire.
1850—⊙ Feb. 12, 6½ h. m., c. Caffraria, Java, Ladrones.
⊙ Aug. 7, 10 h. a., c., Caroline Islands, Owhyhee, S.E.
1851—☾ Jan. 17, 5 h. a., 5½ dig., west of China.
⊙ Feb. 1, 5 h. m., c., Van Diemen's Land, New Zealand.
☾ July 13, 7½ h. m., 8½ dig., California.
⊙ „ 28, 2¼ h. a., c., N.W. America, Iceland, Caspian.
1852—☾ Jan. 7, 6½ h. m., t., 16 dig., Mexico.
☾ July 1, 3 h. a., t., 17½ dig., Japan.
⊙ Dec. 11, 4 h. m., c., Siberia, Japan, Mulgrave Island.
☾ „ 26, 1 h. a., 8 dig., New Caledonia.
1853—⊙ June 6, 8 h. a., c., Society Islands, Gallegos, Peru.
☾ „ 21, 6 h. m., 2¼ dig., Mississippi.
⊙ Nov. 30, 7½ h. a., c., Sandwich Islands, Peru, Rio Janeiro.
1854—☾ May 12, 4 h. a., 3 dig., East of China.
⊙ „ 26, 10 h. a., c., Ladrones, N.W. America, United States.
☾ Nov. 4, 9½ h. a., 1 dig., Russia.
⊙ „ 20, 10½ h. m., c., Paraguay, S.E., v., south of Africa, New Holland.
1855—☾ May 2, 4½ h. m., t., 19¼ dig., Canada.
⊙ „ 16, 2½ h. m., v., north of Asia.
☾ Oct. 25, 8 h. m., t., 18¾ dig., New Albion.
1856—⊙ April 5, 5 h. m., v., New Holland; c., New Zealand.
☾ „ 20, 9½ h. m., Society Islands.
⊙ Sept. 29, 4 h. m., v., all north of Asia.
☾ Oct. 13, 11½ h. a., 11½ dig., France.
1857—⊙ March 25, 11 h. a., c., New South Wales, Pacific, South California.
⊙ Sept. 18, 6 h. m., c., Greece, India, New Guinea.
1858—☾ Feb. 27, 10 h. a., 4 dig., Poland.

1858—⊙ March 15, noon, c., Barbadoes, Spain, St. Petersburg.
☾ Aug. 24, 2½ h. a., 5½ dig., New Guinea.
⊙ Sept. 7, 2½ h. a., c., Chili, S.E.; v., South of Africa.
1859—☾ Feb. 17, 11 h. m., t., 19½ dig., Behring's Straits.
⊙ March 4, 10 h. m., v., Greenland.
⊙ July 29, 9½ h. a., v., north of North America.
☾ Aug. 13, 4½ h. a., t., 19 dig., China.
1860—☾ Feb. 7, 2½ h. m., 9¼ dig., Brazil.
⊙ July 18, 2 h. a., c., New Mexico, Newfoundland, Upper Egypt.
☾ Aug. 1, 5½ h. a., 4¾ dig., Ava.
1861—⊙ Jan. 11, 3½ h. m., c., Isle of France, New Holland, N.E.
⊙ July 8, 2 h. m., c., Java, Caroline Islands, Society Islands.
☾ Dec. 17, 8½ h. m., 2 dig., Nootka.
⊙ „ 31, 2½ h. a., c., United States, Cape Verd, Sicily.
1862—☾ June 12, 6½ h. m., t., 14¼ dig., Mexico.
⊙ „ 26, 7 h. m., v., Cape of Good Hope, Van Diemen's Land.
☾ Dec. 6, 8 h. m., t., 17½ dig., New Albion.
⊙ „ 21, 5½ h. m., v. all north of Asia.
1863—⊙ May 17, 5 h. a., v., North America and Europe, small at south.
☾ June 1, midnight, t., 14¼ dig., London.
☾ Nov. 25, 9 h. m., 11 dig., Pitcairn's Island.
1864—⊙ May 6, 0½ h. m., c., Borneo, Sandwich Islands.
⊙ Oct. 30, 3½ h. a., c., Gallipagos, Rio Janeiro, Cape of Good Hope.
1865—☾ April 11, 5 h. m., 1½ dig., Jamaica.
⊙ „ 25, 3 h. a., c., South Pacific, Brazil, Cape of Good Hope.
☾ Oct. 4, 11 h. a., 3¾ dig., Italy.
⊙ „ 19, 5 h. a., c., Slave Lake, United States, Cape Verd.
1866—⊙ March 16, 10 h. a., v. N.E. of Asia, N.W. of America.
☾ „ 31, 5 h. m., t. 16 dig., Jamaica.
☾ Sept. 24, 2½ h. a., t. 19 dig., Van Diemen's Land.
⊙ Oct. 8, 5 h. a., v., north of America, N.W. of Europe.
1867—⊙ March 6, 10 h. m., c., Cape Verd Islands, France, Tobolski.
☾ „ 20, 9 h. m., 9¼ dig., Pitcairn's Island.
⊙ Aug. 29, 1 h. a., c., Buenos Ayres, S.E.
☾ Sept. 14, 1 h. m., 8 dig., Canaries.
1868—⊙ Feb. 23, 2½ h. a., c., South Pacific, Guiana, N.E. of Africa.
⊙ Aug. 18, 5½ h. m., c., Egypt, India, Caroline Isles.
1869—☾ Jan. 28, 1½ h. m., 5½ dig., Cape Verd.
⊙ Feb. 11, noon, v., South Africa, Madagascar.
☾ July 23, 2 h. a., 6¾ dig., New South Wales.
⊙ Aug. 7, 10 h. a., c., Manchoo Tartary, New Albion, Mexico
1870—☾ Jan. 17, 3 h. a., t. 15 dig., Japan.
☾ July 12, 11 h. a., t. 19 dig., Italy.
⊙ Dec. 22, 0½ h. a., c., Mexico, Spain, Black Sea.
1871—☾ Jan. 6, 9½ h. a., 8 dig., Russia.
⊙ June 18, 2½ h. m., c., Java, New Guinea, Friendly Islands.
☾ July 2, 1½ h. a., 4 dig., Kamchatka.
⊙ Dec. 12, 4½ h. m., c., Persian Gulf, north of New Holland, Mulgrave Island.
1872—☾ May 22, 11½ h. a., 1½ dig., France.
⊙ June 6, 3½ h. m., c., Laccadives, Pekin, Sandwich Islands.
⊙ Nov. 30, 7 h. a., c., Friendly Islands, Cape Horn, S.E.
1873—☾ May 12, 11½ h. m., t. 17½ dig., Friendly Islands.
⊙ May 26, 9½ h. m., v., North Atlantic, north of Europe and Asia.
☾ Nov. 4, 4½ h. a., t. 18 dig., China.
1874—⊙ April 16, 1½ h. a., v., Cape of Good Hope.
☾ May 1, 4½ h. a., 9¾ dig., China.

1874—⊙ Oct. 10, 11½ h. m., c., Baffin's Bay, Norway, Tobolski.

☾ „ 25, 8 h. m., t. 12 dig., New Albion.

1875—⊙ April 6, 7 h. m., c., Caffraria, Maldives, Philippines.

⊙ Sept. 29, 1½ h. a., c., United States, Sierra Leone, Mozambique.

1876—☾ March 10, 6½ h. m., 3½ dig., Mexico.

⊙ „ 25, 8 h. a., c., Mulgrave's Island, Nootka, Greenland.

☾ Sept. 3, 9½ h. a., 4 dig., Russia.

⊙ „ 17, 10 h. a., c., New Guinea, Cape Horn.

1877—☾ Feb. 27, 7½ h. a., t. 19¼ dig., east of Persia.

⊙ March 15, 3h. m., v., north of Asia.

⊙ Aug. 9, 5h. m., v., north of Asia and America.

☾ „ 23, 11½ h. a., t. central, France.

1878—☾ Feb. 17, 11 h. m., 9½ dig., Behring's Straits.

⊙ July 29, 9½ h. a., c., Manchoo Tartary, Behring's Straits, United States.

☾ Aug. 12, midnight, 6½ dig., London.

1879—⊙ Jan. 22, noon, c., Peru, St. Helena, Maldives.

⊙ July 19, 9 h. m., c., Guinea, Abyssinia, N.W. of New Holland.

☾ Dec 28, 4½ h. a., 1¾ dig., China.

1880—⊙ Jan. 11, 11 h. a., c., Pellew Island, Scarboro's, California.

☾ June 22, 2 h. a., t. 12¾ dig., New South Wales.

⊙ July 7, 1 h. a., v., Cape of Good Hope.

☾ Dec. 16, 4 h. a., t. 16¾ dig., east of China.

⊙ „ 31, 2 h. a., v., North America and Europe, small at south.

1881—⊙ May 27, midnight, v., East of Asia, N.W. of America.

☾ June 12, 7 h. m., t. 15¾ dig., New Mexico.

☾ Dec. 5, 5½ h. a., 11½ dig., Ava.

1882—⊙ May 17, 8 h. m., c., Guinea, Persia, China.

⊙ Nov. 10, midnight, c., Borneo, Norfolk Island, Easter Island.

1883—⊙ May 6, 11½ h. a., c., Philippines, Tonga Island, Pitcairn's Island.

☾ Oct. 16, 7½ h. m., 3 dig., California.

⊙ „ 30, midnight, c., north of Japan, Owhyhee, S.E.

1884—⊙ March 27, 6 h. m., v., N.E. of Europe, north of Asia, small at East.

☾ April 10, noon, t. 15 dig., New Zealand.

☾ Oct. 4, 10½ h. a., t. 18½ dig., Greece.

⊙ „ 19, 1 h. m., v., east of Asia, North America.

1885—⊙ March 16, 6 h. a., c., North Pacific, Slave Lake, Baffin's Bay.

☾ „ 30, 5 h. a., 10 dig., west of China.

⊙ Sept. 8, 9 h. a., c., Sidney, New Zealand, S.E.

☾ „ 24, 8½ h. m., 9 dig., Nootka.

1886—⊙ March 5, 10 h. a., c., Torres Straits, Christmas Island, Gulf of Mexico.

⊙ Aug. 29, 1½ h. a., c., Honduras, Ascension, Caffraria.

1887—☾ Feb. 8, 10½ h. m., 5¼ dig., Sandwich Islands.

⊙ „ 22, 8 h. a., v., New South Wales, c., South America.

☾ Aug. 3, 9 h. a., 5 dig., Armenia.

⊙ „ 19, 6 h. m., c., Norway, Lake Baikal, North Pacific.

1888—☾ Jan. 28, 11½ h. a., t. 14 dig., France.

☾ July 23, 6 h. m., t. central, Mississippi.

1889—⊙ Jan. 1, 9 h. a., c., Behring's Straits, Nootka, Hudson's Bay.

☾ Jan. 17, 5½ h. m., 8¼ dig., United States.

⊙ June 28, 9 h. m., c., south of Africa, Madagascar, S.E.

☾ July 12, 9 h. a., 5½ dig., Armenia.

⊙ Dec 22, 1 h. a., c., Carthagena, St. Helena, Abyssinia.

1890—⊙ June 17, 10 h. m., c., Cape Verd Islands, Smyrna, Pegu.

1890—⊙ Dec. 12, 3 h. m., c., Mauritius, New Zealand, Otaheite.
1891—☾ May 23, 7 h. a., t. 15¾ dig., India.
⊙ June 6, 4½ h. a., c., N.W. America, North Pole, Russia.
☾ Nov. 16, 0½ h. m., t. 17¼ dig., Ireland.
1892—⊙ April 26, 10 h. a., c., South Pacific.
☾ May 11, 11½ h. a., 11¼ dig., France.
⊙ Oct. 20, 7 h. a., v., North America.
☾ Nov. 4, 4½ h. a., t. 12½ dig., China.
1893—⊙ April 16, 3 h. a., c., Easter Island, Guiana, N.E. Africa.
⊙ Oct. 9, 9 h. a., c., Sandwich Islands, Peru.
1894—☾ March 21, 2½ h. a., 3 dig., New Guinea.
⊙ April 6, 4½ h. m., c., Egypt, China, Pacific.
☾ Sept. 15, 4½ h. m., 2½ dig., Canada.
⊙ „ 29, 5½ h. m., c., Madagascar, New South Wales, New Zealand.
1895—☾ March 11, 4 h. m., t. 18¾ dig., Barbadoes.
⊙ „ 26, 10 h. m., v., Atlantic, Europe, north of Asia.
⊙ Aug. 20, 0½ h. a., v., north of Asia.
☾ Sept. 4, 6 h. m., t. 18½ dig., Mississippi.
1896—☾ Feb. 28, 8 h. a., 10 dig., east of Persia.
⊙ Aug. 9, 4½ h. m., c., Prussia, E. Siberia, Pacific.
☾ „ 23, 7 h. m., 8 dig., New Mexico.
1897—⊙ Feb. 1, 8 h. a., c., New Caledonia, Easter Island, Guiana.
⊙ July 29, 4 h. a., c., Gallipagos, Barbadoes, Guinea.
1898—☾ Jan. 7, midnight, 1½ dig., London.
⊙ „ 22, 8 h. m., c., Fezzan, Socotra, north of China.
☾ July 3, 9½ h. a., 11 dig., Russia.
⊙ „ 18, 7 h. a., v., South America.
☾ Dec. 27, midnight, t. 16 dig., London.
1899—⊙ Jan. 11, 11 h. a., v., east of Asia, North America.
⊙ June 8, 7 h. m., v., north of Europe and Asia.
☾ „ 23, 2½ h. a., t. 18 dig., New Guinea.
☾ Dec. 17, 1½ h. m., 11½ dig., Cape Verd.
1900—⊙ May 28, 3 h. a., c., Mexico, Azores, Egypt.
⊙ Nov. 22, 8 h. m., c., Benin, Madagascar, New South Wales.

COMPANION TO THE ALMANAC,

FOR

1833.

PART I.

INFORMATION CONNECTED WITH THE CALENDAR AND THE NATURAL PHENOMENA OF THE THE YEAR; AND WITH NATURAL HISTORY AND PUBLIC HEALTH.

I.—ON COMETS.

THE year which has just passed away has been distinguished by the predicted appearance* of two comets, the most remarkable which have yet fallen under the notice of astronomers. These are what are commonly called the comets of Encke† and Biela‡. The latter has been an object of fear to many on account of the nearness with which it has approached, not the earth, but a point of the earth's path. As public attention has thus been turned to this subject in an unusual degree, we seize this opportunity of laying before our readers a slight account of the present state of cometary astronomy, distinguishing that which we really know of these bodies from the many surmises to which they have given rise.

The signification of the word comet has varied, as new bodies have appeared which analogy has led astronomers to include under that name. It was first given, as the word denotes, to bodies which appeared in the heavens with a train of light, or tail, and thus included some of the meteors which belong to our own atmosphere. We now apply the word to those heavenly bodies, without the limits of our own atmosphere, which are nebulous in their appearance, and with or without a tail. We may divide all

* Mr. Henderson has observed Encke's comet at the Cape of Good Hope, and Sir John Herschel that of Biela. We mention these facts here, as neither body is visible to the naked eye, and many of our readers may not be aware of their having been seen by any one.

† First discovered by M. Pons, November 26, 1818, but justly named by astronomers after Professor Encke, from his success in detecting its orbit, motion, and perturbations.

‡ First discovered by M. Biela, an Austrian officer, February 28th, 1826.

B 3

which have been observed into three classes: 1. Those whose returns have been predicted, and the prediction verified by the fact. These are three in number, viz., the celebrated comet of Halley, observed by him in 1682, which returned, according to his conjecture (for it could then hardly be called more) in 1759, and will appear again in 1835; its time of revolution is about seventy-six years. The other two are those above-mentioned, of Encke and Biela, which perform their revolutions respectively in about three years fifteen weeks, and six years thirty-eight weeks. 2. Comets whose return has been predicted unsuccessfully. Of these there is only one of any note, viz. that which appeared in 1770. This, it was found, should have returned in five years and a-half, if the observations made of it were correct; however, it never could be found again. This phenomenon threw doubt upon the return of comets, until the success of Laplace in devising methods for the calculation of the effects arising from the mutual attractions of our system, recalled the attention of astronomers to this almost forgotten failure. It had been found that the comet of 1770, in its approach to the Sun, had passed so near to Jupiter, that, on the theory of gravitation, the attraction of the latter was 200 times as great as that of the former. On applying the methods of Laplace to this case, it was found that, in 1767, while the comet was describing an orbit of more than 50 years, its motion was changed by the action of Jupiter so that it described the orbit observed in 1770; and that in 1779, it came again so near Jupiter that the preceding effect was reversed, and the orbit was again changed into one of long duration. 3. Comets which have been observed, the predicted return of which is yet to be expected. The most remarkable of these is the one observed by Olbers in 1815, which we may now safely say will return in 1887. 4. Comets which were observed at a time when neither theory nor observation was in a state sufficiently perfect to enable the observers to say whether they would return or not; and others, the orbits of which are uncertain, owing to the weather or other accidents not permitting them to be sufficiently well observed. Of these there are a great many, some of which may yet be recovered. For, long before the time of revolution of a comet could be found, astronomers knew how to determine, 1. The magnitude and position of its least distance from the sun. 2. Where its orbit cut the ecliptic. 3. The inclination of its orbit to the ecliptic. If a future comet should strongly resemble any one already observed in these particulars, and if its time of revolution as hereafter determined, should permit of its having been seen about the time of the former comet, we shall have sufficient reason to conclude that the two are one and the same. We must not, however, expect that the accounts of ancient writers on this subject with regard to the form of comets will ever be verified; such, for example, as those which describe comets in the shape of a sword, or surrounded with a shaggy mane. To say nothing of our never having observed such appearances in the course of the last century and a half, we must recollect the well-known fact, that comets were formerly considered as warnings of impending evils, or, at least, of remarkable changes. Thus Bodin, who died in 1596, gives it as his opinion that they are the souls of illustrious men, who having remained many ages upon the earth in the capacity of guardian angels, (for so the context must be interpreted,) are called to heaven in the shape of flaming stars. He attributes the plagues, famines, &c., which were supposed to follow, to the want of the prayers of these superannuated intercessors. Pope Calixtus III., in the fifteenth century, directed the thunders of the church, not only against the Turks, who had gained some successes, but also against a comet, which was supposed to have had some hand in, or at least to have foretold, them. When such impressions prevailed, it was natural that the appearance of the warning body should be somewhat exaggerated.

If from all that has been said upon comets, we take that which we certainly know, we shall have left a mass of conjectures of every grade of probability, from the one which may be considered as nearly proved, to those which, in point of evidence, might be placed side by side with the opinions of Bodin or Calixtus. We shall try to give some notion of the manner in which we come to know that which we do know, and some reasons for the most probable among the conjectures. Those who would read more of the history of surmises on this point, are recommended to consult the *Annuaire* of the French Board of Longitude for 1831, in which will be found a most amusing, as well as instructive, article on this subject, by M. Arago; an English version of which appeared in the *Times* newspaper some months ago.

That a comet is a material body is proved by the same sort of reasoning which is applied to the planets. Firstly, it either reflects the light of the sun, or shines by its own light; which of the two has never been distinctly proved; perhaps both suppositions may be true. Matter is always present where light is either emitted or reflected, at least on our globe. Secondly, comets are found to be acted upon by the laws of gravitation exactly as all other material bodies are, they are attracted by the sun, and move (so far, at least, as we can make out) in ellipses, or other conic sections, and this motion is disturbed, or, technically speaking, *perturbed* by the attraction of the planets, especially by the larger planets, Jupiter or Saturn. In this manner they have furnished one of the most decisive proofs of the Newtonian theory of gravitation. We have already mentioned the comet of 1770; but this, it may be said by those who cannot examine the calculations for themselves, was a trick of the astronomers, to account for their own failure. We will therefore cite another instance, in which the effects of planetary perturbation were very great, were predicted before the event, and verified by it. The comet of 1682, or of Halley, it is well known, was predicted by him as likely to appear in 1757. This he concluded from observing that a comet with a similar orbit had appeared in 1531 and 1607. He however remarked, that as the comet would, if his supposition were true, pass near to Jupiter and Saturn, some alteration might be expected from the attractions of these planets. In 1757, while astronomers were beginning to look

B 4

for the expected body, with no very great hopes of its reappearance, Lalande proposed to Clairaut to undertake the computation of the effect of the planets upon the comet. These names may not be so well known to our readers as to mathematicians and astronomers; we will, therefore, inform them, that Lalande was a practical astronomer of great eminence, and that Clairaut was a mathematician and natural philosopher of even greater celebrity. So little wedded were these men to the system of gravitation, that the first discarded, or, at least, threw doubt upon, the theory of the return of comets, on account of the non-appearance of that of 1770, already mentioned; while the second, on account of some unexplained phenomena, imagined that Newton had mistaken the law according to which the mutual attractions of planets depend upon their distance. The two undertook the enormous labour above-mentioned; and the result was, that Clairaut announced, in the year 1758, that the revolution which was actually taking place, would be 618 days longer than the preceding one, that is, the one which took place between 1607 and 1682. At the same time, he observed, that the methods of calculation were yet so incomplete, that the result could not be depended upon within thirty days. If his conclusion had been quite correct, the comet would have come to its perihelion, or nearest point to the sun, about the middle of April, 1759; and it did arrive there on the 13th of March of the same year, within the thirty days which had been allowed for errors. We may further remark, that the comets of 1832, of which that of Encke has once before appeared, according to prediction, and that of Biela* has been already observed by Sir J. Herschel, both very near their predicted places, could not have had their tables constructed without a strict attention to the planetary perturbations. From such facts we are justified in assuming that comets are material bodies, subject, like the planets, to the attraction of the sun and other bodies of our system, and describing an elliptic orbit round the sun *nearly*, the difference being attributable to the action of the planets, or, perhaps, in some degree, to a resisting medium.

The next question is, comets being material, what is their quantity of matter: that is, if brought to the earth without alteration of their dimensions, would they be light or heavy in proportion to their size. On this point we have sufficient evidence, not as to the actual quantity of matter in any comet, but as to limits below which it must fall, at least in all the comets of which the times of revolution are known. It results from the theory of gravitation, that of two bodies, the first cannot affect the second, without being itself more or less affected by the second. And of two bodies,

* On the subject of this comet, we derive our information from a communication made by Sir John Herschel to the Royal Astronomical Society, and read at the meeting of that body on the 9th of November last. An account of this interesting paper may be found in the Monthly Notice of the transactions of the Society for November, published by Priestley and Weale, Holborn.

one of which is very great compared with the other, the effect which the smaller produces upon the greater is small, compared with that which the greater produces upon the less. This is analogous (though the two phenomena must not be confounded) to a fact of every day observation, that a light body striking against a heavy one, though with great velocity, produces, nevertheless, but a small change in the velocity of the greater one, and *vice versâ*. For example, in the motion of Jupiter and Saturn it is observed, that the average velocity of Jupiter is accelerated, while that of Saturn is retarded more than twice as much. And it is shown, by a process independent of this observation, that Jupiter contains more than twice the quantity of matter of Saturn. After some ages, the motion of Jupiter will cease to be accelerated, and that of Saturn to be retarded. After which, that of Jupiter will begin to be *retarded*, while that of Saturn will begin to be *accelerated*. Hence, if a comet so large, or rather so heavy, as to bear an appreciable proportion to the mass of a planet, were to be disturbed by the latter in any considerable degree, the comet itself would produce a degree of disturbance in the motion of the planet, which would be perceptible to our instruments. Thus, if Halley's comet, which was retarded between 1682 and 1579, more than 500 days by the action of Jupiter, had been only the twenty-thousandth part of the mass of Jupiter, its effect upon the latter would have been even then most distinctly perceptible by good instruments. The same thing would take place now if the mass of that comet were very much less, and yet, in the former case, it would be less than one sixtieth part of the earth. But there are two much more conclusive arguments. Laplace found, that if the comet of 1770 had only been the five-thousandth part of the earth, it would have lengthened our year by three seconds. No such alteration has taken place, and the comet must, therefore, have been less than the five-thousandth part of the earth. The same body passed between the satellites of Jupiter in 1779 without producing any effect; a very little quantity of matter, much less than the five-thousandth part of the earth, would have been sufficient to derange that system perceptibly.

But, it may be asked, are we certain that we know the length of the year with such accuracy, that a difference of three seconds would be of sufficient magnitude to be discoverable by our instruments? To give an idea of the possibility of this, we will state the following fact. Some years ago, Professor Airy of Cambridge, proposed a method of determining the moon's mass, which required accurate observations of Venus near her conjunction. An ephemeris of this planet was accordingly prepared, containing the computed time at which the planet should pass the meridian daily, for that part of the year 1830, in which the conjunction of Venus happened; this was forwarded to different astronomers, English and continental, with a request that they would observe the real time of the meridian passage at their various observatories. Among the observations which were made in consequence, those of Professor Santini, of Padua, were so arranged as to show how

B 5

much they differed from the ephemeris. The difference was, in only a very few instances, so great as one second, and was, for the most part, nearer to half a second. And this result is not considered as anything remarkable.

The appearances of comets are, as far as appearances can be, proofs of their very small mass. The phenomenon of their tails, adopt what explanation we may, can only be accounted for on the supposition that the comets themselves are of very small density. But even the nebulous head of the comet has often been so rare, that small stars, which a fog of moderate intensity would hide, have been seen through its most central parts. Thus Seneca mentions the fact of stars having been seen through comets; Sir W. Herschel saw a star of the sixth magnitude through the centre of the comet of 1795; Professor Struve saw one of the eleventh through that of Encke; and Sir John Herschel, in the Memoir already cited, (in col. 2, note,) informs us, that on the evening of the 23rd of September last, he saw a whole cluster of stars of the *sixteenth* magnitude, almost through the very centre of Biela's comet, the light of which, according to Sir J. Herschel, could not have passed through less than 50,000 miles of the matter of the comet. As neither of the gentlemen above quoted saw any effects of refraction which would have been very apparent had the cometic matter been even many times rarer than our atmosphere, (if, indeed, they could have been seen at all through such a mass, which may fairly be doubted,) we are entitled to conclude that those comets, at least, which are best known to us, are of a rarity far exceeding that of any matter as it exists at the surface of our globe. If any man should assert that the largest comet ever seen, including its millions of miles of tail, contained no more matter than is to be found in the New River Head, he might justly be blamed for asserting more than he knew, but certainly any one who positively denied the fact would deserve the same censure.

As we are not writing for the scientific part of the community, we will say a few words on a very general fear which prevails—namely, that the near approach of a comet would break our planet in pieces, or at least produce a great accession of heat, sufficient perhaps to destroy animal and vegetable life, if not to burn the world altogether. The argument seems to have originated in a notion, that because heat produces expansion, therefore very highly expanded bodies must needs be very hot. It would be as good an argument to say, that because expansion by any other means except heat, produces cold, that therefore all comets must be very cold; and neither argument would, in the least degree, afford matter even for a rational conjecture. We can form so little idea of what the state of a planet of vapour, it may be consisting only of one sort of matter, would be, that we might with as much reason speculate upon the possible organization of the possible animalculæ which swim in that vapour, as try, in the present state of our knowledge, to ascertain whether any and what degree of danger awaits us from such a source. A comet *may* certainly strike the earth in the next century; not one of these which are known, unless the laws of nature be singularly altered, but some one or other yet to come. It has been shown, but by considerations of so high a nature that the result cannot be expected to bring much conviction to any but a mathematician, that if a comet were launched at hazard into our system, for one orbit in which it could strike the earth there are 281 millions in which no such thing could take place as the laws of nature stand at present. The *advocates* of cometary interference (we have met with some whose manner of expressing their opinion on the subject almost entitles them to that name) usually suppose a special interposition of the Divine power, which, (resting on their own interpretation of certain Scriptural prophecies,) they suppose will bring a comet on the earth. They are usually people of some religious feeling, and would act more consistently with the idea they ought to have of their own ignorance and the Divine power, if they ceased to prescribe to the Creator in what way it should please him to alter the course of events which it has hitherto been his will to arrange. It is impossible to produce any other argument on the subject, consistently with the design of this paper; the province of natural philosophy is to collect and compare facts, and to say what *will* be, if things continue as they *have* been; it never presumes even to conjecture what shall be, when the power which has hitherto disposed events in one manner, shall judge it right to ordain a different arrangement.

There are many who, without going the length of fearing danger from the shock of a comet, nevertheless imagine that any unusually hot weather which happens while such a body is visible, or going to be visible, is caused by it in some measure at least. To such a circumstance the fine vintage of 1811 was attributed, and many, even among the educated classes, imagined that the heats of last September and August were occasioned by the approach of Biela's comet. We can certainly re-echo, from this side of the channel, the complaint which M. Arago makes, in the *Annuaire* for 1832, already alluded to, of the scarcity of the meanest knowledge of scientific facts among the middle ranks of society. With a burning sun over head, we have heard those, who might have known better, accusing the comet in the manner aforesaid.

It appears, however, from the table of M. Arago, in which the mean temperature of every year, from 1803 to 1831 inclusive, is placed side by side with the number of comets observed in that year, that there is no visible connexion between the one and the other. Thus 1806 and 1811 were both hot years, the first however hotter than the second, though the first had one comet only of no note, and the second had two, one of which was the most brilliant which the present generation has seen. Again, the year 1826, with its five comets, was not so hot as 1831, which had only one. That hot years in general have more comets than cold ones is very true, and for this simple reason, that the former, generally giving a finer sky, are more favourable for their discovery. We must not

B 6

forget that the greater number of such bodies are not visible to the naked eye. Thus all the years between 1803 and 1831 inclusive, the temperature of which exceeded the average, mustered twenty-nine comets between them; and the remaining, or cold years, only fifteen. We must therefore say, not that the comets brought the heat, but rather that the heat brought the weather which made the comets visible. In the period above-mentioned there were forty-four comets observed, counting distinct appearances of the same comet as different; of which only two were in the least remarkable for brilliancy—those of 1811 and 1823.

Having shown that some comets are bodies in the highest state of tenuity, and conjecturing, with a great degree of probability, that the same is true of all, we may mention a phenomenon which has been several times remarked by different observers, viz., that in their approach to the sun they appear to contract their dimensions, or the nebulous head of the body diminishes in apparent diameter. As they recede from the sun they begin to dilate again. To explain this phenomenon, some have had recourse to the highly elastic fluid or ether, which, as we shall presently see, has been supposed to fill the solar system at least. If this ether, say they, be denser as we approach nearer the sun, we must expect that the comet will be more compressed by it as it approaches its perihelion, and will therefore be confined within smaller limits. To this it is answered, and justly, that such an explanation might suffice, if the comet had an exterior case, which, not being incompressible itself, should hinder the ether from penetrating the light body of vapour. In the memoir of Sir John Herschel already quoted, three distinct possible causes are suggested, two of which are entirely independent of an ethereal fluid, and all so probable, that it may be the phenomenon is partly due to every one of them. In the first place, on account of the great rarity of cometic matter, it may be that what we call cohesion exists only in a very trifling degree, so that perhaps we ought to consider the motion of the several parts of the comet independently of the others. For example, if the diurnal rotation of the earth were suddenly stopped, and it continued in that state to move round the sun, the parts nearest to the sun, being more attracted by it, would, if they were free to move by themselves, describe an orbit differing in a slight degree from that of the parts which are farthest from the sun. But as, owing to the cohesion of the various parts of the earth, they must all move together, the orbit really described by the earth's centre lies between those which would be described by the parts nearest to and farthest from the sun. We have hitherto considered the comet as one mass of matter, the motion of every part of which influences that of the rest. If, however, it should consist of particles so little bound together by cohesion, as to allow of each particle describing, or nearly describing, its own independent orbit, the consequence would be just the phenomenon observed—namely that it would contract as it approached the sun, and dilate as it receded again from it. To illustrate this, draw several ellipses about the same focus, very near to one another, and let one particle move upon each from the perihelion. It will be evident that, as the particles increase their distance from the sun, they increase their distance from one another, and *vice versâ*.

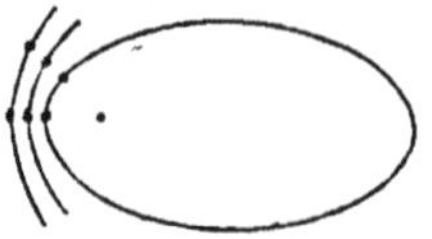

The second explanation proposed by Sir John Herschel is that, perhaps, by the motion of the comet from the sun, and its consequent appearance in a darker part of the heavens, some layers of nebulous matter may become visible, which were not so before on account of their yielding too little light. The third is, that the cometary matter may consist, like a fog, of small particles of moisture floating in a transparent fluid, and which the resisting medium, being hotter near to the sun, renders invisible, by raising their temperature, and turning them into vapour. If this were the case, it is evident, that as the comet approached the sun, the fog at the edges, so to speak, would be cleared up, and consequently the apparent part of the comet rendered less, and *vice versâ*. It is of course impossible to decide between these very ingenious explanations, so as to say which is the more probable; if our notions of the cometary state be just, we cannot deny that the two first must produce some effect: but the greater part of the phenomenon may still be due to the third or some other cause.

The only remaining point of any magnitude, connected with the known facts of comets, is the question, whether there is any fluid medium in space, of such specific gravity, as to offer a sensible resistance to their motion. The question of resistance, or no resistance, is one of great importance, as the *stability* of our system depends in a great measure upon it. The solar system would be said to be *stable*, strictly speaking, if it were so constructed that its motions might continue for ever in the manner now observed, without any such change, arising from the mutual attractions, as would endanger the safety of any one planet. If a number of planets were launched into space, without any particular arrangement of their motions, the chances for the stability of such a system would be very small. We observe in the solar system the following facts, which mathematical analysis shows us are all necessary to its stability, as far as the planets are concerned:—1. The sun is very much greater than any one of them, and the planets are placed at such distances from one another, that the attraction of the sun upon each is always very much greater than that of the other planets. 2. They all move in the same direction round the sun. 3. The orbits are very nearly circular, and are inclined to one another at small angles. From these circumstances, and the law of gravitation, it has been deduced that the average distances of the planets from the sun, and also the average motions, are invariable, or at least will continue the same for a number of ages, which, to our limited ideas, give the notion of eternity. But all this is on the supposition, that there is no fluid which offers any resistance to the planetary motions; if such a fluid exist, however small its density may be, it can be shown that it continually decreases the mean distances, and increases the mean motions. Observation teaches us, that the mean distances of the planets are invariable, or at least that, if they do change at all, the variation is so small, that it has not become perceptible to our best instruments in hundreds of years. The resisting fluid, therefore, if it exists, is of an extremely small density. Comets offer the only chance left to us at present of settling the question; being of very little density themselves, their motion will encounter more resistance than that of the planets, for the same reason that a feather will fall more slowly to the ground than the same weight of iron. If there be a resisting medium, it will certainly diminish the mean distance of the comet, and increase its mean motion; and this phenomenon has been observed in the comet of Encke, which is the one with which we are the best acquainted. Professor Encke says*—"If I may be permitted to express my opinion on a subject which for twelve years has incessantly occupied me, in treating which I have avoided no method, however circuitous, no kind of verification, in order to reach the truth as far as it lay in my power; I cannot consider it otherwise than completely established, that an extraordinary correction is necessary for Pons' comet"—that is, the one which all the world, except Professor Encke, calls Encke's comet—"and equally certain, that the principal part of it consists in an increase of the mean motion proportionate to the time." Professor Airy adds, "I cannot but express my belief, that the principal point of the theory, namely, an effect exactly similar to that which a resisting medium would produce, is perfectly established by the reasoning in Encke's memoir." If this conclusion be correct, we may predict that, in time, this comet and every other will fall into the sun; we know, however, that the medium, if it exists, cannot sensibly affect the planetary motions for a great number of centuries.

There seems to be some cause in operation by which the brilliancy of comets is continually diminishing. That of Halley, in one of its preceding revolutions, is described as giving a degree of light certainly superior to that which it gave in 1682 and 1759. Sir John Herschel could only see Biela's comet through a reflecting telescope of twenty feet in length, an instrument of enormous power in the collection of light; and though he afterwards found it with a refracting telescope, he asserts that he never should have succeeded with the latter, unless he had previously known where to look for it. If the parts of the comet have so little cohesion, as has been, with great probability, conjectured, it may easily lose a part of its substance as it passes through a resisting medium. We have however as yet but little specific information on this subject.

* In a dissertation which appeared in the *Astronomische Nachrichten*, and has been translated into English by Professor Airy, with an Appendix in which the latter gentleman fully coincides in the conclusion of Professor Encke. Those who are acquainted with the present state of science will give great weight to these authorities, to say nothing of their calculations being before the world.

With regard to the cause of the tails of comets, we can say nothing with certainty. Their existence affords a strong presumption for the very little density of the nuclei. They were at one time considered as being in a continuation of the line drawn from the sun to the comet; it has, however, been shown, that they always fall a little behind this line with respect to the comet's path, and have sometimes been even perpendicular to it. That of 1680 was 90° of the heavens in length, so that part of it might have been in the observer's zenith when the comet was setting. It was 141 millions of miles in length. Some comets have had what we may call a succession of tails, one succeeding another, with a vacant space between every two. The conjectures as to the nature and formation of these singular attendants are entitled to very little attention.

As to the multitude of idle theories with which, for want of better information, this part of astronomy has been loaded, such as that the planetary system was formed by matter struck off from the sun by one comet; that another caused the deluge; that the four small planets were formerly one, which was broken in pieces by a third; that the moon was originally a comet, and the like;—we would willingly amuse our readers by an account of them, if our limits permitted. They will however find them all, handsomely exposed by M. Arago, in the *Annuaire* already cited. If any, or all of them, should be hereafter proved to be true, it will be no excuse for those who first made them; for a result produced on insufficient evidence is bad, whether true or false. As the science of astronomy approaches towards perfection, we shall doubtless add some important and interesting facts to our knowledge of comets.

Elements of the Orbits of the three Comets, which have appeared according to prediction, taken from the work of Professor Littrow,—"Ueber den gefurchteten Cometen des gegenwärtigen Jahres 1832, &c. Vienna, 1832."

	Halley.	*Encke.*	*Biela.*
Longitude of the ascending node	54°	335°	249°
Inclination of the Orbit to the Ecliptic . .	162°	13°	13°
Longitude of the perihelion	303°	157°	108°
Greatest semi-diameter, that of the earth being called 1.	18	2·2	3·6
Least semi-diameter	4·6	1·2	2·4
Time of revolution in years	76	3·29	6·74
Time of the perihelion passage	Nov. 16 1835	May 4 1832	Nov. 27 1832

The comets of Encke and Biela move according to the order of the signs of the zodiac, or have their motions *direct;* the motion of that of Halley is *retrograde.*

COMPANION TO THE ALMANAC,

FOR

1834.

PART I.

INFORMATION CONNECTED WITH THE CALENDAR AND THE NATURAL PHENOMENA OF THE YEAR; AND WITH NATURAL HISTORY AND PUBLIC HEALTH.

I. ON THE MOON'S ORBIT.

In this Paper it is our intention to confine ourselves strictly to the question—What is the path described by the moon in absolute space, and what are the difficulties in the way of finding it? Many of our readers will imagine that we are agitating a question which has already been well worn in popular works; and will think they can in few words sum up all the information they can expect to derive from any but a mathematical discussion of the lunar orbit, by saying that the moon moves round the earth in a circle, or ellipse at least, while the earth is moving round the sun. It is our object to show how very rough an approximation to the truth is contained in the preceding notion.

Astronomical language abounds in the species of fictions, which, while they are perfectly capable of being made the foundation of hypotheses mathematically true, are not repugnant to common ideas, but the contrary. We need only instance that, in speaking of the sun, it is always supposed to move round the earth, the latter being at rest. No inconvenience arises from such a supposition,* provided the planetary motions are described in strict conformity to it. The same remark may be made in many other cases.

The definition of the moon's orbit, above-mentioned, seems to leave the astronomer no very difficult task, and it cannot but excite surprise in the minds of those who have adopted it, that the lunar theory is always mentioned as the most abstruse part of astronomy, more so than that of the planets, though the moon seems simply to move round the earth, while some of the planets move sometimes in one direction, sometimes in another, and sometimes stand still. Our object in this paper is, therefore, avowedly to complicate the question, by pointing out what the lunar orbit really is, and the irregular manner in which the moon describes it.

The orbit of the moon was known with very little accuracy by

* For matters connected with this subject in general, see the articles on *Motion*, Penny Magazine, vol. i, pp. 346, 358; and on the *Moon*, vol. ii, pp. 236, 262.

B 2

the ancients. The Egyptians and Jews could not predict more than the day of the appearance of the new moon, on which some of their observances depended: the hour they watched for; and, among the latter people, the appearance of the moon's limb was announced* by sound of trumpet. Ptolemy could not, with any prospect of success, have ventured to predict, a month before the phenomenon, at what time the moon should come on the meridian: he would probably have been wrong seven or eight minutes of time. From Halley's tables an error of half a minute might be looked for on this point; from those of Mayer one of four seconds only. The tables of Burg, which were verified by 148 observations made at different observatories, about the year 1802, very rarely gave an error of so much as one second; and from the observations published by Professor Airy, of the Cambridge Observatory, which are particularly valuable, because a direct comparison is made of the predicted and observed place of the moon, it appears that, though in some very rare instances, perhaps in one out of fifty, the error is as great as one second, its average amount is not more than half a second. This may be considered as an index of the present state of astronomy.

There is, however, yet a possible chance of our having underrated our actual knowledge of the lunar motions. A few years since, Baron Damoiseau published the first lunar tables *formed from theory alone.* At the end of the last century the same method produced tables of the satellites of Jupiter, which represented their motions better than those formed from observation. The phrase, however, needs explanation; previously to which we must introduce the following account of the fiction of the *elliptic orbit.*

It is well known that all secondary bodies move round their primaries in ellipses nearly: that is, the arc moved over in a short time, a day or two for example, is very nearly that of an ellipse. The lunar orbit, therefore, for a short space, might be represented as in the accompanying diagram, in which

—·—·— represents a line drawn on the plane of the ecliptic,
———— " " " moon's orbit *above* the plane of the ecliptic,
········ " " " moon's orbit *below* the plane of the ecliptic.

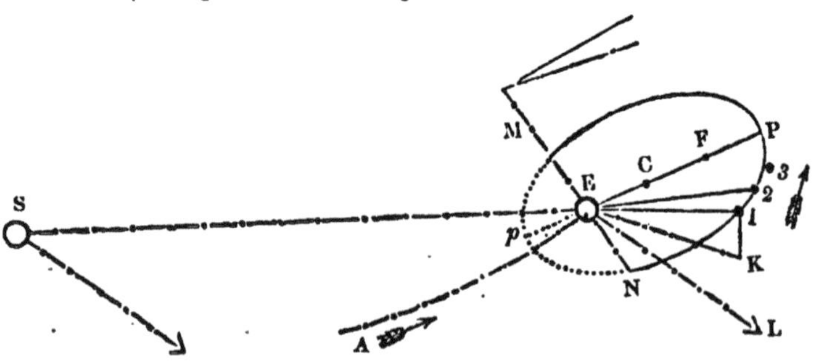

* Montucla, Hist. Math. i. 416. Psalm lxxxi. 3.

S is the sun, E the earth, E A a part of the orbit of the latter round the sun; M P N a plane which, for the time, is that in which the moon moves, and which is carried round with the earth without changing its position in other respects; E L is the line in which the earth's equator cuts the ecliptic, from which the longitude is measured. The moon moves on its orbit from 1 to 2 in a given interval, say a day; and M P N is the ellipse,* of which the arc 1 2 is more nearly the path of the moon than any arc of any other ellipse which can be drawn. This is, for the time, the elliptic orbit of the moon. If a perpendicular 1 K be drawn from any point 1 of this orbit to the plane of the ecliptic, and E K be joined, the angle L E K is the *longitude* of 1, and 1 E K is its *latitude*. The longitudes are measured round in the direction of the earth's motion: thus the longitude of the point N is not the angle L E N, but the *other* opening made by L E and E N, which, as the figure is drawn, is greater than three right angles.

The *elements* of the *momentary* orbit, or the magnitudes which must be found before it can be correctly laid down, are—1. The inclination of the plane M P N to the plane of the ecliptic; that is, the angle made by drawing perpendiculars to N M, one in the plane of the ecliptic, one in the plane M P N, lengthened for the purpose if necessary: this is represented in the figure beyond M. 2. The longitude of one of the nodes M or N; N, the *ascending* node through which the moon would rise to the north side of the ecliptic, is generally chosen: these two elements determine the *position* of the plane of the orbit. 3. The length P*p*, or its half C P, called the *mean distance*, because it is the half sum of E*p* and E P, the greatest and least distances of the moon from the earth: on this element, and this only, depends the average time of a revolution. 4. The *excentricity*, a name given to the proportion which the distance E F between the foci bears to the whole axis P*p*: these two last elements determine the form of the orbit. One focus of the orbit is always at the centre of the primary body; it only remains, then, to know how to place the orbit determined by elements 3 and 4 in the plane determined by 1 and 2, which is known when the position of the point *p* is known. This is the lunar *perigee*, or its point of nearest approach to the earth; and its longitude must be known before the position of the axis P*p* in the plane can be determined.

Let 1 and 2 be the places of the moon at intervals of a day, so that, as above described, the real arc 1 2, described by the moon, may be considered, without any appreciable error, as an arc of the momentary orbit just described. In another day let the moon be at 3. The arc 2 3 will not, strictly speaking, be another arc of the preceding orbit; for, though 3 will be very near to the orbit, it will be possible to draw through 2 3 another ellipse whose arc shall more nearly represent the path of the moon from 2 to 3 than any

* As some of our readers may not know the precise meaning of the word *ellipse*, let them imagine another point F, and C the middle point between E and F. Let a string, twice as long as C P, have its ends fixed at E and F, and let a pencil be drawn round, so as always to keep the string tight. The curve described will be the ellipse, C its centre, E and F its foci, P the point which is farthest from E.

arc of the ellipse M P N. This second ellipse being described, is shown to differ, though slightly, from the ellipse M N P in every circumstance, except that of having one of its foci in the centre of the earth at E. Some of its elements, indeed, do not differ appreciably from those of M N P; such as the excentricity and the mean distance: but others, such as the position of M N, or the longitude of the nodes, and the position of P *p*, undergo very sensible variations. In this sense, then, and this only, can the moon be said to revolve in an ellipse,—namely, that the ellipse itself must be supposed to undergo a continual change. If we were to suppose M P N *p* to be a perfectly flexible tube, always having the form of an ellipse, and contrived to accommodate itself to the moon's path, so that the latter, though always inside it, should never actually come in contact with it, we must then suppose the ellipse alternately to swell and subside, the line M N to be carried round the contrary way to the apparent motion of the sun, with a slow and irregular motion, the line P *p* to follow the moon with another motion, and other phenomena of a similar kind.

We may make a very simple illustration of the same kind of fiction. It is well known to mathematicians, that through any point of a curve a circle may be drawn, which shall be closer to the curve than any other circle; and that, for any small arc of a curve, an arc of a circle may be found, which shall be sufficiently like the arc of the curve for every practical purpose. Let us trace the march of hypothesis upon a distant body, moving in reality along the curve A B C D E, but so slowly, that each of the arcs A B, B C, &c., would be described by the star (as we shall call it)

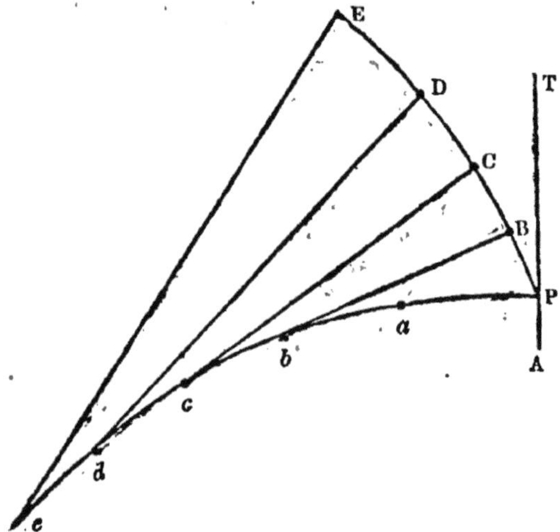

in not less than two or three generations of observers. Let the motion of the star be first suspected when it is at A. Those who first observe it with rough instruments will conclude that it is moving along the straight line A P T, which will continue to be received until better observers will discover a flexure in the direction of the motion a little after P; and will conclude that the real motion is in a circle whose centre is at *a.* This opinion, when its supporters have been sufficiently imprisoned and burnt, will gain ground on the other, and the circular motion will be held to be established. Observers, when the planet is at B, will suspect that their ancestors, though correct as to the circular theory, have not well fixed the position of the centre, which they will place about *b*, instead of *a*. At D, the observers, finding themselves again obliged to change the position of the centre, will see that the circular theory is not correct, unless they suppose the circle itself to have an increasing radius, and a moveable centre. By comparing old observations with their own, they will find the nature of the curve *a b c d*, and will discover the law which links the position of the centre and the length of the radius to the time of observation. From this time, accurate prediction will begin; and, instead of being observed at E, with a view to determine its place, the star will be watched, to ascertain whether its actual place agrees with that laid down in the tables which have been formed.

The first hypothesis made on the lunar orbit was, that it moved uniformly in a circle round the earth. The incorrectness of this supposition soon became apparent; but though several of the irregularities were discovered, and, to a certain extent, tabulated, no great step was made in the determination of an orbit till the time of Kepler, who first proved that the orbits of the moon round the earth, as well as those of the planets round the sun, were more nearly ellipses than circles. But even at this period several circumstances connected with the moon in particular, rendered it necessary to suppose a moveable, not a fixed, ellipse. The Newtonian theory of gravitation, while it accounted for, and connected together various of the inequalities, left much unexplained, until the middle of the last century, when Clairaut, Euler, D'Alembert, Lagrange, and Laplace, availed themselves of the improvements in mathematical analysis; and not only demonstrated that *all* the inequalities discovered by observation were reducible to the same principle, but discovered others, by theory, and pointed out how to make their existence manifest by observation. The result has been, that the lunar orbit can only be considered as an ellipse, on the supposition that every element is, more or less, in a state of perpetual change. These changes are mostly *periodic;* that is, after a certain time, diminution takes the place of increase, and *vice versâ.*

The preceding explanations, though very necessary to every part of our subject, yet grew immediately out of the phrase—*tables formed by theory alone.* A table is said to be formed by theory alone, when nothing is observed except the place of the moon at some one epoch, and the then elements of the moveable orbit; that is, when the nature and magnitude of all the *changes* which take place in the elements, are calculated from the Newtonian law of gravitation, and not from observation. Tables are said to be formed from observation and theory combined, when any one of the *changes* has been observed, and not deduced. For example, though the motion of the line Pp (see fig. page 6) round the point E was determined, by theory alone, so nearly to the result of observation as not to differ from the latter by more than its 440th part, it was customary to prefer the result of observation to that of theory; and the choice was a wise one; for, up to the year 1810 at least, it would have been doubtful whether tables formed from theory alone would have stood the test of observation as well as those already in use. M. Damoiseau, as before stated, has given astronomers the means of judging on this point. With a degree of industry which can never be justly appreciated by one unacquainted with the subject, he has reviewed the whole lunar theory of Laplace, and constructed tables for determining the daily positions of the satellite. These tables have been compared with those of Burckhardt; and the comparison is published in the Nautical Almanac for 1835. It there appears, that the two agree as nearly together as observation agrees with the latter tables; that is, there is very rarely an error of a second of time; and the average error is more near to half a second.

We now come to our main subject. What is the lunar orbit like, independently of all considerations respecting the fictitious ellipse above described? Let E be the earth, and let a circle A q B be

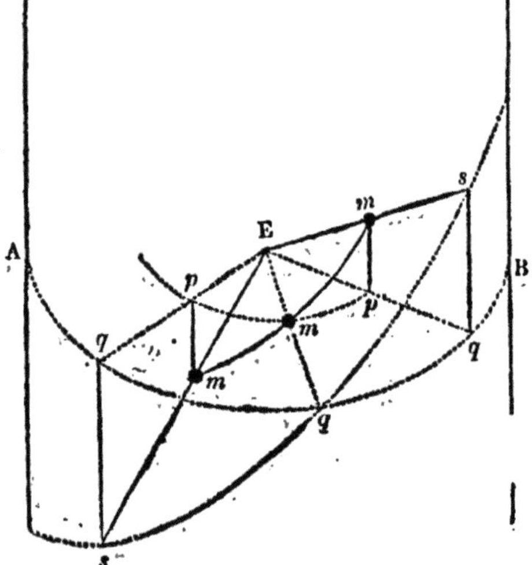

taken on the plane of the ecliptic, greater than the greatest distance of the moon from the earth. On this circle let a cylinder be described; inside which, therefore, the moon will always be. Let *m m m* be the *actual path* of the moon for a time; as we have drawn it, the left hand *m* is below the plane of the ecliptic; the middle *m* on the plane of the ecliptic, or the moon is in her node, and the right-hand *m* is above that plane. Let the motion of the moon be from left to right, that is, the middle *m* is the *ascending* node. Its longitude, on January 1, 1834, is 95½°; so that the line from which longitudes are measured is about 5½° farther round than the line A E, in the invisible half of the cylinder. At every position of m, let mp be let fall perpendicularly on the plane of the ecliptic; and let Em be lengthened to meet the cylinder in s. The point p will trace out a curve on the plane of the ecliptic, always directly under or over the real path; while s will trace a curve on the cylinder. If the eye were placed at a very great* height above E, in the axis of the cylinder, all the real path above the ecliptic would hide its corresponding portion of the curve pp; and the rest of the curve pp would hide the part of the real path below the plane of the ecliptic. But if the eye were placed at E, the real path would entirely hide the curve ss. We may consider the curves pp and ss as a sort of ground-plan and elevation of the moon's real path, which latter cannot be laid down on a plane. We call them the projection on the ecliptic; and the projection on the cylinder. Any point of either projection being given, the moon's corresponding place can be found. For, s being given, draw sq, the side of the cylinder, join E, q, and E, s, and, having found p the intersecting point of E q with pp, draw pm perpendicular to the plane of the ecliptic, which will meet E s in the required point m.

The following curve represents the *species* to which pp, the projection on the ecliptic, belongs. We say the *species* only, because

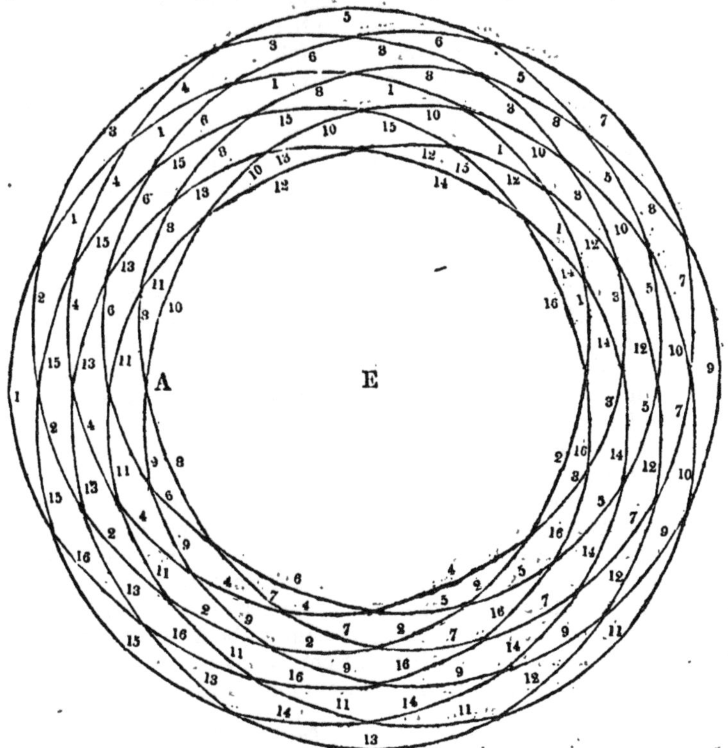

* Mathematically speaking, at an infinite height; but practically, the two curves would combine at any very great height.

the irregularity on which its form depends is very much magnified. We here see how the moon's orbit would appear, to an eye placed at a great height over the earth; and as the height of the moon above the plane of the ecliptic, *pm*, is never so much as one-tenth of E*m*, her distance from the earth, the principal features of her motion *round* the earth are preserved in the design before us. The earth is in the centre of the figure, at E, and the moon is a little above or below the curve of the orbit, which we shall neglect for the present, and suppose it in the projection on the ecliptic.

Let the moon be at 1, on the outside left of the figure, and at her greatest distance from the earth. She moves in the direction 1 1 1, &c., until she comes to 2, at which point she is at her least distance from the earth. The first 1 is called the *apogee;* the first 2, the *perigee* (*apo*, from; *peri*, near to; *gee*, the earth). We observe that the perigee 2 is not directly opposite to the apogee 1; that is, the moon describes more than half a revolution in moving from her greatest to her least distance; and, by following the branch 2 2 2, &c., till we come to the apogee 3, we also see that more than half a revolution elapses between the perigee and the apogee. By following the order of the figures with a pencil, the whole orbit may be traced out, and the successive perigees and apogees determined. Nine complete revolutions will bring the moon again to the apogee 1, before which it will have been in apogee eight times, and in perigee the same. We must, however, correct the idea here given by the following considerations.

1. In the figure, the greatest distance of the moon is half as much again as the least distance. But the greatest distance does not exceed the least distance by more than two-nineteenths of the latter; that is, 1 A should be only two-nineteenths of E 8. The orbit is, therefore, contained within much narrower limits than the one in the figure.

2. In the figure, the average period between apogee and apogee is greater than a revolution by one-eighth of a revolution. But in fact, the distance from apogee 1 to apogee 3 is only about the 117th part of a revolution. At the 116th apogee from 1, the moon will not be quite upon E 1; and at the 117th, she will have passed it: so that, instead of having a closed curve, as in the figure, we shall have a continuing series of similar orbits *, interlacing, as in the figure, but forming a closer net-work.

3. The distances 1 3, 3 5, &c., from apogee to apogee are not always precisely the same, as in the figure; which will give a somewhat irregular appearance to the orbit, such as may be imagined would arise, if it were formed of wire which shrunk a little in some parts, and expanded in others. We need hardly add that the preceding figure is the curve *p p* of page 10.

The whole of the preceding phenomena are such as would take place if the moon moved in an ellipse, the focus of which was at E, and having vertices distant from E, by lines equal to E 1 and E 8. Let the moon move in this ellipse, while the axis itself of the

* The mathematician will here call to mind the distinction between a hypocycloid made by ovals with commensurable and incommensurable circumferences.

ellipse moves slowly round E, in the same direction. The reason of the perigee 2 not being opposite to the apogee 1 (as it would be if the ellipse were stationary) will now be apparent. So far as this one inequality is concerned, we have here the astronomical fiction of the elliptic motion, reconciled to actual appearances by the supposed *progression of the apogee.*

A similar phenomenon may be made apparent by swinging a watch-chain, with seals attached, in an elongated orbit. The progression of the longest line of the orbit will immediately become visible.

But we must recollect, that, in the preceding figure, we have only that assemblage of points over or under one of which the moon is always to be found. We have yet to consider the rise and fall of the orbit with respect to the plane of the ecliptic; for which purpose, suppose the cylinder in page 10 to be unrolled, which will devolope the curve *s s* upon the plane of the paper. The first of the following diagrams is the development so obtained; the cylinder represents the cylinder before it was unrolled. Any given letter corresponds to the same point in both.

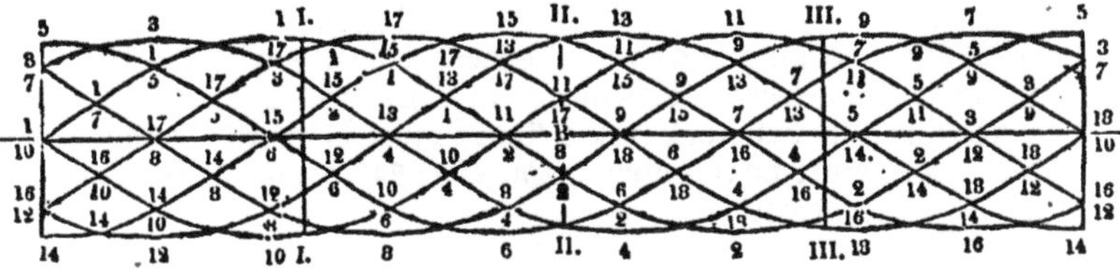

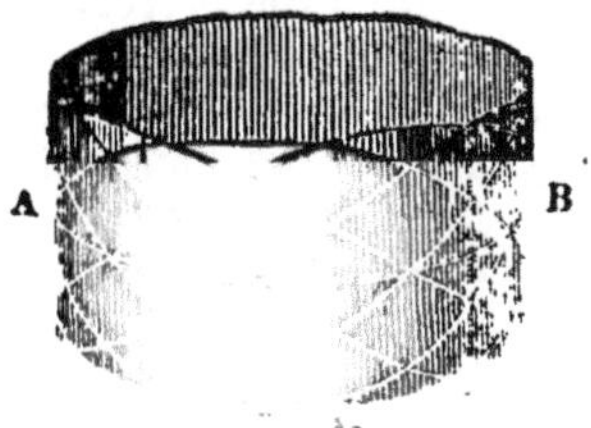

The line A B A is the ecliptic, the letters corresponding to the figure in page 10. The extreme lines 5 14, and 5 14, will unite when the cylinder is restored, and the verticals I., II., III. will then divide the whole revolution round the earth into four quarters. The path may be traced out, as in page 11, by following the course of the figures with a pencil. On looking back to page 10, we see that the latitude of the moon is the angle *s* E *q*; and *s q* the height of *s*, the moon's projection on the cylinder, above the ecliptic, or the perpendicular distance of the orbit above A A in the developed figure, is not the moon's latitude, but what is called, in trigonometry, the *tangent* of the latitude to the *radius* E *q*. Strictly speaking, this does not vary *as* the latitude; for, doubling the latitude will more than double qs; but as the moon's latitude is never more than $5\frac{1}{2}°$, the height sq will vary so nearly with the latitude, that we may suppose sq to vary strictly as the latitude, without sensible error.

Setting out from the ascending node 1, we see that the moon attains her greatest elevation above the ecliptic before she has completed a quarter of a revolution, and comes to the descending node 2, before half a revolution has been finished. She is at the ascending node 3, before the whole revolution is completed, and so on. The rise and fall of the moon is, therefore, completed in less than a revolution; so that, if a star moved uniformly round the ecliptic, always arriving at the moon's node at the same time as the moon, the period between two eclipses of the star would be less than half the time of a lunar revolution. Eight complete revolutions will, in our figure, bring the moon again into the ecliptic at 1, before which she will have been seventeen times at a node. But here we must observe:—

1. In the figure, the height 51 is about the eighth part of A A. It should not be much more than the seventy-second part of A A; that is, the breadth of the zone should be very much diminished.

2. The average time from node to node is less than the whole revolution by the ninth part of the latter; whereas the same should only be about one 249th part of the revolution. Neither the 249th or 250th ascending node would exactly coincide with 1; the first being a little behind, the second a little in advance of it: so that, as in page 11, the net-work of the figure must be more abundant and close, and must be repeated again, after 249 complete revolutions, or 19 years nearly.

3. The intervals from node to node are not exactly equal, as in the figure; but vary—sometimes being greater, sometimes less, than the average drawn from the preceding remark. Neither is the moon's greatest latitude always the same, but subject to a small irregularity, varying, from the average, about one 37th part of the whole. The effect on the appearance of the figure it is needless to describe.

Returning to page 12, the phenomena of the moon's latitude may be explained, by supposing the ellipse, instead of being in the plane of the ecliptic, to be placed partly above, and partly below it, so as to cut the ecliptic in a line passing through E (see fig. page 6). While the moon revolves in this ellipse, and while the ellipse itself revolves in its own plane, so as to carry the apogee and perigee after the moon, as above described, the line M N must revolve slowly round E, in the direction contrary to that of the moon; that is, from east to west. This is what is meant by the *regression* of the line of nodes; and the variation of the greatest latitude may be obtained, by giving the ellipse a small oscillation round the axis M N; that is, an alternate approach to, and recession from, the ecliptic, of about a common month in duration.

It would be difficult to exhibit the actual orbit of the moon in a wood-cut, so as to place before the eye at once the effect of the progression of the apogee, and the regression of the nodes; but the same might be readily done by a coil of wire. If, in the mean time, we suppose the earth E to be carried about the sun, it only remains to consider how the moon must be imagined to move in the ellipse, the motion of which explains the preceding phenomena. If we examine the moon's daily increase of longitude for a few days together, we shall find that it is not *very* different in any one interval of twelve hours, and that which succeeds it. We take the following cases from the Nautical Almanac:—

1834 Jan.		☽'s Longitude.			Increase of Long. in the interval.			
9	Noon . .	283°	25′	9″				
	Midnight .	289	44	22	6°	19′	13″	New.
10	Noon . .	296	0	26	6	16	4	
	Midnight .	302	13	30	6	13	4	
11	Noon . .	308	23	28	6	9	58	
	Midnight .	314	31	0	6	7	32	

Moon in apogee, Jan. 15, 6 in the morning. 1st Quarter Jan. 18, at two in the morning.

18	Noon . .	32	17	54			
	Midnight . .	38	24	50	6	6	56
19	Noon . .	44	35	44	6	10	54
	Midnight . .	50	51	8	6	15	24
20	Noon . .	57	11	35	6	19	27
	Midnight .	63	37	33	6	25	58

We see, then, that the lunar motion in longitude during twelve hours is not *very* different at different times; that is, would not be thought so by an observer without instruments, and unused to astronomy. And so it was considered by those who first observed the heavens, who attributed to our satellite a perfectly uniform motion. They soon discovered that the synodic month, or round of all the phases from full moon to full moon, was about twenty-nine days and a half. But it will appear that the change in the daily motion is *very* considerable, when we take modern astronomy as our test. Suppose that an observer had, on the 9th of January, 1834, come to the conclusion that the moon's motion in longitude was 6° 19′ in twelve hours. If he accordingly predicted the moon's place for the twentieth at midnight, his calculation would be as follows:—

Moon's longitude on January 9, at noon	283° 25′ 9″
Increase of longitude for 23 intervals of 12 hours, at 6° 19′ each...............	145° 17′
Sum, with 360° for a complete revolution, subtracted or his predicted longitude	68° 42′ 9″
Real longitude January 20, at midnight	63° 37′ 33″
His error of prediction	5° 5′ 36″

Now suppose, which might easily have happened, that the moon had been so placed in the heavens as to throw any error made in predicting her longitude almost entire into the time of her coming on the meridian. The error of 5° 5′ would make 20 minutes 20 seconds of difference between this predicted time of coming on the meridian and the real phenomenon. The error of our astronomy would, as we have seen, be about half a second, or at most a second. The error above cited would be 1220 times as great If the observer above supposed persisted sufficiently long in his error, the difference between his theory and observation would continually increase, and he would only be right at the periods when his error happened to be at or near the whole 24 hours. This brings us to the first step of our explanation. If, instead of supposing that the moon's motion in 24 hours was 6° 19′ × 2, or 12° 38′, he had supposed it to be 13·17639639, or 13° 10′ 35″·027, though he would be always wrong in the times of the moon's coming on the meridian, yet this error would never exceed from half to three-quarters of an hour, and would be sometimes on the one side, sometimes on the other; so that if he were to add together the errors of excess for a long period, and compare this with the sum of the errors of defect during the same time, he would find the difference between them to be very small indeed compared with the whole amount of error. To exemplify this, let us see how near we can deduce the longitude on December 31, 1834, at midnight, from that of January 1, 1834, at noon; the interval being 364 days and a half.

Moon's longitude, January 1, at noon	175° 10′ 42″
Motion in 364½ days, at 13° 10′ 35″·027 per day...	4802° 47′ 47″
Sum	4977° 58′ 29″
Deduct whole revolutions, or 13 × 360..............	4680°
Longitude thus predicted..................................	297° 58′ 29″
Real longitude, from the Nautical Almanac.	301° 39′ 39″
Error	3° 41′ 10″

Proportion of the error to the whole arc described, $\frac{1}{1358}$ nearly.

If we were to try the same on any greater period, we should always find the error within limits not much exceeding twice the above error. The fact may be stated thus:—If we were to suppose another satellite moving in the same orbit as the moon, with a uniform daily increase of longitude amounting to 13° 10′ 25″·027, the real moon would never be more than 9° either before or behind the fictitious moon; and, roughly speaking, would be as often before as behind it: so that whatever mistake might be made in *excess*, by substituting the mean moon for the real one at any time, a mistake of equal magnitude would be made in *defect* by making the same substitution at another time.

A certain mathematical affinity between the laws followed by the motion of a pendulum, and the progress of the various inequalities which constitute the whole departure of the real from the mean moon, (which, however, must not be mistaken for an affinity of *principle,*) furnishes an easy and striking illustration of the whole theory of the lunar inequalities.

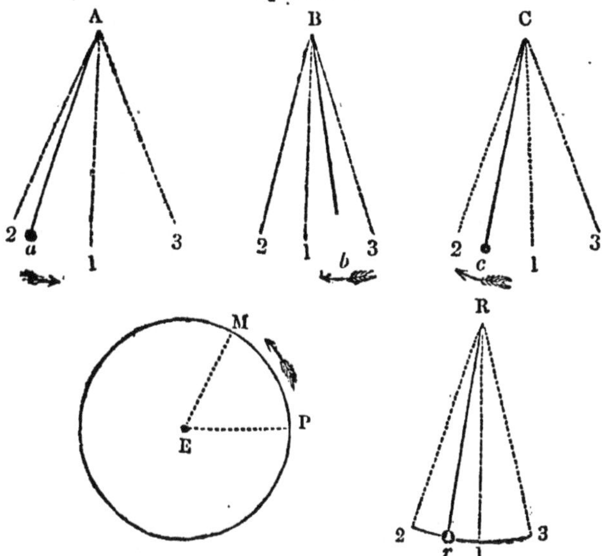

Let A, B, C, &c., be a number of pendulums, not subject to friction, &c.; let A 1, B 1, C 1, &c., be their vertical positions; and A 2 A 3, B 2 B 3, C 2 C 3, &c., their highest points of ascent from the vertical on either side, so that the angles 1 A 2, 1 B 2, &c., are the greatest angular departures from the vertical. We know that, by sufficiently lengthening a pendulum, we can increase its time of vibration to any extent; and that, but for friction and the resistance of the air (which we suppose excluded), any angle of vibration may be maintained as long as we please. The arrows mark the direction of the pendulums for the moment under consideration. Let E represent the earth, round which let the mean moon, already supposed, revolve in the plane of the ecliptic with the daily angular velocity of 13° 10′ 35″·027. By constructing a number of pendulums of certain definite lengths, with certain given extents of vibration, and which reach their highest points at certain given times, not the same for all, it would be possible to form a system from which the moon's real longitude might be deduced from that of the mean moon, in the following way:—Knowing the place of the mean moon at a given epoch, compute the time elapsed between that epoch and the moment for which the moon's real place is required. Calculate the motion of the mean moon for that interval at the rate above given, and thus find its place, as in page 16. Let the pendulums be supposed to be stopped at the moment under consideration, and add together the elongations from the vertical or the angles 1 A *a*, &c., in all those pendulums which are then moving from *right to left.* Do the same in all those which are moving from *left to right;* and, if the first sum exceed the second, put the place of M forward, or increase the angle P E M by that

C 2

excess. But, if the first fall short of the second, put the place of M back, or diminish the angle P E M by that defect. In both cases M will be removed from the mean to the true place of the moon: if the first and second sums should happen to be equal, then, for the moment under consideration, the real moon is in its mean place.

By a *lunar inequality in longitude*, we mean the elongation from the vertical of one of these pendulums, by which it becomes necessary to increase or decrease the moon's *mean* longitude, as part of the process for finding the *true*. By the *period* of the inequality is meant the time of a double vibration, or once backwards and once forwards; and by the *maximum* (or greatest value) of the inequality is meant the angle of greatest elongation from the vertical, or half the extent of the whole vibration.

This general view belongs equally to the motion of the planets and comets round the sun, that of the satellites of Jupiter, &c. round their primaries, that of the ocean and atmosphere of the earth, and in fact to every motion in the solar system which we can watch so well as to detect any irregularities. It is what the mathematician would express by saying that all inequalities yet found can be developed in a series of sines and cosines of angles depending upon the time of observation. We shall try to show how it happens that, instead of being able to consolidate all these inequalities into one, and give a simple method of representing their united effects, we are obliged to point them out separately. The illustration we shall suppose, though utterly inconceivable how it should happen, yet contains no contradiction, and presents the *sort* of difficulty which has caused the planetary theory to be the most complicated part of human knowledge.

Imagine a set of people acquainted with the rule for the multiplication of one fraction by another, but not with that for division, and with the rule of addition, but not of subtraction. Let them observe two phenomena which they suspect to be connected with one another. For example, let the planet A move very slowly backwards and forwards from B to C with an irregular motion, while in the same period the inferior planet *a* moves so as to

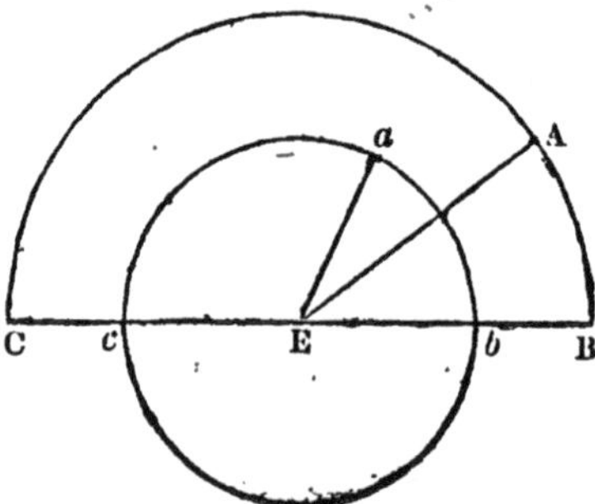

describe the circumference *b a c b*, while A describes B C, and moves back again through *b c a b*, while A returns through C B. Let *a* be observed to be always in advance of A, reckoning from E B, and let all computations be made in fractions of circumferences: thus, let a right angle be $\frac{1}{4}$, a whole circumference 1, and so on. According to the law which we have imagined (but have not yet stated) to connect the motions of the two, observers will soon see, that when *a* begins to be sensibly in advance of A, which will not happen till both are at some little distance from E *b* B, the longitude of *a*, or B E *a* will differ from that of A, or B E A, by that proportion of a revolution which is made by multiplying B E A by itself: for example, that when B E A $= \frac{1}{50}$, B E *a* will be nearly $\frac{1}{50}$ and $\frac{1}{50}$ of $\frac{1}{50}$, or $\frac{51}{2500}$. This will be the first *inequality* discovered. After some time, it will be found that a correction must be applied to the preceding rule, obtained by multiplying the first inequality again by that proportion of a revolution which expresses the longitude of A: for example, if A's longitude be $\frac{1}{20}$, that of *a* will be found to be nearly $\frac{1}{20}$, and $\frac{1}{20}$ of $\frac{1}{20}$, and $\frac{1}{20}$ of $\frac{1}{20}$ of $\frac{1}{20}$, or $\frac{421}{8000}$. Thus, either by theory or observation, they will discover the mathematical law which connects the two to be as follows:—let x be the longitude of A, then that of *a* is $x + xx + xxx +$, &c., *ad infinitum.*

The separate terms they will call inequalities, and will, perhaps, at first, retain $x + xx$ as a simple approximation, similar to the *elliptic orbit* already described for the moon, and will regard the rest as irregularities. Now, had these observers known the subtraction and division of fractions, they could not have failed to discover that the united effect of the whole string of inequalities is contained in the following more simple law. Divide the longitude of A by the remainder left by subtracting it from unity; and the result is the longitude of *a*. Thus, when A's longitude is $\frac{1}{50}$, that of *a* is $\frac{1}{50}$ divided by $\frac{49}{50}$, or $\frac{1}{49}$; which differs from their approximation, $\frac{51}{2500}$, by only $\frac{1}{122500}$. The difficulty arising from the infinite number of terms in the law, as expressed by them, is not felt, because, after a certain point, they become too small to be sensible.

To return now to our case. There is, in the higher part of mathematics, a process known by the name of *integration*, the nature of which it is not here essential to explain. This operation no one has hitherto discovered any method of performing strictly in the case of the planetary motions; nor can any one decide upon the possibility of doing it. But methods have been found by which, when a solution has been obtained which is nearly true, that same solution may be made of use in obtaining one which is still more nearly true, and the second solution may be used in obtaining a third still more near, and so on, till any requisite approach to accuracy has been gained. By observation, the ellipse is found nearly to represent the lunar orbit; and, by using this, a new result is obtained, which adds some *inequalities* to the formulæ which would be required if the ellipse were strictly the orbit. The substitution of the second result produces a third, with more inequalities; but as the magnitude of the inequalities introduced at every step diminishes, they come, in time, to be so small, that it would be useless to retain them, or carry the process any further. With this is joined a method of *expanding* mathematical expressions; that is, of substituting, instead of a formula which is too complicated for any known method of integration, an infinite series of decreasing terms, each of which, by itself, is more manageable, such as $x + xx + xxx +$, &c., above, would be to those who were not in possession of the operations implied in $\frac{x}{1-x}$.

Joining the preceding illustration with that in pages 16 and 17, our reader may conceive that there will be no end of the corrections which must be applied to the mean longitude, in order to obtain the true, but that only a definite number of these corrections will be sufficiently large to be worth taking into account. Where we are to stop, must depend entirely upon the perfection of instruments and the art of observing: since it would be useless to carry prediction to an extent of accuracy which could not be verified by observation. Professor Airy* says, that he has been accustomed to consider the error of a single observation (on seven wires) with his transit instrument as not greater than six-hundredths of a second of time: and Laplace has carried his calculations of the lunar inequalities in longitude down to those which could not, when their effect is greatest, produce an error of more than one-twentieth or five-hundredths of a second. It must be said, therefore, that the approximation to the lunar motions has been fully carried to the point necessary for the present means of observation. But we have seen that the same astronomer makes a difference of one-half a second (one observation with another) between theory and observation. Nevertheless, if we consider (see 'Penny Magazine,' vol. i. p. 286) that the difference between one observer and another may sometimes be three-tenths, or thirty-hundredths of a second, without any assignable cause, it yet remains to see whether the whole half second is really due to a difference between theory and observation. It were much to be wished, not only that more observers would apply themselves to the solar system, but also that they would *reduce* their observations, so as to render them available for the purposes of direct comparison. Our astronomers are too long-sighted a generation; they see nothing nearer than a fixed star.

The inequalities of the moon's distance from the earth may be represented in a manner similar to those of the longitude. To, or from E M (see fig. page 17), the mean distance, the *arcs* of elongation of certain pendulums, such as 1 *r*, must be added or subtracted, according to the direction of the pendulum's motion for the time. The inequalities of the lunar latitude may, many of them, by represented by, and follow from, the various motions attributed to the fictitious elliptic orbit described in pages 6 and 7; but if we dismiss this altogether, we may always represent the latitude as we have done the inequalities in longitude. There is no *average* latitude, or rather, perhaps, we may say the mean latitude is nothing, or the mean moon is always in the plane of the ecliptic. For, by whatever angle of latitude it may be above the ecliptic at one time, it is as much below it at another.

* Mem, Royal Astron. Soc., vol. vi. p. 97.

To give a precise view of the state of theory and observation, as it stood when the *Mécanique Céleste* was published, we have added together the coefficients of thirteen lunar inequalities, neglecting signs, by the side of which we place the sum of all the differences between theory and observation for these inequalities, and also the differences between the tables of Mason and Burg, both deduced, as to *quantity*, from observation*. In neglecting the signs, and in taking the tables of Mason and Burg, as representing the extremes of differences of observation, we have borne very hard upon the theory; but this was our intention. It appears, that out of 3241″ of inequality determined by theory, 85″ do not accord with the tables; but, in the same number of inequalities, different tables differ by 45″: that is, where tables differed from one another by one part out of 72, theory differed from them by one part out of 38.

We shall give some idea of the magnitude of the principal lunar inequalities, which, however, we shall be obliged to abbreviate, as our limits are nearly exhausted. Of those in longitude the largest is the *equation of the centre*. It would not exist if the moon described a perfect circle, and arises from the elliptic form of the orbit. Its *maximum*† is 6° 17′ 20″, and its period the time of the moon's moving from perigee to perigee. It is nothing at the apogee and perigee, and increases the mean longitude when the moon is moving from perigee to apogee, but diminishes it when moving from apogee to perigee. This might all be represented in the illustration of page 17, by letting the pendulum A perform a double vibration in the time of a lunar revolution from apogee to apogee, the angle 1 A 3 be 6° 17′ 13″; and supposing the pendulum so to move, that A *a* is in the position A 1, moving towards A 2, when the mean moon is in apogee. This is the only inequality which the longitude would undergo if the moon were undisturbed by the attraction of the sun and planets. In that case, it can be shown that the radius drawn from the earth to the moon would describe equal areas in equal times; that is, whatever number of square miles might be passed over by the radius in any portion of time, say one day, would also be passed over by the same in any other day. Hence it is evident that, unless the moon describe a perfect circle, the angle through which the moon's radius must move in a day, to describe this area, when the radius is shorter, is greater than that which must be described when the radius is longer. Strictly speaking, this inequality ought to be measured *on the moon's orbit*, and not on the ecliptic; but the error arising therefrom is rectified by the inequalities which depend on the inclination of the lunar orbit of the ecliptic. This inequality was known to the Chaldeans, and of course to Hipparchus and Ptolemy.

* See *Mécanique Céleste*, vol. iii. p. 280. The inequalities taken are those of the third order, with the exception of the two last, in which the inequality determined from theory has not been fully developed. But this makes against the theory, as the differences of the tables for those two inequalities are more than usually great.

† The values of all the maxima are those of M. Damoiseau.

The inequality in longitude of next importance is the *evection*, discovered by Ptolemy. Its effect is, to make the equation of the centre appear too small at the new and full moon, and too large at the quarters. It was soon discovered, however, that the magnitude of the inequality was not always the same, but was nothing when the sun was half way between the moon and either its perigee or apogee, and greatest when the sun is half way between the moon and those points of her orbit which are equally distant from the apogee and perigee. Its maximum is 1° 16′ 30″; its period the time in which 360° is added by the moon's motion to the angle made by subtracting the moon's distance from its perigee from twice the angular distance of the sun and moon. Its cause was suspected by Newton, whose means of analysis were not, however, sufficient to deduce it from his theory. This was first done by Clairaut and D'Alembert.

The *annual equation* was discovered by Tycho Brahé. Its period is what is called an anomalistic year, or the time of the earth's moving from perihelion to perihelion, and it is nothing when the earth is in its aphelion or perihelion. Its maximum is 11′ 14″. It has the effect of making the month a little longer than the average when the earth is nearest to the sun, and *vice versâ*. It was shown to follow from the law of gravitation by Newton.

The *variation* was also discovered by Tycho Brahé. Its period is a *synodic* month, or from new moon to new moon. It increases the moon's velocity when moving to new or full moon, and *vice versâ*. Its *maximum* is 39′ 3″. It was shown to follow from the law of gravitation by Newton.

The *long inequality* was discovered by Laplace. It is remarkable principally for the embarrassment which ignorance of it gave rise to, and the length of its period, which is about 184 years. Its maximum is only 15″. Its period depends upon the united motions of the solar and lunar perigees, and that of the moon's node. It is the time in which twice the longitude of the moon's node, added to that of her perigee, gains 360° upon three times the longitude of the sun's perigee.

We will not go any farther in the mere enumeration of the inequalities. Those of the moon principally depend upon the action of the sun, for that of the planets is very trivial, except in one point, of which we shall presently speak. But there is one more inequality, which is very remarkable. It is well known that the figure of the earth is not exactly a sphere, but much more like the spheroid formed by the revolution of an ellipse about the smaller axis. In an ellipse, the proportion which the difference between the greatest and least axis bears to the greatest axis is called the *compression*. Thus, if we say that the compression of the earth is $\frac{1}{305}$, we mean that the greater axis, diminished by its 305th part, is the length of the lesser axis. Various measurements made upon the surface of the earth have given values of the compression ranging from $\frac{1}{297}$ to $\frac{1}{383}$, though the best of them agree in being between $\frac{1}{300}$ and $\frac{1}{305}$. This non-spherical figure of the earth gives rise, by its attraction, to a sensible inequality in the motion of the moon, both in longitude and latitude: in the former having a maximum of 7″, and in the latter of 8″. Mayer had discovered that the inequality existed before the cause was known; and Laplace connected the result with theory. The compression of the earth, computed from these two independent sources (for so they must be called when it is considered that they follow different laws and require different species of observations), is $\frac{1}{304\frac{1}{2}}$ from the inequality in latitude, and $\frac{1}{305}$ from that in longitude: two results not only according with each other more nearly than others derived from actual measurement, but also agreeing most nearly with the best of the latter.

The direct action of the planets on the moon is, as has been observed, very slight; but there is one of an indirect character, which becomes sensible only in a great number of years. It is a gradual acceleration of the mean motion, so small, however, that the arc now described in a century is only 7′½ more than would have been described in the same time 2500 years ago. Halley first discovered this acceleration by comparison of the recorded times of some ancient eclipses observed by the Chaldeans, B.C. 270, which show that the then longitude of the moon was greater by 1½° than it would have been had the mean motion of the moon continued uniform. Comparison of later observations confirmed this result, and all united in showing that the moon's average velocity was gradually on the increase, so as to increase the arc of longitude described in a century about 10″. This phenomenon drew the attention of all astronomers, and many unsuccessful efforts were made to deduce it from the theory of gravitation. Laplace, at last, discovered that a very slow diminution in the excentricity of the earth's orbit, shown to arise from the action of the planets, must have the effect of accelerating the moon's mean motion. The increase in a century, deduced by theory alone from the known diminution of the excentricity of the earth's orbit, is 10″⅕. Corresponding accelerations have also been shown to exist in the motions of the lunar perigee and nodes, but there are no observations of sufficient age to render the effects of them sensible. An eclipse, the only sort of very ancient observation which can be depended on, though a guide to the knowledge of the moon's place, is none to the elements of her orbit.

In the same manner as ancient eclipses of the moon proved the roundness of the earth, and modern observations of the satellite have confirmed the elliptic form of the primary, the acceleration of the moon's mean motion proves that the time of rotation of the earth has remained sensibly unaltered. If the day had only grown $\frac{1}{100}$th of a second shorter than it was in the time of the Chaldean observations above-mentioned, the *apparent* acceleration of the moon would have been at least a third greater than it has been observed to be. We may also add, that the mean motions of all other planets must have appeared accelerated; but this could not have been well verified from want of sufficiently good and ancient observations.

COMPANION TO THE ALMANAC,

FOR

1835.

PART I.

INFORMATION CONNECTED WITH THE CALENDAR AND THE NATURAL PHENOMENA OF THE YEAR; AND WITH NATURAL HISTORY AND PUBLIC HEALTH.

I.—HALLEY'S COMET.

THE present year will be distinguished by the re-appearance of the first comet whose return was predicted; a body much too remarkable in the history of astronomy to be allowed to pass without notice. Familiar as we now are with periodic comets—having two besides the subject of this article whose predicted return has been verified—the comet of Halley is connected with so many historical associations, that it must always be of pre-eminent interest.

The return of the other periodic comets *, in the year 1832, called the attention of the public so strongly to this part of astronomy, and gave rise to so much description upon comets in general, that we shall have no occasion to touch upon any matters except those which relate to the one before us. We shall therefore proceed to the manner in which the discovery was made.

In the article *On the Moon's Orbit*, in the preceding volume of this work, we pointed out the fiction of the elliptic orbit (see p. 6). It was usual to make a similar fiction or assumption with regard to comets, at the time when Newton published his *Principia* (A.D. 1687). They were supposed to move to and from the sun in parabolas, the parabola not being a curve of revolution, or of an oval kind, but having two infinite branches. In the adjoining figure, let S be the place of the sun, and let MMM be a given straight line. Let the comet P move in such a manner that its distance SP from the sun is always the same as PM, its perpendicular distance from the line MM; it will then describe what is called a *parabola*, of which S is the focus. A comet, describing such a

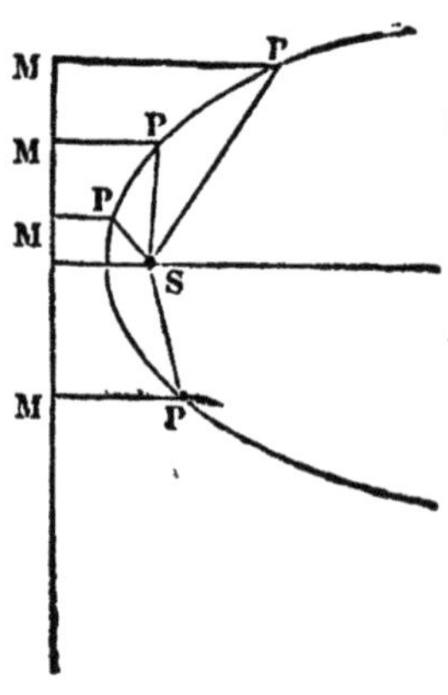

* See the Companion to the Almanac for 1833, pp. 1—15.

path, could never return again to the sun; and, accordingly, the periodic nature of cometary revolutions could never be verified without overturning this supposition. The *parabolic hypothesis* was adopted, because it was found to represent the observed motion of comets with tolerable precision; and it was a considerable step in this part of astronomy. Kepler (A.D. 1618), and Tycho Brahé before him, supposed that comets moved in straight lines. Hevelius (A.D. 1668) mentioned the concavity of the orbit, which notion was verified with regard to the comet of 1680, by Doerfel in 1681, who seems first to have distinctly laid down the parabolic hypothesis. The *Principia* of Newton showed that the same law of attraction under which the planets describe ellipses *nearly* (exactly, it would be, were it not for their attracting one another) would have compelled them to describe parabolas or hyperbolas*, if the velocity with which they were originally projected had been greater. The rule demonstrated by him is as follows. Supposing the planet projected perpendicularly to its distance from the sun, whatever velocity of projection would have made the orbit a perfect circle, anything less would have made it an ellipse, or anything greater under 41 per cent.†: a velocity 41 per cent. greater, would have made the orbit a parabola; and anything still greater, an hyperbola. This rendered the parabolic hypothesis more than dubious; for it naturally seemed very unlikely that so many comets should all have possessed that unique relation between their velocities and distances which is necessary for a parabolic orbit, when anything greater or less would have given an hyperbola or ellipse. Newton himself formed the opinion that the orbits were elongated ellipses, which he expressed in language translated as follows:—

"We have said that comets are a species of planets revolving about the sun in very eccentric orbits. And, in like manner, as among the *planets without tails*, those are less, which revolve in less orbits, and nearer to the sun; so also it seems agreeable to reason, that comets, which in their *perihelia* are still nearer to the sun, should be much less even than the planets, that their attraction might not act too strongly on the sun. But I leave the transverse diameters and times of revolution to be determined by the comparison of comets which return after long periods of time in the same orbits."—*Principia*, Book iii. prop. 41.

At the time when this was written, all comparison was impossible; for the comet of 1680 was the only one of which even the parabolic‡ orbit had been given: and it was no small labour even to deduce so much from the observations.

* An hyperbola is made, by moving a point in such a manner that its distance from one of two fixed points shall always exceed its distance from the other by the same quantity.

† Strictly, any increase less than in the proportion of the square root of 2 to 1. The circle is itself a species of ellipse, made by bringing the two foci together (see last volume, p. 7, note); and if the planet were projected with less than the velocity necessary for a circle, its distance of projection would be its greatest distance from the sun, and *vice versâ*.

‡ A very elongated ellipse is very nearly the same as a parabola in all the parts which are not far from the focus. By the parabolic orbit, is here meant the parabola which most nearly represents the true orbit as long as the comet is visible.

About the year 1700, Halley, afterwards Savilian Professor at Oxford and Astronomer Royal, having in some degree simplified the necessary process, undertook to calculate the orbits of all the comets upon which observations sufficiently definite had been made. Out of more than four hundred such bodies mentioned by ancient and modern writers (that is, 415 up to the year 1665 inclusive), there were not more than twenty-four on which the experiment could be tried. These were the comets of

1337	1556	1590	1652	1672	1683
1472	1577	1596	1661	1677	1684
1531	1580	1607	1664	1680	1686
1532	1585	1618	1665	1682	1698

Halley calculated, for every one of these, the longitude of its ascending node, the inclination of its orbit, the longitude of its perihelion, its nearest distance to the sun, the time of its being in the perihelion, the distance from its perihelion to its node, and the character of its motion, whether direct or retrograde. Having formed his table, he immediately found three comets, the elements of which were very much alike—namely, those of 1531, 1607, and 1682, as follows :—

A	B	C	D	E	F	G	H
1531.	49° 25′	17° 56′	301° 39′	56700	Aug. 24	107° 46′	Retrograde.
1607.	50 21	17 2	302 16	58680	Oct. 16	108 5	„
1682.	51 16	17 56	302 52	58328	Sept. 4	108 23	„

A, year of the comet.
B, longitude of ascending node.
C, inclination of the orbit.
D, longitude of the perihelion.
E, perihelion distance from the sun, that of the earth being 100,000.
F, time of the comet arriving at its perihelion distance.
G, distance from perihelion to ascending node.
H, direction of the comet's motion.

The preceding labour made it, therefore, not difficult to conjecture that these three appearances must have been one and the same comet. With regard to the time of revolution, from August 24, 1531, to September 16, 1607, there is 76 years and 53 days*; while, from October 16, 1607, to September 4, 1682, there is 75 years all but 42 days; about 15 months difference. On this, Halley observes †:—

"All the elements agree, and nothing seems to contradict this my opinion; besides the inequality of the periodic revolutions; which inequality is not so great neither, as that it may not be owing to physical causes. For the motion of Saturn is so disturbed by the rest of the planets, especially Jupiter, that the periodic time of that planet is uncertain for some whole days together. How much more, therefore, will a comet be subject to such like errors, which rises almost four times higher than Saturn, and whose velocity, though increased but a very little, would be sufficient to change its orbit from an elliptical to a parabolical

* In the list of the French Encyclopædia, article *Comète*, the perihelion times of the comets of 1607 and 1682, are given, October 26, and September 14. The same mistake, probably from the same source, has crept into Bailly's History of Astronomy.

† See the English version of the *Cometographia*, in the *Miscellanea Curiosa*, vol. ii, p. 321.

one? This, moreover, confirms me in my opinion of its being the same; that in the year 1456" (that is, seventy-five years before 1531) "in the summer time, a comet was seen passing retrograde, between the earth and the sun, much after the same manner; which, though nobody made observations upon it, yet, from its period, and the manner of its transit, I cannot think different from those I have just now mentioned. Hence I dare venture to foretell that it will return again in the year 1758."

The preceding is a translation from one of the editions of the *Cometographia*, first presented to the Royal Society in 1705. Its author did not then speak with so much confidence as afterwards, when he calls upon posterity to recollect that this discovery was made by an Englishman. Subsequently, he was able to strengthen his conjecture from historical records; for it appeared that comets had been seen in the years 1305 and 1380, as well as 1456; thus giving the complete chain as follows:—

Years .	1305	1380	1456	1531	1607	1682
Intervals	75	76	75	76	75	

Of which the three last, as we have seen, had been *observed*, and the elements of their orbits found to be almost identical. The comet of 1531 had been observed by Apian; that of 1607 by Kepler and Longomontanus; that of 1682 by Halley himself and several others. And it must be remarked, that the orbits of 1607 and 1682, made after the time of Tycho Brahé, agree much better with each other than with the orbit deduced from the observations of Apian, who lived before the great improver of the art of observing.

The opinion of Halley was generally adopted,—namely, that in the year 1758, or thereabouts, a comet might be expected, having nearly the elements above given for the position of its orbit; but, at the same time, it was evident that the verification of the prediction would perhaps never be obtained. Fifteen months of difference had occurred between the period 1531—1607, and 1607—1682; there was reason to believe the comet was diminishing in brilliancy at every successive appearance; and it seemed, therefore, that there was some chance of powerful telescopes being kept constantly in employment for perhaps a couple of years, engaged in *sweeping* for an object which, after all, might never make its appearance. Most of our readers must be aware that a telescope, under a high magnifying power, has a very small field of view, so that the operation of finding a faint object by sweeping for it, is as laborious as finding a little star with the naked eye would be, if a card were pierced with a pin-hole, and fastened over the eye. Various attempts were made to lighten the labour, by calculating presumed orbits for the comet. Klinkenberg, a Dutch astronomer, constructed fourteen different ones; De l'Isle, Lalande, and Pingré, also made attempts of the same sort. But none of these astronomers could take into account the perturbations caused by the attraction of the planets. All they could do was to make different suppositions upon the time of the comet's coming to its perihelion, and to show in each case, where it would be on preceding and succeeding days, on the supposition that the orbit laid down by Halley was the true orbit, but that the place of the comet in its orbit was to be guessed at.

It is important to recollect that the Newtonian theory of attraction was not then by any means established with its present degree of moral certainty. Many phenomena could not be explained by it at all; and of those which were *explained* by it, very few could be *computed* by it. Indeed, it was not till the year 1749 that any additional evidence of decided importance was added to that contained in Newton's Principia. Lalande seems almost to have desponded on the subject of the comet; he says (*Mémoires de Trévoux*, Nov. 1757, cited by Montucla), "We cannot doubt that it will return; and even if astronomers should not see it, they will not be the less persuaded of its return. They know that the faintness of its light, and its great distance, perhaps even bad weather, may keep it from our view; but the public will find it difficult to believe us; they will put this discovery, which has done so much honour to modern philosophy, among the number of predictions made at hazard. We shall see dissertations spring up again in the colleges, contempt among the ignorant, terror among the people, and seventy-six years will elapse before there will be another opportunity of removing all doubt."

Of what kind, our reader may ask, were the difficulties which prevented their doing for the comet, that which was done every day for the sun and moon,—namely, the prediction of its place with at least sufficient nearness to know whereabouts to look for it with the telescope, if it were not visible to the naked eye? We shall try to give an illustration, directing our reader to what we have already said on this subject in the last volume (pp. 18—20).

The orbit of a comet differs from that of a planet in this, that several elements which are small in the latter, are not so in the former. The greatest distance of Juno (the most excentric of all the planets) from the sun is not twice the least distance; and even this is a source of difficulty. The greatest distance of Halley's comet is sixty times its least distance. The inclination of the orbit of Mercury (the most inclined of the *old* planets) is 7°, which much increases the trouble of calculating its perturbations. That of the comet is 17°. We have seen (see last volume, p. 18) that the process of *integration*, for which we are obliged to substitute the summation of a series, is comparatively light when the terms of the series decrease so fast, that the sum of a few of them will be sufficiently correct for practical purposes. But in the present case (owing to the largeness of the above-mentioned quantities), the same series which are serviceable in the case of the planets, *converge* so slowly, as it is called, that is, their terms decrease so slowly, that the use of them is out of the question. Add to this that other series, of which in the case of the planets even the sum of the terms is inconsiderable, become necessary to be taken into the account in the case of a

B 3

comet. The mathematician is therefore obliged to abandon the usual methods almost entirely, and to have recourse to what is called the *method of quadratures*, a process of which the following is an illustration.

The content of a field is easily measured when all its sides are straight; for the measurement of the sides and some of the diagonals (and this not the easiest way) furnishes a ready mathematical rule. The same may be said when some of the sides are circular, elliptical, or part of any curve which follows a not very complicated law. And even where a side is curved, in a manner of which we know nothing, there is little additional difficulty, provided the curvature be very small; for, by means of straight lines and arcs of circles, a figure may be readily constructed, which shall represent an arc of small curvature, with more than the accuracy necessary for practical purposes. But where there is an arc of considerable curvature, and belonging to a curve which either follows no perceptible law, or a very complicated one, land-surveyors use a rough form of the process we wish to describe. Instead of being able, merely by measuring the span and height of the curve, to resort at once to a table or a rule, the curve not being of a known kind, is in exactly the same relation to the measurer as if it were composed of a large number of curves of different kinds. But here a known principle comes to our assistance. A small arc of a curve, of any kind whatsoever, differs very little from a straight line; accordingly, by dividing the base into a great number of parts, and erecting perpendiculars, the curve is divided into a number of figures, the first and last of which are *sensibly* triangles, and the intermediate ones have one curvilinear side each, which however is *sensibly* a straight line. The measurement of these figures, one by one, is necessary to give a sufficiently correct value of the whole area.

Now, to take the case of the comet, and to calculate (let us say) the effect of Jupiter upon it during a whole revolution. This is a summation which cannot be made. With regard to the planets, it can be reduced to the summation of a few terms; even this cannot be done for the comet. But, at the same time, the effect of Jupiter upon the comet's orbit *for a small time* can be calculated with sufficient precision; a difficulty is avoided of precisely the same character as evading the curve by measuring it in such small portions that each may be considered straight; but another is introduced by the vast number of computations which it is necessary to make separately. This is so enormous, that we only know it can be done, because it has been done.

In 1757, Lalande, who took the expected arrival of the comet very warmly to heart, struck by the success of Clairaut in investigating the lunar inequalities, proposed to him to undertake the calculation of the effect of Jupiter's action on the comet during the revolution which was taking place, and that which preceded. Clairaut declined such an undertaking alone, on which Lalande offered to undertake all the astronomical part of the calculations. This enormous labour was actually performed by Lalande, but he also found assistance, and (for this comet is fated to abound in remarkable circumstances) this came from a lady. Madame Lepaute, wife of the celebrated watchmaker of that name, whose astronomical character is not so well known as it deserves to be, engaged in the task with a degree of perseverance which astonished both Clairaut* and Lalande. The latter says, "During six months we calculated from morning till night, sometimes even at meals; the consequence of which was, that I contracted an illness which changed my constitution for the remainder of my life. The assistance rendered by Madame Lepaute † was such, that without her we never could have dared to undertake this enormous labour, where it was necessary to calculate for every degree, and for 150 years, the distance and force of each of the two planets," Jupiter and Saturn, "with respect to the comet."

Clairaut at first imagined that it would be sufficient to confine his attention to the time when Jupiter was near the comet; but he soon saw the necessity of extending his operations. For he not only found that the effect of Jupiter was sensible throughout the whole of its orbit, but he ascertained it to be so great that it was necessary also to examine that of Saturn. The result of the whole enabled him to announce in November 1758, when astronomers had just began to look for the comet, that its revolution was longer by 518 days than it would otherwise have been, on account of the action of Jupiter, and 100 more from that of Saturn ‡. He predicted its arrival at its perihelion for the 13th of April, 1759, but he announced, at the same time, that the imperfection of the method might cause an error as great as a month. He remarked also the possibility of the comet being acted on by some undiscovered planet more distant from the sun than Saturn, a conjecture since so remarkably confirmed.

We now come to the fulfilment of the prediction. The first person who saw the comet was George Palitzch §, a farmer near Dresden, who had cultivated astronomy. This was with an

* Clairaut does not mention her name in his Memoir on the subject. Lalande, their common friend, says that Clairaut had done her full justice, but was induced to suppress it by the jealousy of another lady, to whom he was attached. (*Bibliographic Astron.* p. 677.) The suppression needs something to account for it, for Lalande cites two letters of Clairaut to himself, in which he says, "L'ardeur de Mad. Lepaute est surprenante," and calls her "la savante calculatrice." There was, therefore, a species of poetic justice in the attack of D'Alembert upon Clairaut, who, immediately upon the appearance of the Memoir, divided the opinion of the scientific world upon its merits, by some assertions as to the manner in which the error committed was to be judged of, and considerably lessened, not the merit, but the brilliancy of his labours. Bailly and Bossut have both behaved with extreme unfairness to Lalande; the first by entirely suppressing his name in relation to Halley's comet; the second by merely mentioning that Clairaut was assisted by some "disciples." The reputation of both has sunk as historians, and we conceive neither would now be cited as an authority.

† This lady afterwards assisted Lalande in the calculation of his Ephemerides, till her application weakened her sight too much to allow her to continue. She died 1788, while attending on her husband, who had become insane.

‡ This does not mean that the whole of this quantity was to be added to the supposed revolution, because that had already been conjectured from observations in which similar effects were mixed up.

§ For some account of whom, see *Library of Entertaining Knowledge*, v. 8; *Pursuit of Knowledge under Difficulties*, p. 293.

eight feet telescope*, December 25, 1758. Palitzch announced his discovery to Dr. Hofmann, who immediately went to Palitzch, and saw the comet also on the 27th and 28th December. De l'Isle, and his subordinate Messier †, at Paris, had been on the look-out for many months. On January the 21st, 1759, the latter, with a Newtonian telescope, saw a feeble light in the place he supposed the comet to be near, which he at first took for one of the nebulæ which had so often deceived him. On following it several successive days, he perceived its motion, and soon found it to be the object of which he was in search. He followed it till the 3rd of June, but his observations were kept secret by desire of De l'Isle. The consequence was, that when they came to be made known, they were generally regarded as fictitious. In the mean time it had been seen again at Dresden, Leipsic, Boulogne, Brussels, Lisbon, Cadiz, and other places. The principal observers who saw it were Messier, Maraldi, Cassini de Thury, Hell, and La Caille. The prediction of Clairaut was fulfilled to a degree of precision which must be called accidental coincidence. The comet came to its perihelion on the 13th of March, 1759, just within the month which he had supposed might possibly be the amount of the error. J. A. Euler, the son of the celebrated philosopher of that name, afterwards went over the calculations of Clairaut, and found that by more attention to minute circumstances he might have been within nineteen days of the truth. The actions of Uranus and of the earth have been shown also to affect the comet sensibly.

The following elements of the orbit were deduced from different sets of observations. They furnish the proof that the comet seen in 1759 was really the one observed by Halley; for we need hardly say, that the mere circumstance of a comet having been seen about the time, and even near the place, which was expected, would only have been a remarkable coincidence. The letters stand for the same as in page 7.

A	B	C	D	E	F	H
					h m	
1759	55° 49′	17° 39′	303° 16′	58349	March 12, 13 41	Retrograde.
—	53 46	17 40	303 8	58490	„ 12, 13 59	„
—	53 49	17 35	303 16	58360	„ 12, 12 57	„

The first of these orbits was given by La Caille, the second by Lalande, the third by Maraldi. Their agreement with those of page 7 is sufficiently obvious, as also the effect of perturbation upon the longitudes of the node and perihelion. The only remarkable alteration in the comet, was its decrease in brilliancy. Cassini, in 1682, calls it as round and clear as Jupiter; Messier could only see it faintly in rather powerful reflecting telescopes; and though

* It has been usually stated that Palitzch made the discovery with the naked eye, which is an error. The true account was given in the Dresden *Gelehrten Anzeigen* for 1759.

† A celebrated observer, called *le furet des comètes* by Louis XV. He had discovered twelve comets, every one of which, says Delambre, gained him admission to some foreign academy. While attending to his wife during her last moments, a M. Montagne discovered another comet. This afflicted him to tears, and he exclaimed "Alas, I had discovered twelve, and this M. Montagne has taken away my thirteenth." Then remembering that it was his wife he should mourn for, he began to cry "Ah, la pauvre femme!" and went on deploring his comet.

Palitzch, under very favourable circumstances, and having, we must presume, a remarkably good sight, detected it with the naked eye, we do not read that any one else did the same. In 1456 its tail was sixty degrees in length; in 1607 it was only seven degrees, but brilliant. The same fact—namely, the diminution of light in successive revolutions—has been observed with regard to the other periodic comets; and affords reasonable ground for conjecture that in time they must disappear altogether. Whether this will take place next time, now remains to be seen, but in any case, strong telescopes will most probably be required.

The history of the comet since 1759 consists in the corrections which have been made in the elements then given, not of course by new observations, but by the application of more correct methods to those of 1759. M. Rosenberger * has collected all the observations, and has corrected the elements of 1759; M. Pontécoulant † has also given a corrected result; and Mr. Lubbock ‡ another. We have now a new element introduced, namely, the whole axis of the comet and its eccentricity (see last volume, p. 7). The elements are as follows, the letters being the same as before:—

	Rosenberger.	*Pontécoulant.*	*Lubbock.*
A, year . . .	1759	1759	1759
B, long. asc. node .	53° 47′ 56″	53° 48′	53° 45′
C, inclination . .	17 38 9	17 40	17 36
D, long. perihel. . .	303 10 46	303 14	303 3
E, perih. dist. Earth's mean dist. 100,000 .	58440	58237	58366
F, perih. passage, Paris mean time . .	March 12, 13h 26m	March 13, 3h 50m	March 13, 3h 19m
Eccentricity . .	·96768269	·967705	·9676
Semiaxis Major, that of the earth being unity	18·083275	18·08327	18·0763

These results agree as well as can be expected, considering the discretion which it is necessary to exert in combining observations which are, to a certain extent, discordant.

M. Damoiseau (see last volume, p. 10), about the year 1825, went through the enormous labour of recalculating all the perturbations of the comet through its last and present revolution. M. Pontécoulant has done the same thing; and Mr. Lubbock has applied the perturbations determined by M. Damoiseau to the determination of the elements. The resulting sets of elements are here given; in which, however, the longitude of the perihelion is reckoned *on the orbit* instead of *on the ecliptic.*—See *Nautical Almanac* for 1835.

	Pontécoulant.	*Damoiseau.*	*Lubbock.*
Long. perih. *on the orbit*	304° 31′ 43″	304° 27′ 24″	304° 23′ 39″
Long. asc. node . .	55 30 0	55 9 7	55 3 59
Inclination . . .	17 44 24	17 41 5	17 42 50
Eccentricity . . .	·9675212	·9673055	·967348
Semiaxis Major . .	17·98705	17·9852	17·98355
Perihel. passage, Paris mean time from midnight . . .	1835, Nov. 7·2	1835, Nov. 4·32	1835, Oct. 30·1993

(Copied from the *Nautical Almanac* for 1835.)

* *Ast. Nach.* no. 189. *Phil. Mag.* Jan. 1832. † *Théorie Analytique, &c.* v. ii. p. 140. ‡ *Mem. Ast. Soc.* v. iv. p. 519.

PATH OF HALLEY'S COMET AMONG THE FIXED STARS.

DEGREES OF DECLINATION.

HOURS OF RIGHT ASCENSION.

The accompanying chart, copied from the *Nautical Almanac*, shows the track of the comet through the heavens, on the supposition of the three preceding orbits being correct. The dotted lines are drawn through contemporaneous positions of the comet. Thus it will be observed that on the 6th of October it is in Ursa Major, according to M. Pontécoulant; between that and Bootes according to M. Damoiseau; and in Bootes according to Mr. Lubbock.* These dotted lines represent the directions along which observers with telescopes will sweep on the several nights, and in which there is the greatest probability of its being found.

Small as the chance may be of seeing it with the naked eye, many will doubtless be led by curiosity to attempt to find it. The first ten days of October, if fine, will be favourable for this purpose, as during that time the comet will never set, and will be far from the sun. In the *Nautical Almanac* will be found an ephemeris of the times at which it comes on the meridian for every successive day; but this would be of no use whatever to the observer with the naked eye, as the transit generally takes place in the day-time. The best method will be for the would-be discoverer to make himself perfectly acquainted beforehand with the several stars which lie near the track of the comet. In the neighbourhood of these, he should look carefully at any part of the night when they are very clear, keeping his eye free from any dazzling light (such as candlelight) for some minutes previously. The possessor of a globe should lay down the positions upon it from the *Nautical Almanac*.

It will be observed that the three orbits given are nearly the same, but that the positions assigned to the comet in its orbit are very different. This makes a considerable difference in the time of coming to the perihelion, and will therefore render this comet of little use towards settling the question of a resisting medium (see *Companion* for 1833, p. 13), unless it should be found that one or other of these orbits is very near the truth. Considering the very increased number of observers now ready, the advantages derived from their being more scattered over the world than was the case in 1759, and the comparative ease with which the effects of perturbation are now predicted, we may be justified in supposing that this and every other question now in doubt will need only one more revolution to settle it.

Since writing the above, we have learnt that the track of the comet has been calculated in Germany for some months preceding the beginning of the table in the *Nautical Almanac*, upon the supposition that the comet may become visible about the period of its opposition. Should this happen to be the case, the total appearance of the comet will be nearly a year in duration. The additional tables are published in the 'Philosophical Magazine.'

* See note by Mr. Lubbock, at page 263.

COMPANION TO THE ALMANAC,

FOR

1836.

PART I.

INFORMATION CONNECTED WITH THE CALENDAR AND THE NATURAL PHENOMENA OF THE YEAR; AND WITH NATURAL HISTORY AND PUBLIC HEALTH.

I.—OLD ARGUMENTS AGAINST THE MOTION OF THE EARTH.

EVERY body now knows, or is supposed to know, that the earth moves round the sun, and not the sun round the earth. If we were to ask nine persons out of ten how they came to know this, we should be answered, that they believe it, because those who study the subject assure them it is so, and not otherwise: no bad reason, where a better cannot be found. Or we should be told, that it was taught them in their youth, by means of certain arguments which were used, in which Copernicus was very right, and all who opposed Copernicus very wrong. To the first of these reasons we may say, that, three hundred years ago, an exactly similar argument proved that the earth stood still, and the sun moved. On the second we remark, that not one in a hundred of those who believe in Copernicus ever saw a single sentence of that writer, or has the least idea of the arguments by which his system was either supported or opposed.

In treating of old matters of controversy, it were to be wished that those who write would quote the very words of the earliest advocates of both sides. Firstly, because they may thereby make their readers know that they are not weakening the arguments or exaggerating the absurdities of an opponent. Secondly, because there is a degree of interest which attaches to the actual expressions of by-gone controversialists which seldom can be made to accompany any modern representation or abstract of their opinions. What we now propose to do is, to let an anti-Copernican speak for himself, that such of our readers as do not dabble in old books may say they have seen one of those curious animals, as we now think them, those fossil remains of an extinct theory. The instances we have chosen are—

1. *Thomas Fienus*, doctor of medicine in the University of Louvain, who published, in 1619, his *Disputatio an cœlum moveatur et terra quiescat*, which we shall here give at length. It was reprinted so late as 1670, in London. We choose it because it is a short abstract of the arguments then in use.

2. *Alexander Rosse,* better known by two lines of Hudibras than by all his writings (and they were several) put together. We shall cite passages from the following treatise, published in 1646: '*The new planet no planet, or the earth no wandring star, except in the wandring heads of Galileans.*' The passages from the first (which, put together in their order, make up the whole) are headed F, and those from the second, R. The paragraphs without heading are short remarks upon the arguments employed.

F. "That the heaven moves, and the earth stands still, is proved—firstly, by authority; for besides that Aristotle and Ptolemy have asserted it, and philosophers and mathematicians have followed them by unanimous consent, except Copernicus, Bernardus Patricius, and a very few others, the sacred Scripture manifestly bears witness to it, especially in two places which I have seen; for, in the 10th chapter of Joshua are these words, 'For the sun and the moon stood still, until the people had avenged themselves upon their enemies;' and presently follows, 'So the sun stood still in the midst of heaven, and hasted not to go down for the space of one day, and there was no day so long before it or after it.' In which words the Scripture manifestly speaks of the motion of the *primum mobile,* by which the sun and the moon are carried in their diurnal motion, and by which the day is described, and points out that the heavens and the *primum mobile* are moved. Finally, in the first of Ecclesiastes it is thus said: 'A generation passeth away, and a generation cometh, but the earth standeth always, the sun riseth and goeth down, and returneth to his place.'"

R. "If Solomon had thought otherwise, to wit, that the earth moved, and the sun stood still, he would have said, 'The sunne standeth for ever, the earth ariseth, and the earth goeth downe,' &c. But, for all his knowledge, he was ignorant of this quaint piece of philosophie."

The argument from the Bible has moved many persons, who will at least admit the pious intention and consistent reasoning of Rosse in the following:—

R. "Whereas you say that astronomy serves to confirm the truth of the holy Scriptures: you are very preposterous, for you will have the truth of Scripture confirmed by astronomie, but you will not have the truth of astronomie confirmed by Scripture. Sure one would thinke that astronomicall truths had more need of the Scripture confirmation than the Scripture of them."

But the following quotation will show that this piety was only conditional, and that Rosse had a higher authority than astronomy and Scripture put together, namely, Alexander Rosse himself.

R. "If any booke of Scripture should affirme, as you doe, that the earth moves naturally and circularly, I should verily beleeve that that booke had never been indicted by the Holy Spirit, but rather by a Pythagorean spirit, or by the spirit of Dutch beer."

Let those who in our day are endeavouring to fetter the course of geological induction, by insisting upon a literal interpretation (as they assert) of the first chapter of Genesis, declare whether they will or will not revive the objections to the Copernican system from a similar source. If they will not, their inconsistency can be easily made manifest; if they will, let them "read Alexander Rosse over," and they may not only get powerful arguments, but may express them in all the strength of the older form of our language; to say nothing of proving, if not the sageness of their philosophy, at least its antiquity. Rosse is not by any means the only disputant who has been unable to see that he who appeals to a judge must abide by his decision, whatever it may be: we have seen in the writings of more than one theologian the truth of the Scriptures inferred because a certain doctrine is in them, not the truth of the doctrine because it is in the Bible. The preceding sentence, which would have been indecent in an unbeliever, is intolerably disgusting in Rosse, who professes himself to be altogether a Christian.

F. "Secondly: It is proved by reason. For, first, all the heavens and stars are made for man, and for those terrestrial bodies which are serviceable to men, namely, that they may warm, enlighten, and vivify them, &c. This they cannot do, unless by motion they are applied alternately to the different parts of the world. And it is more likely that they apply themselves by motion to man, and the place in which he lives, than that man should apply himself to them by the motion of his seat or habitation; for they are for the use of man, but man is not for their use; therefore it is more likely that the heavens move, and the earth remains still, than the contrary."

Here a curious question arises; how did they roast meat at the University of Louvain? Did they turn the fire round the meat, or the meat round before the fire? For the fire is lighted for the meat, but the meat is of no use to the fire. Is our modern method of roasting nothing but a vile plagiarism from Copernicus?

F. "Thirdly: No probable argument can be brought from philosophy by which it seems to be proved that the earth is moved, but the heaven is at rest. It cannot be assumed from mathematics. For whether the heavens move and the earth rests (or *vice versâ*), all the phenomena of the heavenly bodies can be kept the same. For like as in optics all things remain the same, whether appearances come from objects to the eye, or rays pass from the eye to the objects; so also in astronomy. Therefore, we should rather remain in the old and common opinion, than receive a new one without cause shown."

All of this, except the first sentence, is correct, and conclusive in one respect; it is surprising that the disputant could feel only one edge of it. It was impossible in that day for either party to give the other absolute demonstration, for the reason just mentioned, that phenomena are the same on either supposition.

F. "Fourthly: The earth is the centre of the universe, and all celestial bodies seem to move about it; therefore itself ought to be immovable; for, whatever moves, it seems, ought to move about or over something which is immovable."

This was a stronger argument against a Copernican of that day than it would be now; for the latter did, themselves, attach some mystical notions to what they called the centre of the universe. The question between them was, whether the earth or the sun was this central body; and, in the preceding, Fienus appeals to admitted notions. The Copernican of that day would move, as an amendment, that for the two first words should be read, "The sun." There is also here an assumption of the question; for whereas the preliminary and undisputed position in the second clause is, that the bodies "seem" to move round the earth; in the fourth clause, this becomes, that they "do" move. Now whether this seeming motion is or is not a real motion, is the point in question.

F. "Fifthly: If the earth move in a circle, it is either moved naturally or violently; either by itself, and of its own nature, or by something else. It is not moved by its own nature or by itself; for the motion natural to it is one in a right line from up to down. Therefore, circular motion cannot be natural to it; for the earth is a simple body; but to one simple body there cannot be two natural motions differing in species or genus. Also it is not moved by any other body; for, by what can it be moved? It must either be asserted to be moved by the sun, or by some other heavenly body; and this cannot be said, because that sun or other body must be asserted either to be at rest or in motion. If it be said to rest, then it cannot impart motion to another; but if it be said to move, neither thus can it move the earth; for it ought to move it either by a motion similar to its own, or contrary. Not with a similar motion, for then neither would be perceived to move; as, when two ships move with the same motion, they do not seem (to each other) to move, but to rest; not with a contrary motion, because nothing can give a motion contrary to its own motion. And since Galileo seems to say, as I have understood from you, that the earth is moved by the sun, I prove at once that this is not true; because the motions of the sun and earth are upon contrary and distinct poles. But the sun cannot be the cause of such a motion, as should take place on different poles. Lastly, the earth does not follow the motion of any other celestial body (not the sun); because if it were moved, it would be turned round in twenty-four hours, but all the other celestial bodies only move in the space of days, months, and years; therefore, &c. Lastly, if the earth were moved by any other body, its motion would be violent. But this is absurd; for no violent motion can be ordinary, or perpetual."

Perhaps in all time there never were so many unproved (to say nothing of untrue) assumptions made in the same quantity of writing. Many of the things asserted were more or less admitted at the time; and probably the answer of any but Galileo would have been a match for the preceding in assumption of premises.

This was written about the time of the prohibition of Copernicus's doctrine by the Inquisition, and before the proceedings against Galileo, of whom Rosse afterwards speaks thus:—

R. "Galileus fell off from you, being both ashamed and sorry that he had been so long bewitched with so ridiculous an opinion; which was proved to him, both by Cardinal Bellarmine, and by other grave and learned men, that it was contrary both to Scripture, divinitie, and philosophie; therefore Galilie, on his knees, did abjure, execrate, and detest, both by word and writ, his errour, which you maintaine, and promised, with his hand on the holy Evangil, never to maintaine it againe."

E pur si muove! Rosse forgets to state the exact nature of the arguments employed by the cardinal and his grave and learned friends; but Rosse's poet gives a hint about those

> "Who proved their doctrine orthodox
> By apostolic blows and knocks."

F. "Sixthly: Even though the earth should be supposed to move, nevertheless it must be confessed that either the planets move, or their orbs; for, otherwise, the diversity of the planetary aspects cannot be explained; nor can a reason be given why the moon does, and the sun does not, depart from the ecliptic; nor how a planet can be stationary, retrograde, high or low, and an infinite number of other phenomena. And hence those who have said the earth moves, as Bernardus Patricius and others, have (also) said that the *primum mobile* is at rest, and the earth moves in its place; but they have by no means been able to deny that the planets move, but have admitted it. And this is the reason why ancient and modern mathematicians have been obliged to lay down and admit a motion of the planets themselves, besides the motion of the *primum mobile.* If therefore it be to be confessed, which is certain it must be, that the stars and heavenly bodies move, therefore it is more likely that all the motion which is perceived in the universe, rather belongs to the heavenly bodies than the earth; for if motion be ascribed to all other bodies, why should not the diurnal motion be ascribed for the same reason, rather to the *primum mobile* than to the earth, especially seeing that our sight seems to draw the same conclusion, which, although it is sometimes deceived in judging of similar motions, yet it is not likely that it should be deceived for ever, or in judging the motion of its own principal object, namely, the celestial lights."

The argument of Fienus here is, that because, on any supposition, all other bodies do move, therefore it is most likely that the earth does not move. To which it might appear a proper answer, that because all other bodies do move, that therefore it is most likely that the earth does move. The *primum mobile* was a very large sphere, invented to make all the stars move round the earth, much as, in a common globe, a *primum mobile* of pasteboard and paper makes the pictures of the stars move at any rate which may be thought necessary. Many of the Copernicans admitted the *primum mobile,* even when they had dismissed it from office, and

B 3

they were therefore hampered with it, as in the preceding argument.

F. "Seventhly: It is proved by experience. For if the earth moved, when an arrow is shot directly upwards it never could fall again upon the place from whence it was shot, but ought to fall upon a spot at many miles distance. But this is not the case; therefore the earth does not move. It may be, and usually is, answered, that this does not follow, because the air is carried with the earth; and thus, since the air which carries the arrow has the same motion with the earth, thence the arrow also is carried with it, and so falls upon the same spot. But this is a mere evasion, and worth nothing as an answer, for many reasons. First, because it appears false that the air is thus moved, and with the same motion as the earth. For what should move it? For if, indeed, the air be moved in the same way as the earth, it must either be moved by the earth itself, or by that which moves the earth, or by itself. Not by itself; because it has another motion natural to it, namely, the rectilinear motion; and also since it has nature, and essence, and qualities all different from the nature and essence of the earth, it cannot by nature have the same motion as the earth, but must necessarily have a different sort of motion. Again, it is not moved by that which moves the earth, because that which moves the earth cannot give exactly the same motion to the air. For since the air differs from the earth in essence, in *active* and *motive* qualities, and *in modo substantiæ*, it cannot receive the moving force of that agent, or the force impressed upon it, in the same manner as the earth, and so, cannot receive the same motion. For the properties of things which act and cause motion are differently received by different bodies, according to their different dispositions. And it cannot be moved by the earth, because if so, it must be said to be moved by carriage; but such a motion appears impossible; for if the air moved the earth by carrying it, the air ought to be more quickly moved than the earth, since the air is the greater body; for that which is without is greater than that which is within. But when that which is greater, being without, is carried round equally quick (in angular velocity) with that which is less, being within, then the former must be moved the more swiftly (in actual velocity); and thus it is certain, that the heaven of Saturn, in its diurnal motions, is much more quickly moved than that of the moon. But it is impossible that the body carried can be moved more quickly than that which carries; therefore the air is not moved by the carrying power of the earth. Let it be that the air moves with the earth, either by itself, or by the carriage of the earth; even here the force of the first argument remains; for its motion cannot be in all things conformable to the motion of the earth, as I have shown; because the air is different from the earth in the consistence of its substance, in its qualities, and essence: but the air should be moved more slowly; which, being laid down, it follows that the arrow shot up could not return to the same point; for the earth in its motion would leave behind it both the air, which moves more slowly, and the arrow, which is carried by the air. It may be added, that if the air move more slowly than the earth, a man in a very high tower should always, however still the air may be, feel a very great wind and agitation of the air. For since mountains and towers move with the earth, and the air does not follow them with equal speed, it must be that they pass through the air by cleaving, and penetrating, and furrowing it; from which passage and penetration a great wind should be perceived."

Most of the above we leave the reader to deal with. The first argument, about the arrow, is one which the Copernicans did not know how to answer, as is evident from their explanation about the air; and this one argument, if it had been good, would surely have been quite enough for the purpose. When the mayor of Dijon, or some other town, excused himself to Henry IV., or some other king, for not firing a salute, alleging that he had twenty good reasons—firstly, that there were no cannon in the town—the king excused him the other nineteen. Kepler himself was obliged to admit the wind, asserted in the preceding paragraph, as a necessary consequence of the motion of the earth, but he thinks it is not sensible near the earth on account of its smallness, nor at very great heights on account of the thinness of the atmosphere. To which Fromond replies, with great justice, that wind is easily felt at the tops of the highest mountains, and should be always felt, by Kepler's admission.

In the ancient Hindoo philosophy, the great question was, not whether the earth moved, but whether it could possibly stand still, and what it could stand upon. The earth, said the vulgar, stands upon the elephant, and the elephant upon the tortoise; but unless the latter were made a *primum stabile* for the occasion, there was no way out of the difficulty. The philosophers asserted that the earth must be always falling, falling, falling, for ever and ever, and with it the sun, planets, and stars. For, said they, what can sustain them? But this same argument of the shooting up of an arrow gave their opponents a terrible advantage. For, said the latter, when you shoot an arrow upwards, the earth gets a start, which it cannot lose. The arrow may in time begin to fall, but how is it to overtake the earth, which is also falling?

The answer of a Copernican to the arguments in the last quotation would probably have been of just the same character as the arguments themselves. Before the time of Galileo, in our opinion, every Copernican was an ingenious theorizer, supporting a system which, though simple and possible, was met by unanswerable and *crucial* arguments, mixed with others derived from pure assumptions common to both parties.

The notion that the air, by its motion, kept bodies in their proper places with respect to the earth, was advanced against Rosse, and thus answered by him.

R. "But what a monstrous absurdity doe you tell us: That if a violent winde be able to drive ships, throw downe towers, turne up trees, much more may the diurnall motion of the aire (which doth so far exceede in swiftnesse the most tempestuous winde) be able to carry with it the bodies of birdes. If the diurnall motion of the aire exceed the windes in impetuosity, how comes it that it doth not the same effects that the winde doth? Why doe we not feele its force? Surely if the aire did move with that violence from east to west that a tempestuous winde doth, we should never have any ships come from the west Eastward, nor ships bound Westward should stay for a winde, seeing the motion of the aire at all times would carrie them with a witnesse. If we should have occasion to saile to New England, we should be there quickly, but no hopes ever to returne thence; how should we be able to walke or sit on horseback, travelling against the motion of the aire, if it did move with that violence you speak of? much less could birds in their flight resiste such a force: not the great bird Ruck (that I may fit you with a bird somewhat proportionable to your conceits), whose wings are twelve paces long, and snatches up elephants (as if they were but mice) in his talons a great way in the aire."

F. "Eighthly: If any one should stand on a high tower, of one or two miles high, and should look down from that tower to a point perpendicularly under the eye, and should let fall a heavy stone in that perpendicular line, it is most certain that that stone would fall upon the point so looked down upon, and lying in the perpendicular. But if the earth moved it would be impossible that the stone should fall upon that point. This I prove, First: Because either the air does not move equally fast with the earth, or it does so move; if not so fast, then it is certain that the stone could not fall upon that point, because the motion of the earth would outstrip the air which carries the stone. If as fast, still the stone cannot fall upon the point below; for though the air be moved equally fast in itself, yet it cannot therefore carry with it as fast and bear forward the stone which is falling through it; because the stone, tending to the centre by its own gravity, resists being carried forward by the air. You will say, like as the earth is moved in a circle, so also will all its parts; whence the falling stone will not only be moved in a circle by the carrying power of the air, but also by its own nature, as an existing part of the earth and having the same motion with it. But the answer is not good; for even if the stone did of its own nature, like the earth, turn in a circle, yet its natural gravity would at once hinder it from moving so fast as the air or the earth, which latter is in its natural place, and which therefore does not gravitate, as does the stone falling from a height. And even if the stone should move in a circle, like the earth, by its own nature, it would nevertheless not be carried so quickly as the whole earth. For just as a stone of one pound weight would fall down from the very heaven to the earth, in a straight line to the centre, because it is a part of the earth; so also would [a stone as big as] the whole earth fall, and yet the first would not fall so quick as the second. In like manner also, though the stone were carried in a circle like that of the earth, because the stone is a part of the earth, yet would not the stone be carried so quickly as the whole earth. And thus it may on every ground be asserted, that the motion of the earth ought always to outstrip the stone and leave it many furlongs behind; and thus it never could descend to the point directly looked down upon by the eye. But this last is false, therefore the earth does not move."

This argument has much the same bearing as one of the preceding; but we see a copious sprinkling of the peculiar phrases which were made substitutes for confessions of ignorance. When we now say that there is gravitation, we mean that, be the cause what it may, bodies do descend to the earth. But Fienus and his predecessors would have a reason for it; it was taken to be very clear that, by the very nature of bodies, a part taken from the whole must strive to return to *its proper place.* Thus, at the time of which we are speaking, and before the discovery of the weight of the air, if ever it were asked why the air did not press, the answer would be, because it is in its proper place. If a bucket were dipped into a river and filled, the weight of the contents not being felt till the bucket was drawn out, would be attributed to the fluid having no weight in its proper place.

But let it be granted that the place in which a body rests should be styled its proper place, which would not, properly understood, be a bad substitute for the more learned phrase, "position of equilibrium:" the preceding would not then be absurd; for, by the implied meaning of proper place, the position that all bodies seek their proper places would be a consequence of the meaning of the last words, just as "the whole is greater than its part" is no more than a necessary consequence of the meaning of *whole, part,* and *greater.* But many of the assumptions herein before contained cannot, by any such convention, be drilled into truths. Such are notions of the comparative excellency or dignity of different parts of creation, from which consequences and analogies were drawn by both parties. As in the following extract from Rosse:—

R. "The reason which you alledge from Pythagoras is also weak, for though the sunne in respect of his light were the most excellent body and the center the most excellent place, yet it will not followe that he is there; for we see that the most excellent creatures are not placed still next the center or in it, but farthest from it: as man is placed in the superficies or circumference of the earth, and not in the center of it, the heart is not in the midst of the body; if the middle or center were alwayes the fittest place for a luminous body, God would have commanded Moses to set the candlesticke with the lamps in the middle of the tabernacle, and not in the side of it; our eyes had been placed in our navels, not in our heads. And albeit Plato say, that the soule of the world resides in the innermost place of it, yet I hope you doe not by this understand the sunne, and you did well to alledge Macrobius against yourselfe, in comparing the sunne in the world to the heart in a living creature; for as the heart is not in the centre of the body, neither is the sun in the centre of the world.

But you give us a profound reason why in living creatures the chiefest part is not alwayes placed in the midst, *because they are not of an orbicular forme as the world is;* then it seems that the outward figure is the cause why the best part is not placed in the midst. What thinke you of a hedge hog when he wraps himselfe up in his prickles, as round as a bowle; is the best part then more in the middle of his body than it was before? Or hath the earth which is of a round forme better things in the center then in the superficies? What difference is there betweene the middle and out-side of a round stone? Againe, you say, the center is not the worst place, although Aristotle proves it from the dignity of the thing containing over that which is contained; and your reason is, *That though the center be contained, yet it is one of the* termini *or limits of a round body, as well as the circumference:* but I reply, that though it be one of the limits, yet it is contained, and therefore more ignoble than that which containeth it; so you have but offered to answer this argument, and indeed you know not how to answer it."

F. "Ninthly: If the earth moved in a circle, it ought to move from west through south to east; and consequently the air should move in the same way. And if this were so, then if anyone should shoot towards the east, the arrow ought to go much further than if he shot towards the west. Because when he shoots eastwards the arrow flies in the direction of the natural motion of the air, and has that motion of the air assisting it. But it is certain that a body should move faster and farther which has the motion of the air with it, than one which has the same against it, as appears in darts thrown with the wind."

"Other arguments might be produced upon this topic; but they are not of equal efficacy with the preceding in demonstration. These seem to me of such force, that I do not see how anyone can rightly answer them."

Such were the arguments by which the system of Copernicus was attacked and defended before the time of Galileo. We presume the reader to be aware of what is now called Newton's second law of motion—a result derived from experiment, which is a sufficient answer to all that was at one time unanswerable. We remark in the pamphlet of Fienus, that he hardly seems to recognise any difference between Copernicus and his followers in point of ability, and in one place he speaks of Bernardus Patricius (now unknown) as if he were as much the proper patron of the opinion of the earth's motion as Copernicus. But in a few years we find another view taken of the subject. Morin, the last of the astrologers, that is, the last man of extensive acquirements who wrote in favour of astrology, and who also wrote against the motion of the earth, speaks as follows:—

Morin. "The opinion of the earth's motion was refuted by Aristotle, Ptolemy, and others, and revived by the great Copernicus in the last century, with such force of talent, that no one at this time (A.D. 1631) is thought worthy of the name of learned who rejects his opinion of the rotation of the earth." And of Copernicus himself he speaks as follows:—" In the year 1473, on the 19th February, was born Nicolas Copernicus of Thorn, in a most happy position of the heavens for talent, as may be seen in his nativity given by Junctinus. He therefore being by nature born a mathematićian....solved all the celestial phenomena with so much skill and elegance by his own hypothesis, that he forced admiration from all astronomers, and obtained the title of 'great.'"

Both systems, the Ptolemaic and the Copernican or Pythagorean, equally well accounted for the phenomena—the former by a complication of movements, the latter by the most simple and apparently the most probable combinations, subject always to the mechanical difficulties already alluded to. On these Copernicus himself touches but slightly; he introduces the hypothesis of the air's motion, and, according to his commentator*, is the first who attended to this part of the hypothesis. His followers, in answering the Ptolemaists, while they had all the advantage of the simplicity of their system, were not behind their opponents in appealing to natures, essences, dignities, and all the tribe of axioms; they even found out some texts of Scripture to place against those which we have cited in Fienus. But here it must be owned they dealt captiously with their opponents, and unfairly with their cause; for supposing it once admitted, that questions of natural philosophy can or may be decided by the plain and literal meaning of any words of the Bible, the question is settled against Copernicus. All the texts cited by his followers clearly imply simply that the power of God shakes the earth, stable as it is. For instance, Job ix. 6, "which shaketh the earth out of *her place*, and the pillars thereof tremble." Still, however, the Copernicans against the Ptolemaists were no worse off than their opponents in any respect, and better in several.

But the modification which Tycho Brahé made in the system of Copernicus, while it preserved the simplicity of the latter, was open to none of the mechanical objections, which could not be answered, against the motion of the earth. But this must be taken with respect to the *annual* motion only, and not the *diurnal*. It consists in making all the planets move about the sun, while the sun itself carried them and their orbits about the earth. Suppose the planets moving according to the notion of Copernicus, on a sheet of paper, while the paper itself receives such a motion contrary to that of the earth that the earth is always over the same point of the table, then the motion of the system with respect to the table will represent the hypothesis of Tycho Brahé. Against this hypothesis the Copernican could bring no argument of any weight; and consequently was obliged to have resort to those of another kind. First, it was complained that the Tychonic system was borrowed from the Copernican, which was not denied. Secondly, they observed that the orbit of Mars must, in Tycho's system, cut the orbit of the sun; which, without giving a reason,

* The question had been previously discussed; the school of Pythagoras was for the earth's motion, those of Aristotle and Ptolemy against. If, as asserted by Aristotle, the Pythagoreans placed the sun in the midst of the universe, because they thought fire the most excellent of the four elements, they have much the same sort of right to be the predecessors of Galileo which the Epicureans have to be those of Dalton.

they assumed to be impossible of any superior planet. It was answered that, in the system of Copernicus, the moon's orbit cut that of the earth in the same way. Thirdly, they remarked that in the Tychonic system, an epicycle (circle which moves with its centre in another circle) was greater than the deferent (circle in which the epicycle moves) but without giving any reason why this should not be. Fourthly, they said that the earth was better in the midst of the planets than in the centre of the universe; because in the first case man was nearer to the throne of God. To which it was answered, that as they themselves placed the fixed stars at a most enormous distance and the throne of God beyond them, the difference was but small. Fifthly, that the earth was placed in the midst of the planets, in order that, by measuring the orbits of Jupiter, &c., and the fixed stars, men might be certain that they would one day have the use and enjoyment of the heaven they knew so well how to measure. To which it was replied, that, after all, man could only measure the lower heavens, while all parties admitted that the third heaven, or residence of beatified spirits, was beyond them all. Sixthly, it was said, that man who is an inhabitant of and looker on the universe could not be supposed to be kept in one place as if he were shut up in a den. To which it was answered, that as the stars and planets came to him, his going round among them was of little consequence. Such were the arguments employed to settle the relative merits of the Tychonic and Copernican systems.

The last of the Anticopernicans, who may be said to belong to the old school, is the Jesuit Riccioli, whose *Almagestum Novum* is a most enormous monument of reading and industry. His attack upon the Copernican system alone consists of more than two hundred double column folio pages; and being at such length it is not easy to pick out any quotations sufficiently complete to be intelligible by themselves. He endeavours to turn the discoveries of Galileo against himself, by trying to show that the descent of a heavy body, according to the law discovered by the last-named philosopher, would be impossible if the earth were in motion. His argument shows that he did not comprehend the law of motion already referred to. He admits the very great merit of the Copernican system, and its applicability to the explanation of all astronomical phenomena; and one of his remarks is, in beginning to show how the motion of the earth's axis explains the precession of the equinoxes:—"We have not yet exhausted the depth of the Copernican hypothesis, in which the further we go, the more shall we find of talent and valuable sagacity." Riccioli takes as much pains to develop the Copernican system in a favourable light, before he proceeds to refute it, as Copernicus himself, and a good deal more space. It has even been suspected that Riccioli was in heart a Copernican, but unable, as a Roman Catholic and a Jesuit, to declare himself.

The church of Rome, or the court it may be, for no council was called on the subject, stopped the mouth of Galileo by means of the Inquisition, as all readers are aware (A.D. 1633). The first actual prohibition of the Copernican system was by the five Cardinals who had the superintendence of the *Index Expurgatorius.* These prelates suspended the work of Copernicus until its errors were corrected (which must have been either ignorance or irony, for the heresy runs from beginning to end), and entirely prohibited that of Foscarini, a Carmelite, who must be considered as the introducer of the doctrine into Italy. Up to this time the contest had been carried on, the times considered, with something like moderation. The tone of contempt with which the orthodox party set out subsided into admiration of the beauty of the system. Indeed, examples are not wanting in which the opponents of the now received system were the more moderate and gentlemanlike of the two. Witness Morin (by no means a man of quiet temper in a personal dispute) who, after admitting the talents of Copernicus and his followers, cites the following from the justly celebrated Kepler :—"The vulgar herd of learned men, not much wiser than the illiterate, produce authorities.... blind in their ignorance....&c." Which remark Morin quotes, not to complain or retaliate, but to observe—"This evidently shows that they have taken up this doctrine, not so much for the sake of dispute and exercise, as because they actually wish to promote the belief of it."

One of our principal objects in writing this article has been to bring some of the arguments of the two parties into contact with the notions our readers may have formed of their relative merits from the popular works most in vogue. We are told, and implicitly believe, that truth is always moderate and argumentative, error violent and spiteful; that in particular the doctrine of Copernicus was truth supported by reason; that of Ptolemy falsehood backed by hypocrisy, stupidity, and malice. The sophisms of those whom a future age, and not they themselves, has shown to be right, are allowed to sink; those of their opponents are preserved and commented upon for ever. We remember, that is, it is remembered for us, that the Ptolemaist attributed gravity to the gravitating body being out of its proper place; but we are not to remember that Copernicus said that circular motion is that of a whole, rectilinear motion is that of a part separated from its whole, and that therefore "we may say that circular motion exists with rectilinear, in the same manner as the notion of animal exists in a horse."

The system of Newton overturned both the Ptolemaic, the Copernican, and the Tychonic, in the sense in which they were asserted by their various supporters. The first and third assumed the absolute stability of the earth, the second that of the sun. Those who are at all acquainted with the nature of relative motion will see that we might (not without inconvenience, but without inaccuracy) assume any one point of the universe we please for a fixed point, provided we give all other points, not their absolute motions, but the motions which they have relatively to the centre chosen. A satellite of Jupiter, a point in Saturn's ring, a cloud in the atmosphere of the earth, a shooting star in its descent, might either of them be assumed to be fixed, provided the proper relative motions were given to all other bodies. The result of Newton's system may be expressed as follows :—

All the primary planets describe ellipses (nearly) about a point in the sun, and all satellites describe ellipses (nearly) about points in or near their primaries; in the meanwhile the centre of gravity of the whole system may be (probably is) in motion towards some point of the heavens, depending upon the impulse originally given to it, and with it the whole system. This motion of the centre of gravity will be in a straight line, unless the attraction of the fixed stars be sufficient to alter it sensibly.

It might happen then that, for a moment, the Tychonic, or even the Ptolemaic, system might be absolutely true as far as the orbital motion of the earth is concerned. This would be the case if ever the orbital motion of the earth were equal in amount, and contrary in direction, to that with which the whole system is moving forward. Similarly the Copernican system might for a moment be an absolute truth.

If the whole system be moving towards any star, the consequence will be that, in course of time, the stars to which the earth draws near will appear farther apart, and those from which it recedes nearer to each other. Sir W. Herschel at one time suspected that the constellation Hercules contained the point towards which our system is moving; but subsequent observations have not confirmed this idea, and it may be ages before any satisfactory conclusion is attained.

Some of our readers may be surprised at our saying that Newton overturned the Copernican system. But we mean the system which actually was promulgated by the man named Copernik, and found in his book published in 1543, under the title *de Revolutionibus orbium cœlestium.* This book considers the sun as the *medium mundi*, or middle of the universe, a term of the time which always implied a fixed point, and there is no reason to suppose that Copernicus used it in any other than the common sense. At the same time it has become customary to call the system of the Universe, as amended by Newton and his successors, Copernican, so far as the actual motions are considered, independently of their cause.

There is not in the whole of astronomy, properly so called, any argument in favour of the motion of the earth which is absolutely and demonstrably conclusive. The argument which admits of no answer is derived from what is called the *aberration of light.* As very few readers, except those who have studied mathematics, have any idea of the cause of this phenomenon, or indeed in what the phenomenon itself consists, we shall endeavour to give an illustration of it.

Suppose a person trying to throw a bullet through the windows of a carriage in rapid motion, so as to pass through both windows, say through the centre of both.

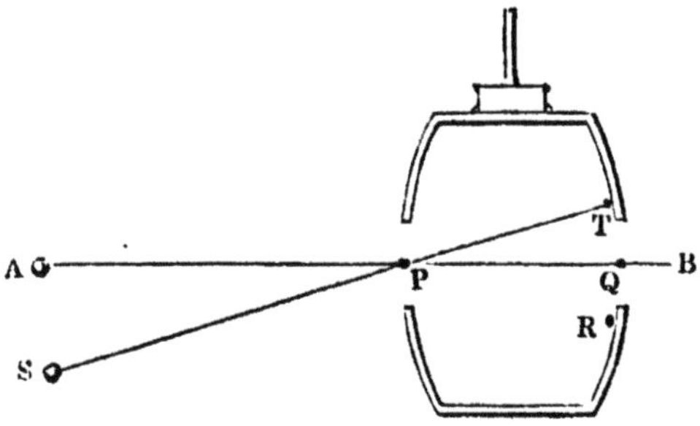

It is plain, that if it be thrown in the line A B, so that it would pass through both windows of the carriage at rest, the thing required will not be done ; for supposing that the bullet strikes the glass at P at the proper time, the forward motion of the carriage will evidently cause the bullet to strike, not the opposite window, but a point R in the hinder pannel, and the passengers in the carriages will suppose that the stone moves in the direction from P to R. But let the bullet be thrown obliquely in the line S P T (so that though it would, if the carriage were at rest, strike the foremost pannel at T), and with such a velocity that the motion of the carriage may bring Q to T, in the same time as the bullet moves from P to T, and the bullet will then pass through the centre of both windows, and will appear to the passengers to be thrown directly through the carriage.

A ray of light, thrown into a telescope from a star, will in the same manner have its apparent direction altered by the motion of the earth, if that be large enough to bear a perceptible proportion to 200,000 miles per second, which is the velocity of light, as ascertained from other sources. Every star in the heavens is found to show exactly such an effect as would be produced by the motion of the earth. Their distance is so great, that the whole yearly circuit of the earth produces no change in their apparent positions, except that each makes a small yearly revolution through a very minute circle or oval. The detail of the evidence upon this point must be reserved for those who understand geometrical reasoning ; but the alternative derived from the result of that reasoning is as follows :—

If the earth do not move, then every star, in what part soever of the heavens it may be situated, at what distance soever from the earth, has light which depends for its laws of motion upon the position and motion of the sun. Even if it were imagined to be possible, which certainly might be the case, that the course of a ray of light would depend (independently of the atmosphere) upon the manner in which it approached the earth, whether horizontally, obliquely, or vertically, this would not affect the present argument. For what is shown to be universally true is, that let a ray of light fall in any manner upon the earth, from any star or planet in the heavens, its course is found to be altered in a manner which can be, and is predicted, when the place of the sun is known, and then only. If all the perceptible motion of the sun arise from a motion of the earth (that is, if the Copernican system be true), then this phenomenon can be, and is explained; but if the earth be still, and the sun only move, it follows that all the light in the universe, as it comes near the earth, undergoes a small variation in its course, which depends on the sun's place and on the quantity and direction of the sun's motion at the time. Even if all the stars derived their light from the sun, this phenomenon would not admit of explanation; but if the stars give light independently of the sun, it would be much more easy to show Tenterden steeple to be the cause of the Goodwin Sands, than to derive from experiment or analogy any connexion between the sun's place and that of a star, except what arises from supposing the motion of the earth. If the earth's motion be not thereby demonstrated, then nothing can demonstrate it; for even if the spectator were removed away from the earth, and saw it move, as he thinks, he could not know whether the motion were in himself, or in the earth. There is no possible way of demonstrating absolute motion mathematically in any one given body.

II.—NOTICES OF ENGLISH MATHEMATICAL AND ASTRONOMICAL WRITERS BETWEEN THE NORMAN CONQUEST AND THE YEAR 1600.

In the 'Companion to the Almanac' for 1836 we gave some idea of the disputes with which the Copernican hypothesis was ushered into notice. We now intend to put together a few historical remembrances of a still earlier state of science in this country, being a part of its history which is as yet altogether unwritten in a connected form. Far from having such a work as those of Montucla or Delambre in our language, we have not even a chronological compendium like that of Weidler, Heilbronner, or Gerard Vossius. In selecting the earliest part of our history, with a view merely to present that which is most striking in mathematics and astronomy, we are thrown upon the interesting period at which works began to be written in English (1500—1600); and the materials for the preceding centuries are so scanty, that it will add but little to our labour to take in the whole period from the Conquest. After the year 1600 English science began again to acquire some note in the other parts of Europe, and we may refer to general histories for the principal names at least. But before that period nothing exists in a collected form; and, imperfect (even for a sketch) as our account must necessarily be, we have the satisfaction of knowing that it must be better than anything of as special a character which exists, to the full extent in which it is better than nothing.

There are various different methods in which an account of English science might be given, with regard for instance to epochs, to places, and to subjects. Immediately succeeding the Conquest, and reaching to the year 1250 or thereabouts, comes a period in which it is difficult to distinguish the Norman from the Anglo-Norman; and the writers (priests of course) may have been only connected with this country by their appointment to abbeys or bishoprics in it. From 1250 to the beginning of the wars of the Roses in 1450, we have a more decidedly English school, due for the most part to the University of Oxford. During the civil, followed by the religious, troubles, we have a period of quiescence which lasted about a century; while finally, from 1550, we see a new era commence, namely, that of writers *in English*. With regard to places, the literary history of this country requires separate and minute accounts of the rise of science in Oxford, in Cambridge, and in the north of England, which should severally end (if it might be no later) with Wallis, Newton, and Thomas Simpson. As to subjects, the writers of whom we treat offer little variety till we come to the last period. The study of the Greek authors formed the whole of the regular course; and except in astronomy, and very much for the sake of astrology, there are but few original writings.

In this present article, however, we shall confine ourselves simply to authors, in order of time, giving a general account of our authorities at the end.

Previously to the Conquest, we find the names of Adelm, about A. D. 680; Walfred of Rippon, about 690; Bede, about 730; Alcuinus, about 760; and Adelbold, about 1000—his correspondence* with Gerbert (Sylvester II.) being preserved in the Vatican library; and John Garland, who wrote on alchymy and minerals. These knew neither Euclid nor Ptolemy, and their principal study was Boethius. It is certain that, in the period succeeding the Conquest, the knowledge of the Greeks was obtained of the Saracens by those who travelled into Arabia and Spain. These "heathen savages," to choose one of the mildest terms with which the benefit they conferred was rewarded, lighted the lamp of knowledge in Europe by communicating Euclid, Ptolemy, and Aristotle.

The following is the list of all the names we have been able to find, which we doubt not might be much enlarged, particularly as regards the latter century.

1060 Oliver of Malmsbury.
1095 Herbert Losinga.
1130 Adelard of Bath.
1132 Richard of York.
1140 William Shelley (de Conchis).
1143 Robert of Reading.
1160 John of Hexham.
1164 Simeon of Durham.
1170 Roger of Hereford.
1170 Clement Langton.
1170 John of Salisbury.
1190 Daniel Morley.
1200 Geoffrey Winesauf.
1210 Gilbert Legley.
1224 Johannes Ægidius.
1230 Alexander Hales.
1242 St. Edmund, Archbishop of Canterbury.
1248 Richard Fisacre.
1253 Robert Greathead.
1253 Adam Marsh.
1255 William Shirewood.
*1255 Roger Bacon.
1256 John of Halifax.
1260 John Peccam.
1270 Odinton of Evesham.
1290 Michael Scot.
1304 Duns Scotus.
1316 Walter Evesham.
1320 Nicholas Trivet.
*1320 John de Dumbleton.
1320 John Baconthorp.
*1326 Richard Wallingford.
1329 John Canon.
1330 Walter Catton.
*1337 Walter Burley.
1340 John Barwick.
1340 Robert Holcoth.
1340 Godfrey de Meldis.
*1342 John Mandovich.
*1347 Nicholas Ockham.
*1349 Thomas de Bradwardyn.
*1350 Roger Swineshead.
*1350 William Grizaunt.
1359 Clinton Langley.
*1360 John Killingworth.
1360 Louis of Caerlion.
*1365 John Estwood.
1370 Richard Lavingham.
1370 Simon Bredon.
*1370 Nicholas of Lynn.
*1385 William Rede.
*1390 John Chylmark.
1390 John Somer.
*1390 Walter Bryte.
1392 Richard II.
1400 Geoffrey Chaucer.
1410 John Walter.
1410 William Batecumb.
1434 Richard Monke.
*1435 Thomas de Rudbourne.
1440 Humphrey, Duke of Gloucester.
*1445 John Killingworth, jun.
1460 William Wircester, or Botoner.
1482 John Shirwood.
*1489 Thomas Kent.
1490 John Erghom.

* One (of Gerbert) *de Causâ Diversitatis Arearum in Trigono Æquilatero.* (What does this mean?) Another, *de Sphæræ Constructione.* There is one of Adelbold, *de Ratione inveniendi Crassitudinem Sphæræ.* It must be remembered, however, that we have only the name by which we can infer that Adelbold was a Saxon.

1524 Thomas Linaker.
1530 John Robins.
1540 Andrew Borde.
1546 William Billingsley.
1552 Anthony Askham.
1555 Richard Eden.
1555 John Adams.
1555 Frederic van Brunswike.
1555 Richard Candish.
1555 John Danke.
1556 John Field.
15..? Robert Norman.
15..? ——— Borough.
15..? Humfrey Baker.
15..? Richard de Benese.
15..? — Sanderson.
15..? — Molineux.
1558 Robert Recorde.
1559 Cuthbert Tonstall.
1559 William Cunningham.
1572 John Dee.
1574 Leonard Digges.
1584 William Boorne.
1585 John Blagrave.
1590 Thomas Hood.
1594 W. Hartgill.
1594 — Blundeville.
1595 Thomas Digges.
1599 Edward Wright.
1599 Thomas Hill.

Oliver of Malmsbury, and Herbert (or Robert) Lozing, Losinga, Lotharingius, or of Lorraine, under which names he appears in different places, were both astrological writers, and formed tables. Athelard translated Euclid from Arabic into Latin, of which there is a MS. in the Bodleian. He translated several Arabic works, and wrote on the astrolabe. It is related that he travelled in the East. This version of Euclid is most likely prior to that of Campanus. Richard of York, abbot of St. Alban's, wrote on the tables of Arzachel, which therefore soon found their way to England. Robert of Reading, and William Shelley, both travelled into Spain in search of books: the former translated the Koran into Latin. A MS. (according to Wallis) containing a particular account of their travels, with mention of those of Adelard, was stolen from Oxford. John of Hexham, and Simeon of Durham, both wrote on Comets; the latter is also known as an historian. Roger of Hereford wrote a theoretical work on Astronomy (MSS. in Bodl.); Clement Langton (or Lanthoniensis, from Lanthony, near Gloucester, says Sherburne) also wrote on Astronomy. Daniel Morley travelled into the East, and into Spain, and on his return wrote *Principia Mathematices.* Several circumstances lead us to believe that at this period the communication between Oxford and Toledo was much greater than now, and Morley was an Oxonian. Geoffrey Winesauf, the historian, also wrote on fruit trees and wine, (whence his name, it is said.) Gilbert Legley, a physician, wrote a compendium of Astronomy. Ægidius, of St. Alban's, was physician to the King of France, and wrote on Astronomy. Alexander Hales commented on the Metaphysics of Aristotle. Of Edmund of Canterbury we only know that he wrote. This is all we have to say on the first period, of which the translation of Euclid by Adelard is certainly the most remarkable feature.

The second period owes a great many of those who have most illustrated it to Merton College, Oxford, a foundation of which Wood remarks, that there was no other for two centuries, either in Oxford or Paris, which could at all come near it in the successful cultivation of the sciences. But he goes on to say that large chests full of the works of writers of this college were allowed to remain untouched by their successors, for fear of the magic which was supposed to be contained in them. Nevertheless, it is not difficult to trace the liberalizing effect of scientific study upon the University in general, and Merton College in particular. The spirit of resistance to foreign usurpation was frequently displayed there, in particular by Robert Greathead, presently to be mentioned. But through Merton College came the blow from which the claim to dictate opinions to others never recovered, given by the hand of William Wickliff, himself a student of philosophy, though we cannot learn that he wrote anything on this subject. Of the immediate followers of Wickliff, several were his contemporaries and fellow collegians. Those who belonged to this distinguished foundation have their names preceded by an asterisk. There is, however, a doubt in the case of Roger Bacon, whether he did not belong to Brasennose College. However this may be, let it be remembered (and need is there that it should be) that to the cultivation of the mind at Oxford we owe almost all the literary celebrity of England in the middle ages.

Richard Fisacre and William Shirewood were friends of Roger Bacon, and their public lectures were held in high estimation. Of Robert Greathead, or Grostête, Bishop of Lincoln, much has been written. Several manuscripts of his exist in the Bodleian, and one treatise, *Compendium Sphæræ*, was published by Gauricus in 1531 (according to Heilbronner). Among other things, he wrote on the Quadrature of the Circle, and commentaries on Aristotle. Adam Marsh is mentioned with honour by Greathead. Of Roger Bacon it is comparatively unnecessary to speak, as several very full accounts of him are published, (Dr. Jebb's preface; 'Biogr. Brit.;' 'Penny Cyclop.') We have little reason to suppose that his writings were much read out of his own university. But, to those who will study them, there is even at this day a combination of simplicity of style and independence of thought altogether unusual in his time; and these induce us to place him far above any writer of the middle ages whose works we have seen. He is the only original investigator of phenomena of whom we shall have occasion to speak. But we purposely pass to those who are now less known.

John of Halifax is the celebrated *Johannes de Sacro-bosco*, better known on the continent of Europe than any other subject of this paper. His work on the Sphere was for centuries almost as general an object of study as Euclid himself, and many commentaries on it were published. The remaining MSS. of this work are very numerous. He also wrote a work on the Calendar, and an *algorithmus* or introduction to arithmetic, of which it has been asserted, but without reason, that it was the first work in which the Arabic numerals were employed. (Peacock, 'Arithmetic, Encyc. Metrop.') It was printed at Venice in 1523.

John Peccam, a friend of Roger Bacon, was Archbishop of Canterbury. He is called John of London by Bacon. He wrote on Optics (Perspectiva), particularly on reflection and refraction. Odinton of Evesham wrote on the motions of the planets, and '*de Mutatione Æris*.' Walter Evesham, mentioned directly after, is, we suppose, the same person; he wrote on the Sphere. Michael Scot was one of the first translators of Aristotle from the Arabic, and commented Sacrobosco. Duns Scotus has a place in this list from his commentaries on the Physics of Aristotle and his Meteorology. Nicholas Trivet and John de Dumbleton are mentioned only by Wood. Richard Wallingford was both a writer on astronomy and an observer. Manuscript tables of eclipses remain, by Louis Caerlion, *Secundum Diametros R. Wallingforde*. There is also a work of his *De Chordâ rectâ et versâ*, and another entitled *Ars componendi Rectangulum et operandi cum eo.* A portrait of him is preserved among the Cottonian MSS. John Baconthorp (sometimes called Bacon, and confounded with Roger Bacon) wrote on the Sphere, and is cited by Recorde as among those who have treated the subject best. Catton, Burley, and Barwick, are cited by Wood: the first and third wrote *against* Astrology.* Holcoth wrote on the Motions of the Stars. G. de Meldis wrote on several comets, the MSS. of which works were in the library of Pembroke College, Cambridge. Mandovich, called also Maudith, gave Tables of the Fixed Stars. Ockham is mentioned by Wood; this writer is also called Nicholas Occam. He was afterwards professor at Paris, and was excommunicated by John XXI., but was protected by the Emperor (Louis IV.), in whose behalf he had resisted the pope's temporal claims. He left two treatises on the errors of this pope, a treatise on the authority of the popes generally, and commentaries on the Physics of Aristotle, &c. Bradwardyn (Bravardinus, Bernardinus, Bragadinus?) wrote on Optics, and also on Proportion. His *Geometria Speculativa* was published at Paris in 1496. Swineshead is mentioned by Wood. William Grizaunt, a physician, who lived first at Oxford, emigrated to Marseilles, where he died. His son was pope, under the title of Urban the Fifth. He wrote on the Quadrature of the Circle, on the Magnitude of the Sun, and on Astrology; and was duly suspected of magic. Langley wrote on Astrology. Killingworth wrote on Arithmetic, Astrology, and Astronomy, and particularly on Twilight and on the Height of Clouds. Louis of Caerlion calculated large numbers of eclipses. Estwood, called also Eschuid, was well known by his *Summa Judicialis*, a mixture of astrology and physics, printed at Venice in 1489. It may give a notion of the liability to confusion which exists in this recapitulation, when we mention that Estwood is also called Estwyde, Eshwood, Eshwid, Eschuyde, de Ashenden, Eshenden, Ashenton, Asyden, Estemdeene. Lavingham wrote on the Distances of the Planets. He was killed, at the same time as the Archbishop of Canterbury, in a tumult in 1381. Bredon wrote on the Almagest, on the Equations of the Planets, (MSS. in Lib. St. Peter's, Camb.) and on the Principles of Arithmetic, (MS. in Bodl.) Nicholas of Lynn gave Astronomical Tables, and

* It must be remembered that the church of Rome never professed to encourage astrology, though it appears to have been obliged to connive at its prosecution. There was no doubt it was discouraged, (with full belief of its efficacy,) as being connected with magic. Hence we may be certified of the universality of the belief in it, by observing how few wrote against it. Bacon was what we may call a corporeal astrologer; he admitted the influence of the heavenly bodies on the mind *through the body*

C

was rather a voluminous writer on these subjects. William Rede gave Tables of the Mean Motions and Equations of the Planets. Chylmark wrote on Astrology. John Somer or Somur made Ephemerides for the meridian of Oxford; he is said also to have written on the Length of the Year. Walter Bryte, the friend of Wickliff, wrote on the Sphere and the Theory of the Planets. Richard II. is asserted by Sherburne to have written "something in Astronomy or Astrology, now (A. D. 1675) extant in his Majesty's library at St. James's." Geoffrey Chaucer, the poet, wrote a treatise on the Astrolabe for his son, which is the earliest* English treatise we have met with on any scientific subject. It was not completed, and the apologies which he makes to his own child for writing in English are curious, while his inference that his son should therefore "pray God save the king that is lord of this langage," is, at least, as loyal as logical. John Walter wrote astronomical tables for the purposes of astrology, and appears to have been more esteemed than we might now suppose would have been the case. William Batecomb, also called Bedecon, is recommended by Recorde as a writer on the Sphere: he wrote on the Astrolabe, and on the *Concave Sphere* and *Solid Sphere* (astronomical instruments).

Richard Monke wrote on the reformation of the Calendar. Thomas de Rudbourne is mentioned by Wood. The good Duke of Gloucester is cited by Sherburne, from Bale, as having set forth "Tables of directions" (astrological) "of his own composing," and Heilbronner mentions a manuscript in the Cambridge Library, entitled "Humphridi ducis de Glocestriâ Tabula de judiciis artis Geometricæ." John Killingworth the younger is mentioned by Wood; but we have no means of distinguishing between his writings and those of his father. W. Wircester, or Botoner (probably W. Botoner of Worcester), wrote on Astrology. There is a manuscript of the Alphonsine Tables by him in the Bodleian. There is in the same library a manuscript of John Shirwood *de ludo Arithmomachiæ*. Thomas Kent, or Kayleg, (Kayleg of Kent?) wrote Astronomical Tables. Linaker, the founder of medical lectures at Oxford and Cambridge, is conspicuous by his editions of Greek authors, particularly of Proclus (Venice, 1500), which is used by Recorde. John Robins, or Robyns, gave Astronomical Tables. Dr. Andrew Boorde wrote 'Principles of Astronomye,' published before 1540.

It may be worth while to notice the selection of English writers which Dr. Bernard (the originator of the editions of the Greek mathematicians which were printed at Oxford in the beginning of the last century) would have recommended; taken from his *Synopsis Veterum Mathematicorum*. We cite those works which have authors' names only, omitting anonymous manuscripts.

Adelard, translation of Euclid. 'Alchindus de sex Quantitatibus. Selections from the Arithmetical Writings of Bede, Bredon, Suisset,† Wallingford, Bradwardin, and Peccam.' 'Wallingford

* The preface to this production is given at length in the *Book of Table Talk*. C. Knight, 1836.

† The arithmetical writings of Suisset (to which we cannot fix a date) were printed in 1488; and several editions followed.

libri IV. de Sinubus ac Proportionibus, cum Scholiis D. Lenys.' 'Bredon, super demonstrationes aliquas Almagesti.' 'Swinshed de Motu Cœli.' 'Calendarium Ric. Monke.' Rede, 'Table of Mean Motions,' 'Oxford Almanack,' 'Canon of Fixed Stars,' and 'Toledan Tables.' 'Grostête's Calendar.' 'Bredon's Theorica.' Wallingford, 'Account of an Astronomical Instrument called *Albion*.' Monk, 'Equations of the Planets and Length of the Year.' Wyrcestre, 'Canon of Fixed Stars.' Evesham 'de Motu Octavæ Sphere.' Astronomical Observations of Estwood, Bredon, Batecomb, Killingworth, Caerleon, Wallingford, and Read. Rob. of Lincoln and Alkindius 'de Prognosticationibus et Observationibus Aeris.' Alkindius 'de Pluviis per Planetas.' Sacrobosco's 'Sphere.' Rob. Lincoln 'de Refractionibus.' Bacon and Peccam, 'Optics.' Bacon 'de Situ Orbis.' Worcester 'on Latitude and Longitude of places in England.' Burroughs 'on Geography.' Grostête 'on the Music and Arithmetic of Boethius.' Chylmead (Chylmark?) 'de Musica Vetera.'

We have now brought our chain of writers down to the beginning of the reign of Edward VI., from which the English (or the *in-English*) epoch dates. On what precedes we must observe that, though the art of printing had been in full play for nearly sixty years, very few English writers had gone through the press. The Greek writers had almost all been printed, either in Greek, as Archimedes and Proclus, or in Latin, as Euclid and Ptolemy. But nothing was done in England towards distributing any of those manuscripts which the educated of the day held in most estimation; and 'The Game of Chess,' 'Troilus and Cressida,' 'The Ship of Fools,' and such like, were the products of our national press; while Ratdolt, at Venice, conquering the difficulty of diagrams by the application of woodcuts, presented Euclid complete to the Latin reader. As the question who first printed in England has been pretty well settled, we recommend the antiquary to turn his attention to the question, Who printed the first useful book in England?

The University of Cambridge appears to have acquired no scientific distinction in the middle ages. Taking as a test the acquisition of celebrity on the continent, we find that Bacon, Sacrobosco, Greathead, Estwood, &c., were all of Oxford. The latter university had its morning of scientific splendor, while Cambridge was comparatively unknown, and (with regard at least to definite college foundations) hardly beginning to exist: it had also its noon-day, at a later period than we shall here have occasion to consider, illustrated by the names of such men as Briggs, Wren, Wallis, Halley, and Bradley. The age of science at Cambridge has been said to have begun with Francis Bacon, and but that we think much of the difference between him and his celebrated namesake lies more in time and circumstances than in talents or feelings, we would rather date from 1600 with the former, than from 1250 with the latter. Praise or blame on either side is out of the question, seeing that the earlier foundation of Oxford, and its superiority in pecuniary means, rendered all that took place highly probable. We rejoice in the recollections by the production of which we are enabled to

C 2

show that this country held a conspicuous rank in the philosophy of the middle ages, and we cheerfully and gratefully remember that, to the best of our knowledge and belief, we are in a great measure indebted for the liberty of writing our thoughts to the cultivation of the liberalizing sciences at Oxford in the *dark* ages.

With regard to the University of Cambridge, for a long time there hardly existed the materials of any proper instruction, even to the extent of pointing out what books should be read by a student desirous of cultivating astronomy. Of this we have a remarkable instance.

Jeremiah Horrox, who is well known to astronomers as having made a greater step towards the amendment of the lunar theory than any Englishman before Newton, and whose name might be well known to every reader but that he died at the age of 23, was at Cambridge in 1633—1635. From the age of boyhood he had been wholly given to the desire of making himself an astronomer. "But many impediments presented themselves: the tedious difficulty of the study itself deterred a mind not yet formed; the want of means oppressed, and still oppresses, the aspirations of my mind: but that which gave me most concern was *that there was no one who could instruct me in the art, who could even help my endeavours by joining me in the study; such was the sloth and languor which had seized all.*"—"I found that books must be used instead of teachers."—"When, therefore, I sought aids for study, and particularly astronomical books, and among these only the best, I happened to light upon a treatise by H. Gellibrand, professor of astronomy in London," (Gresham College,) "in which he greatly praised Lansberg," &c. Whence it appears that the Cambridge student of 1635 could not obtain in the University the means even of knowing to what books he should direct his attention. Nor were the books themselves which he (having but small means, and desiring only the very best) afterwards bought, in any one instance that we can discover, printed in England. The following list is curious, as showing the selection which he made under such circumstances:

Albategnius.
Alfraganus.
J. Capitolinus.
Clavii Apolog. Cal. Rom.
Clavii Comm. in Sacroboscum.
Copernici Revolutiones.
Cleomedes.
Julius Firmicus.
Gassendi Exerc. Epist. in Phil. Fluddanam.
Gemmæ Frisii Radius Astronomicus.
Cornelii Gemmæ Cosmocritice.
Herodoti Historia.
J. Kepleri Astron. Optica.
——— Epit. Astron. Copern.
——— Comm. de Motu Martis.
J. Kepleri Tabulæ Rudolphinæ.
Lansbergii Progymn. de Motu Solis.
Longomontani Astron. Danica.
Magini Secunda Mobilia.
Mercatoris Chronologia.
Plinii Hist. Naturalis.
Ptolemæi Magnum Opus.
Regiomontani Epitome.
——— Torquetum.
——— Observata.
Rheinoldi Tab. Prutenicæ.
——— Comm. in Theor. Purbachii.
Theonis Comm. in Ptolom.
Tyc. Brahæi Progymnasmata.
——— Epist. Astron.
Waltheri Observata.

From the time of the commencement of printing in English, we shall only take such works as we have before us, including those of most note. There were doubtless many others, and there is a printed catalogue extant of the date of 1595, or thereabouts, devoted to books printed in English upon the mathematical sciences; but we have never seen a copy. It must be remembered that 1543 is the era of the publication of Copernicus, which will render it a matter of interest to inquire into the manner in which his opinions were received in England.

To begin with those of dubious date. Robert Norman was a writer on Navigation, as was also Borough, (cited by Blundeville,) who was comptroller of the navy under Queen Elizabeth. Humfrey Baker translated into English a treatise entitled 'The Rules and right ample Documentes touchinge the Use and Practise of the Common Almanackes.' It appears that it was customary to translate almanacks; Henry Van Brunswike, for instance, translated in 1555 'A ryghte excellente Treatise of Astronomie, made in the Thuscane or Italian tongue, by Maister Antonius de Montulm,' promising all sorts of predictions for 1554 and 1555. Richard de Benese wrote 'The Boke of Measuring of Lande, as well of Woodland as Plowland, and Pasture in the Feelde; to compt the true Nombre of Acres of the same.' This is the earliest book on surveying which we have seen. (It was printed* between 1562 and 1575.) The author is called in the preface a *canon of Marton;* the instruments employed in it are a rod of a perch in length, and a waxed string. It contains the most common rules for surveying. Sanderson and Molyneux were among the earliest makers of celestial and terrestrial globes. William Billingsley is called an alchymist, but we do not know whether this is or is not a mistake of the name for Henry Billingsley, of whom presently. Antony Askham, 1552, published 'A lytel treatyse of Astronomy,' showing that in time December would be in the middle of summer, &c. There does not seem to be any proposal to reform the calendar in this work, but it appears merely to have been a work to catch the attention of the curious, as were most of the almanacks of this time, which contained very little except astrology. We have before us a work entitled 'The Compost of Ptolomæus,' full of nothing else. Of translations from the Italian we have traced several, and two from the Spanish of some note; namely, a work of Michael de Coignet either on navigation or surveying, and a work of navigation by Martin Cortez, translated by Richard Eden. There is also mention of a Spanish work on navigation, by Medina. Of John Adams and John Danke we only find the names: and of Richard Candish, the simple assertion that he translated Euclid.† But as we can find nothing elsewhere, and a translation of Euclid could not well be entirely lost, we suppose either that the name of Candalla must have been confounded with that of Candish, or some such mistake. Of John Field we shall presently speak.

* This we know from the printer (Colwell). But there is a 'Boke of Surveying,' between 1530 and 1534, printed by Berthelet.

† In extensive tables of chronology, called *Saturni Ephemerides,* by Henry Isaacson, 1633.

Cuthbert Tonstall died in prison in 1559, having been successively Bishop of London and Durham. He had been a bishop since 1522, and at first approved of Henry's divorce, but afterwards changed his opinion and fell into disgrace. He was ejected in the time of Edward, restored in that of Mary, and again ejected by Elizabeth. His Arithmetic, '*de Arte Supputandi,*' was first printed by Pynson in 1522; by R. Stephens, at Paris, in 1529 and 1535; several times at Strasburg, &c. In a dedication to his friend Sir Thomas More, he states that almost every nation in Europe had books of arithmetic in its own tongue, but of a very low order. His work contains examples of the most simple kind treated at considerable length, rather developing the rules than the principles on which they are founded. In point of simplicity, however, this work stands alone in its age, and is perfectly free from all the extraneous matter which was often introduced into the scientific works of the day.

The founder of the school of English writers (to any useful or sensible purpose) is Robert Recorde, the physician, a man whose memory deserves a much larger portion of fame than it has met with, on several accounts. He was the first who wrote on arithmetic in English (that is, anything of a higher cast than the works mentioned by Tonstall); the first who wrote on Geometry in English: the first who introduced Algebra into England; the first who wrote on Astronomy and the doctrine of the Sphere in English; and finally the first Englishman (in all probability) who adopted the system of Copernicus. According to Wood, his family was Welsh, and he himself was a Fellow of All Souls College, Oxford, in 1531; he died in 1558 in the King's Bench prison, where he was confined for debt. Some have said he was physician to Edward VI. and Mary, to whom his books are mostly dedicated.

The works of Recorde are all written in dialogue between master and scholar, in the rude English of the time. They are enumerated by the author himself in verse (for he, in common with most others,* continually breaks out into poetry in his prefaces and introductions) as follows, at the end of the preface to the 'Castle of Knowledge.'

AN ADMONITION FOR THE

Orderly trade of studye in the Authors woorkes, appertaynyng to the mathematicalles.

The grounde is thought that steddye staye,
Where no foote faileth that well was pyghte:
Whereon who walketh by certaine waye,
His pase is lyke to prosper ryghte.

* A very common practice in his age. Vieta, at the end of his *Variorum de Rebus Mathematicis Responsorum*, after a chapter of quiet description of the calendar, ends abruptly with—sed

Eheu! quis unctum chrismate mystico
Necare regem, sacrilegâ manu
Ausus cucullatus sodalis
In numerum colitur deorum! &c.

This alludes to Jacques Clement, who, after his assassination of Henry III. of France, was regarded as a saint by his party.

1. The *Grounde of Artes* who hathe well tredd,
   And noted well the slyppery slabbes,
   That may him force to slyde or faile,
   He hathe a staffe to staye withall.
2. Then if he trade that *Pathwaye* pure
   That unto Knowledge leadeth sure:
   He maye be bolde tapproche *The Gate*
3. *Of Knowledge* and passe in thereat.
   Where if with *Measure* he doo well treate:
4. To *Knowledges Castle* he maye soone get.
   There if he trauaile and quainte him well.
5. The *Treasure of Knowledge* is his eche deale.
5. This *Treasure* though that some wold haue,
3. Which *Measures* friendshippe do not craue,
2. Nor walke the *Patthe* that leadeth the waye,
1. Nor in *Artes Grounde* haue made their staye
   Thoughe bragge they maye, and get false fame
4. In *Knowledges courte* thei neuer came.

Of 'The Gate of Knowledge,' which appears to have been on Mensuration, and the 'Treasure of Knowledge,' in all probability a (projected) work on the higher part of Astronomy, we can get no information; the other works, and the 'Whetstone of Witte,' published very shortly before the author's death, have their titles as follows:

The 'Grounde of Artes, teachinge the worke and practise of Arithmetike, both in whole numbers and fractions, *after a more easier and exacter sort than any lyke hathe hitherto been set forth.*" The words in italics are perhaps the addition of John Dee, in his edition, the earliest we have seen (1573). The first edition of this work was, in 1551, printed by Reynold Wolfe.

The 'Pathway to Knowledg, containing the firste principles of Geometrie, as they may moste aptly be applied unto practise, both for use of Instrumentes Geometricall and Astronomicall, and also for projection of plattes in everye kinde, and therfore much necessary for all sortes of men.'

" Geometries verdicte
All fresshe fine wittes by me are filed
All grosse dull wittes wish me exiled
Though no mannes witte reject will I
Yet as they be, I wyll them trye.

London 1551, Reynold Wolfe."

The 'Castle of Knowledge.' The title-page of this work is a device representing a castle on a hill, at the bottom of which stand figures of Destiny on a cube holding a sphere, 'whose governour is knowledge,' and Fortune on a ball turning a wheel, 'whose ruler is ignoraunce.' On and below the castle are verses, London, 1556, Reginalde Wolfe.

The 'Whetstone of Witte, which is the seconde parte* of

* 'She returned for answer that she knew of no other books in the house than her young mistress's bible, which the owner would not lend; and her master's "Whetstone of Witte, being the Second Part of Arithmetic, by Robert Recorde, with the Cossike Practise and Rule of Equation."'—Walter Scott, *Fortunes of Nigel.* If Shakspeare makes Lord Say quote Cæsar's Commentaries to Jack Cade's mob, Walter Scott

Arithmetike; containyng thextraction of Rootes. The Cossike practise, with the rule of equation: and the woorkes of Surde Nombers.

> " Though many stones do beare greate price
> The whetstone is for exersice
> * * * * * *
> * * * * * *
> Now proue, and praise, as you doe finde,
> And to yourself be not unkinde.
>
> London 1557, Jhon Kyngston."

The 'Grounde of Artes' was many times republished, and remained in common use till some time after the publication of 'Cocker's Arithmetic,' (1677.) The last edition which we can find is that of Edward Hatton, 1699. The original work was dedicated to Edward VI. The advantages of number are set forth in two distinct ways: we cite these things to show that (to us) singular composition of childish argument and good sense which characterizes so many of the earlier writers in this and other countries. Recorde first asserts that the art of numbering is the 'chiefe pointe (in manner) whereby men differ from all brute beastes'— 'and in manner particularlye, sith that in many thinges they excell us againe.'

> " The Foxe in crafty witte exceedeth moste men,
> A Dogge in smelling hath no man his peere,
> To foresight of weather if you looke then,
> Many beastes excell man, this is cleere.
> The wittinesse of Elephantes doth letters attayne,
> But what cunning doth there in the Bee remayne?
> The Emmet foreseing the hardnes of winter,
> Prouideth vitailes in tyme of summer.
> The Nightingale the Linet, the thrushe, the larke,
> In Musicall harmony passe many à Clerke
> The Hedg hogge of Astronomy seemeth to know
> And stoppeth his caue wher the winde doth blow.
> The Spider in weauing suche arte doeth showe
> No man can him mende, nor follow I trowe
> When a house will fall, the Myse right quicke
> Flee thence before, can man do the like."

Whence Recorde infers that number 'is the onelie thing (almost) that seperateth man from beastes. Hee therefore that shall contempne numbre, he declareth himselfe as brutishe as à beaste, and unworthy to be counted in the felowshippe of men. But I truste there is no man so foule ouerseene, though manie right smallye do it regarde.' Again, the work opens with a dialogue on the advantages of number, of which the following is an extract:

" *Mayster.* If Numbre were so vyle a thinge as you did esteeme it, then neede it not to bee used so muche in mens communication. Exclude Numbre and aunswere me to this question. How many yeares olde are you?

may make a treatise on *algebra* the whole library of an old usurer; and further, two centuries hence, if any novelist shut up a gentleman of our time to pass a rainy Sunday in a country inn, the waiter ought, *pari ratione,* to bring him a volume of the *Mécanique Céleste,* or Mrs. Somerville's translation, at the very least.

Scholer. Mum.

Mayster. How many daies in a weeke? howe many weekes in a yeare? What laudes hathe youre father? How many men dothe he keepe? Howe longe is it sythe you came from him to mee?

Scholer. Mum.

Mayster. So that if Numbre wante, you aunswere all by Mummes: How many myle to London?

Scholer. A poke full of Plummes.

Mayster. Why, thus you may see, what rule numbre beareth and that if Numbre be lackinge, it maketh men dombe, so that to moste questions, they must aunswere Mum.

Scholer. This is the cause Syr, that I judged it so vyle, bycause it is so common in talking euery while: For plenty is not deinty, as the common sayeng is.

Mayster. No, nor Store is no sore: perceaue you this? &c."

The work contains numeration, addition, subtraction, multiplication, division, reduction, progression, the golden rule; a treatise on reckoning by counters, on a principle much resembling that of the Chinese abacus; a system of representing numbers by the hand, like the alphabet for the deaf and dumb; a repetition of all the rules for fractions, with the rules of Alligation, Fellowship, and Falsehood (false position). On the latter rule he remarks that he was in the habit of astonishing his friends by proposing difficult questions, and working the true result by taking the chance answers of "suche children or ydeotes as happened to be in the place."

It may, perhaps, be a question whether this work was not published originally so far back as 1540: for when John Dee (in his edition) comes to the table of coins, he reminds the reader that the table was of coins such as they were when the author first published his book; after which he heads the *table*, 'A table for Englishe Coynes. Anno 1540.' It may not be amiss to give this list of coins, which we do in modern spelling:—

Gold Coins.—A sovereign was two royals, three angels, 4½ crowns, or 22*s.* 6*d.*; half a sovereign, or a royal; the half-royal and the quarter royal; an old noble, called a Henry, was two crowns; half an old noble; an angel was 7*s.* 6*d.*; half an angel; a George noble, 6*s.* 8*d.*; half a noble; a quarter of a noble, 'which in the old statutes is called a farthing;' a crown, or 5*s.*; a half-crown; another crown of 4*s.* 6*d.*, 'known by the rose side, for the rose hath no crown over it.'

Silver Coins.—The groat of 4*d.*; another groat, called a harp, of 3*d.*; the *penny of twopence;* the dandiprat of 1½*d*; the penny, the halfpenny, and the farthing. . . .

The pound of 20*s.*, the mark of 13*s.* 4*d.*, and the shilling of 12*d.* were not *coins*, 'yet there is no name more in use than they.'

Laugh as we may at the 'Grounde of Artes,' we heartily wish it had been our own first book of arithmetic, seeing that it is better than the miserable mercantile compendiums with which the road to mathematics was opened or blocked up, as the case might be. 'Cocker's Arithmetic' is the model on which all these 'Tutor's Assistants,' or 'Pupil's Hinderances' were formed, and all the

C 3

imitations only differ from it in leaving out a few evidences that the author understood Latin, and in curtailing the explanations so as to render the rules unintelligible. But all those who still contend that the only use of arithmetic is to learn how to count money, should combine* to have Cocker reprinted, as they would thereby not only learn the creed from their prophet himself, but would find it more clearly explained than by any of his disciples.

Of the 'Pathway to Knowledge' we have little to say. Probably the translations of Euclid drove it out of the market. It contains —1. A method of working the various questions of practical geometry; 2. A description, not a demonstration, of the theorems in the first four books of Euclid: the whole in a highly useful form.

The 'Castle of Knowledge' is a more remarkable work. It is dedicated in English to Queen Mary, and in Latin to Cardinal Pole. From the preface to the reader, we gather that Recorde had not abandoned astrology. It begins with an account of the Ptolemaic system. All that is cited from Euclid and Proclus is in Greek and Latin, usually both, and Linaker's edition of Proclus is referred to; but the edition of Euclid is not mentioned.

In the 'Pathway to Knowledge,' we observe a strong tendency to turn all Greek names into Saxon-English; but in the present work we find many such English renderings given, though the terms of Greek etymology are preferred. The meridian is called the noon-steede circle; the zodiac, the thwarte circle; a sphere is said to be a round and sound (perfectly enclosed) figure; the Pleiades† (or seven stars) are called the brood hen; the belt of Orion, the golden yard; the milky way, Watling-street; antipodes are called counterfooted. The astronomical instruments in use are described with the following names:—'The Astrolabe, the plaine sphere, the Saphey, the quadrante of diverse sorts, the Chylynder, Ptolome his rules, Hipparchus rules, Tunsteedes rules, the Albion, the Torquete, the Astronomers staffe, the Astronomers ringe, the Astronomers shippe, and a greate numbre more.' We learn that these instruments were described in the 'Gate of Knowledge.' There is a rough determination of the magnitude of the earth, which is summed up in the following table:—

The places.	The elevation of the Pole.	The difference in degrees.	The distaunce in myles.
Southehampton . .	51 0	0 0	000
Newcastell . . .	55 0	4 0	240
Edynburghe . . .	57 0	2 0	120
Catnesse pointe . .	62 0	5 0	300
The summe of all . .		11 0	660

* What has become of Cocker's Arithmetic? It is not in the British Museum, and, with the exception of a mutilated copy of the *thirty-seventh* edition (A. D. 1720), we never found it in London, either in a shop or on a stall.

† In the 'Explanation of the Maps of the Stars,' published by the Society, it is incorrectly stated that it was the seven stars of the Great Bear to which this name was given.

Giving 60 miles to a degree; but this table has evidently been made for that express purpose.

The whole earth is, therefore, made to be 216,000 miles round, which is considerably too small, if common miles be meant, as we suppose was the case. There is a long detail to show that the earth must be round, and not flat or cubical, &c., and much is expressly taken from Cleomedes. A hint is given not to rely on Ptolemy without demonstration of a stronger character than was then usual, and seems to be intended to pave the way for what we shall presently have to cite:—'No man can worthely praise Ptolemye, his travell being so great, his diligence so exacte in observations, and conference with all nations and all ages, and his reasonable examination of all opinions, with demonstrable confirmation of his owne assertion, yet muste you and all men take heed, that both in him and in al mennes workes, you be not abused by their autoritye, but euermore attend to their reasons, and examine them well, euer regarding more what is saide, and how it is proued, than who saieth it: for autoritie oft times deceaueth many menne, &c.' Recorde afterwardes mentiones the 'Arte of *sines* and Cordes,' which is the first time we have found the word in English. He proceeds to explain the eclipses of the sun and moon, and promises a treatise of Cosmography, of which, by another hand, we shall presently have to speak.

We now come to our assertion that Recorde was a Copernican; which we must couple with his own implied assertion that he did not think the world ripe for any such doctrine. We say, also, that he was at least as early an avowed Copernican as any other Englishman, and very likely before any other. The work of Copernicus was published in 1543. In September, 1556, John Field * published an Ephemeris for 1557, 'juxta Copernici et Reinholdi Canones,' in the preface to which he avows his conviction of the truth of the Copernican theory. Recorde begins this subject by asserting that the earth 'standeth in the myddle of the worlde,' on which he uses Ptolemy's reasons avowedly (contrary to his usual practice), and adds nothing but explanation. He then proceeds as follows:—

'But as for the quietnes of the earth, I neede not to spende any time in prooving of it, syth that opinion is so firmelye fixed in moste mennes headdes, that they accōpt it mere madnes to bring the question in doubt. And therfore it is as muche follye to trauaile to prove that which no man denieth, as it were with great study to disswade that thinge, which no man doth couette nother any manne alloweth: or to blame that which no manne praiseth, nother anye manne lyketh.

'*Scholar.* Yet sometime it chaunceth, that the opinion most generally receaued, is not moste true.

'*Master.* And so doo some men iudge of this matter, for not only Eraclides Ponticus, a great Philosopher, and two great clerkes of Pythagoras schole, Philolaus and Ecphantus, were of the contrary opinion, but also Nicias Syracusius, and Aristarchus Samius, seeme with strong arguments to approve it: but the reasons are to difficulte for this firste Intro-

* See a communication by the Rev. J. Hunter to the Royal Astronomical Society. Monthly Notices, vol. iii. p. 3.

duction, and therfore I wil omit them till an other time. And so will I do the reasons that Ptolemy, Theon, and others, do alleage, to prooue the earthe to bee without motion: and the rather bycause those reasons doo not proceede so demonstrablye, but they may be answered fully of him that holdeth the contrarye. I meane, concerning circularre motion: marye direct motion out of the center of the world, seemeth more easy to be confuted, and that by the same reasons, whiche were before alleaged for prouing the earthe to be in the middle and centre of the worlde.

'*Scholar.* I perceaue it well: for as if the earthe were alwayes oute of the centre of the worlde, those former absurdities would at all tymes appeare: so if at anye time the earthe should mooue oute of his place, those inconueniences would then appeare.

'*Master.* That is trulye to be gathered: howe bee it, Copernicus a man of greate learninge, of muche experience, and of wondrefull diligence in obseruation, hathe renewed the opinion of Aristarchus Samius, and affirmeth that the earthe not only moueth circularlye about his own centre, but also may be, yea and is, continually out of the precise cẽtre 38 hundreth thousand miles: but bicause the vnderstanding of that controuersy dependeth of profounder knowledg then in this introduction may be vttered conueniently, I will let it passe tyll some other time.

'*Scholar.* Nay syr in good faith, I desire not to heare such vaine phantasies, so farre againste common reason, and repugnante to the consente of all the learned multitude of Wryters, and therefore lette it pass for ever, and a daye longer.

'*Master.* You are to yonge to be a good iudge in so great a matter: it passeth farre your learninge, and theirs also that are muche better learned then you, to improue (that is, disprove) his supposition by good argumentes, and therefore you were best to condemne no thinge that you do not well vnderstand: but an other time, as I sayd, I will so declare his supposition, that you shall not only wonder to hear it, but also peraduenture be as earnest then to credite it, as you are now to condemne it.'

It appears, then, that Recorde was as much of a Copernican as any reasonable man could well be at the time; at least as much so (in profession) as was Copernicus himself, who makes no decided declaration of belief in his own system, but says, 'It is by no means necessary that hypotheses should be true, or even probable: it suffices that they make calculation and observation agree.'

The 'Whetstone of Wit' was dedicated to the 'companie of venturers into Moscovia.' After treating of numbers in general, with the formation of powers and rootes, he comes to the consideration of *cossike numbers,* meaning an indeterminate number expressed by a letter, as in what is now called *algebra.* But Recorde does not use this term except to denote the application of cossike numbers to the solution of equations, which he calls the rule of Algeber. In this treatise he appears to have compounded, for the first time, the rule for extracting the square roots of multinomial algebraical quantities, and also to have first used the sign =. In other respects he follows Scheubel, whom he cites, and Stifel, whom he does not cite. There is nothing on cubic equations, nor does he appear to have known anything of the Italian algebraists. But the subject of Recorde's algebra can only be discussed at length and in connexion with the rise of algebra in other countries. This is completely done in Dr. Hutton's tracts, vol. ii. It may be observed, however, that Recorde was one of the first who had a distinct perception of the difference between an algebraical operation and its numerical interpretation, to the extent of seeing that the one is independent of the other; and also he appears to have broken out of the consideration of integer numbers, to a much greater extent than his contemporaries. Pointing out to his pupil (for this work is also in dialogue) that a certain fraction appears to be absurd, as being 'lesse than naughte,' which the pupil admits, he then uses this phrase, 'Yet maie your example serve, to teach and practise multiplication by, as well as any other. And furthermore, I will tell you by this occasion, that I spake to you, more after the opinion of the cõmon nomber of artesmen, than after my owne judmente.' He then goes on to show how the same expression ceases to be absurd when a fraction less than unity is the cossic number, and to distinguish generally between the powers of integers and fractions. In perception of general results, connected with the fundamental notation of algebra, Recorde shows himself superior to others, and even we may say, to Vieta: though of course immeasurably below the latter in the invention of means of expression. He (and Scheubel also) are free from the geometrical phantasms which haunted the inventor of the *specious* notation, and made him invest simple numbers with the character of planes, solids, &c., and even imply that numerical equations were impossible unless their coefficients were thus considered. All his writings considered together, Recorde was no common man. It is evident that he did not write very freely at first in English, but his style improves as he goes on. His writings continued to the end of the century to be those in common use on the subjects in which he wrote, though we must gather this more from the adoption of ideas and notation than from absolute citation.

Of William Boorne we only know that he wrote two works on navigation, entitled the 'Attractive' and the 'Regiment of the Sea,' both frequently cited by Blundeville.

The first English work on Cosmography, an intention which Recorde did not live to fulfil, was the 'Cosmographicall Glasse, conteinyng the pleasant principles of Cosmographie, Geographie, Hydrographie, or Nauigation; compiled by William Cunningham, Doctor in Physike. London, John Day, 1559.' This work is on the model of the 'Castle of Knowledge:' the pupil is made to say that he has read the 'Ground of Arts,' the 'Whetstone of Wit,' and the 'Pathway,' and is further recommended to read Orontius, Scheubel, Euclid, and Theodosius. The work contains a description of the Ptolemaic system (we cannot find the name of Copernicus, or the least hint of the existence of his system), a geographical account of the earth, with descriptions of countries, and latitudes and longitudes; the use of instruments in the determination of latitudes, and some notions on navigation. Except that this is the first English work of the kind, there is nothing about it worth notice.

The Life of John Dee, with all his curious attempts to persuade himself, and others, that he dealt with evil spirits, was written in

Latin by Dr. T. Smith, and in English in the 'Biographia Britannica.' We have nothing here to do with him, except with his writings relative to mathematics, in which he was no mean proficient. He was born in 1527, was educated at St. John's College, Cambridge, was chosen fellow of Trinity College at the foundation of that establishment, and died in 1608.

The principal work of Dee is the translation of Euclid, which appeared with his preface in 1573. We call it his, though it is universally stated (and by himself* among others) to have been made by Sir Henry Billingsley. Considering that he wrote the preface, the notes, and the translation of the book of Mohammed of Badgad on the division of surfaces, which was published with the second edition: that he had lectured on Euclid in various places, and left behind him MSS. on the subject, and in particular 'Instructions and Annotations upon Euclid's Elements;' considering, also, that the name of Sir H. Billingsley is not mentioned in the preface to his own Euclid; we imagine that the translator was a pupil of Dee, who worked under general, if not special, instructions, and executed the more mechanical part of the undertaking. That this was the first English translation we know from several testimonies, and from that of Dee himself, in the poetry at the end of his edition of Recorde's 'Ground of Arts,' (that of 1573.) This translation of Euclid was either made from the Greek, or corrected by the Greek, as is evident from comparing the early Latin versions and the Greek text with it. As there may be some dispute about the degree in which Greek was studied in England at the period of which we write, we shall annex one sentence of comparison, namely, the Greek text of the enunciation of the fourth proposition of the first book, the Latin of Campanus, and the English of Billingsley. That the Latin of Campanus was taken from an Arabic text, there is no doubt whatever, were it only from the insertion of Arabic words, which occurs several times.

Book I. prop. iv. (Gregory's edition.) Ἐὰν δύο τρίγωνα τὰς δύο πλευρὰς ταῖς δυσὶ πλευραῖς ἴσας ἔχῃ, ἑκατέραν ἑκατέρᾳ, καὶ τὴν γωνίαν τῇ γωνίᾳ ἴσην ἔχῃ, τὴν ὑπὸ τῶν ἴσων εὐθειῶν περιεχομένην· καὶ τὴν βάσιν τῇ βάσει ἴσην ἕξει, καὶ το τρίγωνον τῷ τριγώνῳ ἴσον ἔσται, καὶ αἱ λοιπαὶ γωνίαι ταῖς λοιπαῖς γωνίαις ἴσαι ἔσονται, ἑκατέρα ἑκατέρᾳ, ὑφ' ἃς αἱ ἴσαι πλευραὶ ὑποτείνουσιν.

English of Dee's edition (omitting letters of reference).

'If two triangles have two sides of the one equal to two sides of the other, each side to his correspondent side, and the angle contained by the equal right lines of the one be equal to the angle contained by the equal right lines of the other: the base also of the one shall be equal to the base of the other, and the other angles remaining shall be equal to the other angles remaining the one to the other by which equal sides are sub-tended.'

Latin of Campanus (Editions of 1482, 1491, and 1516.)

'Omnium duorum triangulorum quorum duo latera unius duobus lateribus alterius equalia fuerint: duoque anguli eorum illis equilateribus

* In the list of his works, contained in his apologetical letter to the Archbishop of Canterbury.

contenti equales fuerint alter alteri: latera quoque eorum sese respicientia equalia: reliqui vero anguli unius reliquis angulis alterius equales erunt: *ac totus triangulus toti triangulo equalis.*'

The general comparison, and particularly the words in italics, are decisive of the question.

This translation contains the whole of the fifteen books, commonly considered as making up the elements of Euclid, and forms the first body of complete mathematical demonstration which appears in our language. For though the works of Recorde were much less dogmatical than the elementary school books of the eighteenth, and (for the most part) of the present century, yet they partake of the character which they tended perhaps to perpetuate, and in many instances teach rules without demonstration, or with at most a rough kind of illustration. At the same time our ancestors throughout Europe did not fall into the error of admitting arithmetic and algebra (such as they then were) to the name of mathematics and the rank of geometry. So long as they continued to be only methods without rigid investigation, they were *arts*, not *sciences*, and the *science* of arithmetic was sought in the 7th and following books of Euclid. Thus Tartalea (the extender and promoter of Algebra) calls Euclid, in the title-page of his Italian Edition, (the earliest translation into a European language, 1543,) 'solo introduttore delle scientie Mathematice,' and, even in our own private schools in England at this day, it is very common to call Euclid alone by the name of *mathematics*, to the exclusion of arithmetic and algebra. We may thus see how great a debt we owe to the Greeks, the *inventors of demonstration*,* so far as can be *shown*. Had it not been for the writings of Euclid, Ptolemy, Aristotle, Proclus, &c., and judging of what would have been the case in geometry by the practice which became universal in other branches of exact science, we cannot see how anything like demonstration could have been introduced. As to England, the pathway of Robert Recorde was almost entirely dogmatical: the example would no doubt have been imitated, and, out of the universities, Geometry would have become no higher a discipline of the mind than the so called arithmetic and algebra. The appearance of Euclid, in an English form, probably saved the credit of the exact sciences, and in this point of view Dee and Billingsley have exercised a material and beneficial influence upon their favourite pursuits. The published writings of Dee are not numerous, and besides his preface to Euclid, (a most tedious nomenclature of arts and sciences.) and his annotations, together with the Book of Mohammed of Bagdad, there are only two which relate to our subject: a preface to the Ephemeris of John Field for 1557, in which he declares his approbation of the Copernican† system, and his *Parallacticæ Commentationis Praxeosque Nucleus quidam*,

* The Hindoo writings, and also the Arabic (except when taken from the Greeks), are strictly dogmatical. See the Viga Ganita, or Mohammed Ben Musa's Algebra, both now in English. We are at liberty to infer, if we please, that there must have been demonstration to establish or discover the results of the former work; but this is not absolutely certain, and nothing like demonstration has been produced.

† See page 36, note.

1573, a geometrical tract upon the ordinary questions of parallax, in which no particular system is alluded to.

We are now to speak of Leonard and Thomas Digges, father and son, who died in 1574 and 1595. As the son either published posthumously, or republished, with additions, all his father's works, these two writers are one for every point now in question. The works of the two are as follows:

1. 'Tectonicon,' by L. D. 1556, republished by T. D. 1592; again, 1647.

2. 'Pantometria,' begun by L. D., published by T. D. 1591.

3. 'Prognostication everlasting,' (Meteorology,) by L. D. 1555, &c., republished by T. D. in 1592, accompanied by 'a perfect description of the celestial orbs, according to the most ancient doctrine of the Pythagoreans,' their *astrological* doctrines, we presume, not their reputed *Copernican* ones. Not having seen the 'Prognostication' of Leonard Digges with the addition of his son in 1592, we can only further state that Anthony à Wood expressly affirms 'Cui subnectitur *orbium Copernicanorum* accurata descriptio.' Weidler improves upon this, for he cites it 'Cui subnectitur *operum Copernici* accurata descriptio.' But Lalande goes still further; for he attempts to recover the English title page, and accordingly we find in the *Bibl. Astron.* the following: '1592 ....Leonard Digges, Accurate Description of the Copernican System to the Astronomical perpetual Prognostication.' Now the real title page is 'Perfect Description of the Celestial Orbs, according to the most Antient Doctrine of the Pythagoreans.' Seeing some days ago in the newspapers an account of a Welch curate who made very good sermons by translating Tillotson into Welch, and then retranslating into English, without the least suspicion of plagiarism arising, we doubted; but the preceding has disposed us to think the account not so improbable.

4. 'Alæ sive scalæ Mathematicæ,' by T. D. 1573.

5. 'Stratiotikos,' begun by L. D., finished by T. D. 1579 and 1590.

The characteristic of the writings of both father* and son is the application of arithmetical geometry to mensuration, the art of war, &c. The first work is entirely upon mensuration, and also the second, which contains a long, and for the time a very ingenious treatise on the mensuration of solids, the results being expressed in an algebraical form. The Stratiotikos is almost entirely on gunnery, (a translation of the work of Tartalea on this subject was published in 1588, by Cyprian Lucar.) The *Alæ*, &c., is a tract upon parallaxes, undertaken, says the dedication, at the suggestion of Lord Burleigh, in consequence of the appearance of the celebrated star in Cassiopeia. It is the first work of an English writer in which we have noticed anything on spherical trigonometry, and the writings of Copernicus are more than once referred to as the source of this subject. With regard to the system of Copernicus, Digges, in his preface, remarks the excessive complexity

* For the assertions of Digges about his father's optical inventions, see *Penny Cycl.* —ROGER BACON.

and diversity of the Ptolemaic system, which he says is like a set of hands, head, and feet, taken off different men, 'which could not have been the case, if they had assumed true hypotheses. And this was the principal reason why Copernicus, a man of admirable ability, and singular industry, used another hypothesis.' He points to the Copernican hypothesis as possibly affording the true explanation of the vicissitudes of apparent magnitude of the new star, and points out that, so far as appearances are concerned, it matters nothing whether the rotation of the *primum mobile*, or the Copernican revolution of the earth, be assumed. This sentence is remarkable, as leading to one of two suspicions; either that Copernicans were common among those for whom he wrote, or that he himself was a known Copernican, desirous of securing a favourable hearing upon a point which admitted of the same mathematical explanation upon either hypothesis. Again, in the *Operis Conclusio*, he remarks, 'If Copernicus (a man never sufficiently to be praised) had been now alive, as indeed he might have been, since he would now have been not more than 100 years old, (1573,) we might have hoped that, so far as mortal weakness would permit, men would have had absolute knowledge of the celestial system.' From such phrases, we imagine Thomas Digges to have been a believer in the real motion of the earth, and not merely an admirer of the system of Copernicus as an explanatory hypothesis.

We have in the names of Recorde, Dee, and Digges, all the principal associations connected with the real advancement of English science in the sixteenth century. We have yet a few works to notice of the older leaven, which will show how little, so far as published writings are concerned, any effective progress seemed to be making towards the splendour of the period succeeding that of which we write. If we may judge from such printed works as are before us, from 1580 to 1600, we should be inclined to suppose that there had been a decline of science. The writers seem to have abandoned what had been newly introduced, and to have betaken themselves to older authors, and older notions. We begin with John Blagrave, of Reading, a worthy gentleman, who we soon found had something upon his mind while he wrote, as appeared by several hints against designing men and inquirers into other men's titles, which appeared in the midst of propositions of spherical geometry. The preface to his work,* 'The Mathematical Jewel,' 1585, let us into the secret. He had lands with a disputed title, and was opposed by 'packing, shifting, suborning, wresting, seducing, and such diabolical practices.'—'It is a world to speak of al their wicked drifts, which bicause it would be but a glasse, wherein vile people of like disposition might see the whole commonwelth of villanous trechery, I here omit, together with their names: and the rather I do it, for that in conscience I thinke they would rejoyce and glory to be in print, though for egregious knaverie, even like that lewd fellow that set Dianaes temple on fire to get him a lasting name. And

* Printed by Walter Venge, whom we notice, because we never saw any other work of his printing, or can find any account of his having printed in any other year.

yet one of their auncestors in the booke of monuments is in print alredie looking through a pillorie which I cannot helpe.' In a bit of poetry, called 'the Authours dumpe,' he describes students in philosophy as men

'That sit and swigge their bellies full, from Queen Minervaes pappes.'

The work itself is a treatise on a new mathematical instrument, which seems to have been a projection of the sphere, for the construction of problems in astronomy. He gives the Ptolemaic system, and several times alludes to Copernicus as an observer, but does not hint at his cosmical system. He also refers to Copernicus in his propositions of spherical trigonometry, and gives various theories of Gemma Frisius, Stoller, Roias, and other writers on the astrolabe.

Thomas Hood* was the inventor of an astronomical instrument, called Hood's staff, described by Blundeville. He published, in 1590, a very poor and insufficient account of the projection of the sphere, which is the first *English* book in the *Bibliographic Astronomique* of Lalande, and in which he remarks that 'Astronomy was beginning to find its way into England.' This is a sort of misrepresentation such as any country must expect, in which no pains are taken to write its own history. Speaking of Hood's staff, we may observe that no point is more completely untouched by the writers on the use of instruments than the method of obtaining accuracy and strength in the instruments themselves, which forms a prominent part of the modern astronomy. In later times, a Ramsden or a Troughton was also an author, who could describe his methods; but, before the time of Tycho Brahè, we have little evidence that observers cared about the quality of the divisions in the instruments which they used. Digges mentions Richard Chansler, as having improved the method of dividing, and Dr. T. Smith, in his life of Dee, (*Biog. Brit.*) affirms this same Richard Chancellor to have constructed a quadrant of five feet in radius, which belonged to Dee.

In 1594, M. Blundevile published his 'Exercises, containing six Treatises,' namely, on Arithmetic, Cosmography, Description of the Globes, Description of the Universal Map of Plancius, of the Astrolabe of Mr. Blagrave, and of the Principles of Navigation. In the Arithmetic we find what is in Recorde, augmented by a description of the sexagenary tables, and by what we suppose (for fault of better) we must call our first English treatise on Trigonometry, being 'A briefe description of the tables of the three speciall right lines belonging to a circle, called sines, lines tangent, and lines secant.' These tables, which are annexed from Clavius, are to a total sine 10,000,000, or, as we should now say, to seven places of decimals, and to every minute of the quadrant. The book on Cosmography describes the Ptolemaic system, the circles of the sphere, the constellations, &c. Of the motion of the earth, it says, 'Some also deny that the earth is in the middest of the world, and some affirme that it is mooueable, as also Copernicus by way of supposition, and not for that he thoughte so in deede:

* See the Society's *Explanation of the Maps of the Stars*, p. 53.

who affirmeth that the earth turneth about, and that the sunne standeth still in the midst of the heavens, by helpe of which false supposition he hath made truer demonstrations of the motions and revolutions of the celestiall spheares, than ever were made before.'

Perhaps the following is the most distinct recognition of authority in our language. 'How prooue you that there is but one world? By the authoritie of Aristotle, who saieth that if there were any other world out of this, then the earth of that world would mooue towards the centre of this world,' &c.

The work on navigation is full of citations from other authors, as already noticed.

Thomas Hill wrote various works, of which the only one we have seen is the 'Schoole of Skil,' London, 1599 (posthumous). It is an account of the heavens and the surface of the earth, replete with those notions of Astrology and Physics which are not very common in the works of Recorde or Blundevile. Of the Copernican doctrine, he says, that 'Aristarchus Samius . . . tooke the earth from the middle of the world, and placed it in a peculiar orbe, &c. The like argument doth that learned Copernicus apply unto his demonstrations. But overpassing such reasons, least by the newnesse of the arguments they may offend or trouble young students in the Art: wee therefore (by true knowledge of the wise) doe attribute the middle seat of the world to the earth, and appoynte it the center of the whole,' &c.

There are several writers who lived and studied in the latter part of the sixteenth century, but whom we must place at the beginning of the seventeenth century, in respect to the character of their works. Such were Wright, Briggs, Harriot, Gellibrand, (Napier in Scotland,) &c. &c. But these belong to what we may call the period of *mutual* intercourse with foreigners. Hitherto, all was derived from abroad, and nothing was returned. We do not find in the continental histories or bibliographies any decided proof that English works speedily found their way abroad. From the beginning of the seventeenth century, the history of science becomes more European. We shall now conclude this article with some account of the works which have formed the basis for it.

Indirectly, much has been drawn from the catalogues of English writers of Bale and Pits, and from the catalogue of Manuscripts published at Oxford in 1697. The two former works contain names which are not here, and so does the 'Antiq. Univ. Oxon. of Wood:' our rule has been to take only those names to which manuscripts remain, or which are mentioned by mathematical writers: a collection of the titles of all the mathematical and astronomical writers was made with care from the work just cited by Heilbronner in his 'Historia Matheseos Universæ,' Leipsic, 1742, a work of unequalled research on this subject. Next to it comes the 'Chronologia Mathematicorum' of Gerard Vossius (Amsterdam, 1650), which contains a slight notice, with dates, of every noted name. We have then the 'Synopsis Historiæ Litterariæ' of J. N. Eyring, Gottingen, 1783, a chronological notice of every writer from the earliest period to the invention of printing. After these, and far below them, comes the list of mathematical writers in the Appendix to Sir Edward Sherburne's Translation of Manilius (1675), which is a guide rather than an authority.

With regard to books, the unclassed state of the library of the British Museum renders it impossible to investigate the condition of any century by means of it. Our principal assistance has been the 'Bibliotheca Mathematica' of Murhard, Leipsic, 1803, which, however, has served to do little more than to show how very little the English printed works of the sixteenth century have found their way abroad. The 'Biographia Philosophica' of B. Martin, and the 'Mathematical Dictionary' of Dr. Hutton, have been used only where confirmed from other sources: the articles in the 'Biographia Britannica' (principally Roger Bacon, Dee, and Digges) may be depended on. There are none but the most common names in Riccioli, Blancanus, or Weidler, and the summary in the first volume of the 'Cursus Mathematicus' of Dechâles is not to be depended on for correctness as to English matters. The French writers, Lalande, Montucla, and Delambre have been in this matter of no use whatever. Blount's 'Censura' contains some names: Ward's 'History of Gresham College' does not go back far enough. Costard's 'History of Astronomy,' though an English work, does not treat specially of Astronomy in England. The fact is, that the history of science in England, during the period of which we have treated, is unwritten, though the preceding may serve to show that it has not been for want of materials. It would be comparatively easy to give a full account of the sixteenth century only, but those which precede would require access to the manuscripts in various libraries. To the credit of this country be it spoken, the readiest access to a knowledge of the names and works of the *English* writers who lived before the introduction of printing, is through the writings of the *Germans*, Vossius, Heilbronner, and Eyring.

V.—ON CAVENDISH'S EXPERIMENT.

Our object in choosing this particular time to give a somewhat detailed account of the celebrated experiment of Cavendish, is the circumstance of the council of the Royal Astronomical Society* having announced an intention of repeating it immediately, and

* See their Annual Report for 1836.

[Continued next page]

thus of verifying or overturning one of the most remarkable physical investigations which ever was undertaken. We have heard it stated that the government has granted funds for the purpose, and that the construction of the necessary apparatus will soon be in progress; however this may be, we are sure that hardly any expenditure for a scientific purpose could be imagined which would be better justified by the utility of the end proposed.

In looking at our last assertion, we desire the reader to bear in mind that no single *practical* purpose (in the common sense of the word) will be advanced in the smallest degree by the repetition of the proposed experiment. Neither railroad nor canal, steam-engine nor ship of war, will be in the most remote manner advantaged, so far as can be seen. It is certain that useful applications frequently do arise, out of scientific inquiries which seem at first sight entirely barren; but of the present subject it may be said, that though it be closely connected with astronomy, no person professing that science can point out a single page of the Nautical Almanac which has any chance of being rendered more correct.

Why is it, then, that we speak so highly of the utility of the end proposed? Because there is in our world mind as well as matter, and because we conceive that few things would better repay as large an expenditure as the one which is proposed, than the opportunity afforded of turning the attention of the reading part of the community upon a result of science which tends to destroy crude theories, by planting, as it were, in the midst of them, a *numerical* result with which they are incompatible.

There was once a board of longitude, instituted for the purpose of promoting the discovery of methods for the determination of the longitude at sea. If we are not misinformed, a large portion of the time devoted by this board to its general meetings was expended in hearing schemes for finding the longitude miraculously (that is, by means demonstrably insufficient in the common course of nature), for squaring the circle, trisecting the angle, overturning the diurnal motion of the earth, &c. &c. So long as this board remained in being, it was a standing evidence of the existence of a class of speculators which we have reason to know is not yet extinct. If such be the case with astronomy, we may suppose that less accurate sciences still more abound with persons who have begun their career of discovery before they have learnt their subjects. Such a class of investigators lives upon the ignorance of the public; *not as to mere results, but as to the trains of reasoning by which they are obtained.* Now Cavendish's experiment is one of which the rationale may be made tolerably clear; and is therefore well adapted to furnish an instance of sound investigation. That mathematicians and astronomers feel interested in its repetition on purely scientific grounds, does not diminish the interest with which it must be viewed; but our object in this paper is specially the information of the public at large upon the species of investigation employed. We shall add, at the end, a short mathematical account, which will be more easily read by the elementary student than that given by Cavendish himself.

C 2

The result of the experiment is as follows: the earth on which we live, with its various compounds of matter of different densities, is altogether five times and a half as dense as water, or has an average density, roughly speaking, which is about a mean between that of glass and brass, and is about half that of lead. If the globe were entirely composed, from the centre to the surface, of nothing but water, it must be five times and a half as large as it now is, to retain the moon in its orbit: that is, its radius must be increased from 4000 miles to about 7000 miles. This result is of a nature to excite surprise: how could a single individual ascertain such a point? It will appear still more strange when it is said that the apparatus employed was all contained in a small room, and that the experiment, if repeated, will probably be entirely performed (calculations excepted) in a room not exceeding twelve feet square.

The doctrine of the attraction of matter upon matter, of which Newton first demonstrated the actual law, had been suggested from time immemorial by the fall of bodies to the earth, and various other common phenomena. The Newtonian process consisted in showing that the planetary motions *would be* such as they *actually are*, if such a law of attraction prevailed as that of which it treated. Either the force of gravity (varying inversely as the square of the distance) does exist, or something else which produces exactly the phenomena which such a force would produce. It is enough, says Newton, that I have proved that gravity exists and acts: meaning a *cause of motion** which obeys the law laid down. Into the nature of this cause he did not inquire, though there are surmises scattered through his writings which might lead us to suppose his thoughts leaned towards some invisible fluid. Let it be remembered, then, that at the death of Newton all reasonable ground for inferring the attraction of matter upon matter rested on the fact, that the leading characters of the heavenly motions coincided with the necessary results of a simple hypothesis.

Up to the end of the last century the efforts of the followers of Newton added nothing to his main argument, except amplification of its details. If Newton could not explain how the motion of the lunar apogee was a consequence of the sun's attraction, Clairaut did; if Newton did not attempt any deduction of the mutual perturbations of Jupiter and Saturn, Laplace, with more powerful analysis, was successful; and so on. No celestial phenomenon of any magnitude was left unattached to the main theory; and instances had even occurred in which the theory was made to suggest phenomena which, from their smallness, never would have been obtained by direct observation, unless the observer had been taught by theory when and where to look. Further details on this head will be found in the article on the Moon's Orbit, in the volume of this work for 1834.

The question therefore of attraction absolutely considered, remained as Newton left it, with this exception, that the Newtonian *law* was fully established, namely, that of the inverse square of

* Force, in dynamics, means nothing but the *cause of motion*.

the distance. If any notable inequalities had remained unexplained when all the increased powers of mathematics had been expended upon the question, there would have been strong ground of suspicion that the law above-mentioned is not exactly true, and subject to some slight modification, the character of which would be determined by that of the unexplained motions. It was at one time the opinion of Clairaut, that the law expounded by Newton needed an additional term, which should vary as the inverse fourth power of the distance; but his subsequent investigations showed him that the simpler law was sufficient.

But there is another part of the Newtonian law which remained, up to the time of which we have spoken, and which still remains, entirely unsettled. It is sufficiently obvious, that whatever may be the attractive effect of a given amount of a given substance, twice as much must produce twice as great an effect, and so on. The attractive effect of two pounds of *lead* must be twice as great, *cæteris paribus*, as that of one pound of *lead*. But it is not equally evident that the attraction of a pound of lead must be the same thing as that of a pound of any other matter, as iron or granite. So that, though it may be said of different quantities of the same substance, that their attractions are proportional to their masses, there is no certain ground for assuming this of mixtures, or of quantities of different substances. Nevertheless, it is a constituent part of the Newtonian theory that *all* equal masses attract equally: or, measuring masses at the earth's surface by the weight they produce (which is, be it remembered, itself a particular case of the same assumption), that the attraction of portions of matter is in proportion to their weights. Should this hypothesis be untrue, it would not affect any astronomical result of the theory, since the only effect of its being overturned would be this, that the numbers which now are said to represent the masses must still be used under some other name. To make this clearer, we lay down and explain the following table:—

The Sun	100,000,000	Mars	39
Mercury	49	Jupiter	93,400
Venus	246	Saturn	28,500
Earth	282	Uranus	5,600
Moon	4		

The reader must imagine the law of gravitation divided into two parts: the first expressive of the manner in which attraction depends on the distances; the second, of that in which it depends on the masses of the attracting bodies. Assuming the first on grounds at which we have hinted above, it is then possible to demonstrate the following. Let a given mass be placed at a certain distance from the sun's centre, and imagine all the other planets placed in a circle round it at equal distances, so that the mass in question is equi-distant from all. Then, if the sun exerted on it an attractive pressure equivalent to 100,000,000 of grains, Jupiter would cause an attraction of 93,400 grains, and so on; while the moon would cause no more than 4 grains. The whole of the preceding table requires no other premiss than the assumption of the law of the inverse square of the distance, except in the case of Mercury, the number corresponding to which is determined in a manner which Laplace calls very precarious, and which appears to us utterly insufficient.

Now *if we assume* that, *cæteris paribus*, the attraction of different substances is as their masses, without reference to the composition of those masses, it then becomes proper to say that the masses of the planets are in the same proportion as the numbers attached to them in the table; and, *such assumption being made*, the numbers above mentioned *are* considered as expressing the proportion of the masses. It may turn out that this is a mistake, which would be of some importance as to physics in general, though of none whatever as to astronomy. We have no doubt this point will be settled in the repetition of the experiment.

Towards the end of the last century, the question was raised whether the actual attraction of matter upon matter could be discovered, independently of the manner in which its consequences are written in the celestial motions. The first suspicions of such a phenomenon being sensibly displayed, arose, as far as we know, in the visit of Bouguer and La Condamine to Peru (1735-1743) for the measurement of their degree of the meridian. Astronomical observation was of course most essential, and a correct knowledge of the zenith point is a necessary preliminary to accurate results. The plumb-line, used to determine the vertical, is liable (if matter attract matter) to be deflected more or less from true perpendicularity by the attraction of any neighbouring mountain. The observations of Bouguer were carried on near Chimboraço, and he found discrepancies in his results which, though very small in amount, were larger than usual, or than could be explained by any probable instrumental error. He suspected that the mountain attracted the plumb-line, but not knowing the density of the materials which composed the mountain, he was unable to complete his investigation. The discordance observed by him was hardly more than the fifteenth part of what he expected to arise from such a cause, and being only about seven seconds, was not, at that period, large enough to be distinctly assumed as arising from a cause extraneous to the instrument. The result of La Condamine, to use his own expression, was, "que si l'on ne peut rien tirer d'absolument décisif en faveur de l'attraction Newtonienne, encore moins en conclura-t-on rien qui y soit contraire."

In 1774-1776, Dr. Maskelyne, whose astronomical labours are so well known, determined to measure the effect of a mountain upon the plumb-line, and chose Schehallien as his scene of operation. His method was simply observations of the same stars made both at the north and south of the mountain, with an instrument depending for the accuracy of its zenith point upon a plumb-line. If the mountain were to deflect such a plumb-line, it would be in opposite directions at the opposite sides of the mountain; and the difference between the resulting zenith distances of a star, observed north and south of the mountain, would be the united effect of the attractions. The mountain was surveyed, and its material ascertained to the best of the observer's power. How it is that such a process could lead to the knowledge of the earth's density, we shall have occasion to explain when we come to the detail of Cavendish's experiment. The result of Dr. Maskelyne was, that the earth is about $4\frac{1}{2}$ times as dense as water. However, from a subsequent survey of the mountain by Professor Playfair, there were found grounds for suspicion that Maskelyne had wrongly estimated the density of the mountain; Dr. Hutton undertook the calculations necessary for the correction of this error, and found that 5, instead of $4\frac{1}{2}$, was the result of Maskelyne's observations. Newton, who suggested that the attraction of mountains might affect plumb-lines, surmised that the mean density of the earth was between 5 and 6 times that of water.

If we look carefully at Maskelyne's method, we shall see that there is but one circumstance which would be unanswerable in favour of attraction. The difference between the meridian zenith distances of stars observed north and south of the mountain might be easily attributed partly to errors of observation, partly to change in the form of the instrument. The whole discrepancy was only about 11 seconds; and it is also to be remembered that Maskelyne obtained his result from selected observations: he chose a small number out of the whole, such as he considered the best. Baron Zach afterwards reduced the whole of Maskelyne's observations, and found that they gave very nearly the same result as that of the selected ones; but this, however much it may add to the credit of the observer, does not add anything to the conviction that attraction was the cause of the differences observed. The circumstance which justifies us in inferring the attraction of the mountain is, the deflection of the plumb-line being, in both cases, towards the mountain. When north of the mountain, the zenith *of the instrument* was thrown a little towards the northern horizon, and *vice versâ.*

An incredulous person might very well be supposed to say, "let me see matter (neither affected by magnetism nor electricity) move out of its place when other matter is made to approach it, and return to its old station when that other matter is removed, and not till then will I believe in attraction." Cavendish's experiment would have shown such a doubter what he desired to see; and we shall now proceed to its details, neglecting mechanical contrivances, which tend only to convenience, and confining ourselves to points connected with the principle of the experiment.

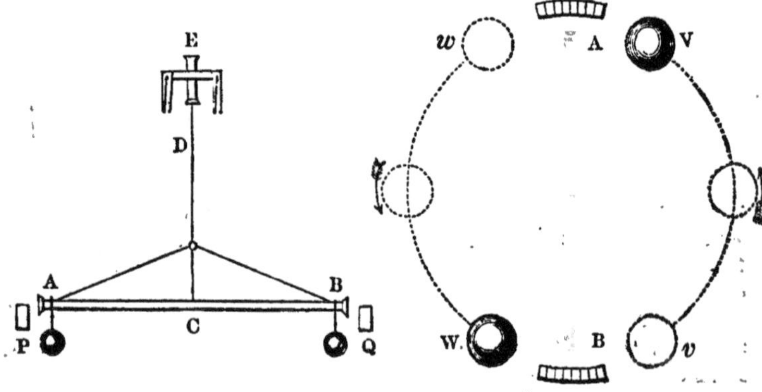

A slight arm, made of deal, A B, seen sideways on the left hand figure and from above on the right, was suspended at the middle by a wire of silvered copper C D, about 40 inches long, and very near each extremity was hung a leaden ball of about 2 inches in diameter. As near as conveniently could be to the ends of the bar were placed two graduated scales, divided to twentieths of an inch (P and Q), by which any quantity of horizontal vibration in the arm might be measured; but instead of providing each end of the arm with an index, a vernier was substituted, by which the fifth part of a division, or the hundredth part of an inch, was read. The whole was then inclosed in a wooden case, made nearly to fit the figure, leaving only slits beyond P and Q, which were stopped with glass. All currents of air were thus prevented. The only communication with the interior of this case was, an apparatus at E, by means of which the wire could be turned round, so as to make the ends of the arm fall opposite to any division of the scale.

Exterior to the case, and revolving by means of apparatus perfectly unconnected with it, were two large balls, V and W (shown only in one of the figures), each weighing nearly 350 pounds. These balls could be brought directly opposite to the smaller balls, either at V and W, or at *v* and *w*. This apparatus was then inclosed in another chamber, having two apertures on each side, opposite to the slits of the first case: one for the insertion of a small telescope for reading from the scale; another to admit the light of a lamp, which was thus thrown on the slits.

Let us now suppose the larger weights brought into the positions V and W. If the wire were without any *torsion*, that is, if the lever could turn horizontally without any effort on the part of the wire to untwist, and if there were also attraction enough in the balls V and W to overcome friction, &c., the bar A B would begin to turn, A towards V, and B towards W, and would continue moving until the ends struck the sides of the inner case. But the effort on the part of the wire to restore itself, or the torsion, which becomes stronger and stronger the more the bar has turned, will gain upon the attraction, and at last become equal to it. Let K H be this position:

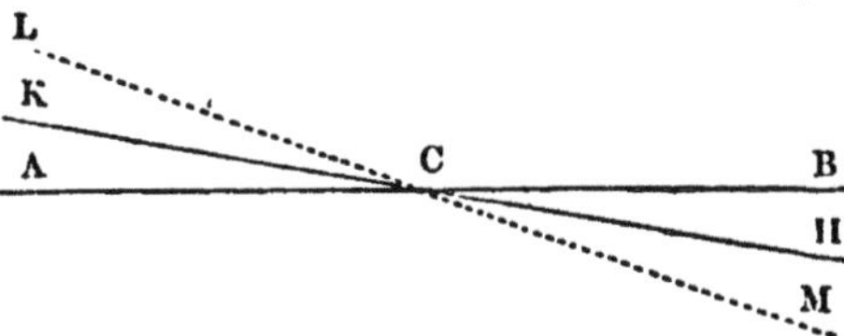

the arm will not stop here, for the velocity acquired by the motion will carry on the rotation, until the excess of torsion over the attraction has destroyed all that velocity. This same excess will then cause the bar to begin turning from L M (the extreme position) towards K H again, after which the excess of attraction over torsion will begin to destroy the velocity acquired in turning from L M to K H. The arm will not reach A B again, for friction and atmospheric resistance will destroy some part of the velocity. Thus the bar will continue to perform vibrations, of less and less extent, about the position K H, in which it will ultimately become stationary.

The fact of attraction was easily established: for on approximating the larger balls to the smaller, this series of oscillations began to take place. Indeed, the first wire which Cavendish used was not sufficiently stiff, so that the smaller balls, on the larger ones being brought near, were made slowly to approach the sides of the inner case, by which they were stopped. Even when a stiffer wire was substituted, the new position of rest was 3 divisions of the scale distant from the former one, or more than an eighth of an inch. In every experiment, therefore, the approach of matter to other matter at rest produced motion in the latter, visible to the naked eye.

The two phenomena which it was necessary to observe were, the time of one of the small vibrations, and the position in which the arm finally rested. To know in what way these phenomena conduce to the settlement of the question, some acquaintance must be obtained with the laws which regulate the motion of a pendulum; and we therefore proceed with this necessary preliminary.

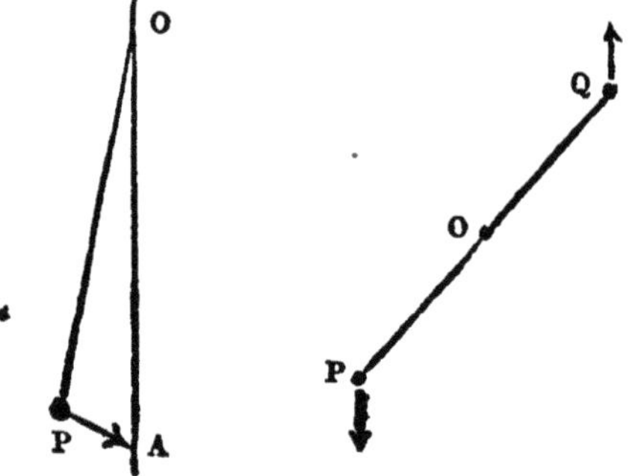

Let O A be the position of rest of a pendulum, and O P the pendulum itself, descending towards O A; let it also be so arranged that the vibration shall be of very small extent. Then the force which is acting upon the mass P is not the whole weight of P (for much the greater part is sustained by the string) but only a minute portion of it, which may be thus found—multiply the weight of P by the number of seconds in the angle A O P, and divide by 206265: or thus, for every 206265 grains (29½ pounds avoirdupois nearly) in the weight there is a grain of propelling force for every second in the angle. *Cæteris paribus*, then, the propelling force changes in the same proportion as the angle A O P. And the same may be said of the retarding force during the ascent of the pendulum on the other side.

The *isochronism* of the oscillations of a pendulum, which is a result both of theory and experiment, is the following: two pen-

C 3

dulums of equal length vibrate in the same time, whatever the extent of the vibrations may be, provided they are small; or rather, that the times of vibration corresponding to different small extents are *so nearly* equal, that the difference requires a great many vibrations to make it perceptible. If a pendulum begin by vibrating through one degree on each side of the position of rest, and be then allowed to oscillate until it is at rest, the time of single vibrations will gradually diminish, but no vibration will be less than the original one, by more than its ten thousandth part. This being the case, we assume that all small vibrations of the same pendulum are performed in the same times, whatever their extents may be; after which, the time of vibration of a pendulum may be ascertained by the following rule. From the square root of the number of inches in the length subtract its thousandth part, and take 16 per cent. of the remainder, which will be the number of seconds in the time of a complete vibration from the highest point on one side to the highest point on the other. Thus, to find the number of seconds which a pendulum 50 inches long will vibrate, proceed as follows:—

Square root of 50 is	7·071
	·007
	7·064
	4
	28·256
	4
	100)113·024
	1·13024

Answer, in 1·130 seconds, or 100 vibrations in 113 seconds.

It must be understood, however, that in the above rule, the pendulum is considered as a mere point, suspended by a string which has no weight. The size of the ball, the weight of the string or rod, the resistance of the air, and the extent of the arc of vibration, all create a necessity for some very minute corrections. The preceding rule will be quite sufficient for our present purpose.

If in the last figure P O were continued to Q, making P O and O Q equal, and if a weight equal to P were placed at Q, the apparatus would cease to be a vibrating pendulum, and would, when put in motion, revolve round and round O, until brought to rest by friction, &c. But if we imagine the weight of Q altered in direction, and made to press Q in a direction contrary to that of the weight of W′ (as indicated by the arrows), then the apparatus is again a vibrating pendulum. And the time of vibration remains the same as in the simple pendulum O P, as also does the proportion of the weight which acts in producing the motion at a given angle from the position of rest (or 1 grain in 29½ pounds for each second, as before stated). This double pendulum, with weights in opposite directions, is in all respects equivalent to the single pendulum, but has the advantage of a conformation resembling that of the *torsion* pendulum, with which it is presently to be compared.

In all that precedes, the force of gravity, or the earth's action as evidenced in the production of *weight*, is the moving power of the pendulum. This force is not quite the same in different latitudes, and in the preceding rules the latitude of London is supposed. But there are other ways of obtaining pendulums, vibrating in all respects after the manner of *gravitation pendulums*, but not with the same degree of force employed. The pendulum of Cavendish already described, performed vibrations solely through the torsion of the suspending wire; the weights of the balls were neutralised, being as ineffective in producing or hindering motion as if they had been suspended at D (page 31). But though the weights of G and H are counter-balanced, the masses of matter composing the balls still remain, and the torsion produces an oscillation, the force of which may be thus computed.

Observe the time of vibration of the pendulum in seconds, multiply this number by itself, and 8073057 by the product: divide the result by the number of inches in the half pendulum (C A), and the result is the number of grains of weight in each ball, to which one grain of propelling power must be applied for every second (as in page 33) to produce the motion actually observed.

Thus, if the time of vibration be 2 seconds, and the length of the half pendulum 5 inches, we have

$$2 \times 2 = 4 \qquad \begin{array}{r} 8073057 \\ 4 \\ \hline 5)32292228 \\ \hline 6458445 \end{array}$$

or the proportion is 1 grain out of 6458445.

Now suppose an extraneous force to retain the pendulum in its position at O P, not allowing it to descend. It is plain that this extraneous force must be equal to the force with which the pendulum would be urged downwards; that is, if the moving power be gravitation, it must be 1 grain for every 206265 grains in the weight, repeated for every second in the angle A O P: again, if we saw such a pendulum as that in the last example permanently deflected 20 seconds out of its position of rest, we should infer that each ball must be kept by a detaining force equivalent to 20 grains out of every 6458445 grains in its weight. We are now in a condition to proceed with our account of Cavendish's experiment.

The weights having been brought to V and W, a series of oscillations commenced, as already described, and the arm A B ultimately came to a position of rest, with its ends nearer to V and W than in the figure. But it was not necessary, nor even desirable, to wait until the arm was at rest; for it is in this experiment of as much importance to measure the time of the arm's vibration as the quantity of its deflection. Now the time of vibration can only be measured while the vibrations are of some extent; if this had been done, and if the observer had then waited until the arm was at rest, it is possible that some disturbing circumstance might have occurred in the interval which would have permanently altered the torsion of the wire, and thus caused the time and the deflection to be measured upon two different pendulums. Considering that in this pendulum the forces to be measured are not more than one grain in fifty millions of the weights used, it will appear by no means certain that the force of torsion in a wire, and all other circumstances, will remain the same for several hours, so nearly as that no one of them shall undergo anything like an alteration amounting to a fifty millionth part of the whole. Experience soon showed that whatever care might be taken, the arm would not retain *precisely* the same position for an hour together. If the vibrations were always the same in extent, it would be easy to determine the position of rest, which would be no other than half way between the extremes of vibration. But seeing that resistances are perpetually retarding the vibration, and diminishing its extent, some other method must be employed. Now it may be assumed, quite correctly enough for this purpose, that the diminutions of extent in two successive vibrations are the same. Let S be the position of one extremity of the arm at the beginning of a vibration, and T the other extremity, as it would be if there were no resistance, so that K, half way between S and T, is the real position of rest. Now let S V be the real extent of the vibration, and V W that of the next, whence S W (the effect of resistance in two vibrations) is double of V T (that in one): or *v* (half way between S and W) and V are equidistant from K. Thus K may be found by bisecting *v* V.

The method of finding the time of a vibration was as follows: as soon as the middle point of a vibration was ascertained, care was taken to note the time at which the end of the arm passed the preceding and following division, whence the time of the middle of the vibration was readily determined. Thus, if the middle point of the vibration were at 25·82 on the scale, and if the times of arriving at 25 and 26 were 8m. 40s. and 8m. 53s., it follows, by a simple proportion, that (on the supposition of the uniform motion of the arm, which is true for so short an interval) 82-hundredths of the 13 seconds employed in moving from 25 to 26 must have elapsed when the arm is at 25·82. By noting the time of the middle of two vibrations, between which several have elapsed, the time of a number of them is known, from which that of a single vibration can be found. This is subject to a slight error, because the time of coming to the middle point (owing to the resistance of the air, &c.) is not half-way between the times of coming to the two extremes: but, as this error affects the first and last middle point, almost in the same manner, its effect will be inappreciable.

It appears in the figure, that the weights may either be placed as at V and W, or as at *v* and *w*, in which two cases the deflections of the arm will be in opposite directions. To distinguish the two cases, Cavendish called that position *positive*, in which the deflection made the arm point to a higher number on the scale; and the other one negative. It was thus in his power to make four distinct experiments: that is, to observe the effect of bringing the weights, 1. From the intermediate to the positive position; 2. From the intermediate to the negative; 3. From the positive to the negative; 4. From the negative to the positive. We now give the results of one experiment (the eleventh) entire.

Extreme Points.	Divisions.	Time.	Position of Rest.	Time of middle of vibration.
		h. m. s.		h. m. s.
	The weights in the positive position.			
34,90				
34,10	..	..	34,47	
34,80	..	..	34,49	
34,25				
The weights brought into the negative position.				
23,30				
	28	9 59 59	..	10 0 8
	29	10 0 27		
33,30	..	..	28,42	
	29	0 6 52	..	0 7 5
	27	0 7 51		
23,80	..	..	28,35	
32,50	..	..	28,30	
24,40				
Not observed.				
24,80				
31,30	..	..	28,17	
	29	0 48 37	..	49 8
	28	0 49 21		
25,30	..	..	28,20	
	28	0 56 8	..	0 56 13
	29	0 56 56		
30,90				

At the beginning of the experiment, the weights being in the positive position, (say at V and W) the arm had nearly come to rest, and was making vibrations of short extent, about a position of rest 34·47 divisions from the beginning of the scale. The method already explained was used to find that position: thus, two consecutive extremes on the same side are marked 34·90 and 34·80, and the point half-way between them was at 34·85. Half-way between this and 34·10 we have 34·475, which may be called 34·47. On the weights being brought into the negative position, (at v and w) the arm is thrown back on the scale, and the extreme is at 23·30. The middle of the vibration it was known, from preceding experiments, would fall somewhere between 28 and 29: accordingly, the time of passing these divisions was attentively marked, and found to be at 9h. 59m. 29s. and 10h. 0m. 27s. On the return of the arm, we must suppose the time of passing 28 was lost, as we see 27 in its place; and this is not a misprint, as we see that the interval (from 6m. 52s. to 7m. 51s.) is about double of the preceding. The two adjacent extremes, 23·30 and 23·80, have their middle point at 23·55; half-way between which and 33·30 gives 28·425, accounting for 28·42, which we see entered as the position of rest. Again, the middle of the vibration is at 28·30, and the whole interval from 28 to 29 being passed over in the 28 seconds next following 9h. 59m. 59s., the arm will be at 28·30 in 30-hundredths of 28 seconds, or 8s·4, and we see 10h. 0m. 8s. put down as the time of arriving at the middle. In this way we proceed through eight complete vibrations, namely, from the middle of that whose extremes are 23·30 and 33·30 to the middle of that whose extremes are 25·30 and 30·90, which eight vibrations occupy from 10h. 0m. 8s. to 10h. 56m. 13s., giving 56m. 5s. for the whole, or 7m. 1s. for each. The point of rest varies during the experiment from 28·42 to 28·20; and, as the balls are on the negative side, this is the effect that would take place if the attraction of the weights became stronger with time (of which hereafter). But, as Cavendish observed that the position of the arm would vary slightly in an hour, without any disturbance from without, he took only the first effect, or that produced immediately after the change of the weights, as the measure of the whole effect, instead of obtaining a new point of rest from the average of those observed. In the present experiment, the position of rest is changed at once from 34·49 to 28·42, or the reversal of the position of the balls causes the arm to move through 6·07 divisions of the scale.

Previously to entering upon the deductions which may be made from this and the other experiments, we must ask whether the preceding effects may not be the consequence of magnetism or electricity. With regard to the latter, the permanent nature of the result, which, however often repeated, was very nearly the same, rendered it impossible to attribute the effects to the varying and accidental disturbances of the electric state of the surrounding bodies. The case of magnetism is rather different; for, though it be sensible only in iron, yet other metals may have a portion sufficient to produce so slight an effect as the one in question. Cavendish provided an apparatus for turning the weights V and W round on their axes, so as to cause them (if they should be magnetical bodies) to present different poles to the smaller balls, in different experiments, or in the course of the same. No change of effect was produced; though, had the phenomenon been magnetical, the attraction ought to have become a repulsion, and *vice versâ*.

Cavendish never suspected that magnetism was the cause of the whole of the phenomena, and the apparatus just mentioned was constructed with reference to the remarkable phenomenon which is visible in the experiment we have quoted: namely, that the attraction of the balls appears to increase a little with the time during which the balls remain in their position. Thus, the point of rest passes, in less than an hour, from 28·42 to 28·20, without any assignable cause. At first it was suspected that the suspending wire might be only imperfectly elastic, and might require time to exhibit the whole effect of the attraction; but, upon fixing the wire for some time in a somewhat twisted position, and then allowing it to go free, nothing was perceived which indicated that the wire had, as the phrase is, taken a set, or lost any of the elasticity necessary to restore the arm to its former position. Magnetism might have explained so small a discrepancy, as the balls might, by remaining in one position, gradually acquire a small degree of polarity. But no effect whatsoever was observed on reversing the balls; and it at last occurred to Cavendish, that the discrepancy might arise from the balls being warmer than the sides of the case, which would produce a current of air towards the weights, and thus tend to bring the arms nearer to them. This explanation turned out to be correct: by warming the weights slightly before bringing them near the ends of the arm, the gradual alteration of the deflection was very much increased, and by cooling the weights with ice, previously to using them, the effect was reversed, that is, after some time the deflection of the arm begun to diminish.

The method of obtaining the earth's density, from the experiments, is as follows. The arm itself is a double pendulum, to the ends of which equal forces are applied, and it may therefore be compared with the modification of the gravitation pendulum, in page 34. The experiments give the time of vibration of the pendulum, and therefore (page 35) the amount of force which is necessary to maintain it deflected by a given angle from its position of rest; but the experiments also give the deflection, that is to say, they inform us what degree of force actually was exerted by the weights. That is, knowing the size and density of the weights, we possesss, in one instance, the attractive effect of a sphere upon a mass of matter placed at a given distance from it.

It is a well-known mathematical result, that, when the forces of attraction of particles on each other are inversely as the squares of their distances, any sphere attracts any other precisely as it would do if all the particles of both were at their several centres. And remembering that spheres of different diameters are as the cubes of their diameters in volume, and that their weights are *cæteris paribus* as their densities, there are all the conditions necessary for the solution of the following problem. What is the relation which must exist between the densities and magnitudes of two spheres, their distance from two given equal weights, and the attractions they exercise upon these weights. Between all these quantities an equation exists, so that if all be known except one, the equation gives that one. Now, in the present instance, we have the following set of quantities. Of the two spheres, one is the whole earth, the other is the ball V or

W. We know the magnitude of V, and its density; and from the torsion-pendulum we ascertain, in the manner preceding, its attractive force on a given ball at a given and known distance. We know the diameter of the earth, and therefore its size, *but not its average density;* we know, also, (from the common pendulum) how to measure its attractive effect on a ball at its surface, that is, at a known distance from its centre. Consequently, with the exception only of the earth's average density, we know all the quantities which enter into such an equation as the one described; and this equation itself, therefore, determines the earth's density. There are one or two very small corrections to be made, such as taking into account the attraction of the rods which carry the weights, &c. &c. But, as the details of these corrections are complicated, and their united effect very small, we shall speak no further of them.

The experiments made by Cavendish, as to their individual results, were as follows. The first set was made with a wire of so little stiffness, that a vibration was not completed in less than fifteen minutes. The results of these experiments gave for the earth's mean density—

5·50, 5·61, 4·88, 5·07, 5·26, 5·55,

the mean* of which is 5·31.

The remaining experiments, made with a stiffer wire, which vibrated in a little more than seven minutes, we shall arrange according to their details: *m*, +, and − signify the intermediate, positive, and negative positions of the weights; and *m* − signifies that, in the column underwritten, the weights were moved from the intermediate position to the negative, &c.

	m+	*m*−	−+	+−
	5·36	5·53	5·58	5·29
		5·29	5·57	5·65
		5·34	5·62	5·10
			5·44	5·27
Mean		5·39	5·79	5·39
			5·42	5·63
			5·47	5·46
			5·34	5·85
			5·30	
			5·75	
			5·68	
Mean			5·54	5·46

Mean of the whole, 5·48.

The nearness of this result to that of Maskelyne is a presumption in favour of the latter: for so much superior in precision must the experiment of Cavendish be considered, that, if the two had differed very materially, there is no question that the one we have described must have been relied on altogether, to the exclusion of the other. The accordance is a satisfactory result; and we have little

* Cavendish says this mean is 5·48, which it would be if for 4·88 we should read 5·38 in the preceding.

doubt that the repetition, which is shortly to take place, will be an additional confirmation.

It seems surprising, however, that the question raised at the commencement of this article should never have suggested itself to Cavendish: never, at least, with force enough to induce him to try a few experiments with pendulum balls of brass, iron, or wood. The ordinary gravitation pendulum may, perhaps, be considered as proving that matter is nearly equally attracted by matter, of whatever substance: but the numerous small discrepancies which still are found in pendulum observations, made at different spots, render it a question not only of curiosity but of interest, to know whether the attraction of different species of matter on each other is precisely the same. This fact may be assumed on strong grounds, but it has never yet been brought to actual measurement by any method so free from extraneous disturbances as that of Cavendish.

Let g be the measure of the accelerating force of gravity (32·19 ft. of velocity per second).

R .. the earth's mean diameter 7915·5 miles in feet.

D .. the earth's mean density.

l .. half the arm of the torsion pendulum (in feet).

r the radius of the sphere V or W (in feet).

δ the density of the same.

n the number of seconds in a vibration of the arm.

k the radius of the circle on which the graduated scale is laid down (in inches).

b the number of divisions in the observed deflection, each division being β of an inch.

h the distance (in inches) of the centre of a weight and its pendulum ball.

W and w the weight of the large and small balls.

a the weight in grains of a cubic foot of water.

If we consider the torsion wire as deprived of its stiffness, and the place supplied by a pressure on the weight acting from the centre of motion always in one direction (in the manner of gravity acting on a common pendulum), we have, g' being the acceleration in question on each of the balls,

$$n = \pi\sqrt{\frac{l}{g'}} \qquad g' = \frac{\pi^2 l}{n^2}\ ;$$

if we decompose this force in the direction perpendicular to the arm, we must multiply the preceding expression by $\sin\theta$, when the pendulum is inclined to its position of rest by an angle θ. But as this angle is always small, we may substitute θ instead of $\sin\theta$, θ being measured by the ratio of the subtending arc to the radius. And g' is to the acceleration produced by gravity (or g) as the pressure produced by it on the whole ball to the pressure produced by gravity on the ball (or its weight); that is—

$$g' \text{ or } \frac{\pi^2 l}{n^2} : g :: \text{pressure in question} : w\ .$$

the pressure is, therefore, $\frac{\pi^2 l}{n^2 g} w$, and $\frac{\pi^2 l}{n^2 g} \times w\,\theta$ is its resolved part. This is then the force, which applied in an opposite direction, will maintain the arm deflected at an angle θ from its position of rest. But θ has an arc of b times β to a radius k, in the results of the experiment: consequently $\frac{\pi^2 l}{n^2 g} \times \frac{b\beta}{k} \times w$ is the force exerted by the larger ball on its ball of the pendulum. From the smallness of the angle θ, it is unnecessary to consider that the action of the larger weights on the balls is not quite perpendicular to the arm in the deflected position.

Now, since the weight of a body is, *cæteris paribus*, as the product of its bulk and density, it follows that $a \times$ bulk $\times$ density is the weight of any body in grains. Hence

$$W = a \times \frac{4}{3}\pi r^3 \times \delta, \text{ or } \frac{4}{3}\pi r^3 \delta = \frac{W}{a}$$

And since the attractive forces of different spheres on bodies at different distances are as their values of

$$\frac{\text{bulk} \times \text{density}}{(\text{Distance of the attracted point})^2},$$

we find that the attraction of W on one of the balls is to the attraction of the earth on it (its weight w) as

$$\frac{\frac{4}{3}\pi r^3 \times \delta}{\left(\frac{h}{12}\right)^2} \text{ to } \frac{\frac{4}{3}\pi R^3 D}{R^2}, \text{ or as } \frac{144\,W}{a\,h^2} \text{ to } \frac{4}{3}\pi R D$$

or as 108 W to $a\pi h^2$ R D. Observe that h must be reduced to feet, or $\frac{h}{12}$ used instead of h. Whence the attraction of W is $\frac{108\,W}{a\pi h^2 R D} \times w$. But this attraction being the force which deflects the lever, we have

$$\frac{\pi^2 l}{n^2 g} \times \frac{b\beta}{k} \times w = \frac{108\,W}{a\pi h^2 R D} \times w$$

$$\text{or } D = \frac{108\,k g W}{a\pi^3 l h^2 \beta R} \times \frac{n^2}{b}$$

Here n and b are determined from the experiment, while the coefficient is composed entirely of known quantities, which may be compared with those of Cavendish, as follows:

k = 38·3 inches.

g is implicitly contained in the length of the second's pendulum, (or, $g \div (3{\cdot}14159)^2$), which he assumes at 39·14 inches.

W = 2439000 grains.

a is implied in his statement, that the preceding is 10·64 spherical feet of water (or spheres of a foot diameter); that is, he makes 2439000 contain $\frac{4}{3} \times 3{\cdot}14159 \times \frac{1}{8} \times a$, 10·64 times.

$\pi = 3{\cdot}14159$ as usual. $h = 8{\cdot}85$ inches.
$l = 36{\cdot}65$ inches. $\beta = \frac{1}{20}$ of an inch.
$R = 20900000$ feet.

and a correction was required to be introduced by a mistake in the apparatus, not worth explaining, which amounted to diminishing W in the proportion of ·9779 to 1.

The paper of Cavendish is in the 'Philosophical Transactions' for 1798; it is translated entire in the tenth volume (17 ième cahier) of the *Journal de l'Ecole Polytechnique*, and is also given, excepting only the details of the actual experiments, in the eighteenth volume of Dr. Hutton's abridgment of the 'Philosophical Transactions.'

VI.—OCCULTATIONS OF PLANETS AND FIXED STARS BY THE MOON, VISIBLE AT GREENWICH.

(From the Nautical Almanac.)

Day of the Month.	Star's Name.	Magnitude.	IMMERSION. Sidereal Time.	Mean Time.	Angle from N. Point.	Angle from Vertex.	EMERSION. Sidereal Time.	Mean Time.	Angle from N. Point.	Angle from Vertex.
1838.			h m	h m	°	°	h m	h m	°	°
Jan. 5	δ Arietis	4	7 16†	12 15	18	58	. .	. .	. .	. .
8	C Tauri	4.5	3 18	8 7	85	46	4 33	9 21	294	269
9	47 Geminorum..	6 1	4 28	19 11	133	168	14 58	19 41	205	238
10	c Geminorum..	6	2 44	7 24	57	14	3 42	8 23	297	253
Feb. 1	40 Arietis......	6	3 7	6 21	129	136	4 21	7 35	288	313
4	C Tauri	4.5	12 36	15 36	100	139	13 27	16 27	252	287
6	c Geminorum..	6	12 53	15 46	92	135	13 47	16 40	233	274
7	λ Cancri.......	6	3 33	6 23	47	4	4 31	7 21	296	254
9	37 Leonis......	6	16 15	18 56	75	114	17 10	19 50	231	268
10	l Leonis	6	8 0	10 38	94	63	9 3	11 41	199	179
Mar. 6	λ Cancri	6	13 15	14 18	88	131	14 11	15 13	229	270
10	σ Leonis	4	10 58	11 45	32	29	12 13	13 0	251	264
16	A² Scorpii	6	12 10†	12 33	334	303	. .	. .	. .	. .
16	(237) Scorpii....	6	16 42	17 4	38	46	17 49	18 12	285	303
Apr. 1	47 Geminorum..	6	14 25	13 45	54	90	15 12	14 32	282	314
7	β Virginis	3.4	13 14†	12 11	321	339	. .	. .	. .	. .
12	χ Libræ	5.6	12 25	11 2	105	78	13 18	11 55	199	180
18	κ Capricorni ...	5	19 0†	17 12	203	180	. .	. .	. .	. .
25	MERCURY ...	.	10 42	8 29	49	85	11 16	9 2	333	8
29	c Geminorum..	6	14 35	12 5	87	125	15 25	12 55	243	276
May 2	37 Leonis	6	17 12	14 30	80	117	18 3‡	15 21	229	262
13	(84) Sagittarii ..	6	17 14†	13 49	185	167	. .	. .	. .	. .
15	33 Capricorni...	6	17 50	14 17	117	88	19 0	15 27	284	263
31	σ Leonis	4	14 55†	10 19	146	181	. .	. .	. .	. .
June 6	A² Scorpii	6	13 33	8 34	338	318	13 38	8 39	330	311
6	(237) Scorpii....	6	18 7	13 7	34	54	19 3	14 3	298	326
8	γ¹ Sagittarii. ...	5	19 28	14 20	11	25	19 38	14 30	355	10
12	x Aquarii......	6	17 55	12 31	107	74	19 2	13 38	301	276
27	χ Leonis	4.5	15 51	9 29	49	87	16 52	10 29	249	288
27	JUPITER.....	.	16 52	10 29	26	65	17 41*	11 19	276	313
July 9	κ Capricorni...	5	18 4	10 54	94	64	19 8	11 59	313	291
12	t Piscium	6	21 11†	13 49	215	155	. .	. .	. .	. .

Day of the Month.	Star's Name.	Magnitude.	Immersion. Sidereal Time.	Immersion. Mean Time.	Immersion. Angle from N. Point.	Immersion. Angle from Vertex.	Emersion. Sidereal Time.	Emersion. Mean Time.	Emersion. Angle from N. Point.	Emersion. Angle from Vertex.
			h m	h m	°	°	h m	h m	°	°
1838.										
13	ζ¹ Piscium	6	19 31	12 6	82	43	20 16	12 50	341	302
18	C Tauri	4.5	20 56§	13 10	109	81	21 43	13 57	266	233
31	m Scorpii	6	17 36	9 0	63	78	18 52	10 15	269	295
Aug. 2	γ¹ Sagittarii	5	16 24	7 41	59	45	17 34	8 50	292	289
6	50 Aquarii......	6	0 54	15 53	170	195	1 38	16 37	258	287
7	χ Aquarii	5.6	23 18†	14 13	37	39	. .	. .	. .	. .
12	f Pleiadum....	5	2 15†	16 50	205	180	. .	. .	. .	. .
Sept. 2	κ Capricorni...	5	17 22	6 36	86	52	18 20	7 34	318	290
Sept. 3	70 Aquarii	6	23 39†	12 49	216	226	. .	. .	. .	. .
4	κ Piscium	5.6	1 56	15 1	156	179	2 51	15 56	276	306
7	ψ Arietis	6	1 53	14 46	178	171	2 32	15 25	246	249
8	ζ Arietis	5	19 19	8 9	125	91	20 6	8 56	279	242
9	χ Tauri	6	23 3	11 49	127	84	0 0	12 45	275	232
13	λ Cancri	6	2 29	14 58	90	48	3 32	16 1	254	211
30	50 Aquarii......	6	1 54	13 17	151	182	2 46	14 9	274	310
Oct. 1	χ Aquarii......	5.6	0 36†	11 55	37	52	. .	. .	. .	. .
8	C Tauri	4.5	21 7	7 59	36	7	21 30	8 22	339	307
25	a Sagittarii	5.6	21 22†	7 7	200	215	. .	. .	. .	. .
28	70 Aquarii......	6	23 53†	9 26	216	229	. .	. .	. .	. .
29	κ Piscium	5.6	3 10	12 39	128	160	4 10	13 38	299	335
Nov. 1	ψ Arietis	6	2 38	11 55	164	169	3 28	12 45	256	274
2	ζ Arietis	5	19 16	4 30	135	102	19 59	5 13	269	232
2	σ¹ Arietis	6	21 48†	7 2	25	343	. .	. .	. .	. .
2	9 Tauri	6	5 20	14 33	115	144	6 29	15 41	280	319
2	d Pleiadum	5	10 3†	19 15	190	229	. .	. .	. .	. .
3	χ Tauri	6	21 49	6 59	178	139	22 6	7 16	219	178
6	c Geminorum..	6	4 35	13 32	49	8	5 37	14 34	295	262
10	χ Leonis	4.5	8 28	17 9	35	8	9 43	18 23	254	238
29	ζ Arietis	5	9 50†	17 15	13	51	. .	. .	. .	. .
30	χ Tauri	6	11 58	19 20	134	168	12 36	19 57	238	268
Dec. 2	C Tauri	4.5	20 31‡	3 47	102	77	21 17	4 32	271	241
3	47 Geminorum..	6	2 25	9 36	72	28	3 27	10 38	287	244
8	H Leonis	6	11 35	18 25	54	56	12 54	19 44	228	245
24	(189) Piscium...	6	1 1	6 50	44	48	1 10	6 59	30	36
25	π Piscium	6	0 7†	5 52	216	198	. .	. .	. .	. .
26	ψ Arietis	6	23 22	5 3	166	131	0 8	5 49	257	227
27	9 Tauri	6	3 33	9 10	103	104	4 46	10 22	300	323
27	d Pleiadum....	5	8 26	14 2	129	172	9 18	14 53	251	293
27	η Tauri	3	9 10	14 46	163	205	9 35	15 11	216	257
27	f Pleiadum....	5	9 38	15 13	128	169	10 25	16 0	250	289
27	h Pleiadum....	5.6	9 46	15 22	151	192	10 19	15 54	227	266
29	C Tauri........	4.5	10 3†	5 30	354	39	. .	. .	. .	. .
31	c Geminorum ..	6	3 30	8 51	137	93	4 10	9 31	211	168

• Star setting.
† A near approach.
‡ Star below the horizon.
§ Star rising.

IV.—NOTICES OF THE PROGRESS OF THE PROBLEM OF EVOLUTION.

In the article Involution and Evolution, in The Penny Cyclopædia (and in various other places mentioned in the course of this paper), will be found a method of obtaining the rational roots, and approximating to the irrational roots, of an equation of any order whatsoever. This process, being a generalization of the method by which the roots of numbers are extracted, has received the name* of *Evolution*, and the meaning of the term *Involution* has received a corresponding extension. We presume the readers of this article to be acquainted with the method, of which we shall here give no further explanation, than that any method which proceeds by finding figure after figure of the root required, making use of the error of the root already found to determine a new figure of the root, is evolution. Thus, the Newtonian method of approximation only differs from evolution in not being conducted by finding the figures of the root successively.

A little controversy has been going on for some years relative to the invention of this method, which, being apparently for the most part conducted without sufficient attention to the previous history of algebra, has been argued upon what appears to us altogether a false basis. If Vieta, Herigone, Harriot, Oughtred, Wallis, and various others, whose writings are not now before us, had never lived nor written, we might not have had this objection to make; but, as it is, we think it will sufficiently appear in the course of this article that it has become necessary to compel the appearance of these parties. The following points will be asserted:—

1. That a method of evolution, exhibited in the performance of division and the extraction of the square and cube roots, was communicated by the Hindoo to the Arab writers, and passed from the latter into Europe.

2. That an extension of this method to the solution of equations in general, laborious but certain, was made by Vieta, and became public before the publication of his collected works.

3. That this method was somewhat improved in form by Harriot, used by Oughtred, Wallis, and the generality at least of English writers, and continued to be a common method up to the time at which the Newtonian approximation appeared in Wallis's Algebra.

4. That the method was so laborious, that its derivation of the root, figure by figure, was not preferable to the use of the Newtonian method, for which it was accordingly abandoned.

5. That before the middle of the year 1819, Mr. W. G. Horner, of Bath, discovered a rule for performing the computations necessary in Vieta's extended method, by which that method was made preferable to the Newtonian method, which simplification was read to the Royal Society July 1, 1819, and published by them December 1, 1819.

* Mr. Peter Nicholson wrote a treatise on the method, under the name of 'Essay n Involution and Evolution.' London, Davis and Dixon, 1820.

6. That, though more than one individual may claim to have rediscovered the method of Vieta, or that method with modifications not materially decreasing its difficulty, and partly derived from the Newtonian approximation, Mr. Horner is the first and (as far as can be proved) the only inventor of all that renders the extension of Vieta, as now used, preferable to the approximation of Newton.

7. That with reference to its practicability, this method is properly called *Horner's method;* but, looked at, altogether, it must be called the united method of Vieta, Newton, and Horner.

On the first point there is no dispute: the Hindoo treatise on arithmetic, called the Liliwati, establishes the knowledge which that nation had of the processes mentioned; and the passage of this knowledge into Europe by means of Mahometan works, professedly indebted to Hindoo writers, is matter of common history.

On the second point there can be no dispute that we know of, except, perhaps, with the reader of Hutton's history of algebra (in the second volume of his tracts), who must have found, and perhaps received, the assertion that Stevinus was the inventor of a general method of solving equations, from which this of Vieta differs little. Now, we have searched the writings of Stevinus* in vain for anything more of the sort in question than the simple extraction of roots. But we find that there was a method of solving equations (so called) which sometimes bore the name of Stevinus (in Ronayne's Algebra, for instance), and which is nothing but successive trial of different numbers. Now, Dr. Hutton says of Vieta's method, that "it is very laborious, and but little more than was before done by Stevinus, depending not a little upon trials." Here, then, we have the secret of his assertion: he knew a little of what some had dignified by the name of Stevinus's *method*, and did not take the trouble to follow Vieta through one of his examples. Seeing a trial at every step (as there is in division, &c.), he set the plan down at once as being that of Stevinus, which is *nothing but trial.* We dwell upon this, because we have no doubt that the carelessness of Dr. Hutton (of which this is not the only instance) is the proximate cause of the present article. His work is the history of algebra which was most accessible to the English reader; if he had faithfully looked into the method of Vieta, and represented it accordingly, the odds are great against a revivification of that method having been promulgated as a new discovery. We shall see more of this before we have done.

Wallis (in the preface of his Algebra, English edition) says of Vieta, "He introduced also his numeral Exegesis of affected equations, extracting the roots of these in numbers; which had before been applied to single equations, such as the extracting the roots of squares, cubes, &c., singly proposed: but had not been applied (or but rarely) to equations affected." The words, "or but rarely," do not indicate that Wallis knew of any predecessor, for he cites none throughout a voluminous history; but they probably

* The collected edition of his works by Albert Girard, two volumes of Hypomnemata, and the arithmetic of 1585.

imply that he considered the extension so natural, that it would be unsafe to say it had never been made. But he always calls the method after Vieta, and frequently after Vieta, Oughtred, and Harriot, putting Vieta first. And in cap. 52 he says expressly, that none, to his knowledge, used it before Vieta.

Leaving the testimony of historians, we take up the tract of Vieta, entitled 'De Numerosâ Potestatum purarum atque adfectarum ad Exegesin Resolutione:' if such a method existed prior to the publication of this tract, it is for those who make the assertion to produce it.

The year of publication of this Exegesis was 1600. Vieta printed his works at his own expense, and distributed them as gifts; which, with the complete collection of his writings, published by Schooten in 1646, has rendered the original editions scarce. It would not be easy for us to say whether this tract was ever matter of much publicity previously to its appearance in Schooten's edition, if it were not for the course of mathematics which Herigone, or Herigonius, published in and about 1634. In the second volume of this course appears the method of solving adfected equations, with a reference to Vieta's name.

We shall now take an example from Vieta, previously giving the same by the method now used. The equation to be resolved is—

$$x^5 - 5\,x^3 + 500\,x = 7905504.$$

0	−5	0	500.	7905504(20+4
20	400	7900	158000	3170000
20	395	7900	158500	4735504
20	800	23900	636000	4735504
40	1195	31800	794500	0
20	1200	47900	389376	
60	2395	79700	1183376	
20	1600	17644		
80	3995	97344		
20	416			
100	4411			
4				
104				

In the following, taken from Vieta, explanations are omitted. The divisor used for obtaining the second figure of the root is not the same as in the modern form, but differs little from it when a few places have been obtained, as in Wallis's example at the end of this article.—

"In notis QC − 5 C + 500 N æquatur 7,905,504, et fit 1 N unitatum quot?"

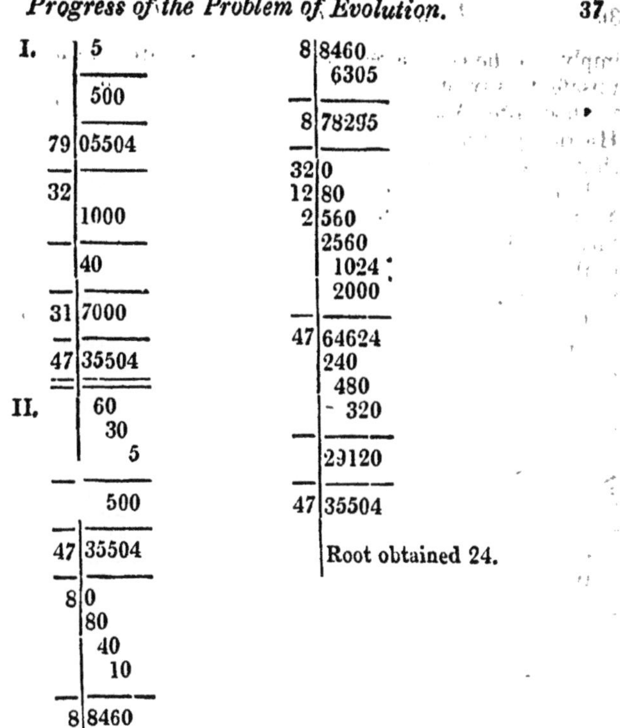

See next Col.

To explain the divisor of Vieta in modern language, we must say that that it is $\phi(a+1) - \phi(a)$,* instead of $\phi' a$, ϕa being the unknown side of the equation, and a the part of the root already found.

We now come to our third point. The Clavis of Oughtred was first published in 1631, but no evolution is found in this edition, except the extraction of ordinary roots. Dr. Wallis says, "The method of Vieta is followed, and much improved, by Mr. Oughtred, in his Clavis (first published in the year 1631), and other treatises of his." This means that Wallis had before him a later edition of the Clavis than the first, and we shall presently dispute the alleged improvement. The 'Artis Analyticæ Praxis' of Harriot (posthumous) also appeared in 1631; and here we find a specific improvement, namely, the formation of only so much of the divisor as is necessary for the determination of the next figure. It is strange that Wallis, who seems in almost every other case to have a bias of exaggeration towards Harriot's works, should, in this, place Oughtred by his side. The following example is from page 156 of the Praxis of Harriot, and contains a root of three places, all the examples of Vieta (in adfected cases at least) containing only

* Or rather a unit less than this, both in Vieta and Oughtred.

two. It should be noticed that Harriot always deprives the equation of its second term, which was then an improvement, though it would not be one now.

Sit æquatio numersosè proposita

$$aaaa - 1024\,aa + 6254\,a = 19633735875.$$

Radix universalis successivè educenda.

	3 7 5
Homogeneum resolvendum *hhhh*	19633735875
ggg	6254
$-ff$	−1024
bbb	27
+	27006254
Divisor A	26903854
gggb	18762
Radix sing. prima $b = 3$, $-ffbb$	−9216
bbbb	81
+	81018762
Ablatitium A *b*	80097162
Radix singularis	3
Homogeneum residuum resolvendum	11624019675
ggg	6254
$-ff$	−1024
$-2ffb$	−61440
Radix sing. decupl. $b = 30$, $4bbb$	108000
$6bb$	5400
$4b$	120
+	113526254
−	62464
Divisor B	112901614
gggc	43778
$-ffcc$	−50176
$-2ffbc$	−430080
Radix sing. secund. $c = 7$ $4bbbc$	756000
$6bbcc$	264600
$4bccc$	41160
cccc	2401
+	1064204778
−	480256
Ablatitium B *c*	1059402218
Radix aucta	3 7
Homogenei residuum resolvendum	1029997495

	ggg	6254
	− *ff*	− 1024
	− 2 *ffb*	− 757760
Radix aucta decupl. *b* = 370,	4 *b bb*	202612000
	6 *bb*	821400
	4 *b*	1480
	+	203441138
	−	758784
	Divisor B	202682354
	gggc	31270
	− *ffcc*	− 2560
	− 2 *ffbc*	− 3788800
Radix sing. tertia *c* = 5	4 *bbbc*	1013060000
	6 *bbcc*	20535000
	4 *bccc*	185000
	cccc	625
	+	1033811895
	−	3814400
	Ablatitium B *c*	1029997495
	Radix universalis completé educata	3 7 5
	Homogenei reliquum finale	000000000

We see, then, that Harriot was the first English writer who used the extended method, which came in England to be sometimes called by his name. The second edition of Oughtred's Clavis was published by himself in 1647, in English, and was called 'The Key of the Mathematicks, new forged and filed; together with a Treatise of the resolution of all kinde of affected Æquations in numbers.' The method is that of Vieta, as distinguished from that of Harriot, even to the omission of the unit in the divisor, mentioned in a preceding note; but the names of Vieta and Harriot are entirely suppressed. The following is an example from Oughtred —

$$1\,qc - 15\,qq + 160\,c - 1250\,q + 6480\,l = 170304782.$$

$170\dot{3}$	$0478\dot{2}$	(47
1	5	− B
1	250	− D *q*
1	60	C *q*
	6480	F *q q*
1024		A *q c*
102	40	C *q* A *c*
2	5920	F *q q* A
+ 1128	9920	
384	0	− B A *q q*
20	000	− D c A *q*

− 404	000	
724	9920	*Ablatit.*
R 978	05582	
128	0	5 A *q q*
6	40	10 A *c*
	160	10 A *q*
	20	5 A
7	680	C *q* 3 A *q*
	1920	C *q* 3 A
	160	C *q*
	4680	F *q q*

+ 142	50040	
38	40	− B 4 A c
1	440	− B 6 A q
	240	− B 4 A
	15	− B
1	0000	− D c 2 A
	1250	− D c
− 40	87665	
+ 101	62375	Divisor.
896	0	5 A q q E
313	60	10 A c E q
54	880	10 A q E c

4	8020	5 A E q q
	16807	E q c
53	760	C q 3 A q E
9	4080	C q 3 A E q
	54880	C q E c
	45360	F q q E
+1333	62047	
268	80	− B 4 A c E
70	560	− B 6 A q E q
8	2320	− B 4 A E c
	36015	− B E q q
7	0000	− D c 2 A E
	61250	− D c E q
− 355	56465	
+ 978	05582	*Ablatit.*

In Wallis's Algebra, chapter 62, we find a 'numerose Exegesis,' as he calls it, in which appears, for the first time, the important improvement of the contraction of the divisors and remainders; by which means he has extracted the root of a biquadratic equation to 17 places. The work, as written down, occupies a little more than two of the small folio pages of the time, in double column, with Oughtred's marks of explanation; and, from various expressions used by Wallis, it appears that the method was commonly known and practised. Nor was it locked up in so large a work as Wallis's Algebra; for so late as 1702 an edition of Oughtred's Clavis (in English) was published by Halley, with a laudatory preface, and 'recommended by Mr. E. Halley, Fellow of the Royal Society,' conspicuous in the title-page. And this is the case in favour of our third point.

We now proceed to the fourth point. The method of approximation given by Newton was first published in Wallis's Algebra, cap. 94, and applied to the case of $y^3 - 2y = 5$, which is to this day the example commonly given of this method. Wallis himself does not give any particular approbation to it, but only states that it "is very different from that of Vieta, Oughtred, and Harriot, which is commonly received." After this period English writings begin to give Newton's method to the exclusion of that which preceded. The reason why the Newtonian approximation is less laborious than its predecessor, arises from the earlier method containing no provision for making the steps of one part of the process facilitate those which succeed. The manipulation of algebraical quantities was in a very rude state in the hands of Oughtred; and the reader will detect in the preceding example such a process as $a\,b + a\,c$, instead of $a\,(b+c)$. That the preceding method declined, and that Newton, and subsequently others, supplied its place by *new* processes, is very well known to every reader of mathematical history. But the old method was not so entirely forgotten that it could never come to the knowledge of a modern reader: for example, in Barlow's Mathematical Tables (1814), page xxiii., the whole history of the problem is summed up (from Lagrange) in 12 lines, as clearly and correctly as it could possibly be done in that space; with the exception only that Pell is mentioned where Wallis should have been written, which probably arises from Wallis's improvement being introduced in that part of his history of algebra in which he treats of Pell's writings. No more is necessary on the fourth point.

We do not mention the method of Raphson, which is not properly evolution, though it throws the Newtonian approximation into a form somewhat resembling it. If this method were made to proceed by single figures of the root, it would much resemble the method of Mr. Atkinson, presently to be noticed; and this was actually proposed by one Richard Sault,* between 1690 and 1694.

We should incline to think that the "numerose Exegesis" did not make much way on the Continent, compared with the progress it made in this country, where mathematical writers have always been greatly attached to numerical applications. In the course of mathematics of Dechâles† (published in 1690) we find Vieta's form applied only to the most trivial examples.

The fifth point is matter of notoriety; and as Mr. Horner is the first who *published* even so much as a revival of the method as Wallis left it, he certainly has the claim of discovery for the additions, whether there be independent inventors or not.

There are two stages between Wallis and Horner, for the method of the former will allow of some reduction by the use of these processes, which are now familiar to every algebraist. But a person who revives an old method, merely changing its rude form into one of a more simple character, to which he has been led by the use of the same methods of simplification in other cases, has no claim to the title of a discoverer. Suppose, for instance, a person should revive a problem of Vieta, using the binomial theorem instead of the long process of multiplication, he would certainly not be held to have been the independent author of an improved method. In looking back, therefore, with the present state of the problem in our hands, to its former history, we must ask, what new thing is it, to the invention and application of which we owe the present state aforesaid?

We now come to the sixth‡ point. In or after the month of

* In a work called 'A New Treatise of Algebra, with a converging Series for all manner of adfected Equations.' It has no date, but it refers to Raphson (whose first edition was 1690), and an old possessor's autograph (in our copy) has the date 1694. But so evidently does this reduction of Raphson's method (so called) to the determination of figure after figure bring the process back to the old 'numerose Exegesis,' that the possessor above mentioned (or some other person of his time) has inserted the single word 'numerose' in the title-page, opposite to the author's announcement about converging series.

† But this work is behind its time in many respects. Its readers were introduced to Newton only by the information that Isaac *Nouton*, Professor of Mathematics at Cambridge, had edited the Geography of Varenius.

‡ The number of inventors, as well as that of persons supposing themselves to be inventors, may be larger than that which we have given. Sherburne mentions one Milbourne, who solved an equation of the fifth degree approximately, before he had

June, 1820, a tract was published by Mr. Theophilus Holdred, entitled, 'A new method of solving Equations,' &c., with a supplement. These two must be discussed as if they were two separate works; the first being the old method in a modern form, and with no more difference than the application of the usual modern forms and the use of Newton's divisor instead of Vieta's; the second being neither more nor less than the method which Mr. Horner had put in print more than six months before.

We take an example from the *work;* observing that Mr. Holdred was acquainted with Harriot's method, which he mentions in the preface.

EXAMPLE 6.

Let the equation $x^4 + 5x^3 + 7x^2 + 3x = 91672020000$ be proposed.

By comparing this with the general equation, we have $h = 5$, $i = 7$, and $k = 3$.

Make $r = 500$.

$r^4 =$	62500000000	$4r^3 =$	500000000
$hr^3 =$	625000000	$3hr^2 =$	3750000
$ir^2 =$	1750000	$2ir =$	7000
$kr =$	1500	$k =$	3
A =	63126751500	B =	503757003
		${}^2B = B + 2Ca + 3Da^2 + 4a^3$	

$6r^2 =$	1500000	$4r =$	2000
$3hr =$	7500	$h =$	5
$i =$	7	D =	2005
C =	1507507	${}_2D = D + 4a$	
${}_2C = C + 3Da + 6a^2$			

$$N = 91672020000(500 =$$
$$A = 63126751500$$

$$B = 503757003) \qquad 28545268500(\ 40 = a$$
$$C = 1507507 : \times 40 = 60300280 = Ca$$
$$D = 2005,\ 3Da = 240600 \qquad 3208000 = Da^2$$
$$Da = 80200,\ 4a = 160,\ 6a^2 = 9600 \qquad 64000 = a^3$$
$$a^2 = 1600 \qquad \text{Divisor } 567329283 : \times 40 = 22693171320 \text{ subtrahend}$$

$${}_2B = 634237563 \qquad 5852097180(9 = b$$
$${}^2C = 1757707 : \times 9 = 15819363 = {}_2C\,b$$
$${}^2D = 2165 \qquad 175365 = {}_2D\,b^2$$
$${}_2D\,b = 19485 \qquad 729 = b^3$$
$$a^2 = 81 \qquad \text{Divisor } 650233020 : \times 9 = 5852097180 \text{ subtrahend}$$
$$0$$

In the preceding method, the calculation of ${}_2B$, ${}_2C$, and ${}_2D$, which forms no small item of the whole, is omitted, and the results only are stated. If this were done and all the operation were

seen Harriot's work. (*See* Penny Cyclopædia, article HORROCKS.) And in our day we have seen a treatise on Algebra, of the most moderate claims in other respects, the author of which claims a general method of solving equations, which is no other than the Newtonian method in its most usual form.

stated in Harriot's way, it would be found to differ in nothing from the old method, except in the employment of the Newtonian divisor. Mr. Holdred knew of that method, since he mentions it, for the purpose of depreciation, in his preface: and his own method has the mark of Harriot's fingers strongly impressed upon it. The reader who looks back to the preceding examples will see the same redundancy of ciphers in both, distinguishing both from the examples of Vieta and Oughtred.

We now come to the Supplement, premising that, though we began this article under the impression that Mr. Holdred was an independent inventor of the old method, we are now convinced of the contrary. The Supplement is a curious production: Mr. Holdred, having been, as he states, for 40 years in possession of his method, was about to publish it, when a subscriber (not named) sent him an equation to solve. An accidental mistake led him, after some weeks of delay, to the method of his Supplement, which caused a corresponding delay in the publication. This lucky accident happened precisely in that six months of the 40 years during which Mr. Horner's improvement was in print. All this may have been; but it is needless to say that, when the claims of another person are in question, such evidence must be rejected, particularly as coming from a person who had already called Harriot's method his own; nor can we allow ourselves to infer from it that Mr. Holdred was an independent inventor of the method, or of any part of it, new or old.

The next claimant is Mr. Peter Nicholson, who, as far as we can see, has intended to treat his predecessors with all fairness, but who is evidently under a most singularly erroneous impression, namely, that he was the first publisher of Mr. Horner's simplification. Mr. Holdred communicated what he called his methods to Mr. Nicholson, who states that he suggested some improvements, which Mr. Holdred declined to publish, unless * he were allowed to pass them as his own. On the appearance of Mr. Horner's paper, Mr. Nicholson published his Essay on Involution and Evolution, which also appeared in 1820 (date of preface May 1). Mr. Nicholson says of Horner's paper, "I perceived, however, that the paper contained the substance of what I had previously† written and published, and that, in addition thereto, he had shown the means of performing the operation without using the figurate factors." (Preface, p. vii.) And again (page 24), "I cannot help observing the singular coincidence of the rule which I have given in my Combinatorial Essays for the decomposition of algebraic products with the non-figurate method of extracting the roots of equations." The *non-figurate* method means Mr. Horner's rule.

So far Mr. Nicholson's assertion is rather vague: we have searched the Combinatorial Essays without finding anything which more resembles Mr. Horner's process than would any other in which such a succession of processes occurs as $A=P+Qa$,

* We do not know that this statement was ever answered.

† Where? Where? Where?

B=R+Aa, C=S+Ba, &c. And here we should be obliged to leave this claim, were it not for a more precise repetition of the statement which occurs in 'A practical System of Algebra, &c., by Peter Nicholson and J. Rowbotham,' London, 1837 (third edition). In a note to page 196 the following statement is made with reference to Mr. Horner's rule.

"The very same process was published in Mr. Nicholson's Increments, in the year 1817, and was derived from the doctrine of combinations. Examples may be seen in the introduction, pages xxxv. and xxxviii., and in the body of the work, pages 4 and 38. Page 126 contains an improvement upon the former, and is the same as that now used. The author did not then think of applying these rules to finding the roots of equations, though they were far superior to the methods discovered, three years afterwards, by Messrs. Holdred and Horner, in point of brevity, facility, and perspicuity."

The precision of this paragraph is good evidence of the fairness of its intention; but the last sentence would alone settle the question. Mr. Nicholson had the means in his hands, but did not *then* think of applying them. Did he do so *afterwards?* His own words, previously quoted, show that the "non-figurate method" was first revealed to him by Mr. Horner's paper. The whole of this part of the claim is unintelligible; and we pass to the more specific portion, which is nothing more than a total mistake.

Is it true that the "very same process" was thus published? We have examined the pages alluded to, and can find there an ingenious process for forming the co-efficients of $(x+a)(x+b)\ldots$; but not the rule in question. However, since it is asserted that "p, 126" contains an "improvement," "the same as that now used," we will cite here the process contained in it entire.

"In order to show the superiority of this method I shall again give another solution of example xiv., page 70.

"To find the sum of the series—

$$1^3_{\cdot}\,2^3 + 3^3_{\cdot}\,4^3 + 5^3_{\cdot}\,6^3 + \ldots\ldots\ldots(2x-1)^3 \times (2x)^3.$$

Let $z = 2x - 1$; then will $z^3(z+1)^3 = (2x-1)^3 \times (2x)^3$

A	B		Method of Decomposition.			
2	0,0,1,1,1	− 2	2 ·2 = 4	4·1 = 4	4·1 = 4	4·1 = 4
4	0,1,1,1	− 4	6 ·3 = 18	22·3 = 66	70·3 = 210	214 *aff.*
6	1,1,1	− 5	11·5 = 55	77·5 = 385	455 *neg.*	
8	1,1	− 7	18·7 = 126	203 *aff.*		
10	1	− 9	27 *neg.*			

$$\therefore z^3(z+1)^3 = z^{6|2} - 27\,z^{5|2} + 203\,z^{4|2} - 455\,z^{3|2} + 214\,z^{2|2} - 4\,z^{1|2}\text{"}$$

The preceding then is a very good process for turning a product of the form $(x+a)(x+b)\ldots$. where a, b, &c., are not in arithmetical progression, into the sum of other products in which the substitutes for a, b, &c., *are* in arithmetical progression. But Mr. Horner's rule is for the purpose of finding the *development* of $(x+a+k)(x+b+k)\ldots$. by means of the *development* of

$(x+a)$ $(x+b)$.... The two rules* have different objects, and different modes of proceeding; different things done in different ways: how can two rules differ more?

We now come to the assertion that Mr. Nicholson's methods are superior in "brevity, facility, and perspicuity." It is to be remembered that an arithmetical operation is not to be judged by the space it occupies in a printed book; nor is it wanted to save paper, but time and risk of error. If then the writing down of two or three figures be avoided by a process which will involve a moment's thought, it may very easily happen that there is nothing gained by the omission.

The following instance is the most complicated, we believe, which occurs in the elementary treatise from which our citation is made:—

$$x^4 + 5x^3 - 3x^2 + 15x - 25 = 0$$

5	− 3	15+····	−25 (1·2165
6 ×1=.... 6	+ 3 ×1=	3·······	+18×1=
7 ×1=.... 7	10 ×1=	10	+18
8 ×1= ... 8	1800	28000+····	− 70000 (·2
92 ×2= ..184	1984 ×2=	3968+····	31968×2=
94 ×2= ..188	2172 ×2=	4344	63936
96 ×2= ..192	236400	36312000+····	− 60640000 (1
981×1= ..981	237381×1=	237381+····	36549381×1=
982×1= ..982	238363×1=	238363	36549381
983×1= 983	239346	36787744+····	− 24090619 (6
9846 59076	2399 ×6=	14396 +····	3693170×6=
9852 59112	2405 ×6=	14432	22159020
	2411	3707602	− 1931599(5208

The process, by the method we recommend, is as follows:—

First Coefficient 1.

5	−3	+15	25 (1·216520810
6	6	3	18
7			
8	3	18	70000
90	7	10	63936
92			
94	10	28000	60640000
96	8	3968	36549381
980	1800	31968	24090619
981	184	4344	22159008
982			
983	1984	36312000	1931611
984	188	237381	1854405
	2172	36549381	77206
	192	238363	74200
	236400	36787744	3006
	981	14394	2968

* We have not explained the details of Mr. Nicholson's process; for it either is or is not the "very same" as that now used: and, in the first case, explanation would be useless, and in the second irrelevant.

237381	3693168	38
982	· 14430	37
238363	3707598	1
983	121	
239346	370881	
6	121	
2399	371002	
6		
2405		
6		
2411		

On comparing the preceding processes, we see that Mr. Nicholson has attained what he states, in a subsequent note, to have been his object, namely, the placing of corresponding processes in the same horizontal line, by sacrificing a much more important element of arithmetical accuracy. Numbers which are to be added together should always be written under one another; whereas, in Mr. Nicholson's form, this is very frequently not the case. We think that the brevity of his method can profit no one but a very expert arithmetician indeed; that its facility is considerably below that of the form given by Mr. Horner; and that, in point of perspicuity, it is hardly as good. But the only way of deciding such a question is by trial; and this the reader must do for himself. The introduction of the ciphers is a decided improvement.

The next claim is contained in a posthumous tract of the late Mr. Henry Atkinson of Newcastle, entitled, 'A new Method of extracting the Roots of Equations; read at the Literary and Philosophical Society of Newcastle-upon-Tyne, August 1, 1809.' Newcastle, 1831. Mr. Atkinson appears to have been wholly unacquainted with the writings of the older algebraists, and his method is that of Vieta, somewhat like the form in which Wallis left it; but with the use of the Newtonian divisor, and the use of Newton's approximation at the conclusion. There is but one wrong assumption pervading Mr. Atkinson's assertion of his claim,—namely, that the whole of the method used by Mr. Horner was supposed to be Mr. Horner's property. We suspect that Mr. Horner himself was under the same impression. He (Horner) says, "The whole process may be defined, *division by a variable divisor*. For an accurate illustration of this idea, as discoverable in the existing practice of arithmeticians, we cannot, however, refer to the mode of extracting any root except that of the square; and to this only in its most recently improved state."* Had the

* We do not quite see the meaning of this. The present method of extracting the square root is as old as European algebra; at least we find it, precisely as performed at present, in P. A. Cataldi's 'Trattato del modo brevissimo di trovare la radice quadra,' &c. Bologna, 1613.

In the fourth volume of Leybourn's Repository will be found a paper by Mr. Horner "on the popular methods of approximation," written a little while after his paper in the Philosophical Transactions. In this paper he notices Newton, Halley, Raphson, De Lagny; but no one of older date.

author known Oughtred's, &c., writings, he would here, most probably, have referred to them.

Acting upon this wrong notion, Mr. Atkinson claims as much of the whole method as would amount to adding the use of Newton's divisor to the ancient process. But he abandons all pretension to the "non-figurate method," which he terms a "capital improvement." It would not have taken an hour to show Mr. Atkinson, whose view seems to have been perfectly fair, that though, up to a certain point, he had all the merit, he had not the priority, which is necessary to a claim of discovery, except in the points presently noticed.

We now give an example from Mr. Atkinson's work, in which it will be seen that the steps which fill up nearly the whole of Oughtred's process are omitted; that is, the results only are stated.

$$2\,x^3 + 4\,x^2 - 10 = 60000.$$

$$\text{N} = 60000(30\cdot4722476 \text{ the value of } x$$

$$\text{A}\,a + \text{B}\,a^2 + \text{C}\,a^3 = 57300$$

		2700·000
$\text{A}+2\,\text{B}\,a+3\,\text{C}\,a^2$	5630	2281·568
$\{\text{B}+3\,\text{C}\,(a+b)\}c$	73·6	
$\text{C}\,c^2$	·32	418·432000
		405·385246
	5703·92	
	73·6	13·046754000
·32×2	·64	11·609318096
	5778·16	1·437435904
$\{\text{B}+3\,\text{C}\,(a+b+c)\}d$	13·048	1·16101
$\text{C}\,d^2$	·0098	
		·27642
	5791·2178	·23220
	13·048	
	·0196	4422
		4064
	5804·2854	
$\{\text{B}+3\,\text{C}\,(a+b+c+d)\}e$	·37364	358
$\text{C}e^2$	8	348
	5804·659048	10
	·37364	
	16	
	5805·032704	

Mr. Atkinson, besides applying Newton's divisor, was the first who suggested the idea of making one divisor help in the formation of the next; insomuch that, had Mr. Horner allowed time, this method must have superseded many others. Perhaps, all things considered, it would be fair to say that Mr. Atkinson's method is Newton's approximation rather than Vieta's, digested so as to furnish figure by figure, until the number of decimals obtained can be doubled in the usual manner, for which the process furnishes the materials. He consequently made a real step on the road, which had not time to become known before it was rendered useless by the "capital improvement" of Mr. Horner.

Mr. Atkinson was induced to suppress his method for many years in consequence of an opinion given by Hutton, who, though he declared it to be "new and ingenious," preferred the rule of position. Now, whether Dr. Hutton, who, as we have seen, could not distinguish the methods of Stevinus and Vieta, saw the real point of novelty which Mr. Atkinson's plan presented, we cannot certainly tell; but, on either of two suppositions, such an opinion is surprising as coming from an algebraist and an historian of algebra. If he thought that the mere evolution was new, what are we to say to the annalist of Vieta, Oughtred, Harriot, and Wallis? but if he detected the new form of the Newtonian approximation, how could he contrast it with the rule of double position, the elements of which are much the same, and which is, in fact, much such a predecessor of Newton's approximation as Vieta's method itself?

We have thus, we submit, established the sixth point, and the seventh is the necessary consequence of the preceding six. To Vieta belongs the extension of the process of evolution; to Newton, the addition of the most convenient divisor; and to Horner, the simplification of the process of finding the divisors, which renders what we may call the digested Newtonian approximation preferable to the form in which it was proposed by Newton himself. Of various modifications of Mr. Horner's process which have been proposed we do not much approve, believing that the tendency to error which they introduce far more than counterbalances any advantage which they may have in other respects.

The only historian who has yet noticed this method is Professor Peacock, in his Report to the British Association upon the present state of Analysis, a work which we have found of almost daily use. But, as Professor Peacock had never seen the tract* of Mr. Holdred (as he states in a note), and as Mr. Atkinson's was probably not published when his paragraph on this subject was written, we do not feel it necessary to controvert some one or two sentences to which we might otherwise have taken some exceptions.

One thing must strike every reader who can follow the details of the preceding article—namely, that any person who announces a discovery should, before he commits himself, pay a little attention to the history of the science of which it purports to be an advancement. If any one of the competitors in this matter had published an account of the previous state of the problem, much unnecessary discussion would have been avoided. It is particularly to be regretted that Mr. Horner did not do this; since, in

* It was sold only by Messrs. Davis and Dickson, and by the author, and is very little known.

such a case, it is more than probable that his honours would not have been claimed by others. It is not to be wondered at that the bias of self-love, and the honourable desire of distinction, should lead to the assertion of claims which are really unfounded, in subjects of which the history * is so little written as that of the mathematical sciences. The quarter to which we look for some improvement in this respect is the British Association, the volumes of which have contained some valuable (because minute and precise) additions to the accounts of preceding discoveries. It is to be hoped that it will continue this part of its career; and that it will not, in the excitement of novel research, lose sight of what appeared, at its first formation, to be an object which was particularly set forth as one of its most prominent pursuits. If, by describing what has already been done, it should save persons possessing the talent of investigation (but ignorant of history) from fruitless labour and ultimate discouragement, and should show them the path in which their exertions may really prove useful, it would benefit the branches of knowledge which its members cultivate as much as if their own exertions had been gifted with double energy.

Perhaps no case that ever happened would more strongly illustrate this position than that of Mr. Atkinson, whose whole account of his labours carries with it every possible mark of truth and fairness, so far as his knowledge enabled him to tell what his claims might be. "It was," he says, "during the summer of 1801 that the first idea of the method occurred to me; and, so impatient was I to know the result to which it would lead, that I actually sat down by the side of the road, on which I was then walking, and, upon some waste paper that happened to be in my pocket, worked out the general formula, so far as to satisfy myself that it would answer. Being then but a young mathematician, in a remote part of the country, without connexions, and unaccustomed to write, the idea of publishing my discovery never entered my mind." Some years afterwards Mr. Atkinson presented his memoir to the Philosophical Society of Newcastle, but he does not appear, to the day of his death, to have met with one person † who could suggest a comparison of his method with that of Oughtred or Wallis, were it only to see if they differed or not.

Since the removal of the injurious and unprofitable tax ‡ on almanacs, two well-known publications, the Ladies' and Gentle-

* We speak particularly of our own country; but the whole of the scientific community has reason to wish for a complete history of mathematics, done in the manner of Professor Peacock's report.

† A friend, who knew Mr. Atkinson well, informs us that he may "state with certainty that, at the time he (Mr. Atkinson) wrote the paper, he knew nothing of the works of Harriot, Oughtred, and the other elder mathematicians, except what might be gathered from the memoirs of their lives in Hutton's Dictionary, or from the Encyclopædia Britannica. The chief mathematical works to which he had previously had access were those of Emerson and Simpson."

‡ It now needs no very sanguine mind to see that the great incubus of all, the tax on communication between persons who do not live in the same place, will be reduced to its fair proportions; and the time must come when the public, beginning to see the benefits of the removal of a restriction which is all but a stamp-duty on speech, will wonder at the stupid apathy with which it allowed itself to be muzzled from year to year.

D

man's Diaries, which have for more than a century been a medium of communication between the members of a large class of mathematical students, have commenced series of articles which cannot fail, if properly continued, to do much towards awakening the spirit of historical inquiry. In the Ladies' Diary for 1838 is a reprint of Horner's paper.

In diverging to the apparently irrelevant subject of the restrictions laid upon communication by heavy taxes, we do not consider that we have lapsed into politics. The case before us will distinctly show how much even the abstract and speculative sciences may have suffered from the enormous imposts which have prevented the communication of thought. The precision with which the facts of mathematical history can be recorded may enable us to point out, more specifically than the nature of other subjects will allow, how much of industry is thrown away, and of talent rendered unavailing, by unnatural obstacles placed in the way of information and discussion. Who can tell the extent to which this has gone, whether in matters of science and literature, or in the progress of the arts of life?

We subjoin a list of all the works, of which we know anything, in which any account of Horner's method occurs:—1. Mr. Horner's paper in the Phil. Trans. for 1819, reprinted, as stated before, in the Ladies' Diary for 1838. 2. Nicholson on Involution and Evolution, London, 1820. 3. Holdred on the Solution of Equations, London, 1820. 4. Young's Algebra, London, 1826. 5. Young's* Solution of Equations, London, 1835. 6. Nicholson's and Rowbotham's Algebra, London, 1837. 7. The Penny Cyclopædia, article INVOLUTION and EVOLUTION.†

Example of the complete method proposed by Mr. Horner, applied to the equation solved by Wallis‡ (hereinbefore mentioned), and to finding the same number of places of the root:—

$$x^4 - 80\,x^3 + 1998\,x^2 - 14937\,x + 5000 = 0.$$

It is supposed to be found that one positive root lies between 12 and 13.

* This work contains a more complete account of the actual state of the theory of equations than any English work which we can name.

† It is proposed in that article to annex ciphers (as had been done by Mr. Nicholson) in the proper parts of the process, after the manner adopted in the extractions of the square and cube roots. If the beginner will try this method, rejecting all the abbreviations by which it has been attempted to be simplified, he will soon be in a condition to use it with much more facility (whether he afterwards abbreviate or not) than if he had commenced with one of the condensed forms. But, to avoid the appearance of producing two testimonies, where there is only one, the author states that he was also the author of the article cited. He was not then aware that Mr. Nicholson or any one else had proposed the use of ciphers: had he known that this belonged to another person, he would have stated it, adding his opinion that it is the step which completes the analogy of Horner's process with that of common evolution.

‡ This is the equation resulting from the celebrated problem proposed by Dr. Pell to Colonel Silas Titus, and by him to Dr. Wallis, who seems to have considered it as a sort of challenge from Pell, and has discussed it in a most masterly manner. A great deal may be found on the subject of this very equation in the Scriptores Logarithmici and other publications of Baron Maseres.

		First co-efficient, 1.	
−80	+1998	−14937	−5000(12·7564417944430744
−68	− 816	+14184	−9036
−56			
−44	+1182	− 753	+40360000
−320	− 672	+ 6120	37373441
−313	+ 510	+ 5367000	29865590000
−306	− 528	− 27937	26459730625
−299			
−2920	−1800	+ 5339063	34058593750000
	−2191	− 42931	31723000237296
−2915			
−2910	−3991	+ 5296132000	2335593512704
−2905	−2142	− 4185875	2114644002200
−29000	−6133	+ 5291946125	220949510504
	−2093	− 4258625	211462866012
−28994			
−28988	−822600	+ 5287687500000	9486644492
−28982	− 14575	− 520793784	5286568076
−28976			
	−837175	+ 5287166706216	4200076416
	− 14550	− 521837352	3700597551
	−851725	+ 5286644868864	499478365
	− 14525	− 3486336	475791108
	−86625000	+ 528661000550	23687757
	− 173964	− 3486800	21146271
	−86798964	+ 528657513750	2541486
	− 173928	− 34872	2114627
	−86972892	+ 52865716503	426859
	− 173892	− 34872	422925
	−87146784	+ 52865681631	3934
	− 116	− 87	3700
	−871584	+ 5286568076	234
	− 116	− 87	211
	−871700	+ 5286567989	23
	− 116	− 6	21
	−871816	+ 528656793	2
		− 6	
		+ 528656787	

The whole of the last column is to be found in different parts of Wallis's process; but the divisors in the preceding column differ until the middle of the process. Wallis's divisors are as follow—

5364881 instead of 5367000
5295306481 " 5296132000
5287600846001 " 5287687500000;

and they afterwards are only abbreviations of those exhibited above. Surely it is now hardly necessary to say that, in *principle*, the

D 2

method of Horner, so called, has been known two hundred years; this process of Wallis being in fact that of Vieta (who died in 1603). But when we come to speak of the *practicability* of the method, we have no longer any right to say that Horner had a predecessor. The preceding process is complete: it omits no process except such as are usually omitted in common multiplication and division, and the number of figures is under 730. The process in Wallis has more than 1100 figures *written down;* and, if all those additions were made which would render it as complete as the preceding, we imagine that the 1100 figures would be at least trebled in number.

If any other proposed method, among those which have come to our knowledge, were applied to the preceding example, *with every necessary figure put down*, we repeat it would be found that the number of their figures would not very materially fall short of those required in the process of Wallis (made complete). At any rate, the following challenge is fair: let any method, other than that of Horner, and completely exhibited, cost less than 1500 figures, and we are ready to admit that method as being a greater improvement upon the ancient process than we had imagined to exist independently of the one to which we have given the preference.

Since the preceding pages were written, we have found the following in Halley's paper on the solution of equations, in the Philosophical Transactions:—"That great discoverer and restorer of the modern Algebra, Francis Vieta, about 100 years since, showed a general method for extracting the roots of any equations, which he published under the title of 'A numerical Resolution of Powers,' &c. Harriot, Oughtred, and others, as well of our own country as foreigners, ought to acknowledge whatever they have written upon this subject as taken from Vieta." This paper of Halley's was republished both in the 'Miscellanea Curiosa' and in Dr. Harris's 'Lexicon Technicum.' It would almost seem as if the method of solving equations were fated to give rise to more unfounded claims of invention than any other. In the same Lexicon Technicum, edited by a person exceedingly well versed in mathematics, and whose work is still unsurpassed as an exhibition of the state of science in a particular country at a particular time, we find the Newtonian method, with no more alteration than writing $(a\,\varphi' a - \varphi a) \div \varphi' a$, instead of $a-(\varphi a \div \varphi' a)$, proposed as a novelty, and attributed to a Mr. Wastel of the Navy Office, who published an edition of a work called 'Parson's Arithmetic.'

COMPANION TO THE ALMANAC

FOR

1840.

PART I.

INFORMATION CONNECTED WITH THE CALENDAR AND THE NATURAL PHENOMENA OF THE YEAR; AND WITH NATURAL HISTORY AND PUBLIC HEALTH.

I.—ON THE CALCULATION OF SINGLE LIFE CONTINGENCIES.

It is the object of the present article to put together a number of formulæ which it may be useful to the actuary to find in one place. At the same time, it may show all persons who possess an elementary knowledge of algebra, that they may, with no great amount of tables, and processes of very easy application, learn to compute the value of any benefit in which the duration of one life only is concerned. The same principles, with more extensive tables, apply to cases in which two or more lives are involved.

A sketch of the history of the subject will be found in the Library of Useful Knowledge, Treatise on *Probability;* and more fully in Mr. Milne's articles on *Mortality* and *Annuities* in the new edition of the Encyclopædia Britannica. The articles *Annuity*, *Interest*, *Mortality*, and *Reversions* (when the latter appears) in the Penny Cyclopædia, may also be consulted.

About thirty years ago, a Mr. George Barrett presented to the Royal Society a method by which the calculation of life contingencies was very materially facilitated. This method the society did not think worthy of publication; and it was accordingly given to the world by Mr. Francis Baily, in the appendix to his well-known work on Annuities, with some severe remarks on the omission just alluded to. It was certainly an unfortunate want either of examination or of judgment which caused the Philosophical Transactions, the depositary of the writings of many eminent inquirers on this particular subject, to miss a contribution which would have done honour to any one of them. This method of Mr. Barrett was rendered still more commodious, and we believe extended, by Mr. Griffith Davies, in his tables of life contingencies (1825), a work now unfortunately out of print: it is Mr. Barrett's method, as improved by Mr. Davies, which we propose to present to the reader, with some extension of notation and generalization of processes.

Let it be the law of mortality that of a_0 persons born alive, a_1 are alive at the end of a year, a_2 at the end of two years, a_x at the end of x years. Let v be the present value of £1, to be received at the end of a year, which depends entirely on the interest of money; if r be the interest of £1 for one year, we have

$$v=\frac{1}{1+r}:$$

thus, at 3 per cent., $r=\cdot 03$ and $v=\cdot 970874$. The present values of £1 to be received at the end of 2, 3,....x years are v^2, v^3,.... v^x. The best way to find these powers will be to take the logarithms* from a larger table, such as that in the Penny Cyclopædia, article Interest. Thus the logarithm of $1+r$ being

$$\cdot 0128372247\text{, that of } v \text{ is } 9\cdot 9871627753-10$$

Multiples of this logarithm being formed up to 104 times, we have all the logarithms which will be wanted, in the use of the Carlisle Table.

Persons not used to computation should remember that the easiest way of forming a set of multiples is to write the quantity to be added each time at the bottom of a card, and to make each addition by holding the card so that the writing on it may stand over the last result. In this way it will not take many minutes to form a hundred multiples of the preceding, and a verification of the last multiple should be made by actual multiplication.

Obtain as many logarithms as are wanted of the powers of v in the preceding manner, allowing as much space between the lines as will† contain four lines of figures. Take the table of mortality which is to be used, say the Carlisle table, and under the logarithm of v write that of a_1, the number surviving a year; under that of v^2, write the logarithm of a_2; and so on up to the end of life. The Carlisle table will be found in Mr. Milne's work on Annuities; it is also in the Penny Cyclopædia, article‡ *Mortality*. Under the last logarithms, write in succession the logarithms of a_0-a_1, a_1-a_2, &c., the numbers who die in the first, second, &c. years: as follows, in which 3 per cent. is supposed. Five decimal places are taken, merely as an example; but it will be almost as easy to use seven, as there are no interpolations.

	10000			
1		$\log v$	$=9\cdot 98716-10$	
2	8461	$\log a_1$	$=3\cdot 92742$	
3	1539	$\log(a_0-a_1)$	$=3\cdot 18724$	
4			$3\cdot 91458$	$a_1 v=8214\cdot 5$
5			$3\cdot 17440$	$(a_0-a_1)v=1494\cdot 2$
1		$\log v^2$	$=9\cdot 97433-10$	
2	7779	$\log a_2$	$=3\cdot 89092$	
3	682	$\log(a_1-a_2)$	$=2\cdot 83378$	
4			$3\cdot 86525$	$a_2 v^2=7332\cdot 5$
5			$2\cdot 80811$	$(a_1-a_2)v^2=642.85$
	&c.	&c.	&c.	&c. &c.

* We have re-examined this table, and find no error, by Hutton's Tables, p. 386.]
† It will be desirable to have the spaces equal.
‡ We have re-examined this reprint, and find it correct.

Those who cannot easily add one line to another which is separated by a third should now cut pieces out of a card in such manner that being laid upon one of the compartments of the table, the parts cut open will show the first and third line, and the rest hide the second and fourth. The fourth line is then formed by adding the first and second, and the fifth line by adding the first and third, covering the second and fourth. We thus obtain two successions of results.

$$a_0 \qquad a_1 v \qquad a_2 v^2 \qquad a_3 v^3, \qquad \&c.$$

$$(a_0 - a_1) v \qquad (a_1 - a_2) v^2 \qquad (a_2 - a_3) v^3 \qquad (a_3 - a_4) v^4, \qquad \&c.$$

Let these be denominated D_0, D_1, D_2, &c., and C_0, C_1, C_2, &c., so that

$$D_0 = a_0, \qquad D_1 = a_1 v, \ldots\ldots \ldots D_x = a_x v^x$$

$$C_0 = (a_0 - a_1) v, \quad C_1 = (a_1 - a_2) v^2, \ldots\ldots C_x = (a_x - a_{x+1}) v^{x+1}.$$

The following table is then to be completed in the manner which will be described.

Age.	D_x	N_x	S_x	C_x	M_x	R_x	Age.
0	D_0	N_0	S_0	C_0	M_0	R_0	0
1	D_1	N_1	S_1	C_1	M_1	R_1	1
2	D_2	N_2	S_2	C_2	M_2	R_2	2
3	D_3	N_3	S_3	C_3	M_3	R_3	3
&c.	&c.	&c.	&c.	&c.	&c.	&c.	&c.

The columns D and C have been described; the rest are formed from them as follows. The last of the column N is nothing, and N_x is always the sum of those in column D beginning with D_{x+1}, and continuing to the end: thus any one D and its N added together give the preceding N. Or

$$N_x = D_{x+1} + D_{x+2} + D_{x+3} + \ldots\text{(to the end)}$$

$$N_x = D_{x+1} + N_{x+1}.$$

The last S is nothing, and the column S is formed from N as N was formed from D in every point except this, that each S begins with its own N instead of the one after: thus

$$S_x = N_x + N_{x+1} + N_{x+2} + \ldots\text{(to the end)}$$

$$S_x = N_x + S_{x+1}.$$

The last M is the last C, and M is formed from C precisely as S from N: thus

$$M_x = C_x + C_{x+1} + C_{x+2} + \ldots\text{(to the end)}$$

$$M_x = C_x + M_{x+1}.$$

Lastly, R is formed from M as M from C: thus

$$R_x = M_x + M_{x+1} + M_{x+2} + \ldots\text{(to the end)}$$

$$R_x = M_x + R_{x+1}.$$

We here give, as a specimen, the first five, the last five, and an intermediate five years, from the Carlisle table at 3 per cent., keeping only four significant figures.

Age.	D.	N.	S.	C.	M.	R.	Age.	No. Living.	Dying.
0	10000	173200	3702000	1494·0	4664	70040	0	10000	1539
1	8215	165000	3529000	642·9	3170	65372	1	8461	682
2	7332	157700	3364000	462·1	2527	62202	2	7779	505
3	6657	151000	3206000	245·2	2065	59674	3	7274	276
4	6218	144800	3055000	173·4	1820	57610	4	6998	
⋮	⋮	⋮	⋮	⋮	⋮	⋮	⋮	⋮	⋮
30	2324	45460	732100	22·80	932·7	25070	30	5642	57
31	2234	43220	686600	22·14	909·9	24140	31	5585	57
32	2147	41080	643400	21·11	887·8	23230	32	5528	56
33	2063	39020	602300	20·13	866·6	22340	33	5472	55
34	1983	37030	563300	19·55	846·5	21470	34	5417	
⋮	⋮	⋮	⋮	⋮	⋮	⋮	⋮	⋮	⋮
100	·4683	·7879	1·4580	·10100	·43170	1·17700	100	9	2
101	·3536	·4343	·6696	·09809	·33070	·74550	101	7	2
102	·2452	·1891	·2353	·09524	·23260	·41480	102	5	2
103	·1429	·0462	·0462	·09246	·13730	·18220	103	3	2
104	·0462	·0000	·0000	·04488	·04488	·04488	104	1	

That the beginner may attach definite ideas to the several columns, we subjoin an explanation of each. The assumption is that of 10000 individuals born alive, the numbers surviving to each age, and dying in each year, are as in the last two columns.

D. £2324 invested at 3 per cent. at the birth of 10,000 persons, will, improved at compound interest, yield every survivor £1 at the age of 30: the number of survivors being 5642.

N. £45460 invested at the birth of 10,000 individuals will produce, by the time they attain the age of 30, enough to guarantee to each person then surviving an annuity of £1 on his life, the first payment being made when he attains the age of 31.

S. £732,100 will in the same case produce enough to guarantee an increasing annuity, paying £1 to each at the age of 31, £2 at that of 32, &c.

C. £22·8 invested at 3 per cent. at the birth of 10,000 persons, will, improved at compound interest until the survivors are 31 years of age, yield £1 for each of those who died between 30 and 31.

M. £932·7 similarly invested, would yield to each one who reaches 30, £1 at the end of the year in which he dies.

R. £25070 similarly invested, would yield to each one who

lives to be 30, £1 if he die in the 31st year, £2 if in the 32nd, and so on.

The preceding list simply enunciates the method of constructing the tables: the following shows the use to which the mere inspection may be put. Take any two ages, say 30 and 34, and transpose the numbers opposite each age to the other age: then, whatever may be the present age (less than 30);—

D. A person might now give up £1983, due at the age of 30, to receive £2324, if he live to be 34.

N. A person might now give up an annuity of £37030, to be granted at the age of 30, to receive in return another of £45460, to be granted at the age of 34, if he should live so long.

S. A person might now give up a uniformly increasing annuity of £563,300 the first year, twice as much the second, &c. to be entered upon at the age of 30, to receive another annuity of the same kind, beginning with £732,100, to be entered upon at the age of 34, if he should live so long.

C. £19·55 secured to a person in the event of his dying between 30 and 31, is now of the same value as £22·80, secured to the same person in the event of his dying between 34 and 35.

M. £846·5 secured to a person at the end of the year in which he dies, if after attaining 30, is of the same value as £932·7 secured to the same person at the same period, if after attaining 34.

R. An increasing assurance, to be £21470, if a person die in his 31st year, twice as much if in his 32nd, &c., is now of the same value as another, to be £25070 if he die in his 35th year, twice as much if in his 36th, &c.

These properties are independent of the present age of the party, and show that the most simple indication of the tables is the proportion in which a benefit due at one age ought to be changed, so as to retain the same value and be due at another age. They might, therefore, with great propriety be called *commutation tables.*

The following formulæ and additional notation will be found useful.

I. $N_x = D_{x+1} + D_{x+2} + D_{x+3} + \ldots$ to the end of life.

$M_x = C_x + C_{x+1} + C_{x+2} + \ldots \quad . \quad . \quad . \quad . \quad .$

$S_x = N_x + N_{x+1} + N_{x+2} + \ldots = D_{x+1} + 2D_{x+2} + 3D_{x+3} + \ldots$

$R_x = M_x + M_{x+1} + M_{x+2} + \ldots = C_x + 2C_{x+1} + 3C_{x+2} + \ldots$

II. $N_x - N_{x+1} = D_{x+1}$ $\qquad S_x - S_{x+1} = N_x$

$M_x - M_{x+1} = C_x$ $\qquad R_x - R_{x+1} = M_x$

III. Let $N_{x,y} = N_x - N_{x+y}$ $\quad\Big|\quad$ $S_{x,y} = S_x - S_{x+y} - yN_{x+y-1}$

$M_{x,y} = M - M_{x+y}$ $\quad\Big|\quad$ $R_{x,y} = R_x - R_{x+y} - yM_{x+y-1}$

B 3

IV. Then $N_{x,y} = D_{x+1} + D_{x+2} + \ldots + D_{x+y} \quad \ldots (y \text{ terms})$

$$M_{x,y} = C_x + C_{x+1} + \ldots + C_{x+y-1} \quad \ldots (y \text{ terms})$$

$$S_{x,y} = D_{x+1} + 2D_{x+2} + \ldots + (y-1)\, D_{x+y-1} \ldots (y-1 \text{ terms})$$

$$R_{x,y} = C_x + 2C_{x+1} + \ldots + (y-1)\, C_{x+y-2} \ldots (y-1 \text{ terms}).$$

V.

$$D_x + \ldots + D_y = N_{x-1,\, y-x+1}$$

$$C_x + \ldots + C_y = M_{x,\, y-x+1}$$

$$D_x + \ldots + (y-x+1)\, D_y = S_{x-1,\, y-x+2}$$

$$C_x + \ldots + (y-x+1)\, C_y = R_{x,\, y-x+2}$$

VI.

$$N_{x,y} + S_{x+1,y} = S_{x,y+1} = S_x - S_{x+y} - yN_{x+y}$$

$$M_{x,y} + R_{x+1,y} = R_{x,y+1} = R_x - R_{x+y} - yM_{x+y}$$

VII.

$$N_{x,y} - \frac{1}{y} S_{x+1,y} = N_x - \frac{1}{y}(S_{x+1} - S_{x+y+1})$$

$$M_{x,y} - \frac{1}{y} R_{x+1,y} = M_x - \frac{1}{y}(R_{x+1} - R_{x+y+1})$$

VIII.

$$C_x = v\, D_x - D_{x+1}, \quad M_x = vN_{x-1} - N_x$$

$$R_x = v\, S_{x-1} - S_x$$

$$M_{x,y} = v\, N_{x-1,y} - N_{x,y}, \quad R_{x,y} = v\, S_{x-1,y} - S_{x,y}$$

The formulæ VIII. will be useful in verifying the tables.

All that will be found of demonstration in the present article is intended for those who are familiar with the subject, being meant to give a method of dealing with the more complicated cases, and particularly a method by which the succeeding formulæ may be verified. This will be followed by a collection of preparations for formulæ which may be easily used even by a person unacquainted with the demonstration.

The present value of £1 to be received by a person now aged x, if he live to attain $x+k$, is $D_{x+k} \div D_x$.

The present value of £1 to be received by the representatives of a person now aged x, if he die between the ages of $x+k$ and $x+k+1$, is $C_{x+k} \div D_x$.

The following problem will include every case we have yet seen proposed of annuities, whether for the whole life, or temporary, or deferred, increasing or decreasing uniformly; and also of insurances: with every manner yet proposed of paying the premium. It matters nothing that it involves payment of premium after the benefits begin to be received; since every application of it will require the part of the premium so paid to be made equal to nothing.

Problem. A person now aged x is to receive s (pounds sterling) if he attain $x+k$. In the n following years he is, if he live, to receive $a, a+h, \ldots a+(n-1)h$ pounds at the end of the successive years, and ever afterwards during his life, t pounds at the end of each year. But if he die during the n years, he is to have A, A+H .... A+$(n-1)$ H, according as he dies in the first, second, &c. year, and T at the end of any subsequent year in which he dies. Besides this, he is, if he die in w years, to have a return of part of the premiums presently described.

For this he is to pay at once, σ, and a premium ϖ, which he is to pay l times; the next m premiums are to be ϖ, $\varpi(1+\mu)+\beta$, $\varpi(1+2\mu)+2\beta, \ldots \varpi(1+\overline{m-1}\,\mu)+(m-1)\beta$; after which the premium is always τ. But if he should die before he attains $x+w$ years, he is to receive at the end of the year of his death $\varrho+\nu\varpi$ if he die in the first year, $\varrho+\theta+2\nu\varpi$ if in the second, and, finally, $\varrho+(w-1)\,\theta+w\nu\varpi$ if in the wth year. Required the equation that must exist among these quantities to make the receipts and payments of equal value.

The first set of receipts has the value of the following expression divided by D_x.

$$aD_{x+k+1}+(a+h)\,D_{x+k+2}+\ldots\ldots+(a+\overline{n-1}\,h)\,D_{x+k+n}$$
$$+AC_{x+k}+(A+H)\,C_{x+k+1}+\ldots\ldots+(A+\overline{n-1}\,h)\,C_{x+k+n-1}$$
$$+sD_{x+k}+t\,(D_{x+k+n+1}+\ldots\ldots)\quad +T\,(C_{x+k+n}+\ldots\ldots),$$

which, by the preceding formulæ, is

$$\left.\begin{matrix} aN_{x+k,n}+hS_{x+k+1,n}+AM_{x+k,n}+HR_{x+k+1,n} \\ +sD_{x+k}+tN_{x+k+n}+TM_{x+k+n} \end{matrix}\right\}\ldots\ldots(A).$$

The balance of the premiums and returns is the following divided by D_x:

$$(\sigma+\varpi)\,D_x+\varpi\,(D_{x+1}+\ldots+D_{x+l-1})+\varpi D_{x+l}+(\varpi\,\overline{1+\mu}+\beta)\,D_{x+l+1}$$
$$+\ldots\ldots+\{\varpi\,(1+\overline{m-1}\,\mu)+\overline{m-1}\,\beta\}\,D_{x+l+m-1}+\tau\,(D_{x+l+m}+\ldots\ldots)$$
$$-(\varrho+\nu\varpi)\,C_x-(\varrho+\theta+2\nu\varpi)\,C_{x+1}-\ldots-(\varrho+\overline{w-1}\,\theta+w\nu\varpi)\,C_{x+w-1},$$

the value of which, by the same formulæ, is

$$\left.\begin{matrix} \sigma D_x+\varpi N_{x-1,l+m}+\varpi\mu S_{x+l,m}+\beta S_{x+l,m}+\tau N_{x+l+m-1} \\ -\varrho M_{x,w}-\theta R_{x+1,w}-\nu\varpi R_{x,w+1} \end{matrix}\right\}\ldots\ldots(B).$$

The equation of the results (A) and (B) gives

$$\left.\begin{matrix} \varpi\{N_{x-1,l+m}+\mu S_{x+l,m}-\nu R_{x,w+1}\} \\ +\sigma D_x+\beta S_{x+l,m}+\tau N_{x+l+m-1} \\ -\rho M_{x,w}-\theta R_{x+1,w} \end{matrix}\right\}=\left\{\begin{matrix} sD_{x+k} \\ +aN_{x+k,n}+hS_{x+k+1,n} \\ +AM_{x+k,n}+HR_{x+k+1,n} \\ +tN_{x+k+n}+TM_{x+k+n}. \end{matrix}\right.$$

This problem contains the circumstances of all which are proposed, and is here introduced that any question may have the result of the common investigation compared with that deduced by considering it as a particular case of the preceding. Most of the letters will be $=0$ in any question which occurs; and the following list will serve to remind the calculator what letters enter into the case before him.

1	k	s	Occurs in questions which comprise a fixed sum at a certain age; an endowment.
2	k, n	a	An annuity, present or deferred, &c.
3	k, n	h	An increasing or decreasing annuity.
4	k, n	A	A fixed assurance of any kind.
5	k, n	H	An increasing or decreasing assurance.
6	k, n	t	An increasing or decreasing annuity, with an end to the increase, &c.
7	k, n	T	An increasing or decreasing assurance with an end to the increase, &c.
8	—	σ	Present value of any kind.
9	l, m, w	ϖ	Fixed premiums of any kind.
10	l, m	μ	Premiums increasing or diminishing by a proportion of the first premium.
11	l, m	β	Premiums increasing or decreasing by a sum independent of the first premium.
12	l, m	τ	Premiums increasing or decreasing with an end to the increase, &c.
13	w	ϱ	Return of a sum in case of death.
14	w	θ	Return of increasing or decreasing absolute sum in case of death.
15	w	ν	Return of increasing or decreasing proportions of the first premium in case of death.

The third column contains the letters indicating benefits or payments, and the second column shows the terms of years with which the benefits, &c. are particularly connected in the general problem. By attention to the conditions, the solution of any case can be readily picked out of the general equation, as in the following instances.

What is the premium to be paid for an insurance of £A on a life aged x, accompanied by a return at death of all the premiums paid?

Here all the letters of the third column vanish except A, ϖ, and ν, and m, n, and w are to be extended beyond the possible term of life, while $l=0$, $k=0$, and $\nu=1$. Again, when the age $x+y$ extends beyond the term of life, $N_{x,y} = N_x$, &c. Consequently, the equation gives

$$\varpi(N_{x-1} - R_x) = AM_x.$$

Again, what is the present value of an assurance of £A, with which the sum paid is to be returned? Here only σ, ϱ, and A have value, and $\sigma=\varrho$, the letters of the first column being as before. Hence

$$\sigma(D_x - M_x) = AM_x.$$

If D_x be ever less than M_x this problem is impossible· but D_x is necessarily greater than M_x, being $C_x(1+r)+C_{x+1}(1+r)^2+\ldots$, while M_x is $C_x+C_{x+1}+\ldots$.

Thirdly, an annuity of £a to commence when a life aged x attains $x+k$ is to be bought by a premium regularly diminishing, so as to be last paid when the annuity begins (that is at $x+k$), and a year before a payment of the annuity is made. Here only ϖ, μ, and a have value, k is given, $l=0$, m is $k+1$, n outruns the term of life, and $w=0$. And μ must be taken negatively, as $-1\div(k+1)$. Hence

$$\varpi\left(N_{x-1\,k+1,}-\frac{1}{k+1}S_{x,\,k+1}\right)=a.N_{x+k},$$

or (VII.) $$\varpi=\frac{aN_{x+k}}{N_{x-1}-(S_x-S_{x+k+1})\div(k+1)}.$$

Before we proceed further, it may increase the interest attached to the formulæ if we remark that the principle of these commutation tables (as we have called them) can be extended from the case of life contingencies to that of interests certain, in such manner that every formula which gives the value of one of the former, may, by going to a different table, be applied to the corresponding one of the latter. That is to say, the mathematical treatment of the hypothesis that a life is to last for ever does not differ from that of a table of *mortality*. We should imagine, that in questions of instalments particularly, increasing or decreasing, such tables would be of very great use.

To construct them, proceed as follows:

$$D_x=v^x,\quad N_x=\frac{v^{x+1}}{1-v},\quad S_x=\frac{v^{x+1}}{(1-v)^2}.$$

The remaining quantities are useless, being always $=0$. $N_{x,y}$ and $S_{x,y}$ may be exhibited as before in the forms

$$N_{x,y}=N_x-N_{x+y},\quad S_{x,y}=S_x-S_{x+y}-yN_{x+y-1}.$$

Suppose, for instance, we take the last question, and require the value of the annuity for the *whole life* (which here means perpetuity) of a after the expiration of k years, to be bought by regularly diminishing instalments one paid now, &c. The formula then becomes

$$\varpi=\frac{aN_k}{N_{-1}-(S_0-S_{k+1})\div\overline{k+1}},$$

and $N_{-1}=\dfrac{1}{1-v}$, $S_0=\dfrac{v}{(1-v)^2}$. If the value of the tabular quantities be restored, the preceding (cleared of fractions) is

$$\frac{a\,(k+1)\,v^{k+1}\,(1-v)}{(k+1)(1-v)-(v-v^{k+2})};$$

the same as would be obtained by common methods. The first is (with tables) as easily calculated as the second by the common tables; or, if any thing, somewhat more easily.

The particular uses of such commutation tables for certain interests would be, 1. That those who can use the life tables more readily than the usual tables of interest (many, perhaps most, actuaries) would at once be able to apply their facility in the former to the new form of the latter. 2. That whenever the formula is given in terms of D_y, N_y, and S_y, it is indifferent at what age the perpetual life of the problem is supposed to begin, so that a repetition of the simple process upon another age verifies the computation.

We now come to the classification of problems, and the presentation of their results. In all cases, one of the quantities in the third column is to be unknown, and found from the equation. Two cases arise. 1. Where all the quantities of the third column are independent of each other. 2. Where one is to be a simple fraction of another. Thus τ, the premium remaining over, or what we might call the *residual* premium, might be required to be $\gamma\varpi$, a given fraction of ϖ; and the problem might as easily be solved if all except, say a, (and the indicators of fractions already existing, as μ, ν) were to be made given fractions or multiples of ϖ.

If $(\nu\varpi)$, in parentheses, be taken as an abbreviation of the phrase 'coefficient of $\nu\varpi$,' &c., we may, for purposes of general consideration, write the equation

$$\varpi\left\{(\varpi) + (\mu\varpi).\mu - (\nu\varpi).\nu\right\} + \&c. = (s).s + (a).a + \&c.$$

the first side entirely depending upon the mode of offering payment, and the second upon the nature and amount of the benefit. Hence it is useless to combine each of the different benefits with all the modes of paying for it; for, as cannot but have been observed by those who have used these tables, a given benefit must be always calculated by the same numerator, whatever the single mode of payment may be; and the payment by the same denominator, whatever the benefit may be. Thus if a simple deferred annuity be bought by uniform premiums, we have

$$(\varpi).\varpi = (a).a, \quad \text{or} \quad \varpi = \frac{(a)}{(\varpi)}.a.$$

But if the single value be paid for the same, we have

$$(\sigma)\,\sigma = (a).a, \quad \text{or} \quad \sigma = \frac{(a)}{(\sigma)}.a.$$

But if the mode of payment be double, say partly by a single value, partly by a succession of uniform premiums, we have

$$(\varpi).\varpi + (\sigma)\,\sigma = (a).a\,;$$

from which ϖ may be given, and σ be found, or vice versâ. If σ is to be a given fraction of ϖ, the *payment* part of the calculation is wholly in the denominator. The following rules will be found useful as preservatives from error.

1. When no part of the benefit is to depend upon the unknown

item of payment,* no function of the benefit can be in the denominator; and the contrary.

2. When all the items of payment are fractions of one among them, no function of the payment can be in the numerator, and the contrary: but when there are parts of the payment not so connected, those which are known are found in the numerator.

3. In all the cases not before specified, the numerator is entirely a function of the benefit and the denominator of the payment.

If we call the two sides of the equation the payment side and the benefit side, and if taking twenty different cases of each, we write down the corresponding sides of the equation, we have the materials for solving instantaneously any one out of four hundred problems, out of which all will be practically useful, in which the conditions of the problem make all payments cease at or before the time when the benefit begins. This we proceed to do. The equation is

$$\left.\begin{array}{c} \sigma D_x - \varrho M_{x,w} \\ \varpi\{N_{x-1,l+m} + \mu S_{x+l,m} - \nu R_{x,w+1}\} \\ \varsigma N_{x+l+m-1} + \beta S_{x+l,m} - \theta R_{x+1,w} \end{array}\right\} = \left\{\begin{array}{c} sD_{x+k} \\ + aN_{x+k,n} + hS_{x+k+1,n} + tN_{x+k+n} \\ + AM_{x+k,n} + HR_{x+k+1,n} + TM_{x+k+n} \end{array}\right.$$

$$N_{x,y} = N_x - N_{x+y} \qquad S_{x,y} = S_x - S_{x+y} - yN_{x+y-1}$$

$$M_{x,y} = M_x - M_{x+y} \qquad R_{x,y} = R_x - R_{x+y} - yM_{x+y-1}$$

The following are the principal cases of benefits to be bought, and under each is written the benefit side of the equation in which it enters. The age of the life is x throughout.

Benefit Terms. Annuities.

1. *Endowment.* £s to be received in k years if the party be then alive,

$$sD_{x+k}.$$

2. *Life Annuity* of £a. First payable in one year, continuing through life,

$$aN_x.$$

3. *Deferred Life Annuity.* Deferred for k years, makes payment in $k+1$ years,

$$aN_{x+k}.$$

4. *Temporary Annuity.* Makes no payment after n years though the annuitant continue alive,

$$a(N_x - N_{x+n}).$$

5. *Temporary deferred Annuity.* Deferred k, continues n, years, first payment in $k+1$ years,

$$a(N_{x+k} - N_{x+k+n}).$$

* With the payment class any returns of payment in case of the conditions of benefit ceasing to exist before it becomes due.

6_1. *Increasing or decreasing Life Annuity.* Differs from (2.) in the successive payments being a, $a \pm h$, $a \pm 2h$, &c.,

$$aN_x \pm hS_{x+1}.$$

6_2. When $h = a$ for the *increasing* annuity,*

$$aS_x.$$

7_1. *Deferred increasing or decreasing Annuity.* Deferred k years, first payment a, in $k+1$ years, second $a \pm h$ in $k+2$ years, &c.

$$aN_{x+k} \pm hS_{x+k+1}.$$

7_2. When $h = a$, for the *increasing* annuity,

$$aS_{x+k}.$$

8_1. *Temporary increasing or decreasing Annuity.* Lasts n years only.

$$a(N_x - N_{x+n}) \pm h(S_{x+1} - S_{x+n+1} - nN_{x+n})$$

8_2. When $h = a$, for the *increasing* annuity,

$$a(S_x - S_{x+n} - nN_{x+n}).$$

9_1. *Temporary deferred increasing or decreasing Annuity.* Deferred k years, continues n years,

$$a(N_{x+k} - N_{x+k+n}) \pm h(S_{x+k+1} - S_{x+k+n+1} - nN_{x+k+n}).$$

9_2. When $h = a$, for the *increasing* annuity,

$$a(S_{x+k} - S_{x+k+n} - nN_{x+k+n}).$$

10. *Decreasing Annuity, temporary by extinction.* That is, it lasts n years, and each payment is less than the preceding by 1-nth part of a ($hn = a$),

$$a\left\{N_x - \frac{1}{n}(S_{x+1} - S_{x+n+1})\right\}.$$

11. *Deferred decreasing Annuity, temporary by extinction.* Deferred k years, expires after $k+n$ years,

$$a\left\{N_{x+k} - \frac{1}{n}(S_{x+k+1} - S_{x+k+n+1})\right\}.$$

12. *Arrested increasing or decreasing Annuity.* Here after n years, when the annuity would begin to pay $a \pm nh$, $a \pm (n+1)h$, &c., the increase or decrease is *arrested*, and it pays $a + (n-1)h$ for the rest of life,

$$aN_x \pm h(S_{x+1} - S_{x+n}).$$

13. *Increasing or decreasing Annuity, deferred and arrested.* The period of deferment is k years, and the increase or decrease continues n years, after which as in the last,

$$aN_{x+k} \pm h(S_{x+k+1} - S_{x+k+n}).$$

14_1. *Temporary Annuity, continued by increase or decrease.* Here the annuity is a for $n+1$ years,† after which it increases or decreases by h for $p-1$ years, and then stops.

$$a(N_x - N_{x+n+p}) \pm h(S_{x+n+1} - S_{x+n+p+1} - pN_{x+n+p}).$$

* This case can be easily calculated from the common life tables, by a method given by the author of this article in his Essay on Probabilities. (Cabinet Cyclopædia.)

† Namely, for n years from the first part, and one year of the continuation.

14_2. The same *continued to extinction*, ($hp = a$).

$$a\left\{N_x - \frac{1}{p}(S_{x+n+1} - S_{x+n+p+1})\right\}.$$

15_1. *Deferred temporary Annuity, continued by increase or decrease.* Here, after k years, a is paid for $n+1$ years, and $a \pm h$, $a \pm 2h$, &c. during $p-1$ years more,

$$a(N_{x+k} - N_{x+k+n+p}) \pm h(S_{x+k+n+1} - S_{x+k+n+p+1} - pN_{x+k+n+p}).$$

15_2. The same *continued to extinction*, ($hp = a$).

$$a\left\{N_{x+k} - \frac{1}{p}(S_{x+k+n+1} - S_{x+k+n+p+1})\right\}.$$

In the preceding list it is obvious that any benefit there described is converted into another of the same kind, but deferred for k years, simply by changing x into $x+k$. In like manner, the benefit might be anticipated a year, by writing $x-1$ for x, which would make all the immediate annuities become due, or would alter their technical character from *annuities* to *premiums*. Similarly, if we compare the meanings of $D_{x+k} \div D_x$ and $C_{x+k} \div D_x$, we see that

The first is the value of £1, to be received if the person *begin* his $(x+k+1)$th year, whether he live through it or not.	The second is the value of £1, to be received if the person *begin* his $(x+k+1)$th year, *and do not live to finish it.*

If, then, we change D_x into C_x, &c., in any problem of annuities, and alter the benefit side of the equation accordingly, we make a change of benefits as follows: at every period at which the claimant, being alive, should receive a sum of £1, let him receive it a year later, but only if he die within the year, and let it be forfeited if he live. Consequently, an annuity to be paid, say at the seventh, eighth, and ninth birthday from the present time, would thus be turned into an assurance to be paid at the eighth, ninth, or tenth birthday, if the party should die in either of these years. But since M_x was made to begin a year earlier than N_x, or $M_x = C_x + \ldots.$ and $N_x = D_{x+1} + \ldots.$, this change of conditions as to time is compensated by the structure of the tables; and any one of the preceding annuity benefits is converted into its corresponding assurance benefit, so far as the benefit side of the equation is concerned, by changing N_x into M_x and S_x into R_x. But if D_x ever occur, we must change it into C_{x-1} if the time of payment is to be the same in both.

We might thus dispense with the following list, but, in an article of reference, it is desirable, were it only to avoid the necessity of looking under one head, while thinking of another.

Benefit Terms. Assurances.

1. *Endowment Assurance.* £S to be received in k years, if the person now aged x died in the preceding year,

$$SC_{x-1}.$$

2. *Life Assurance* of £A.* Payable at the end of the year of death,

$$AM_x.$$

3. *Deferred Assurance.* Payable at death, if more than k years hence,

$$AM_{x+k}.$$

4. *Temporary Assurance.* Payable at death if within n years,

$$A(M_x-M_{x+n}).$$

5. *Temporary deferred Assurance.* Payable at death, if between the ages of $x+k$ and $x+k+n$,

$$A(M_{x+k}-M_{x+k+n}).$$

6_1. *Increasing or decreasing Life Assurance.* Payable at death, A if in the first year, $A\pm H$ if in the second, $A\pm 2H$ if in the third, &c.

$$AM_x\pm HR_{x+1}.$$

6_2. When $H=A$, for the *increasing* assurance.

$$AR_x.$$

7_1. *Deferred increasing or decreasing Assurance.* Deferred k years, A if death in $(k+1)$th year, &c.

$$AM_{x+k}\pm HR_{x+k+1}.$$

7_2. When $H=A$ for the *increasing* assurance

$$AR_{x+k}.$$

8_1. *Temporary increasing or decreasing* Assurance.* If death take place in n years.

$$A(M_x-M_{x+n})\pm H(R_{x+1}-R_{x+n+1}-nM_{x+n}).$$

8_2. When $H=A$ for the *increasing* assurance,

$$A(R_x-R_{x+n}-nM_{x+n}).$$

9_1. *Temporary deferred increasing or decreasing Assurance.* Deferred k years, continues n years,

$$A(M_{x+k}-M_{x+k+n})\pm H(R_{x+k+1}-R_{x+k+n+1}-nM_{x+k+n}).$$

9_2. When $H=A$ for the *increasing* assurance,

$$A(R_{x+k}-R_{x+k+n}-nM_{x+k+n})$$

10. *Decreasing Assurance, temporary by extinction.* Payable at death, A, if in the first year, $(n-1)$-nths of A if in the second, &c.,

$$A\left\{M_x-\frac{1}{n}(R_{x+1}-R_{x+n+1})\right\}.$$

11. *Deferred decreasing Assurance, temporary by extinction.* Payable A, if death take place in the $(k+1)$th year, &c.,

$$A\left\{M_{x+k}-\frac{1}{n}(R_{x+k+1}-R_{x+k+n+}).\right.$$

* Actuaries say *assurance*, and others *insurance*. The difference may be made useful in remembering (what the courts of law have not yet found out) that a life *assurance* and a fire *insurance* are very different things.

12. ***Arrested increasing or decreasing Assurance.*** Here, after n years, when the sum payable should be $A \pm nH$, &c., the increase or decrease is arrested, and $A \pm (n-1)H$ is the assurance for the rest of life.

$$AM_x \pm H(R_{x+1} - R_{x+n}).$$

13. ***Increasing or decreasing Assurance, deferred and arrested.*** Deferment k years, increase or decrease n years, after which as in the last.

$$AM_{x+k} \pm H(R_{x+k+1} - R_{x+k+n}).$$

14_1. ***Temporary Assurance, continued by increase or decrease.*** Here the assurance is A for $n+1$ years, after which it increases or decreases by H for $p-1$ years. and then stops,

$$A(M_x - M_{x+n+p}) \pm H(R_{x+n+1} - R_{x+n+p+1} - pM_{x+n+p}).$$

14_2. The same *continued to extinction*, ($Hp = A$),

$$A\left\{M_x - \frac{1}{p}(R_{x+n+1} - R_{x+n+p+1})\right\}.$$

15_1. ***Deferred temporary Assurance, continued by increase or decrease.*** Here, after k years, A is the assurance for $n+1$ years, and $A \pm H$, $A \pm 2H$, &c. for $p-1$ years more.

$$A(M_{x+k} - M_{x+k+n+p}) \pm H(R_{x+k+n+1} - R_{x+k+n+p+1} - pM_{x+k+n+p}).$$

15_2. The same, *continued to extinction*, ($Hp = A$),

$$A\left\{M_{x+k} - \frac{1}{p}(R_{x+k+n+1} - R_{x+k+n+p+1}).\right.$$

We now come to the enumeration of the different cases of the payment side of the equation. This we shall divide into two tables, one expressing the terms dependent on the premiums to be paid, the other the returns (where there are any) to be made in the event of no benefits becoming due.

Payment Terms.

1. ***Single Premium.*** The whole present value of the benefit, σ, paid at once,

$$\sigma D_x.$$

2. ***Life Premium.*** £ϖ now, and the same at the end of every year during life,

$$\varpi N_{x-1}.$$

3. ***Temporary premium.*** £ϖ now, and $l-1$ more times, l times in all,

$$\varpi(N_{x-1} - N_{x+l-1}).$$

4_1. ***Life Premium, increasing or decreasing by a proportion.*** £ϖ now, and $(1 \pm \mu)\varpi$, $(1 \pm 2\mu)\varpi$, &c. in succeeding years,

$$\varpi(N_{x-1} \pm \mu S_x).$$

4_2. When $\mu = 1$, for the *increasing premium*,

$$\varpi S_{x-1}.$$

5. *Temporary Premium, increasing or decreasing by a proportion.* To last only m years, last premium $(1\pm\overline{m-1}\,\mu)\varpi$,

$$\varpi\{N_{x-1}-N_{x+m-1}\pm\mu(S_x-S_{x+m}-mN_{x+m-1})\}.$$

6. *Premium temporary by extinction.* Here $m\mu=1$, and the extinction takes place after m premiums,

$$\varpi\left\{N_{x-1}-\frac{1}{m}(S_x-S_{x+m})\right\}.$$

7. *Arrested proportionally increasing or decreasing Premium.* The premiums of m years are $\varpi,\ldots.\varpi(1\pm\overline{m-1}\,\mu)$, at which they afterwards remain,

$$\varpi\{N_{x-1}\pm\mu(S_x-S_{x+m-1})\}.$$

8_1. *Temporary Premium, continued by proportional increase or decrease.* Here $l+1$ premiums ϖ are to be paid; afterwards $m-1$ premiums $\varpi(1\pm\mu)$, $\varpi(1\pm2\mu)\ldots.\varpi\{1\pm(m-1)\mu\}$,

$$\varpi\{N_{x-1}-N_{x+l+m-1}\pm\mu(S_{x+l}-S_{x+l+m}-mN_{x+l+m-1})\}.$$

8_2. The same *continued to extinction* ($\mu m=1$),

$$\varpi\left\{N_{x-1}-\frac{1}{m}(S_{x+l}-S_{x+l+m})\right\}.$$

9. *Life Premium, increasing or diminishing by an absolute sum.* Premium ϖ, $\varpi\pm\beta$, $\varpi\pm2\beta$, &c.

$$\varpi N_{x-1}\pm\beta S_x.$$

10. *Temporary Premium, increasing or diminishing absolutely.* To last m years, last premium $\varpi\pm(m-1)\beta$,

$$\varpi(N_{x-1}-N_{x+m-1})\pm\beta(S_x-S_{x+m}-mN_{x+m-1}).$$

11. *Arrested absolutely increasing or diminishing Premium.* The first m premiums are ϖ, $\varpi\pm\beta,\ldots.\varpi\pm(m-1)\beta$, at which they afterwards remain,

$$\varpi N_{x-1}\pm\beta(S_x-S_{x+m-1}).$$

12. *Temporary Premiums continued by absolute increase or decrease.* Here $l+1$ premiums ϖ are to be paid; afterwards $m-1$ premiums $\varpi\pm\beta$, $\varpi\pm2\beta\ldots.\varpi\pm(m-1)\beta$,

$$\varpi(N_{x-1}-N_{x+l+m-1})\pm\beta(S_{x+l}-S_{x+l+m}-mN_{x+l+m-1}.)$$

The following is the table of modes of returning a portion of the premiums.

N.B. These tables do not suffice to calculate the effect of the return of a given proportion of varying premiums. The quantities following are *positive* when put on the benefit side, and negative when on the payment side. It must, of course, be obvious that this is only another table of the values of assurances, described so as to meet the form in which problems are usually given.

Return Terms.

1. *Fixed Return at death.* A fixed sum ρ returned whenever the death may take place,

$$\rho M_x.$$

2. *A fixed Return at death, if before w years have elapsed,*

$$\rho\ (M_x - M_{x+w}).$$

3. *Return at death of a proportion of fixed premiums.* That is, $\nu\varpi$, if the death take place in the first year, $2\nu\varpi$ if in the second, &c.

$$\nu\varpi R_x.$$

4. *Return at death of a temporary proportion of fixed premiums,* if the death take place within w years.

$$\nu\varpi\ (R_x - R_{x+w} - wM_{x+w}).$$

5. *An increasing or decreasing sum returned at death;* ϱ if in the first year, $\varrho \pm \theta$, if in the second, &c.

$$\rho M_x \pm \theta R_{x+1}.$$

6. *The same if the death take place in w years,*

$$\rho\ (M_x - M_{x+w}) \pm \theta\ (R_{x+1} - R_{x+w+1} - wM_{x+w}).$$

7. *An arrested increasing or decreasing sum returned at death.* ρ if in the first year, $\varrho \pm (w-1)\ \theta$ if in the wth, or any following year,

$$\rho M_x \pm \theta\ (R_{x+1} - R_{x+w}).$$

8. *A fixed sum,* ϱ, *or a fixed proportion of premiums* returned if the life continue w years,* ρD_{x+w}.

A few general rules will be readily collected from the preceding, and may be simply demonstrated. They might be made the foundation of a synthetical view of the subject.

1. Every thing depends on the fundamental calculation of the various cases of the benefit side of the equation.

2. The benefit side of the equation being found for the whole life, that for the same benefit deferred k years is found by writing $x+k$ for x.

3. And that for the same benefit to last n years is found by changing

$$N_x \text{ into } N_{x,n},\ M_x \text{ into } M_{x,n}.$$

With regard to S_x and M_x the change must be regulated by the following consideration. When their exponent is *one more* than the present age, or the age at a term mentioned in the problem, change S_x into $S_{x,n}$ and R_x into $R_{x,n}$. But whenever S_x or R_x has the exponent of the present age, or of that at the beginning of a term, change S_x into $S_{x,n+1}$, and R_x into $R_{x,n+1}$, (compare (page 16,) the transition from 6_1 to 8_1 with that from 6_2 to 8_2). If no simplifications were allowed, that is, if N or M were always retained for the permanent portion of an annuity or insurance, and S or R for the term depending on the value of the incremental portion, the first rule would be sufficient.

4. All the cases are then derived from the following:

N_x on the benefit side of the equation, an annuity of £1, £1, £1, &c.
M_x . . . . assurance of £1, £1, £1, &c.
S_x . . . . annuity of £1, £2, £3, &c.
R_x . . . . assurance of £1, £2, £3, &c.

* For a given year a proportion of the premiums paid by that time is simply a fixed sum.

5. Every formula in which any particular relations exist should be carefully looked at with a view to simplification.

If we wish to make a benefit begin k years *earlier*, we write $x-k$ for x. In the case of an immediate annuity, this is intelligible enough when $k=1$, but N_{x-k} when k is greater than 1, is the impossible quantity of this branch of algebra. Its meaning is as follows: suppose a person has been k years in the enjoyment of a benefit, or of the chance of a benefit, for which he ought to have paid when such enjoyment began; suppose also that, had he died during the k years, the claimant of the payment would have had no means of recovering his rights. According to the principles which regulate these transactions, the holder of the unbought benefit ought to pay the claimant not only all arrears with compound interest, if any, but also compensation for the chance of loss which he has run. The value of such compensation is found by writing $x-k$ for x in the value of the benefit as reckoned from the present time. Thus, $N_{30} \div D_{50}$ is what a person now aged 50 should pay for the past and the future, who has been in unbought possession of an annuity of £1 for 20 years; and $C_{30} \div D_{50}$ is what a person aged 50 should now pay for the unbought chance of having formerly received £1, if he had died between 30 and 31.

This last consideration will be particularly important in its application to the commutation tables for interests certain, as we may thus find the value of all arrears, or may solve a case in which partly arrears and partly prospects are to be valued. For instance, a person engaged to pay a decreasing rent for certain tenements, £a at the end of the first year, $a-h$ of the second, &c., and $a-(n-1)h$ at the end of the nth and last: k years elapse during which he pays no rent, and then his affairs pass into the hands of assignees, who are desirous of paying the arrears and buying the remaining term for one sum. Here at the commencement, the payment side of the equation is σD_x, and the benefit side is $aN_{x,n} - hS_{x+1,n}$: put the last back k years (assuming the age, which is indifferent,* to be k) and we have

$$\sigma = \frac{a(N_0 - N_n) - h(S_1 - S_{1+n} - nN_n)}{D_k},$$

which is the sum to be demanded of the assignees.

It now only remains to show an example of the mode of proceeding with the registered cases.

An assurance, to commence in k years, and to be £A, A+H, ...A+$(n-1)$H in the following n years, at which last sum it is arrested, together with an annuity of £a, to begin at the same term, and to last n years, is to be bought by present payment of a sum σ, and also of a premium which is extinguished after k years, or in $k+1$ payments, on condition that the sum σ shall be returned, with simple interest, if the life drop during the k years. Required the first premium ϖ.

* In the tables for interests certain it will do equally well to put the payment sid forward k years.

Benefit terms. $\begin{cases}\text{The assurance (13), } AM_{x+k}+H(R_{x+k+1}-R_{x+k+n}) & =V \\ \text{The annuity (5), } a(N_{x+k}-N_{x+k+n}) & =W\end{cases}$

Payment terms. $\begin{cases}\text{The fixed sum } \sigma \text{ (1), } \sigma D_x & =X \\ \text{Premium (6), } \varpi\left\{N_{x-1}-\dfrac{1}{k+1}(S_x - S_{x+k+1})\right\} & =Y\end{cases}$

Return terms. $\begin{cases}\text{Return of } \sigma(1+r) \text{ in the first year, \&c.} \\ \sigma(1+r)(M_x - M_{x+k})+r\sigma(R_{x+1}-R_{x+k+1}-kM_{x+k})=Z \\ \text{or } \sigma(M_x - M_{x+k})+r\sigma(R_x - R_{x+k}-kM_{x+k}).\end{cases}$

As no further simplification suggests itself, each term had better be calculated for the particular case wanted: we have then

$$X+Y\varpi-Z=V+W, \quad \varpi=\frac{V+W+Z-X}{Y}.$$

When it becomes necessary to return proportions of decreasing premiums, new tables must be constructed from S_x and R_x, say Z_x and Y_x, so that

$$Z_x = S^x + S_{x+1} + S_{x+2} + \ldots.$$
$$Y_x = R_x + R_{x+1} + R_{x+2} + \ldots.$$

These, divided by D_x, will give the values of the annuity or assurance £1, 3, 6,...., or $n\dfrac{n+1}{2}$ in the nth year: and the effect of these tables, combined with the others, will be to give the value of an annuity or assurance which is $a+hn+qn^2$ in the nth year.)

In Mr. Barrett's original method, which is still followed by some actuaries, are three columns only, answering to D, N, and S, which by aid of the first three formulæ VIII., give C, M, and R. The *great principle* of the method, namely, the formation of tables by which deferred, temporary, and increasing benefits are as easily calculated as those for the whole life, belongs to Mr. Barrett as much as the invention and construction of logarithms to Napier. On the other hand, Mr. Griffith Davies, by the alteration presently noted, and the separate exhibition of M and R (he has not given C, which is of little use in practice, though essential to the theory) has increased the utility and extended the power of the method to an extent, of which its inventor had not the least idea; and has all the rest of the claim in the matter which is made for Briggs in the adaptation of logarithms to practical use. Nor must it be forgotten, that in all probability this most expeditious mode of conducting operations would not have been now in existence, but for the sagacity of Mr. Baily, who, as we have seen, saw further into its merits than the Royal Society. In Mr. Barrett's form, there are three columns, A, B, and C: and A_x is not $a_x v^x$, but $a_x(1+r)^{w-x}$, where w is the greatest age any individual can attain. Also B_x is not $A_{x+1}+\ldots$, but $A_x+\ldots$., so that the value of an annuity on a life aged x is $B_{x+1} : A_x$. Again, C_x is $B_x + B_{x+1} + \ldots.$ The following comparisons may be useful to those who are habituated to Barrett's original form, remembering that C_x now means Barrett's third column, and not what it has hitherto stood for :

$$A_x = D_x v^{-w}, \qquad D_x = A_x v^w,$$
$$B_x = N_{x-1} v^{-w}, \qquad N_x = B_{x+1} v^w,$$
$$C_x = S_{x-1} v^{-w}, \qquad S_x = C_{x+1} v^w.$$

Since, then, Barrett's form is that of Mr. G. Davies multiplied by a constant factor, the former are also *commutation* tables, using the life a year older than the given life in the second and third columns.

In conclusion, we may mention that when a whole table is to be calculated, it may happen that it is better to dispense with the assurance columns by means of VIII. Thus in the case of the premium of assurance for a term of years (as noted by Mr. G. Davies)

$$\frac{M_x - M_{x+n}}{N_{x-1} - N_{x+n-1}} \text{ is not so convenient as } v - \frac{N_x - N_{x+n}}{N_{x-1} - N_{x+n-1}}.$$

The only works of which the writer is aware, in which the preceding method, whether called by the name of Barrett or Davies, is treated, are the Appendix to Mr. Baily's Treatise on Life Annuities, &c.; the French translation of the same; Mr. G. Davies' work, already cited; a Note in the Appendix to Mr. Babbage's Treatise on Life Assurance; and the treatise on Life Annuities, &c. by Mr. Jones, now in course of publication in the Library of Useful Knowledge.

A. DE MORGAN.

University College, London,
October 1, 1839.

Since writing the above, it has struck me that it would be more convenient to make the calculations in page 6 by writing the logarithms of a_x and $a_x - a_{x+1}$, one above, and the other below, the logarithm of v^x, than by writing both of the former below the latter.

COMPANION TO THE ALMANAC

FOR

1841.

PART I.

INFORMATION CONNECTED WITH THE CALENDAR AND THE NATURAL PHENOMENA OF THE YEAR; AND WITH NATURAL HISTORY AND PUBLIC HEALTH.

I.—ON THE USE OF SMALL TABLES OF LOGARITHMS IN COMMERCIAL CALCULATIONS, AND ON THE PRACTICABILITY OF A DECIMAL COINAGE.

THE invention of logarithms is now two hundred years old, during which time the labour of computation has been reduced, for the mathematician, to about a tenth part of its previous expense of time and trouble. The mathematical computer, however, has had them all to himself; they have never been applied, except indirectly, to the purposes of commerce, or the wants of life Some account of the reason of this is necessary to a proper explanation of the object of this article.

The tables of logarithms which first came into general use were constructed for seven places of figures; and these still continue in use for the most accurate calculations. Such tables, when for example the square root of 2 is required, give 1·414214, almost as easily as 1·414: or, which is the same thing, require the computer to whom 1·414 is a result of sufficient accuracy to take nearly as much trouble as would give him 1·414214, which is more than he wants. It is true that tables of six places were soon constructed, but they did not find much favour until, in the last century, the tables of Lalande showed astronomers that five places were sufficient for many, perhaps most, purposes. These tables of Lalande found their way into use in England, and a little while ago, a reprint,* much superior to the French work in appearance, was published in London, under the sanction of the Useful Knowledge Society. This was soon followed by a better table of six-figure† logarithms than had previously existed; and this again by a table of four-figure logarithms.‡ It is to these last that our attention is now to be directed; we assert, and shall prove, that they are sufficient for a great many purposes, and we think that they might be advantageously introduced in instruction.

Much of the trouble of using logarithms arises, 1. From the necessity of turning leaf after leaf of the book to arrive at the

* Tables of Logarithms. Taylor and Walton.
† Tables of six-figure Logarithms. Longman and Co.
‡ Four-figure Logarithms, on a card. Taylor and Walton.

proper page; and 2. in a minor degree, from the process of finding the number to a logarithm being somewhat more difficult than that of finding the logarithm to a number. Attention and practice conquer both these difficulties, and, as in other cases, make them appear trivial on retrospection. In the four-figure logarithms they are both avoided, since the whole table of logarithms to numbers is on one side of a card, and the smallness of the table enables another table, as extensive, of numbers to logarithms, to be entered on the other side of the card.

These four-figure logarithms were first printed, we believe, some years ago, for private circulation among practical astronomers, to be used in the reduction of the observed places of stars to the mean. They were afterwards introduced into the treatise on algebra in the Library of Useful Knowledge.

It may be, however, that much of the monopoly which mathematicians have had of the use of logarithms has arisen from the want of a decimal system of weights and measures; which not only renders the tables difficult of use in the case of fractions, but enables a commercial arithmetician (unless indeed he aspire to the higher questions of interest and annuities) to get on, after a fashion, without any knowledge of decimal fractions. Much as a decimal coinage is to be wished for in almost every respect, we doubt if any advantage accruing from it would equal that of its indirect consequence, the forcing of the attention of people in general to the subject of decimal fractions.

So far as the subdivisions of the pound sterling are concerned, an approximate reduction to decimals is very easily made, the absence of which from our common works on arithmetic is part of a system, namely, the rejection of every rule by which an unnecessary degree of accuracy is avoided and a more simple mode of operation substituted for the tedious rules which give the uttermost fraction of a farthing. As much of the application of logarithms to commercial arithmetic depends upon the approximate rule* just mentioned, we must begin by explaining it.

Take the fraction ·28417 of a pound, meaning 2-tenths, 8-hundredths, 4-thousandths, &c. It is required to find what shillings, pence, and farthings this fraction contains. The common method is as here written, by which three multiplications show that £·28417 is 5 shillings, 8 pence, and 8032 parts out of ten thousand of a farthing. Taking the next farthing above, we have 5*s.* 8¼*d.* If this degree of accuracy would have been sufficient; that is, if it were not material to know that ·8032 of a farthing was the precise fraction remaining, the three multiplications might have been avoided by a very simple rule, to which the following is the key, the columns standing for decimal places.

```
 ·28417
     20
 -------
5·68340
     12
 -------
8·2008
      4
 -------
 ·8032
```

* Till a few years ago this rule was the property of actuaries, to whom it saved many a tedious reduction, and much risk of error. It has of late years been introduced into elementary works, but there are still but few who know how much labour may be saved by it, for which reason we repeat it here.

Two shillings	1	0	0	
One shilling		5	0	
One farthing			1	Abating 1 at and above 25.

Thus in ·284, the 2 stands for 4 shillings, 5 out of 8 for another shilling, and the remaining 3 with the 4, or 34, must be diminished by 1 as being above 25. This gives 33 (farthings), so that 5 shillings and 33 farthings, or 5*s.* 8¼*d.* is the answer. Similarly, ·005 is 1¼*d.*; ·025 is 6*d.*; ·036 is 8¾*d.*; ·058 is 1*s.* 2*d.*; ·067 is 1*s.* 4¼*d.*; ·094 is 1*s.* 10¾*d.*; ·118 is 2*s.* 4½*d.*; ·139 is 2*s.* 9½*d.*; ·188 is 3*s.* 9¼*d.*; ·427 is 8*s.* 6½*d.*; ·602 is 12*s.* 0½*d.*; ·659 is 13*s.* 2½*d.*; ·777 is 15*s.* 6½*d.*

To turn the shillings, &c. into the fraction of a pound, reverse the preceding process; that is, allow a unit in the first place for every pair of shillings, 5 in the second place for the odd shilling, if there be one, and fill up the second and third places with the number of farthings in the pence and farthings, increased by 1 if it be 6*d.* or upwards. Thus 2*s.* 4½*d.* is ·118; 3*s.* 4½*d.* is ·168; 3*s.* 7¾*d.* is ·182; 12*s.* 8½*d.* is ·635; 13*s.* 8½*d.* is ·685: and so on.

If a fourth decimal place should be wanted, put in the fourth place the nearest ten to four times the number of farthings, or four times the number above sixpence. Thus in 3*s.* 4½*d.*, the first three places of which are ·168, the number of farthings is 18, and 4×18 is 72, the nearest number of tens being 7; hence ·1687 is the fraction required. But the following table will, with the preceding rule, enable any one to find the decimal fraction to any number of places immediately.*

¼*d.*	04166	2¼*d.*	37500	4¼*d.*	70833
½*d.*	08333	2½*d.*	41666	4½*d.*	75000
¾*d.*	12500	2¾*d.*	45833	4¾*d.*	79166
1*d.*	16666	3*d.*	50000	5*d.*	83333
1¼*d.*	20833	3¼*d.*	54166	5¼*d.*	87500
1½*d.*	25000	3½*d.*	58333	5½*d.*	91666
1¾*d.*	29166	3¾*d.*	62500	5¾*d.*	95833
2*d.*	33333	4*d.*	66666	6*d.*	00000

In every one of these it must be understood that the last figure may be repeated any number of times, according to the degree of exactness required. Thus for 5*s.* 3½*d.* the preceding rule gives ·264, and opposite to 3½*d.* we see ·5833333....: whence 5*s.* 3½*d.* is ·2645833333....., being a circulating decimal, as it is called, of which every figure after the fifth is 3. By so very slight a process as this we are enabled to multiply 5*s.* 3½*d* by 10, 100, 1000, &c. by mere inspection, and with perfect accuracy: thus ten times this sum is £2 12*s.* 11*d.*, one hundred times is £26. 9*s.* 2*d.* one thousand times is £264 11*s.* 8*d*, ten thousand times is £2645 16*s.* 8*d.*, and so on. Again, 14*s.* 9¾*d.*, by the preceding rule, is ·740; annex the part answering to 3¾*d.* in the preceding table, and this becomes ·740625.

* A complete table, including every possible case, will be found in the Library of Useful Knowledge. Value of Annuities, &c., p. 66.

The preceding rule takes advantage of the accident that one farthing differs little from the thousandth part of a pound, being £·00104166.... Another, depending on the year consisting of 365 days, may be seen in the Penny Cyclopædia, article INTEREST. A third is as follows:

Given the price of a hundred weight in pounds sterling, required that of a pound in pence. The rule is, take the pounds $2\frac{1}{7}$ times. Thus £7 a hundred weight is 15*d.* a pound. Again, if a hundred weight cost £39. 13*s.* 9*d.*, or 39·687, the price of 1lb. in pence is thus found, and the answer is 85*d.* and about $\frac{1}{23}$ of a penny.

39·687
2
79·374
5·669
85·043

The inverse question, given the price of 1lb. in pence required that of 1 cwt. in pounds, is solved by the following rule: take half the price, and one thirtieth of it; the difference is the number of pounds required. Thus 90*d.* per lb. is 45−3 pounds sterling per cwt., or £42 per cwt. Again, 1*s.* 7¾*d.* per lb. gives £9. 4*s.* 4*d.* per cwt., as in the adjoining calculation.

19·75	pence
9·875	the half
·6583	the thirtieth.
9·2167=£9. 4*s.* 4*d.*	

We now go on to the method of using logarithms, with particular reference to the four-figure table already described.

1. The logarithm of a number or fraction is another number or fraction which represents the first in calculation, in a manner of which it is not our business here to explain the theory, but to describe the use.

2. The logarithm to a number or the number to a logarithm is to be found partly by rule, partly by means of a table presently to be described.

3. The absolute logarithm of a number would consist of figures going on for ever: consequently no table can give all the figures of a logarithm, but only a certain number of the first figures: and a table which gives five figures of any logarithm is called a table to *five places*, and so on. The general rule is, that so many places as a table of logarithms is carried to, so many figures of the correct answer to a calculation can be obtained by it. Thus in the table of four places which we intend to use, if the real answer were 37196472, we should find the nearest round number of four places, followed by ciphers, or 3720,0000: similarly, if the answer were 166·219, we should find 166·2. The degree of incorrectness necessarily belonging to tables of four places may be stated at less than one part in a thousand of the result: for example, if the answer were in pounds and decimals of a pound, the error would very seldom indeed amount to a farthing in a pound.

4 A logarithm is an algebraical number, positive or negative, and it consists, like other mixed quantities, of an integer and a fractional part.

5. The integer part is found by rule, and not from the tables; it is called the *characteristic:* the fractional part is found from the table, and may be called the *decimal part.* Thus the logarithm of 2000 being 3·3010, the characteristic is 3, and the decimal part ·3010.

6. The characteristic may be positive or negative, which means as follows. A positive characteristic is one which is always to be used in the same manner as its decimal part, and is denoted in the usual way: thus in 3·3010, when ·3010 is to be added, 3 is to be added, and so on. A negative characteristic is one which is always to be used in the contrary manner to the decimal part, and is denoted by a small bar placed over the figure: thus the logarithm of ·0002 is $\bar{4}$·3010, and it means that when ·3010 is added 4 is subtracted, and when ·3010 is subtracted 4 is added. Here $\bar{4}$ is called a *negative* characteristic.

7. The use of such mixed numbers as $\bar{4}$·3010 does not differ in any very difficult points from common arithmetical operations: the following rules contain all that is peculiar to them.*

Addition and Subtraction. The whole of these rules will be best remembered by the fiction that the positive characteristic shall stand for a credit, and the negative one for a debt. Thus the taking away of a debt makes a credit larger, and a larger debt less. The following will then be easily understood.

4 increased by 3 is 7	4 diminished by 3 is 1
4 . . . . . . $\bar{3}$ is 1	4 . . . . $\bar{3}$ is 7
$\bar{4}$ . . . . . . 3 is $\bar{1}$	$\bar{4}$ . . . . 3 is $\bar{7}$
$\bar{4}$ . . . . . . $\bar{3}$ is $\bar{7}$	$\bar{4}$ . . . . $\bar{3}$ is $\bar{1}$

3 diminished by 4 is $\bar{1}$
3 . . . . $\bar{4}$ is 7
$\bar{3}$ . . . . 4 is $\bar{7}$
$\bar{3}$ . . . . $\bar{4}$ is 1

'Increased by $\bar{a}$' means 'diminished by a'.

'Diminished by $\bar{a}$' means 'increased by a'.

In all carriages, the carried part is reckoned as positive; thus when 1 is carried to $\bar{4}$, it makes it $\bar{3}$.

Add.				Subtract.			
6·2	$\bar{3}$·1	$\bar{3}$·6	$\bar{4}$·1	4·8	$\bar{2}$·9	$\bar{2}$·6	$\bar{4}$·2
$\bar{1}$·8	2·5	$\bar{2}$·5	6·9	$\bar{2}$·1	3·7	3·8	$\bar{3}$·9
6·0	$\bar{1}$·6	$\bar{4}$·1	3·0	6·7	$\bar{5}$·2	$\bar{6}$·8	$\bar{2}$·3

Multiplication and Division. In multiplication the only addi-

* The student who has a little acquaintance with algebra is supposed not to need these rules.

tional rule is that the negative characteristic, when it comes to be multiplied, must have its product taken negatively, while the part carried must be taken positively. Thus $\bar{2}\cdot616\times4$ gives $\bar{6}\cdot464$ whereas $2\cdot616\times4$ is $10\cdot464$.

In division, instead of taking the multiple of the divisor next below the characteristic to make the first quotient figure, let the multiple next above be taken, and let the excess thereby added to the characteristic be carried in the usual way. The following is a comparison of the first step in the two kinds of division.

$7)\ 39\cdot414\ldots\ldots$ $5\cdot630\ldots\ldots$	$7)\ \overline{39}\cdot414\ldots\ldots$ $\bar{6}\cdot487\ldots\ldots$
The nearest multiple of 7 *below* 39 is 35, which is four *below* 39; put down 5 and carry 4; proceed similarly with 44, &c.	The nearest multiple of 7 *above* $\overline{39}$ is $\overline{42}$, which is three *above* 39; put down 6 and carry 3; proceed in the common way with 34, &c.
$\bar{3}\cdot414 \times 10 = \overline{26}\cdot140$ $\bar{1}\cdot618 \times 9 = \bar{4}\cdot562$	$2)\ \bar{3}\cdot728$ gives $\bar{2}\cdot864$ $5)\ \overline{12}\cdot999$ gives $\bar{3}\cdot799$

8. The manner in which the characteristic is to be taken is as follows: if the number be a simple digit followed by decimals, as 9·616, 1·312, the characteristic is 0; in every other case reckon the number of digits from the units place exclusive to the figure of highest value: the number so reckoned is the characteristic, *positive* when the original number is greater than unity, *negative* when it is less. In the following list the characteristic is written after the number to which it belongs: thus the characteristic of 327·81 is 2.

9·616	0	327·81	2	·1	$\bar{1}$	·0246	$\bar{2}$
4·7219	0	14623	4	·1276	$\bar{1}$	·03792	$\bar{2}$
21·3	1	291·0874	2	·34729	$\bar{1}$	·0004168	$\bar{4}$
38·1127	1	7761812	6	·77	$\bar{1}$	·0017296	$\bar{3}$

The characteristic neither depends on the number of figures nor on their value, but solely on the position of the leading figure with respect to the decimal point (which is always supposed to be written after the unit's place, even where no fractional digits follow.)

9. And in like manner as the characteristic or integer part of the logarithm depends only on the position of the decimal point, so the rest of the logarithm, or its fractional part, does not depend at all on the position of the decimal point, but only on the digits of the number and their order. Thus 1234, 123·4, 12·34, 1·234,

·1234, ·01234, &c., all have the same decimal part in their logarithm, with the characteristics 3, 2, 1, 0, $\bar{1}$, $\bar{2}$, &c. This decimal portion happens to be ·0913, to be found as presently shown: so that we have

log^m of 1234 is 3·0913	log^m of 12·34 is 1·0913
. . . . 123·4 is 2·0913	. . . . 1·234 is 0·0913

log^m of ·1234 is $\bar{1}$·0913
. . . . ·01234 is $\bar{2}$·0913.

10. We now show how to find the decimal portion of the logarithm. Every table consists of numbers which are made the references to other numbers, and this mode of speaking may be altered, since the second mentioned numbers may be made references to the first. Thus, in a common interest table, we may either ask, what is the interest of £—— for — days, (in which case we look for the days, and by them find the sum required,) or we may ask how many days does it take £—— to make £——, (in which case we look for the sum and find the number of days to it.) In every case, that which is first looked for is called the argument, and we shall call that which is found the *result*. Thus, when we find the logarithm to a number, the number is the argument, or thing looked for, the logarithm is the result.

11. The tables previously cited, namely, those to four places, consist of two pages. The first, headed *logarithms*, is so arranged as to be most easily used when the number is the argument, and its logarithm the result. The second, headed *antilogarithms*, is arranged for ready use when the logarithm is the argument, and its number the result. But either table may be inverted: thus the logarithm may be made the argument in the first, and the number in the second.

When the argument is in the leading column, and the result in the body of the table, the table may be said to be *directly* used; when the argument is in the body of the table, and the result in the leading column, *inversely*.

To find the logarithm of a number use the logarithms directly
. . . . antilogarithms inversely
To find the number of a logarithm use the logarithms inversely
. . . . antilogarithms directly.

12. In every use of the table, neglect entirely the decimal point of a number, and the characteristic of a logarithm. When we speak of the first four figures of an argument, we mean the first four significant figures of a number, and the first four decimal figures of a logarithm: and these may be called *the* arguments.

13. To use either table directly.

In the row which begins with the *first two* figures of the argument and to the result which is in the column headed by the *third*, add that number from the side-table whose column is headed by the *fourth*.

Direct use of 'Logarithms.'		Direct use of 'Antilogarithms.'	
Logarithm of 261·7?		Number to logarithm 2·1493.	
Argument 2617.		Argument 1493.	
In row 26, under 1 . .	4166	In row 14, under 9 . .	1409
Side-table under 7 . .	11	Side-table, under 3 . .	1
Decimals of log. . . .	4177	First 4 figures of number	1410
(8) Characteristic . .	2	Which (charac. being 2) must be written . . .	141·0
Log. 261·7 =	2·4177		

14. To use either table inversely.

In the body of the table find the result which is next beneath the argument, and note the unattained part.

To the two figures of the row in which the result is found annex the figure at the head, and the figure at the head of the side column in which the unattained part (or the nearest to it) is found.

Inverse use of 'Logarithms.'		Inverse use of 'Antilogarithms.'	
Number to logarithm 2·1493.		Logarithm of 261·7.	
	·1493		2617
Row 14, col. 1	1492	Row 41, col. 7	2612
Unattained part* .	1	Unattained part .	5
Answer 1410.		·4178, or ·4179; charact. 2.	
By charact. 141·0, as before.		Answer 2·4178, or 2·4179.	

15. Thus it appears that the results of the two tables may agree entirely, or may differ slightly in the last figure. This is a consequence of the logarithms not being exact, and belongs to all tables; the last figure can never be found in all cases with perfect accuracy. The error may often be corrected slightly by the following addition to the preceding rules.

Take the difference between the result under use and the next one, and take its half, or if an odd number, take the whole number next above the half. Take care to use a row of the side-table which has this last mentioned whole number in the middle. Thus, in (13), the result under use is 4166, the next being 4183, the difference 17, and the whole number required 9. Now 9 is not in the middle of the side table, row 26; but in the middle of the one above it, which use. Here under 7 falls 12 instead of 11, and the result is 2·4178, which is correct.

This trifling correction may be deferred until the student is well practised in everything else.

16. The properties of logarithms are as follows.

* Here the unattained part indicates that 0 is the fourth figure, because by the side able, 1 would require 3 unattained.

The logarithm of a product is the sum of the logarithms of the factors.

The logarithm of a quotient, ratio, or fraction, is the excess of the logarithm of the dividend, antecedent, or numerator, over the logarithm of the divisor, consequent, or denominator.

The logarithm of the square, cube, fourth, &c. power of a number is 2, 3, 4, &c. times the logarithm of the number.

The logarithm of the square, cube, fourth, &c. root is the half third, fourth, &c. part of the logarithm of the number

Hence the operations for finding the logarithm of a result are always of an easier order than those for finding the result itself: in which the advantage of logarithmic calculation consists.

The following are examples of the first rule:

1737	3·2397	·8183	$\bar{1}$·9130	64380	4·8087
×2346	3·3703	×·1147	$\bar{1}$.0595	×·009194	$\bar{3}$·9635
4074000	6·6100	·09387	$\bar{2}$·9725	591·9	2·7722

The following is the description at length of the first process. The logs. of 1737 and 2346 have each the characteristic 3, because the highest figure of each is 3 places to the left of the unit's place. In the table of logarithms, row 17, col. 3, is found 2380, and under 7 in the side table is 17; 2380+17=2397: again, row 23, col. 4, gives 3692, with 11 in column 6 of the side table; 3692+11=3703. The two logarithms being 3·2397 and 3·3703, their sum is 6·6100, indicating a number whose highest figure is six from the unit's place. Since, then, only four figures are attainable by our table, the vacant places must be filled by ciphers. In the table of antilogarithms, row 61, col. 0, is found 4074, and the fourth figure of 6100 being 0, we have no use for the side table. Hence the number 4074000 is the approximation to the product of 1737 and 2346. The real answer is 4075002, and the error is 1000 out of 4075002, or less than 1 out of 4075, being about one quarter of a farthing in one pound.

Let us now see whether by attention to (15) we could have made the fourth figure correct. Since 173 gives 2380, and 174 gives 2405, of which the difference is 25, and the whole number next greater than the half, 13, and since the middle column of the side table at row 17 has only 12, we must use the next higher row of the side table, which gives 18 under 7 instead of 17. Hence 3·2398 is the most correct log. of 1737. That of 2346 is 3·3703, as before, the half-diff. and middle column of the side-table agreeing. Hence the sum of the logarithms is 6·6101, and in the antilogarithms, row 61, col. 0, we find 4075, while in col. 1 of the side table we find 1, which gives 4075, or 4075000 is the most correct answer the tables will give.

The following are examples of the raising of powers:

$(\cdot 6167)^4$		$(142\cdot 7)^2$		$(22\cdot 66)^8$	
·6167	$\bar{1}\cdot 7901$	142·7	2·1544	22·66	1·3553
	4		2		8
·1446	$\bar{1}\cdot 1604$	20360	4·3088	69,560,000,000	10·8424

The following are examples of division:

6143	3·7884	·1176	$\bar{1}\cdot 0705$	22748*	4·3570
÷22·14	1·3452	÷414·9	2·6179	÷·06295	$\bar{2}\cdot 7990$
277·4	2·4432	·0002835	$\bar{4}\cdot 4526$	361400	5·5580

The following are examples in the extraction of roots:

$\sqrt[3]{(159\cdot 9)}$	$\sqrt{13689}$
3) 2·2039	2) 4·1364
0·7346	2·0682
Ans. 5·428.	117·0
$\sqrt[5]{\cdot 001727}$	$\sqrt[12]{\cdot 00001192}$
5) $\bar{3}\cdot 2372$	12) $\bar{5}\cdot 0763$
$\bar{1}\cdot 4474$	$\bar{1}\cdot 5897$
·2802	·3888

57·916	1·7629
·0475	$\bar{2}\cdot 6767$
192	2·2833
	2·7229
365	2·5623
1·447	0·1606
£1 8*s.* 11½*d.*	

Example 1. What is the interest on the sum of £57. 18*s.* 4*d.*, at 4¾ per cent., and for 192 days. Here the annual fraction of the sum which arises from interest is ·0475, or 4·75 in 100. The sum is £57·916, or, for our present purpose, £57·92. This being taken ·0475 of a time, or multiplied by ·0475, the year's interest is found, of which 192-365ths is to be taken, or this result is to be multiplied by 192, and divided by 365. The answer is £1 8*s* 11½*d.*, which is within a farthing of the truth.

Example 2. Three men subscribed £12 13*s.* 4*d.*, £18 9*s.* 11½*d.*, and £10 17*s.* 5¾*d.* towards an adventure which yielded £181 18*s.*; in what proportions ought it to be divided among them.

12·666	181·9	2·2598
18·497	42·04	1·6236
10·873		0·6362
42·036		

* 22750 must be used here, as being the nearest number with four significant places.

12·67	1·1028		10·87	1·0363
	·6362			6362
54·83	1·7390		47·04	1·6725
18·50	1·2672		£54·83	
	6362		£80·05	the several shares.
			£47·04	
80·05	1·9034		181·92 = £181 18*s.* 5*d*	

Example 3. If 1 cwt. 2 qrs 21 lbs. cost £33 17*s.* 7*d.*, how much will 6 cwt. 3 qrs. 15 lbs. cost.

To turn quarters and pounds into decimals of a hundred-weight, write the proper decimal for the quarters, (·25, ·5, or ·75,) and for the odd pounds write down their number in the first and second places of decimals, subtracting twelve times the number in the third and fourth places. The difference must be taken to three places, and a unit added in the last place for every 7 pounds, or quarter of a quarter.

1 qr 24 lbs.	2 qrs. 13 lbs	3 qrs. 9 lbs
·24	·13	·09
·0288	·0156	·0108
·2112	·1144	·0792
·253	·502	·751
·464 Ans.	·616 Ans	·830 Ans.

Thus in the question before us, 2 qrs. 21 lbs is ·687 cwt., and 3 qrs. 15 lbs. is ·884 cwt. The question then is, if 1·687 cost 33·879, how much will 6·884 cost. The answer is £138 4*s.*, as found by the accompanying process. A calculator who has frequent occasion for questions involving fractions of hundred-weights, should provide himself with a table in which every pound up to 111 is registered, and its fraction of a hundred weight exhibited. Supposing such a table, and one of logarithms, every figure of the work which need be written down is here before us; it will be worth while to put down the solution of the question in the shortest way in which it can be accurately done without any tables, by a person familiar with decimal fractions.

6·884	0·8379
33·88	1·5299
	2·3678
1·687	0·2271
138·2	2·1407

1 cwt. 2 qrs. 21 lbs.	: 6 cwt. 3 qrs. 15 lbs.	:: £33 17*s.* 7*d*
6 qrs. 21 lb.	27 qrs. 15 lbs.	4·079365
28	7	£ 30
168	189	122·38095
21	4	£3. 12·23810
189	756	15*s.* 3 05953
	15	2*s.* 6*d.* ·50992
		1*d.* ·01699
	189) 771 (4·079365	138·20549
	756	Ans. £138 4*s.* 1¼*d*
	1500	
	1323	
	1770	
	1701	
	690	
	567	
	1230	
	1134	
	96	

With such a knowledge of decimals as is supposed in the manner of working which has just been given, logarithms might be easily learnt: but the calculator who can only proceed in the common way must turn the first and second terms into pounds avoirdupois, the third into pence, and will then have the satisfaction of multiplying 8131 by 771, dividing the product by 189, and turning the quotient into pounds, shillings, and pence.

91·88	1·9632
3·5	0·5441
	2·5073
3	0·4771
107·2	2·0302

Example 4. When the three per cents. are at 91⅞, what is the proper price of three and a half per cents.? Multiply 91⅞ or 91·88 by 3·5, and divide by 3, which gives 107⅕ as the price of the three and a half per cents.

13·18	1·1199
100	2·0000
	3·1199
84·54	1·9271
15·59	1·1928

Example 5. What per centage is £13 3*s.* 6*d.* of £84 10*s.* 9*d.* The first being multiplied by 100, and divided by the second, gives 15·59, or £15 12*s.* per cent.

Example 6. The exchange between England and France being 25 francs, 7 sous, to the pound sterling, what English money is 2074 francs worth? Here 25 francs, 7 sous is 25·35 francs, by which dividing 2074 we have 81·81, the number of pounds required. This is £81 16*s.* 2*d.* nearly. The correct answer is £81 16*s.* 3½*d.*

2074	3·3168
25·35	1·4040
81·81	1·9128

In many questions, approach to exactness may be attained by taking out a large part of the answer by simple perception, as in the following.

Example 7. Sold goods to the amount of £3619, on which 27½ per cent. was gained; what was the prime cost? Here £3000 of prime cost would have made 3000+30×27½, or £3825, which being over the mark by £206, we have to deduct from £3000 the prime cost which would produce £206, or £161. 10*s.* This gives £2838 10*s.*, the real answer being £2838 8*s.* 7½*d.* If we had begun with the whole £3619, the answer obtained by the logarithmic process would have been £2839.

206	2·3139
100	2·0000
	4·3139
127·5	2·1055
161·5	2·2084

Example 8. A population doubles itself in 26 years, what is the yearly rate of increase. It is here required to find the 26th root of 2. The logarithm of 2 is ·3010, the 26th part of which is ·0116, the number to which is 1·027. Hence the increase is 27 in 1000 yearly, or $2\frac{7}{10}$ per cent.

Example 9. At an increase of $4\frac{3}{10}$ per cent. per annum, how long will a population be in becoming five times as great as it now is. Here 1 at the end of a year becomes 1·043, the logarithm of which is ·0182, which is contained in the logarithm of 5, or ·6990, 38 times and a fraction. The answer then is, that the population will be a little less than five times its present amount in 38 years, and a little greater in 39.

Example 10. What will be the state of a population in 42 years which increases yearly by 6 in 1000? The question here is to find the 42nd power of 1·006. The logarithm of 1·006 is ·0025, which taken 42 times is ·1050, which is the logarithm of 1·274. The answer then is, that for every 1000 inhabitants there will in 42 years be about 1274.

The preceding examples will be sufficient to show the facility with which logarithmic methods may be applied. If this article should induce any one to learn the use of the four-figure table, we should recommend him next to turn his attention to those of five and seven places. We do not recommend six-figures, except for the more exact trigonometrical purposes: in the tables of simple logarithms, we believe it will be found, that owing to the completeness with which what are called the proportional parts are given, the seven-figure tables will be more readily used than the six-figure ones.

We now make some remarks upon a subject which grows out of the preceding, though not immediately connected with the subject of logarithms: namely, the state of our coinage. We have seen that, owing to an accidental circumstance, the reduction of shillings, &c. to decimals of a pound is easily done, approximately at least. The question then arises, could not a slight alteration be made, which should make the preceding approximation perfectly exact? And could this be done without introducing permanent confusion, injustice, or error?

There are those who think that any alteration of established weights, measures, or coins must be injurious, whatever may be the abstract merit of the proposed innovation. We cannot lean to this opinion ourselves, further than to the extent of admitting that the alteration should be glaringly advantageous, and that this should be made obvious to a very large proportion of the educated community. Others again doubt the practicability of introducing any changes without a long period of confusion, and the conquest of a large mass of resistance. We agree with them as to weights and measures, but not as to coinage, in the matter of which it is in the power of any government to make any amount of change as speedily as it pleases.

The French have succeeded in establishing a purely decimal coinage. Their franc now consists of 10 decimes, each decime being 10 centimes. The sou, or semi-decime, though still a coin, is not used in accounts. Thus 3·76 francs means 3 francs, 7 decimes, and 6 centimes, or 3 francs, 15 sous, and 1 centime. The inconvenience and even absurdity of their system is, that it subdivides the franc too far: the centime is hardly in use, and all merchants' accounts present nothing but sous at lowest, so that nothing under 5 centimes appears.

We have amongst ourselves advocates of an alteration, who may be divided into two classes; the first advocating a purely decimal coinage, and the second an approximation to it. We hold entirely with the first: being satisfied that nothing short of the complete advantage of a purely decimal coinage will ever repay the trouble and difficulty of the change.

Again, the arguments against a decimal coinage are frequently put in a manner which it is not our intention to meet. Persons already well practised in the existing system ask what advantage *they* are to derive from the change. We answer, perhaps none at all, the conveniences and inconveniences being fairly balanced. But the next generation, which is educated in the new system, will have the former without the latter: and a decimal coinage, if ever it be introduced, will be a sacrifice on the part of some of those who make the change for the benefit of their successors. Those who bequeath such a national debt as ours may not unreasonably be expected also to bequeath the readiest mode of reckoning it.

The inconveniences of a change will be much alleviated, if it can be brought about so gradually that no one shall be able to say exactly where the new system begins, and the old one ends. This we think might be done in the following manner.

As the matter stands, the pound (which should remain the same quantity, both of gold and silver, as at present) is subdivided into 20 parts, each of which is again subdivided into 12 parts, each of which is again subdivided into 4 parts. It is required to alter it, so that each part shall be subdivided into 10 others. That we may give names, let us call these parts royals, new groats, and new farthings: thus a pound is 10 royals, a royal 10 new groats, and a new groat is 10 new farthings. Hence it follows immediately that

The royal is two shillings.
The new groat is $2\frac{4}{10}$ of a penny, or $2\frac{1}{2}d.$ very nearly.
The new farthing is $\frac{96}{100}$ of the old one.

First, let pieces of two shillings each be coined, with a distinct name, and let no more half-crowns nor crowns be issued, either from the Mint or the Bank. The old crowns and half-crowns would thus be gradually withdrawn from circulation, and by what we have called the royal, the first step towards the decimal division would be established.

This first step having been made, no other need be taken, until the new coin has become familiar, which would require perhaps a couple of years. The next process must be, to issue copper or mixed coins of $2\frac{1}{2}d.$ each. These should be made so portable as to be convenient, which might easily be attained, as in the small French pieces of two sous each, for instance. This coin of $2\frac{1}{2}d.$ would in the end stand for what we have called the new groat; that is, would count as $2\frac{4}{10}d.$ of the old money. This alteration of four per cent. in the value of a copper coin would be of no consequence whatever: since the daily fluctuations in the price of copper amount to much more. But no pence nor half-pence should be withdrawn from circulation, nor any reference made to the new system, in the first issue of groats. These new coins would be immediately nicknamed *copper half-crowns*, which would assist the comprehension of them materially, accustomed as persons are already to $2\frac{1}{2}$ in shillings, by means of the silver half-crown.

Since 10 of these new groats would make 2*s.* 1*d.*, they would immediately be reckoned as having the value of five to the shilling, in all purchases of three shillings or upwards, when change is given. Articles would be priced in shillings and groats; and in all large transactions it would become customary to keep accounts in royals and groats, neglecting the fractions of the latter. Thus £17·83 would stand for 17 pounds, 8 royals (16 shillings), and 3 groats. And those who were more particular would drop the penny, and divide the groat into five half-pence, or even into ten farthings, which would complete the decimal system. However this might be, it is clear that while all the present coinage (except the half-crown) continued in existence, with the royal of two shillings, and the groat of $2\frac{1}{2}d.$ superadded, those who liked would have the means of introducing the decimal system into their accounts; while those who did not would be able to retain the old one.

The last step would be to coin farthings of ten to the groat, which would never differ practically from the common farthings. The only positive measure now requisite to make the system perfectly decimal, would be to enact that five groats should be legal tender for one shilling, and ten groats for two shillings: and this would practically be true previous to the enactment.

The money used in accounts would now be

1	Pound.			
10	1	Royal.		
100	10	1	Groat.	
1000	100	10	1	Farthing.

10 farthings are 1 groat.
10 groats are 1 royal.
10 royals are 1 pound or sovereign.

The coins in use would be, the farthing, the half-penny, penny, *groat*, sixpence, shilling, *royal*, half-sovereign, and sovereign. Before the half-pence and pence were gradually withdrawn, the only new rule relative to an old coin would be that the silver sixpence would be sixpence farthing in copper, and the silver shilling twelve pence half-penny. The final coinage should be the farthing, half-groat, groat, quarter-royal (old sixpence), half-royal (old shilling), royal, quarter-pound, half-pound, and pound, of course under more convenient names, the existing ones being taken as far as they go.

If the preceding scheme were looked at all at once, it would appear liable to a great deal of confusion, as must be the case in any scheme which proposes to change a reckoning by dozens into one by tens at one stroke. But in order to judge of it fairly, it must necessarily be looked at in detail, and it must be asked, how much of this plan is in operation at once? What quantity of change is required to be learned at any one time; that is, in any one year?

We will now state a few of the advantages of a perfectly decimal coinage.

1. All computations would be performed by the same rules as in the arithmetic of whole numbers.

2. An extended multiplication table would be a better interest table than any which has yet been constructed.

3. The application of logarithms would be materially facilitated, and would become universal, as also that of the sliding rule.

4. The number of good commercial computers would soon be many times greater than at present.

5. All decimal tables, as those of compound interest, &c., would be popular tables, instead of being mathematical mysteries,

6. The old coinage would be reduced to the new by the simple rule given at the beginning of this article. Thus any person would see at once, after a moderate degree of practice in that rule, that £14 17*s*. 9½*d*. (old coinage) is £14, 8 royals, 8 groats, 9 farthings, of the new coinage, at least within a farthing: this would be written £14·889. Again, £23·614 of the new coinage, or £23, 6 royals, 1 groat, 4 farthings would be seen by the same rule to be £23 12*s*. 3½*d*. (old coinage).

7. When the decimal coinage came to be completely established the introduction of a decimal system of weights and measures would be very much facilitated, and its advantages would be seen.

The preceding advantages are sufficiently obvious; but the question is, do they overbalance the disadvantages? These are

1. The period of confusion attending all changes.

2. The existence of a class of persons (adults) who cannot by any process master any difficulty of an arithmetical kind.

3. The loss of the subdivision of the shilling into twelve parts; twelve having more divisors than 10.

There is no question in our own minds as to the side on which the scale predominates: but perhaps practised arithmeticians have hardly a right to give a positive opinion on this question, which must be settled by the feeling of the large majority who would reconcile themselves to the change with more or less of difficulty. Those who consent to face this difficulty will deserve the thanks of posterity: and we cannot but think that there are few who, looking at the gradual and easy manner in which the new system could be introduced, would count their own share of the necessary inconvenience too much to pay for a real and lasting benefit to society.

A. DE MORGAN.

University College,
July 22, 1840.

The rule given in this article for turning shillings, &c., into decimals of a pound, and *vice versâ*, occurs, for the first time that we know of, in Kersey's edition of Wingate's Arithmetic, 1673 (p. 191). It was given again by Hawkins, in what he calls Cocker's Decimal Arithmetic, 1685 (see *Penny Cyclop.*, article COCKER, as to whether this, which is not the well-known work of Cocker, be really his or Hawkins's). It was claimed as his own (probably under colour of a trivial alteration for the better) by one William Weston, who published (without date, but subsequent to 1762) a work called "Specimens of Abbreviated Numbers." Owing to not appearing in the ordinary works on arithmetic, it fell entirely out of use, though it is frequently a material simplification of operations, until it was revived, we believe, by Mr. Milne, in his work on Annuities, published in 1815. We are told that it was also revived in the same year or thereabouts, by the author of a "Catechism of Arithmetic."

COMPANION TO THE ALMANAC

FOR

1842.

PART I.

INFORMATION CONNECTED WITH THE CALENDAR AND THE NATURAL PHENOMENA OF THE YEAR; AND WITH NATURAL HISTORY AND PUBLIC HEALTH.

I.—ON LIFE CONTINGENCIES. No. II.

THE present paper is a continuation of that in the Companion for 1840, on the application of Barrett's method of calculating life contingencies. No tables were then published for contingencies involving two lives; but since that time Mr. Jones's tables have appeared, in his work on the subject in the Library of Useful Knowledge. It may therefore be worth while to collect and describe the formulæ relative to two lives which are most likely to be useful in practice; with some other matters, which may partly interest the actuary and partly the commencing student.

The method described in the former paper is extended by calculating every case of

$$D_{m,n}=a_m\,a_n\,v^k,\ \ D_{m+1,\,n+1}=a_{m+1}\,a_{n+1}\,v^{k+1},\ \&c.$$

where a_x always means the number living at the age x, v^x the present value of £1 due in x years, and k is any convenient number depending on m or n, or both. In the tables published, k is taken to be the greater of the two, m or n; the consequence of which is, that two formulæ are necessary in every case in which the order of survivorship is a part of the question, according as the life, on the survivorship of which a benefit depends, is the elder or younger of the two. This might have been avoided by choosing $\frac{1}{2}(m+n)$ for k, and on such an assumption we shall first exhibit the formulæ which the different cases require; pointing out, at the same time, how to adapt the results to the published tables. Such a course will not only be the most simple with respect to the theory, but will always enable the reader to detect a misprint, if any should exist.

We begin, then, as follows:—

$$D_{x,y}=a_x\,a_y\,v^{\frac{1}{2}(x+y)},\ \ N_{x,y}=D_{x+1,\,y+1}+D_{x+2,\,y+2}+\ldots.$$

If the tables were complete we should then have

$$C_{x,y}=(a_x\,a_y-a_{x+1}\,a_{y+1})\,v^{\frac{1}{2}(x+y)+1}$$

$$M_{x,y}=C_{x,y}+C_{x+1,\,y+1}+\ldots.$$

but as these tables have not been published, we must express M in terms of N as follows:—

$$M_{x,y}=vN_{x-1,\,y-1}-N_{x,y}=D_{x,y}-(1-v)\,N_{x-1,\,y-1}$$

B

In the former paper $N_{x,y}$ stood for $N_x - N_{x+y}$; in the present one let

$N_{x,y} - N_{x+z,y+z}$ be denoted* by $N_{x,y|z}$.

In all questions in which the joint duration of two lives takes the place of one life, the formulæ are obtained from those of the former paper by simply substituting $N_{x,y}$ for N_x, and so on. The following are instances, (x) and (y) representing individuals aged x and y years:—

1. The value of an annuity of £1 on the joint lives of (x) and (y) is

$N_{x,y}$ divided by $D_{x,y}$

2. The same, for n years, if both parties live so long, is

$N_{x,y|n}$ or $N_{x,y} - N_{x+n,y+n}$, divided by $D_{x,y}$

3. The value of £1 to be received in k years, if both (x) and (y) be then alive, is

$D_{x+k,y+k}$ divided by $D_{x,y}$

4. The value of an annuity of £1 deferred for k years is

$N_{x+k,y+k}$ divided by $D_{x,y}$

5. The premium for the same (k premiums in all) is

$N_{x+k,y+k}$ divided by $N_{x-1,y-1} - N_{x+k-1,y+k-1}$

Or, to pay for it in l premiums, take $N_{x-1,y-1} - N_{x+l-1,y+l-1}$ for a divisor.

6. Present value of an assurance of £1 on the death of the first of (x) and (y).

$$\frac{M_{x,y}}{D_{x,y}} \text{ or } \frac{vN_{x-1,y-1} - N_{x,y}}{D_{x,y}} \text{ or } \frac{D_{x,y} - (1-v)\,N_{x,y}}{D_{x,y}}$$

7. The same, if the first death take place within n years.

$$\frac{M_{x,y|n}}{D_{x,y}} \text{ or } \frac{v(N_{x-1,y-1} - N_{x+n-1,y+n-1}) - (N_{x,y} - N_{x+n,y+n})}{D_{x,y}}$$

$$\text{or } \frac{D_{x,y} - D_{x+n,y+n} - (1-v)\,(N_{x-1,y-1} - N_{x+n-1,y+n-1})}{D_{x,y}}$$

8. To pay for either of the preceding in premiums, write for the divisor

$N_{x-1,y-1}$ or $N_{x-1,y-1} - N_{x+l-1,y+l-1}$

according as the premium is for life or for l years.

These are the most common cases, and any others which arise can as easily be found from the preceding paper. The usual cases of reversionary annuities, annuities on the longest liver, &c., may

* It was not contemplated in the last paper that any one would have had the courage to compute tables for two lives. Had it been thought likely, what was there denoted by $N_{x,y}$ would have been denoted by $N_{x|y}$.

be omitted here, as they are more easily calculated by the common tables.

In cases which involve the order of survivorship, tables in which $\frac{1}{2}(x+y)$ is the exponent of v in $D_{x,y}$, will introduce the square root of v. The method of adapting their results to the case in which that exponent is the *greater of the two*, x or y, must be described as follows:—

Change $D_{x,y}$ into $D_{x,y}\,v^{-\frac{1}{2}(x-y)}$ or $D_{x,y}\,v^{-\frac{1}{2}(y-x)}$ according as (x) or (y) is the older life; and the same for $N_{x,y}$.

In the following instances the first result given is that in which $D_{x,y}$ stands for $a_x\,a_y\,v^{\frac{1}{2}(x+y)}$; the second, that in which (x) is the older and $D_{x,y}$ denotes $a_x\,a_y\,v^x$; the third, that in which (y) is the older and $D_{x,y}$ denotes $a_x\,a_y\,v^y$.

9. Required the present value and premium of an assurance of £1 on the life of (x), provided that (y) survive (x). The first divisor is for the present value, the second for the premium.

$$\frac{v\,N_{x-1,\,y-1}-N_{x,\,y}+\sqrt{v}\,(N_{x-1,\,y}-N_{x,\,y-1})}{2D_{x,\,y}\text{ or }2N_{x-1,\,y-1}}$$

$$(x>y)\qquad \frac{v\,(N_{x-1,\,y-1}+N_{x-1,\,y})-(N_{x,\,y-1}+N_{x,\,y})}{2D_{x,y}\text{ or }2N_{x-1,\,y-1}}$$

$$(y>x)\qquad \frac{v\,(N_{x-1,\,y-1}-N_{x,\,y-1})+N_{x-1,\,y}-N_{x,\,y}}{2D_{x,\,y}\text{ or }2N_{x-1,\,y-1}}$$

$$(y=x)\qquad \tfrac{1}{2}\text{ result of (6)}$$

10. If the risk of the preceding question be not to begin until k years have elapsed, write $x+k$ and $y+k$ instead of x and y *in the numerators only*, which then become

$$(x>y)\qquad v\,(N_{x+k-1,\,y+k-1}+N_{x+k-1,\,y+k})-(N_{x+k,\,y+k-1}+N_{x+k,\,y+k})$$

$$(y>x)\qquad v\,(N_{x+k-1,\,y+k-1}-N_{x+k,\,y+k-1})+N_{x+k-1,\,y+k}-N_{x+k,\,y+k}$$

11. And in all cases in which premiums are to be paid t times, instead of during the whole joint duration, the divisor has

$$N_{x-1,\,y-1\,|\,t}\text{ or }N_{x-1,\,y-1}-N_{x+t-1,\,y+t-1},\text{ instead of }N_{x-1,\,y-1}.$$

It will very much simplify both the use and theory of these formulæ if the student make himself perfectly master of the symbol $N_{x,y\,|\,t}$, which means the value of N at the ages x and y cut off, as it were, at the end of t years, or diminished by all the terms arising from the period beyond t years. Thus

$$\begin{aligned}
N_{x,\,y} &= D_{x+1,\,y+1}+\ldots\ldots+D_{y+t,\,y+t}+D_{x+t+1,\,y+t+1}+\ldots\ldots\\
N_{x+t,\,y+t} &= \qquad\qquad\qquad\qquad\qquad\; D_{x+t+1,\,y+t+1}+\ldots\ldots\\
N_{x,\,y}-N_{x+t,\,y+t} &= D_{x+1,\,y+1}+\ldots+D_{x+t,\,y+t}\\
&= N_{x,\,y\,|\,t}
\end{aligned}$$

12. Required the present value and premium (payable t times) of £1 to be paid at the death of (x), if within n years, and if (y) be then alive.

B 2

Cut off every term in the numerator of (9) at n years, and use the divisor for l premiums, when necessary:

$$(x>y) \qquad \frac{v\,(N_{x-1,\,y-1\,|\,n}+N_{x-1,\,y\,|\,n})-(N_{x,\,y-1\,|\,n}+N_{x,\,y\,|\,n}}{2D_{x,\,y} \text{ or } 2N_{x-1,\,y-1\,|\,l}}$$

$$(y>x) \qquad \frac{v\,(N_{x-1,\,y-1\,|\,n}-N_{x,\,y-1\,|\,n})+N_{x-1,\,y\,|\,n}-N_{x,\,y\,|\,n}}{2D_{x,\,y} \text{ or } 2N_{x-1,\,y-1\,|\,l}}$$

13. To find the corresponding formulæ in the case in which the assurance is for the death of (x), provided y be then *dead*, calculate the present value of the assurance on the life of (x) absolutely (by the former paper), and subtract the present value of the proper one of the preceding formulæ; the remainder is the present value required. Find the premium by the usual method, or else by multiplying by $D_{x,y}$ and dividing by the proper divisor.

14. The present value of a life annuity of n payments on (y), the first being made at the $(k+1)$th anniversary of the present time which follows the death of (x).

N.B. If $k=0$ and n outrun the term of life, this is a common reversionary annuity on (y) after (x).

$$\frac{N_{y+k}-N_{y+k+n}}{D_y}-\frac{N_{x,\,y+k}v^{\frac{1}{2}k}-N_{x,\,y+k+n}v^{\frac{1}{2}(k+n)}}{D_{x,\,y}}$$

To suit the published tables the second numerator is to be altered as follows, all the rest remaining:—

$x<$ or $=y$	into	$N_{x,\,y+k}-N_{x,\,y+k+n}$
$x>y<$ or $=y+k$	,,	$(N_{x,\,y+k}-N_{x,\,y+k+n})\,v^{x-y}$
$x>y+k<$ or $=y+k+n$	,,	$N_{x,\,y+k}v^{k}-N_{x,\,y+k+n}v^{x-y}$
$x>y+k+n$		$N_{x,\,y+k}v^{k}-N_{x,\,y+k+n}v^{k+n}$

15. The present value of £1 to be paid at the death of (x), if he die before (y) or within n years after (y).

$$\frac{M_x-M_{x+n}}{D_x}+$$

$$v^{\frac{1}{2}n}\,\frac{N_{x+n-1,\,y-1}v-N_{x+n,\,y}+v^{\frac{1}{2}}(N_{x+n-1,\,y}-N_{x+n,\,y-1})}{2D_{x,\,y}}$$

To suit the published tables the second numerator ($v^{\frac{1}{2}n}$ included) must be

$$(x+n<y)$$

$$(N_{x+n-1,\,y-1}-N_{x+n,\,y-1})v^{n+1}+(N_{x+n-1,\,y}-N_{x+n,\,y})v^{n}$$

$$(x<y,\,x+n>y)$$

$$\{(N_{x+n-1,\,y-1}+N_{x+n-1,\,y})v-(N_{x+n,\,y}+N_{x+n,\,y-1})\}\,v^{y-x}$$

$$x>y \qquad (N_{x+n-1,\,y-1}+N_{x+n-1,\,y})v-(N_{x+n,\,y}+N_{x+n,\,y-1})$$

If this risk be to continue only for $n+p$ years, all the Ns must be stopped after p years, or $N_{x+n-1,\, y-1|p}$ must be written for $N_{x+n-1,\, y-1}$, and so on.

If $x+n=y$, the second numerator must be

$$\{(N_{x+n-1,\, y-1}v+N_{x+n-1,\, y})-(N_{x+n,\, y}+N_{x+n,\, y-1})\}v^n$$

16. The present value of £1 to be paid on the death of (x), provided he die within n years after y.

From the preceding result subtract the present value of £1 payable at the death of (x), if (y) survive. (See No. 9.)

17. The present value of £(1) to be paid at the death of (x), if he survive (y).

From the present value of an assurance on the life of (x) subtract the result of (9), or the value of £1 to be paid at his death, if y survive.

18. The present value of £1 to be paid at the death of (x), if more than n years after the death of (y).

From the present value of an assurance of £1 on the life of (x) subtract the result of (15). This gives $M_{x+n} \div D_x$ for the first term.

I have not thought it necessary to write down the formulæ connected with the last few cases, as they seldom occur.

There are, we are told, still some persons who cannot see how much shorter this method of Barrett's is than that in previous use. It is certain that any new method does not become so easy as the old and well-known one until a few examples have been tried; and there are dispositions which will not give a new method a fair trial. But, for those who may be commencing the subject, and to whom, therefore, all methods stand entirely on their own merits, we subjoin the example of a simple temporary assurance on two joint lives, worked by both methods.

Required the present value of an assurance for the next ten years only, on the joint continuance of two lives, now aged 45 and 50 years: interest 5 per cent. (Milne, p. 341.)

Old Method.

a_{65}	3·60991	1 − ·43827	
a_{60}	3·56146	= ·56173	$\bar{1}$·74953
a^{-1}_{45}	$\bar{4}$·32541	v	$\bar{1}$·97881
a^{-1}_{50}	$\bar{4}$·35684		
v^{10}	$\bar{1}$·78811	·53498	$\bar{1}$·72834
·43827	$\bar{1}$·64173	6·623	·82105
	·85163	$1-v$	$\bar{2}$·67778
3·114	·49336	·31538	$\bar{1}$·49883
9.737		·53498	
6·623		·21960 present value of £1.	

Barrett's Method.

$N_{44,49}$ 19460646			$N_{45,50}$ 17648150	
$N_{54,59}$ 6438764			$N_{55,60}$ 5644407	
13021882	7·11468		12003743	
$1-v$	$\bar{2}$·67778			
620100	5·79246			
12401782				
12003743				
398039	5·59992			5·59992
$D_{45,50}$ 1812496	6·25828		13021882	7·11467
Present value ·21960	$\bar{1}$·34164		Premium ·030567	$\bar{2}$·48525
			Milne, p. 346, ·030565	

If we were to look at the number of figures only there would seem to be little difference between the two methods, the first containing about 130 figures and the second 120. But if we look at the number of entries of different tables in the two, we find it as follows:—

Old Method.		New Method.	
No. of pages in Milne which must be consulted for log a_{55}, &c., v^{10}, 9·737, v and $1-v$	3	No. of pages in Jones to be consulted for $N_{41,59}$.. ..$D_{45,50}$	1
Logarithms to be taken out	6	Logarithms taken out .	4
Numbers to logarithms .	4	Numbers to logarithms .	2
	13		7

In multiplying by v in the above process, the formula taken is

$$av=a-(1-v)a,$$

and $(1-v)a$ is found by logarithms and subtracted from a. The reason is that v is so near unity that it would take tables of logarithms to seven places to get five places of the answer correct, if av were to be directly found.

As another example, take the following (Milne, p. 360): Required (at 5 per cent.) the annual premium for £1, payable at the death of (50), if within 10 years, and (45) be then alive; premiums being payable for ten years, if the joint lives endure so long. The first formula of No. 12 must be employed:

$N_{49,44}$	19460646	$N_{50,44}$	18016514	$N_{49,45}$	19057899	$N_{50,45}$	17648150
$N_{59,54}$	6438763	$N_{60,54}$	5834103	$N_{59,55}$	6227029	$N_{60,55}$	5644407
	13021883		12182411		12830870		12003743
	12830870						12182411
	25852753						24186154

25852753	7·4125067
$1-v$	$\bar{2}$·6777803
1231082	6·0902870
24621671	
24186154	
435517	5·6390051
13021883	7·1146737
2)·03344502	$\bar{2}$·5243314
.01672251 required premium.	

Let the problem be now changed by simply inverting the ages. Required the annual premium for £1 payable at the death of (45) if within ten years, and (50) be then alive. The second formula of No. 12 must now be used, but our preceding process contains all the data:

$N_{44,49\mid 10}$	13021883		$N_{44,50\mid 10}$	12182411
$N_{45,49\mid 10}$	12830870		$N_{45,50\mid 10}$	12003743
	191013	5·2810629		178668
v		$\bar{1}$·9788107		
	181917	5·2598736		
	178668			
	360585	5·5570076		
	13021883	7·1146737		
	2)·02769069	$\bar{2}$·4423339		
	·01384534	present value required.		
	·01672251	result of the preceding.		
	·03056785			

This last result ·03056785 is the premium for the assurance on the joint lives, and agrees both with the above result and with Mr. Milne. But Mr. Milne, instead of ·01672251, has (p. 361) ·015714, differing materially from the truth. The reason is, as those who are used to the common method are well aware, that the old tables of annuities, calculated to three decimal places, are not sufficiently exact for survivorship questions, the results of which depend upon the differences of nearly equal annuities, particularly where, owing to the differences of ages given being only every five years, the defective interpolation by first differences is employed. Mr. Milne's process, it is hardly necessary to say, is as good as, with his tables, it could have been: but the above examples show, not only that Barrett's method is much the more easy of the two, but also the more exact, the tables actually published* being used. All the above results are true to six significant figures.

We now proceed to consider one or two questions in which approximations are usually made, with a view to state the results of an examination of the approximations, and to propose some amendments upon them.

1. The values of life annuities are usually calculated on the supposition that interest can be converted into principal only once a year, the truth being that such conversion always takes place twice a year at least. Mr. Milne has given a rule (p. 367) for making the requisite correction, but I believe the following will be found to be more easy of application: it is given for the case of half-yearly conversions only. Let the value of an annuity (for lives or a term of years) be A at c per cent., A_1 at $c+1$ per cent., A_2 at $c+2$ per cent., and so on. Take the differences of A, A_1, &c., as follows:—

$$\begin{array}{ccccc} A & & & & \\ & B & & & \\ A_1 & & C & & \\ & B_1 & & D & \\ A_2 & & C_1 & & \\ & B_2 & & & \\ A_3 & & & & \end{array} \qquad \begin{array}{lll} B=A-A_1 & B_1=A_1-A_2 & B_2=A_2-A_3 \\ C=B-B_1 & C_1=B_1-B_2 & \\ D=C-C_1 & & \end{array}$$

Then the value of the same annuity at c per cent., interest being convertible twice a year, is

$$A-\frac{c^2}{400}\left(B+\frac{C}{2}+\frac{D}{3}-\frac{c^2(C+D)}{800}\right), \text{ very nearly.}$$

Example 1. Required the value of a perpetual annuity of £1 at 4 per cent., payable yearly, on the supposition that money is convertible half yearly.

$$\begin{array}{llll} A\ = 25\cdot000 & & & \\ & 5\cdot000 & & \\ A_1 = 20\cdot000 & & 1\cdot667 & \\ & 3\cdot333 & & \cdot715 \\ A_2 = 16\cdot667 & & \cdot952 & \\ & 2\cdot381 & & \\ A_3 = 14\cdot286 & & & \end{array}$$

$$(1\cdot667+\cdot715)\frac{16}{800}=\cdot048 \qquad \begin{array}{r} 5\cdot000 \\ \cdot889 \\ \cdot238 \\ \hline 6\cdot127 \\ \cdot048 \\ \hline 6\cdot079 \end{array}$$

$$6\cdot079\times\frac{16}{400}=\cdot243$$

$$25\cdot000-\cdot243=24\cdot757 \text{ answer.}$$

* The superior exactness of the above method, as it stands, is not owing to anything in the method, but solely arises from the greater extent of the tables published.

The real answer is 24·75225, and a perpetual annuity is the severest trial which can be made of the rule.

Example 2. (Milne, p. 367.) Required the value of a life annuity on (60), at 5 per cent., interest being convertible twice a year:

$$\begin{array}{llllll} A & 8\cdot940 & & & & \\ & & \cdot636 & & & \\ A_1 & 8\cdot304 & & \cdot075 & & \\ & & \cdot561 & & \cdot012 & \\ A_2 & 7\cdot743 & & \cdot063 & & \\ & & \cdot498 & & & \\ A_3 & 7\cdot245 & & & & \end{array}$$

$$(\cdot075+\cdot012)\frac{25}{800}=\cdot0027 \qquad \begin{array}{r}\cdot6360\\ \cdot0375\\ \cdot0040\\ \hline \cdot6775\\ \cdot0027\\ \hline \cdot6748\end{array}$$

$$\cdot6748\times\frac{25}{400}=\cdot04218$$

$$8\cdot940-\cdot04218=8\cdot89782$$

Mr. Milne's answer is 8·8977.

2. When an annuity is to be paid, not yearly, but half-yearly or quarterly, &c., interest being convertible into principal only once a year, the common correction of A, the present value of a yearly annuity, is to write $A+\frac{n-1}{2n}$ instead of A, n being the number of times which the annuity is payable yearly. This correction is strictly true when we speak of a perpetual annuity, on the supposition that by interest being convertible once a year, we mean that money may be invested at any one time of the year, so that the grantor of the annuity must be held to lose simple interest for the rest of the year on each of the prepaid portions. When this correction is to be investigated for a life annuity, it would generally be supposed that those who die in each year die at equal intervals. This is an insufficient supposition; for if the mortality of the table be increasing from year to year, it ought to be supposed that the mortality of the latter part of a year is greater than that of the former. We have therefore taken the supposition that the mortality of each year is to be constructed on the hypothesis of constant third differences, which has the effect of distributing the deaths of each year in a manner depending on the mortality of the preceding and following years.

Proceeding on this plan we find the following result. Let the annuity be payable n times a year, let the age be k, and let a_k be the number in the table alive at that age, A_k being the tabular yearly annuity. Also let

$$\beta=\frac{n-1}{2n}, \qquad \beta'=\frac{n^2-1}{72n^2}, \qquad \beta''=\frac{n^4-1}{180n^4}.$$

The value* of the annuity then is, accurately, so far as the suppositions are accurate (r being the interest of £1 for one year)

* As we give no demonstration, we should say that, besides obtaining this expression from demonstration, we have tried several verifications: the simplest is that in which a_k is a constant, and $A_k=1:r$, which gives the perpetual annuity.

B 3

$$\left(1+6\beta'\frac{r^2}{1+r}-(\beta'+\beta'')\frac{r^4}{(1+r)^2}\right)A_k$$

$$+\beta-(8\beta'+2\beta'')\frac{r}{1+r}-(\beta'+\beta'')\frac{r^2}{(1+r)^2}$$

$$-3\beta'\frac{a_{k-1}-a_{k+1}}{a_k}+(\beta'+\beta'')r\left(\frac{a_{k+1}}{ra_k}+\frac{1}{1+r}\frac{a_{k-1}}{a_k}\right).$$

This expression contains a coefficient for A_k depending (n apart) on the rate of interest only; certain terms which depend on the rate of interest only; another depending on the mortality only; and others depending jointly on the rate of interest and the mortality. To form an idea of the utmost value of these corrections, let us take the extreme case of an annuity payable momently (or n infinite) and a high rate of interest, 10 per cent. We have then

$$\beta=\frac{1}{2}, \qquad \beta'=\frac{1}{72}, \qquad \beta''=\frac{1}{180};$$

and the value of the annuity is

$$1{\cdot}00076\ A_k+{\cdot}48873-{\cdot}03989\frac{a_{k-1}}{a_k}+{\cdot}04361\frac{a_{k+1}}{a_k},$$

$A_k+{\cdot}5$ being the common approximation.

We may say then that the common approximation is generally correct to the first place of decimals, but by no means to the second or third. Now, it may fairly be insisted, looking at the manner in which such calculations are usually conducted, that the second place at least in the decimal part of an annuity shall be made as correct as can be conveniently done; not that this is, generally speaking, of very great consequence, but because there are other things by the dozen of no greater consequence, which are always attended to.

3. If in the preceding money make no interest, A_k becomes $(a_{k+1}+a_{k+2}+\ldots)$ divided by a_k, and the annuity represents the mean duration of life, which is therefore (n being infinite)

$$\frac{a_{k+1}+a_{k+2}+\ldots}{a_k}+\frac{1}{2}-\frac{1}{24}\frac{a_{k-1}-a_{k+1}}{a_k},$$

the first two terms are those from which the mean duration is usually deduced: the third is substantially the correction which we have elsewhere* proposed.

4. An annuity for a term of years must be computed in the usual way from the corrected life annuities.

5. No tables have been published for the mean duration of pairs of lives; but we understand that this defect as far as regards the Carlisle Table, will be remedied in Mr. Jones's work. In the

* Cab. Cyclop. Probabilities and Life Contingencies, p. 163.

meantime it may be approximately supplied (see the work last cited, Appendix, p. xxx.), but much better from Mr. Jones's Tables of Annuities at 3, 3½, 4, 4½, and 5 per cent. If these be called A_6, A_7, A_8, A_9, and A_{10}, the approximate formula is

$$105(2A_6+9A_8)+126A_{10}-80(9A_7+7A_9)+\cdot 5.$$

Example 1. Required the mean duration of a life aged 45 (Carlisle Table).

$(2A_6+9A_8=158\cdot671)\times105=$	16660·455	15942·885
$(A_{10}=\ 12\cdot648)\times126=$	1593·648	1521·828
	18254·103	17464·713
$(9A_7+7A_9=227\cdot870)\times\ 80=$	18229·600	17440·160
	24·503	24·553
	·5	·5
	25·003	25·053
Correct answer	24·46	

Example 2. Supposing male and female life not to be of different value, what is the average duration of a marriage between parties aged 30 and 25?

Answer, 25 years, calculated as above on the right.

Example 3. What is the average age at which a son of 7 years succeeds to the estate of a father of 37, and how long on the average does he enjoy it?

The first part of the question means the average age of those sons who do succeed, without any reference to those who do not survive their fathers; and since at 37 the mean duration of life is 30 years, the answer is that such sons succeed, one age with another, at the age of 37. The last part of the question cannot be directly solved by existing tables, if it refer to the actual average period of enjoyment of all those persons who live to come into possession; but it must first be found how many years of enjoyment those sons have, one with another, on the supposition that the actual years of enjoyment are divided among the whole, both those who live to come into possession and those who do not. If L_x be the mean duration of (x), and $L_{x,y}$ that of (x) and (y) jointly, then $L_7-L_{7,37}$ is *approximately* the mean period by which a son of 7 years survives a father of 37. Or, if L_x and $L_{x,y}$ represent the values of the annuities when money makes no interest, $L_x-L_{x,y}+\cdot5$ is the mean period of survivorship; then $L_{x,y}$ is calculated by the above method, omitting the ·5, and $L_x+\cdot5$ is the mean duration of (x) in the table. Calculate $L_{7,37}$, which gives 27·62, and L_7 being 50·30, we have 22·68 (say 23) years for the average term of enjoyment. That is to say, if 1000 sons, each aged 7, and the father of each being aged 37, were to agree to club their future years of possession in such manner that all (or their executors) should take equal shares, whether they lived to come into possession or not, the years of enjoyment of those who do come into possession would yield to each of the 1000 sons 23 years a-piece, or thereabouts; and of course the average term of those who *do* come into possession, nothing being supposed given up, is considerably larger than this; and an approximation (and only an approximation, though very plausible-looking reasoning might be put together to make it appear an exact method) may be made to it as follows: Divide this term of 23 years by the chance which a son of 7 years has of surviving a father aged 37. In the table of chances of survivorship, it appears that ·7837 is the chance of (7) surviving (37), and 23÷·7837=29, which is about the number of years required. This approximation always gives the result too great, perhaps by a year or two; but rough as the method may be, it is, as far as we know, the first which has been proposed—the want of that most essential preliminary, tables of mean survivorship of two lives, having prevented the subject from being considered.

Example 4. What would be the joint duration of twins if they were assumed to be, at birth, lives of the same value as other infants? The calculation gives 21 years.

Example 5. What is the mean duration of the survivor of (25) and (30)? This must be found, approximately, from the formula $L_{25} + L_{30} - L_{25,30} + \cdot 5$, or $37\cdot 4 + 33\cdot 8 - 24\cdot 6 + \cdot 5$, or 47 years.

Example 6. What is the mean joint duration of the three lives (7), (24), and (40)? Take the two elder lives and calculate their mean joint duration, which is found to be 26·5. Find a life which has this mean duration, which is (42), and find the mean joint duration of (7) and (42) or 27 years. This result is absurd, since the mean duration of (42) itself is not quite 27 years; it arises from the error of the process. In the present instance the mean duration of (7) is upwards of 50 years, so that the joint duration of (7) and (42) is very little less than that of (42) itself; so little less, that the error incidental to the process (which is generally in excess) more than makes the difference.

Also the annuities used in the application of this rule must be tolerably exact to three decimals; and if there be any doubt about the interpolated annuities, recourse must be had to those derived from Barrett's tables.

6. Supposing an individual (x) to lay by £1 every year, what may he reasonably count upon having at his death, supposing that he improves the money at compound interest?

Supposing £1 to be laid by now, the question is the same as, how much should a premium of £1 assure at his death? and the answer is;— N_{x-1} divided by M_x in the table, or $(1+r)(1+A)$ divided by $1-rA$ from the common table, A being the value of a life annuity of £1. We mention this here because there is a table in Mr. Morgan's "Treatise on Assurances" (Table III., described in p. 54, and note V.) which was copied in part into the "Penny Cyclopædia," as being what its name would import, "the sum to which an annuity forborne and improved will amount on the extinction of a given life." Mr. Morgan's table gives the answer to the following question; if every one at his death were to receive his annuity (not a premium; exclude therefore the immediate payment) improved to its amount at the beginning of the year, what would persons of the same age at commencement receive one with another, from all the different amounts which they actually do receive? In our notation it is a calculation of the value of

$$\frac{a_{x+1}-a_{x+2}}{a_x}+\frac{a_{x+2}-a_{x+3}}{a_x}(1+\overline{1+r})+\frac{a_{x+3}-a_{x+4}}{a_x}(1+\overline{1+r}+\overline{1+r}^2)+\dots$$

or,
$$\frac{a_{x+1}}{a_x}+\frac{a_{x+2}}{a_x}(1+r)+\frac{a_{x+3}}{a_x}(1+r)^2+\dots.$$

This table may therefore serve (at least by the Northampton table at 4 per cent.) to compare the following cases. A and B are both of the same age: A is to assure his life immediately at a premium of £1, and B will in a year lay by £1, and the same at the end of every subsequent year. Supposing each to get the average of his class at death, what will they severally get? To bring both accounts up to the end of the year of death, at which the common assurance is supposed to be paid, a year's interest must be added to the result in Mr. Morgan's table. The following are the average sums obtained in the two cases, at different ages:—

	A	B
20	49·4	98·3
25	44·7	82·2
30	40·2	68·6
35	35·7	56·8
40	31·3	46·7
45	27·2	38·0
50	23·2	30·7
55	19·7	24·5
60	17·0	19·2

Supposing the Northampton table and 4 per cent. to be facts (we have chosen this table and rate because the only deduced table of the kind we are considering is Mr. Morgan's), it thus appears that the office gets from persons aged 20, one with another, in premiums and interest, £98·3 and the first premium with its interest, from each person, for every £49·4 assured. But many of these persons receive each his £49·4 long before he has earned it by his premiums, and the advances thus made by the office must be repaid with interest by those who live longest.

7. The effect of compound interest, as distinguished from simple, is not very easily made apparent, and is not of great importance, since simple interest, as applied to annuities, is really a fiction, not more worthy of attention than would be the transactions of a man who, having a number of sovereigns in his hand, and others coming in, should invent a rule for separating some from the rest, and locking them up in a cash-box, instead of investing them. Some writers have argued that the *theory* of

annuities at simple interest must be wrong, inasmuch as its consequence is, that an infinite sum is necessary to pay a perpetual annuity; the simple truth being, that if any one were to carry that theory into practice, he would find that the prediction of theory was verified; namely, that no sum, however great, would enable him to pay a perpetual annuity. According to the theory, the present value of an annuity (the rate of interest being r per pound) of £1 for n years is

$$\frac{1}{1+r}+\frac{1}{1+2r}+\frac{1}{1+3r}+\ldots+\frac{1}{1+nr}.$$

If n be considerable, the labour of summation will be materially lessened by using the following approximation.

Let $\frac{1}{1+r}=P_1$, and $\frac{1}{1+nr}=P_n$:

the value of the preceding series is nearly (the logarithm used being the common one)

$$\frac{2\cdot3025851}{r}\log\frac{1+nr}{1+r}+\frac{1}{2}(P_1+P_n)+\frac{r}{12}(P_1^2-P_n^2)-\frac{r^3}{120}(P_1^4-P_n^4).$$

To try this approximation, suppose r as great as $\cdot1$ and $n=10$. We have then $1 : r=10$, $nr = 1$ (the powers of P_1 may be taken from the table of present values of £1).

$$\left(\log\frac{2}{1\cdot1}=\cdot2596373\right)\times\frac{2\cdot3025851}{\cdot1} = 5\cdot978370$$

$$\frac{1}{2}(\cdot909091+\cdot5) = \cdot704546$$

$$\frac{\cdot1}{12}(\cdot826446-\cdot25) = \cdot004804$$

$$-\frac{(\cdot1)^3}{120}(\cdot683013-\cdot0625) = -\cdot000005$$

$$6\cdot687715$$

Verification.		Reciprocal.
	1·1	·90909091
	1·2	·83333333
	1·3	·76923077
	1·4	·71428571
	1·5	·66666667
	1·6	·62500000
	1·7	·58823529
	1·8	·55555556
	1·9	·52631579
	2·0	·50000000
		6·68771403
	Approx.	6·687715
	Error	0·000001

The approximative process would not be more difficult, nor less exact, if the number of terms were a million.

8. A more practical question than the last is derived from the following considerations. It is well known that individuals rarely or never make compound interest of what they lay by; while it is possible that many lay by a sum which increases from year to year equally or nearly so. We have seen (in 6.) that an assurance office seems to offer a greater moral benefit to the old than the young; inasmuch as the latter, in making the office their trustee, seem to give up a more brilliant average prospect. But this is on the supposition that compound interest is made: let us now suppose that the *office* makes compound interest, while the *individual* would only lay by an increasing annual sum. We now introduce the following table. (See p. 16.)

This table is nothing but the table described in our last paper, on the supposition that money makes no interest, so that all results depend merely on the average duration of life. This table, as we shall proceed to show, concerns persons in general more almost than any other of the kind. All persons speculate, more or less, on what they can do with their incomes during their lives, without thinking of interest; that is, they attempt to form notions which this table would enable them to form justly.

Example 1. A person aged x is to lay by £1 at the end of every year from this time. What will the savings amount to at his death, one person with another?

Ans. $\dfrac{N_x}{D_x}$; if $x=20$, $\dfrac{249432}{6090}=£40\cdot96$.

Example 2. A person aged x is to lay by £1 for n years, if he live so long: what will be the average amount?

Ans. $\dfrac{N_x-N_{x+n}}{D_x}$; if $x=20$, $y=10$, $\dfrac{249432-190900}{6090}=£9\cdot6$.

Example 3. What should a person aged x lay by now and at the end of every year, that, one such person with another, there may be £1 at death?

Ans. $\dfrac{D_x}{N_{x-1}}$; $x=20$, $\dfrac{6090}{255522}=£\cdot024$, say 0*s.* 6*d.*

Example 4. A person aged x is to lay by a at the end of a year, $a+h$, $a+2h$, &c., at the end of 2, 3, &c., years; what will be the average savings of such persons, one with another?

Ans. $a\dfrac{N_x}{D_x}+h\dfrac{S_{x+1}}{D_x}$.

This problem represents better than any other in our opinion, the *best* supposition that can be made as to what any individual can do who attempts to be his own life-assurer. If $x=20$, we have $41a+990h$ for the average required.

Suppose then an individual (A) in this way steadily to invest £1 every year, with three per cent. simple interest upon the preceding investments, so that $a=1$, $h=\cdot03$, and the preceding is £71; such individuals, one with another, will have £71 at death;

Age.	D.	N.	S.	Age.	D.	N.	S.
0	10000	382213	12603644	53	4211	77784	967464
1	8461	373752	12221431	54	4143	73641	889680
2	7779	365973	11847679	55	4073	69568	816039
3	7274	358699	11481706	56	4000	65568	746471
4	6998	351701	11123007	57	3924	61644	680903
5	6797	344904	10771306	58	3842	57802	619259
6	6676	338228	10426402	59	3749	54053	561457
7	6594	331634	10088174	60	3643	50410	507404
8	6536	325098	9756540	61	3521	46889	456994
9	6493	318605	9431442	62	3395	43494	410105
10	6460	312145	9112837	63	3268	40226	366611
11	6431	305714	8800692	64	3143	37083	326385
12	6400	299314	8494978	65	3018	34065	289302
13	6368	292946	8195664	66	2894	31171	255237
14	6335	286611	7902718	67	2771	28400	224066
15	6300	280311	7616107	68	2648	25752	195666
16	6261	274050	7335796	69	2525	23227	169914
17	6219	267831	7061746	70	2401	20826	146687
18	6176	261655	6793915	71	2277	18549	125861
19	6133	255522	6532260	72	2143	16406	107312
20	6090	249432	6276738	73	1997	14409	90906
21	6047	243385	6027306	74	1841	12568	76497
22	6005	237380	5783921	75	1675	10893	63929
23	5963	231417	5546541	76	1515	9378	53036
24	5921	225496	5315124	77	1359	8019	43658
25	5879	219617	5089628	78	1213	6806	35639
26	5836	213781	4870011	79	1081	5725	28833
27	5793	207988	4656230	80	953	4772	23108
28	5748	202240	4448242	81	837	3935	18336
29	5698	196542	4246002	82	725	3210	14401
30	5642	190900	4049460	83	623	2587	11191
31	5585	185315	3858560	84	529	2058	8604
32	5528	179787	3673245	85	445	1613	6546
33	5472	174315	3493458	86	367	1246	4933
34	5417	168898	3319143	87	296	950	3687
35	5362	163536	3150245	88	232	718	2737
36	5307	158229	2986709	89	181	537	2019
37	5251	152978	2828480	90	142	395	1482
38	5194	147784	2675502	91	105	290	1087
39	5136	142648	2527718	92	75	215	797
40	5075	137573	2385070	93	54	161	582
41	5009	132564	2247497	94	40	121	421
42	4940	127624	2114933	95	30	91	300
43	4869	122755	1987309	96	23	68	209
44	4798	117957	1864554	97	18	50	141
45	4727	113230	1746597	98	14	36	91
46	4657	108573	1633367	99	11	25	55
47	4588	103985	1524794	100	9	16	30
48	4521	99464	1420809	101	7	9	14
49	4458	95006	1321345	102	5	4	5
50	4397	90609	1226339	103	3	1	1
51	4338	86271	1135730	104	1	0	0
52	4276	81995	1049459				

that is, some will have next to nothing, a majority less (many of them much less) than £71, and a minority more, a few of them much more. Another individual (B) goes with £1 down and £1 yearly to an office which charges the Carlisle premiums at 3 per cent., increased by 25 per cent. for management and fluctuation. Such an office, returning profits, will pay, let us say £120 for every £100 assured, to its customers, one with another. If this be the case it will pay the preceding individual £64 at his death. Let us now see how the parties stand.

A	B
has been gambling, setting a remote chance of more than the office would give against a great risk of leaving next to nothing, or at any rate very much less than the policy. He has had all the trouble of investing his savings, all the risk of being led to engage in some foolish speculation* which should lose them, and if neither his industry nor his prudence should fall short, all the anxiety of leaving too little behind him for many years. *If* he succeed, he may succeed very signally, and he may have the compound interest of his savings, if he know how to make it; but it is much more than an even chance that he does not do nearly so well as B.	has exchanged chance for certainty at a very moderate sacrifice, owing to the office being able to gain compound interest. He has no trouble in investing, beyond that of a walk once a year to the office, and is very nearly rid of temptation to change his investment, by knowing that he can only do it at a loss, and by having felt the comfort of knowing himself secure. He has, from the very beginning, a certainty of having made sure of his average, a state of mind highly conducive to health. His success is moderate, but certain; and it never surprises him, if, any day of any week, he is solicited to aid in saving from starvation the widow and children of A.

9. We shall end this paper with the consideration of a question which is of frequent occurrence, namely, the borrowing of money to be paid by annuity. A person aged x (the annuity on his life being worth A_x year's purchase) borrows £1, for which he is to pay a life annuity. This life annuity ought to be $1 : A_x$ of a pound. Now if p_x be the premium of assurance for £1 at the required age, we have, r being the interest of £1 for one year,

$$\frac{1}{1+A_x} - \frac{r}{1+r} = p_x, \qquad \text{or} \qquad \frac{1}{A_x} = \frac{r+(1+r)p_x}{1-(1+r)p_x}.$$

First, suppose an assurance office to exist, charging the absolute premium of the tables, and nothing more. Let the lender insure the borrower's life, and take from him a yearly payment equivalent both to the interest of the sum lent and the premium. Remember that the lender pays the premium at the beginning of

* In particular, the temptation of spending the *interest* of his savings in making a greater appearance, a folly which has prevented many a competency from being acquired.

the year, while the borrower's annuity payment does not become due till the end, whence the borrower must pay interest for each premium advanced. This he does (with the interest of the loan) as long as he lives, and, at the end of the year of death, the lender claims from the office: 1, the £1 lent; 2, a year's interest on it, just due; 3, the premium last paid; 4, a year's interest due on it. Let S be the sum assured; then S must be made up of 1, r, p_x S (its premium), and rp_xS a year's interest. Hence

$$S=1+r+p_x(1+r)S, \quad \text{or} \quad S=\frac{1+r}{1-p_x(1+r)}=\frac{1}{v-p_x},$$

which is the sum the lender must assure, to keep himself clear of all loss. Now the borrower must pay at the end of each year, 1, the interest just due on £1; 2, the preceding premium advanced for him a year before, with interest, so that his annuity must be

$$r+\frac{(1+r)^2}{1-p_x(1+r)}.p_x, \quad \text{or} \quad \frac{r+(1+r)p_x}{1-(1+r)p_x}, \quad \text{or.} \quad \frac{1}{v-p_x}-1,$$

as found before. And this last process* is true, whether the office charges the tabular premium, or makes an addition.

Let us now examine how far the existence of assurance offices affects the position of a borrower. Formerly, in borrowing on annuity, the usury laws made the position of the lender hardly respectable, for his risk (as no one individual could have many transactions of the kind, so as to secure an average) required what might either be called a high rate of mortality or a high rate of interest, at the pleasure of the lender; for the price which supposed 4 per cent., deduced from one table, might be the price at 8 per cent. deduced from another. It is certain that, with the money-lenders, 10 per cent. on the Northampton table was not uncommon. The life of a man of 35 can be respectably insured for £2. 10*s.* per £100, or £·025 per £1. If he borrow at 5 per cent. in the preceding manner, we have then $r=\cdot05$, $p_x=\cdot025$, and £·0783 is the value of the preceding expression, say, £78. 6*s.* for £1000. Now, if this man had gone to a money-lender who asked 10 per cent., even taking the Carlisle table (which is favourable in this case to the borrower), he must have paid an annuity of £116. 11*s.* for the same accommodation. And in many other ways it might be shown that the existence of assurance-offices is a most valuable protection, not as enabling persons to borrow (an objection we have before now heard urged against them), for borrowing was always practicable, but as enabling them to borrow on reasonable terms, and of respectable people; which the usury laws had rendered almost impracticable.

This transaction must be arranged as follows, when it is desired to pay off the loan in n years. Let p be the premium for assuring £1 for n years, k that for endowing the debtor with £1, if he live

* As far as I know, this exact method is due to Mr. Griffith Davies, and the first proof of it is in Mr. Jones's work (p. 189).

over n years. The lender must then (the loan being supposed £1) insure the debtor's life for $\frac{1+k}{v-p}$, and must also pay a premium for £1, to be paid by the office if the debtor survive n years. This premium to the office will therefore be

$$\frac{1+k}{v-p}p+k, \text{ or } \frac{p+vk}{v-p}.$$

And the debtor's annuity is $\frac{1+k}{v-p}-1$, the first payment being due in a year, and the last at the end of the n years, if he live so long.

A. DE MORGAN.

University College, June 21, 1841.

PART II.

GENERAL INFORMATION ON SUBJECTS OF CHRONOLOGY, GEOGRAPHY, STATISTICS, &c.

IV.—REFERENCES FOR THE HISTORY OF THE MATHEMATICAL SCIENCES.

THERE are three classes of persons who, being mathematical students, require to know something of the history of their pursuit. The first want only a general view of leading points, such as can be furnished by one writer in a few volumes. The second wish to be able to compare the accounts given by different persons, and, up to a certain point, to examine the authorities used by those persons, or at least to keep watch upon their mode of using them. The third are desirous of being the critics of the historians, and of amending their works, if need be.

For persons of the third class there is but one plan, which is to examine closely all the writings in which discovery was or is announced. A political historian must refer ultimately to witnesses of the things which he relates; but, for the most part, the mathematical historian has before him the very book the publication of which is the proof, and the contents the matter, of the discovery which he is to describe. The references in this article may be of some use to this highest species of student, but only as reminding him, not of his authorities, but of the places in which he may search for their names, and get information as to his future proceedings. The second class of students will find in this article, we believe, a large majority of all the works which they can possibly want, noticed under different heads: while those of the first may be enabled to judge what writings ought to be selected to serve their purpose.

In most bibliographical lists many works are contained which the maker of the catalogue has never seen, having copied the titles and dates from some predecessor. The confusion which this has introduced into the subject of mathematical history is well known to all who have attempted to reconcile descriptions of books with each other or with the books themselves. The fault seems to lie here, in part at least: persons desirous of presenting lists of writings on any particular subject feel it beneath their dignity simply to put down what they *know*, for the information of those who know less, and the criticism of those who know more. They must be *complete*, and accordingly they add to that which they can describe of their own knowledge, a copy of the description given by some other person, of that which they cannot. Hence errors are created and perpetuated without end, and books become so variously described, that the more prudent bibliographers, when they come to try to make sense of the various descriptions which they find, are obliged to cut the knot by doubting* of the existence of the works in question. Or

* As happened in the case of Lucas Pacioli's edition of Euclid, which the editor of

perhaps information as to the contents of a work passes from hand to hand until it assumes some shape* very different from the right one. To secure ourselves from mistakes of this sort, *we give in the present article no account of any book, tract, pamphlet, or other printed document whatsoever, unless such book, &c., be before us at the moment of writing down its date, place, and title;* nor any opinion or account of any book whatsoever, unless such as is derived from our own personal acquaintance with its contents. We may therefore leave out many works which ought to be included, describe second editions where one with any bibliography in his soul would wish for the first, and even possibly abridge the number of volumes which a book ought to consist of. All this risk we shall run for the sake of the security which we thereby gain, and the authority which our article will assume; for, thanks to the malpractices of catalogue-makers, common correctness is very rare.

The great source of error in dates is the ease with which one figure may be printed for another, or transposed. Such mistakes can be easily provided against by printing dates in words; thus for 1805 we shall write eighteen-five, for 1628, sixteen-twentyeight, and so on.

We write entirely for the second class of students above mentioned. To make a similar catalogue for the third class would be almost the work of a life; but students of this last-mentioned description will find the list which we give not without its use.

The works of which we treat may be divided into those which are wholly devoted to matters of scientific history, and those which contain the same sort of information in a partial or incidental manner. Of the first class we believe we shall name a large majority of those which are known; but not so of the second. There is no library in London, public or private, which contains every work from which one authoritative statement on matters of science might be made. Among the things which we have almost wholly omitted are the various Transactions of scientific societies, which, all-important as they are to inquirers of the third class, are only incidentally useful to those of the second. A separate article might be made, with much advantage, giving references to the places in which the Transactions aforesaid contain historical collections or memoirs devoted to history, particularly *éloges* of deceased members. But this should be a work by itself, and it would be one of no small labour, requiring examination of upwards of four hundred thick volumes.

Again, the works of which we treat are most naturally subdivided into accounts of the progress of science, or history; of the lives of persons, or biography; and of books themselves, or

Fabricius described as not existing, so various were its dates and titles. But we know that the book does exist, for we have seen it.

* Teissier says that some letters of Vieta are printed among those of Caselius. This last book we found, after much research, and can only account for its being said to contain letters of *Francis Vieta*, from the occurrence in it of some of *Peter Victor:* at least this is the correspondent of Caselius, whose name has most letters in common with that of Vieta.

bibliography. Most of them contain a mixture of these kinds; but, at the same time, one or the other of them predominates in most, in a manner which will render the classification a matter of no difficulty.

The written histories are either on the plan of Montucla, Bossut, &c., in which a general account is framed out of the writers' notes or remembrances of miscellaneous reading; or in that of Delambre, Woodhouse, &c., in which the successive writings of eminent men are examined and described one after the other, so that each chapter or section is a description of the progress of science in the hands of some one person, and is complete in itself. The latter is the plan which is most favourable to accuracy, and most interesting to the inquirer of the third class; the former, while it better suits the first and second class, leaves the writer open to many sorts of error which the latter avoids. But, at the same time, Delambre's plan is rather the discussion of the successive points of history, than history itself; and nothing but a very good index, with very full preliminary dissertations, can entitle the results of it to that name. This index is wanting in all Delambre's writings, and is a serious drawback on their utility: while the dissertations are too short. But the same may be said as to the indexes to the writings of Montucla, Bossut, and others. In fact, historians will not write good works of reference: they think to oblige their readers to go through them from beginning to end, by making this the only way of coming at the contents of their volumes. They are much mistaken; and they might learn from their own mode of dealing with the writings of others, how their own will be used in turn. No writer is so much read as the one who makes a good index, or so much cited.

There is no description nor classification of the various manners in which biography is written; there are biographical dictionaries, short and long, separate works, detached pamphlets, *éloges* in scientific transactions, volumes of correspondence, obituaries in periodical works, &c., &c. We have taken everything of these kinds which came under our notice, and it would not be difficult to find hundreds of works as much to this purpose as some of those which we have inserted. It is hard to say where we might have stopped. Many large works contain a few notices of scientific men; Casaubon's letters, for instance, have one page about Vieta, from the writer's personal knowledge, and nothing more to our present purpose. We have, however, taken no work in which there are not accounts, more or less original, of several writers, unless where there is a formal biography of one or more. Neither have we, for the reason above given relative to philosophical transactions, attempted to collect the names of those of whom *éloges* have been written, except where those *éloges* have been afterwards published in separate collections.

In our bibliographical list we have introduced sale catalogues and shop catalogues. Both these and the professedly scientific catalogues are very liable to be mistaken, but their tendencies to error run in different ways, and are frequently corrective of each other. Add to this, that sale and shop catalogues are always those of books which the catalogue-maker actually sees, which is more than can be said of scientific bibliographies. An entry in an auction-list, or in a bookseller's catalogue, is a better proof of the existence of the book than its announcement by Murhard or Lipenius. Those catalogues are uniformly the best which, like Hutton's and Kloss's, are made for the sale of the books by their owners, being persons who understand them; such catalogues combine both characters. We have left out many lists which we could have put down, because they do not contain any very decided quantity of mathematical works.

The chronological works are accessaries to all history; but those of Riccioli and Calvisius are written by mathematicians, and have a peculiar value for the purposes of this article.

In the specifications of books we have followed one uniform plan, namely, that the first word should always be the name of the place of publication, in the language in which the book is written, and in Italics; and that the last word (that is, of the description of the title) should always be the form of the publication, folio, quarto, &c., with the number of volumes, if more than one, also in Italics. The form of the works is described accurately, which may mislead those who are not acquainted with old books. Thus many an ancient *octavo* (eight leaves to the sheet) is, in size, not so big as a modern *duodecimo.*

I. *History.* [*Mathematics, pure and mixed.*]

London, seventeen-sixteen. John Harris, D.D., 'Lexicon Technicum, an Universal English Dictionary of Arts and Sciences.' Third Edition. There are three volumes, or two volumes and a supplement. In our copy, the first volume has the date above; the second has *seventeen-ten,* the third, *seventeen-fortyfour* (both apparently first editions). There is not much history in the first two volumes, and what little there is is not very good: but they are themselves history, in this manner. There is no existing account of a state of science, in a particular country, at a particular time, so good as is contained in the two volumes (the third is mostly biography and theology). *Three volumes folio.*

Paris, seventeen-thirtyseven. Goujet, 'De L'Etat des Sciences en France, depuis Charlemagne jusqu' à Robert.' *Octavo.*

Paris, seventeen-fiftyeight. Montucla, 'Histoire des Mathématiques.' This is the first edition. *Two volumes quarto.*

Paris, seventeen-ninetynine to eighteen-two. Montucla, 'Histoire des Mathématiques.' Second edition, *four volumes quarto.* The authors whom he mentions in the preface are Proclus, Diogenes Laertius, Plutarch, Stobæus, the anonymous writer of the Φιλοσοφουμενα, a tract which was once attributed to Origen, Achilles Tatius, Baldi, Vossius, Wallis, Weidler, Heilbronner, Wolf, Bailly, Priestley (the History of Optics, with the notes to Klugel's German translation). In his first edition he mentions also Riccioli, Dechâles, Savérien, and '*Institutiones Geometriæ Sublimioris* de M. G. W. Craft (Tub. 1753 in 4to),' which he says contains the best existing history of the higher geometry, though not without error. He had, probably, between the two editions, learnt to think less of those whom he has omitted in the second, and he certainly speaks with more respect of Weidler and Heilbronner in the second than in the first.

Montucla's work itself is spoiled for a work of reference by the badness of the index and the want of separate tables of contents: the index, in particular, is the worst of which we know anywhere. But the work is invaluable: if it sometimes, say often, fail in accuracy, it is no more than might have been expected; while, in all the moral qualities of an historian, few have more claims to respect than Montucla. He died during the publication of this edition, having arrived at volume three, page three hundred and thirtysix: the rest of the work was completed by Lalande, from Montucla's papers, with assistance from various quarters. We shall try to give such a table of contents as may enable an inquirer to go to the part of the work in which what he wants is most likely to be found.

First Volume. Montucla.

First part, First book, pages 1—41. General Considerations. Opinions on use and abuse of mathematics.

——— *second book*, pages 42—99. Origin of the sciences: Chaldeans, Egyptians, Persians, Indians, Greeks.

——— *third book*, pages 100—201. The Ionian and Platonic schools, and Grecian science generally till the foundation of the Alexandrian school.

——— *fourth book*, pages 202—288. The Alexandrians, Euclid, Aristarchus, Archimedes, Eratosthenes, Apollonius, Hipparchus, &c. &c., to the Christian era.

——— *fifth book*, pages 289—350. To the taking of Constantinople. The Theons, Ptolemy, Proclus, Diophantus, &c. The history of magic squares.

Second part, first book, pages 351—414. Arabs, Persians, and Turks: Almamon, Albatenius, Alhazen, Hulagu, Nassireddin, Ulughbeigh, &c. &c.

——— *second book*, pages 415—422. The Jews.

——— *third book*, pages 423—447. The Indians, Siamese, &c.

——— *fourth book*, pages 448—480. The Chinese.

Third part, first book, pages 481—534. The Romans and Western nations to the end of the fourteenth century. The Calendar. Boethius, Bede, Alcuin, Gerbert, Athelard, Alphonso, Roger Bacon, &c.

——— *second book*, pages 535—557. Fifteenth century. Leonard of Pisa, D'Ailly, Cusa, Purbach, Regiomontanus, Walther, Pacioli, &c.

——— *third book*, pages 558—619. Editors and publishers: Cardan, Tartalea, Vieta.

——— *fourth book*, pages 620—687. Astronomy in sixteenth century. Copernicus, Reinhold, Mœstlinus, Tycho Brahé, reformation of the Calendar. *Supplement*, pages 715—739. Gnomonics, ancient and modern.

——— *fifth book*, pages 688—714. Mechanics and optics in sixteenth century. Ubaldi, Tartalea, Porta, Maurolicus, de Dominis, &c.

Second Volume. Montucla.

Fourth part, first book, pages 1—101. Seventeenth century, geometry and other pure mathematics: Snellius; logarithms: Napier; Guldinus; Cavalieri; the cycloid: Roberval; &c.

——— *second book*, pages 102—177. Seventeenth century, algebra: Harriot; Bachet; Girard; Des Cartes; Fermat; Beaune; Schooten; Huyghens; Slusius; &c.

Fourth part, third book, pages 178—220. First half of the seventeenth century, mechanics: Stevinus; Galileo; Castelli; Torricelli; Pascal; Des Cartes; &c.

——— *fourth book,* pages 221—267. First half of the seventeenth century, optics: Kepler; the telescope and microscope: Snellius; Des Cartes; Fermat; de Dominis; &c.

——— *fifth book,* pages 268—346. First half of the seventeenth century, astronomy: Kepler; Galileo; Scheiner; Snell; Norwood; Riccioli; Horrox; Des Cartes; &c.

——— *sixth book,* pages 347—403. Last half of the seventeenth century, mathematics: Wallis; Brouncker; Mercator; Barrow; Newton; James Gregory; Leibnitz; the Bernoullis, &c.

——— *seventh book,* pages 404—500. Last half of the seventeenth century, mechanics: Des Cartes; Huyghens; Newton; &c.

——— *eighth book,* pages 501—546. Last half of the seventeenth century, optics: James Gregory; Barrow; Grimaldi; Newton; Halley; &c.

——— *ninth book,* pages 547—647. Last half of the seventeenth century, astronomy: Huyghens; Cassini; observatories: Auzont; Picard; Richer; Römer; Hook; Wren; Flamstead; Halley; Newton; Hevelius; &c.

Supplement, pages 648—661. Navigation to the commencement of the eighteenth century.

Index and corrections, pages 662—718.

THIRD VOLUME. MONTUCLA.

Fifth part, first book, pages 1—426. Eighteenth century, mathematics: Vieta; Harriot; equations; curves: Newton; Maclaurin; tangents; differential calculus; discussion about its invention: Rolle; Saurin; Robins; integration: Leibnitz; Bernoulli; Cotes; De Moivre; D'Alembert; differential equations: Riccati; series: De Moivre; Stirling; Euler; finite differences: Taylor; Emerson; limits; functions: Lagrange; imaginary quantities; interpolation; continued fractions; trajectories; calculus of variations; logarithms; contested problems; theory of probabilities and its applications.

——— *second book,* pages 427—605. Eighteenth century, optics; application of algebra; images: Barrow; Berkeley; Smith; achromatism: Euler; Dollond; Clairaut; Klingenstierna; Boscovich; W. Herschel; optical instruments; photometry; undulation and emission; &c.

——— *third book,* pages 606—719. Eighteenth century, theoretical mechanics; statics: Galileo; Stevinus; Huyghens; Wallis; dynamics: D'Alembert; Lagrange; Daniel Bernouilli; vis viva: D'Alembert's principle; least action; Maupertuis: tautochrones; vibrating chords; resisting media; Robins: hydrodynamics; rivers; waves.

——— *fourth book,* pages 720—832. Eighteenth century, practical mechanics; force of animals; friction: Coulomb; stiffness of ropes; water-power; steam-engine; wind, water, and hand mills: Lambert; machines employed in the arts; horologery; turning; automata; perpetual motion; various mechanicians and their works.

FOURTH VOLUME. MONTUCLA.

Fifth part, fifth book, pages 1—125. Eighteenth century, system of the world; physical hypothesis in its several parts; stars; catalogues; nebulæ; solar theory; lunar theory and inequalities; gravitation the moon: Newton; Clairaut; Euler; D'Alembert; Mayer; acceleration of the moon; parallax; eclipses; their use in longitudes; transits of Mercury and Venus; theory of principal planets; discovery of Uranus.

Fifth part, sixth book, page 126—300. Eighteenth century, physics* of astronomy; refraction: Ptolemy, &c.; Lacaille; figure of the earth, by measurement; the same from the theory of gravitation; Saturn's ring; aberration; precession and nutation; obliquity of the ecliptic; satellites; comets, on their approach to the earth; moon's libration; tides; Laplace's theory of them.

——— *seventh book,* pages 301—380. Eighteenth century, astronomical calculation; tables; ephemerides; Gregorian calendar; French calendar; astronomical instruments; observatories; astrology.

——— *eighth book,* pages 381—508. Eighteenth century, seamanship; ship-building; form of vessels; oars; sails; management; rudder; resistance; rolling and pitching; *arrimage,* disposition of the cargo; 'Examen maritimo' of Don Georges-Juan; rigging; obtaining fresh water from salt water.

——— *ninth book,* pages 509—584. Eighteenth century, navigation; compass; log; altitude-instruments; longitude; chronometers; lunar method.

First Supplement, pages 585—588; capstan.

Second Supplement, pages 589—618; geography.

Third Supplement, pages 619—643; quadrature of the circle.

Fourth Supplement, pages 644—652; music.

Fifth Supplement, pages 653—658; defence of some ancient philosophers.

Sixth Supplement, pages 659—661; Arbogast's derivations.

Life of Montucla, pages 662—672.

Paris, eighteen-ten. Bossut, 'Histoire générale des Mathématiques.' Much better than Bossut's Preface to the mathematics of the Encyclopédie Méthodique; carries the history farther down than Montucla; exceedingly good for a general view, but not of the highest authority for accuracy; has an imperfect list of mathematicians at the end. There is a translation (See 'Penny Cyclopædia' art. 'Bonnycastle'). *Two volumes octavo.*

Paris, eighteen-ten. Delambre, 'Rapport Historique sur les Progrès des Sciences Mathématiques depuis 1789.' For a report to be presented to the emperor, and without any pretence to give full references, this is good of its kind, as far as relates to France. *Octavo.*

London, eighteen-thirtyfour. Powell, 'View of the Progress of the Physical and Mathematical Sciences.' (Cabinet Cyclopædia.) *Duodecimo.*

II. *History.—Pure Mathematics.*

Oxoniæ, sixteen-twentyone. Savile, 'Prelectiones in Euclidem.' *Quarto*

Francofurti ad Mœnum, sixteen-twentyseven. Peter Ramus, 'Scholæ Mathematicæ,' contains a large number of historical hints and allusions. Third or later edition. *Quarto.*

London, sixteen-eightyfive. Wallis, 'A Treatise of Algebra, both Historical and Practical.' This work contains a very studied and learned history of algebra: its defect is nationality, particularly with reference to Harriot as compared with Vieta; as pointed out by Montucla, who was,

* We do not say *physical astronomy,* because, by an absurd misnomer, those words are applied in this country to the theory of gravitation.

convicted of the same with reference to Vieta as compared with Cardan, &c., by Cossali; and the last may perhaps in process of time be served with a *tu quoque* by some Mohammedan writer. *Folio.*

Oxoniæ, sixteen-ninetythree. Wallis, Works, vol. ii., contains the same algebra in Latin, with twelve additional chapters, and various other augmentations. This is the one which should be referred to. *Folio.*

London, seventeen-fifteen, published also in Latin in the same year. Raphson, 'History of Fluxions.' A very partial history on the Newtonian side of the question. Though it appeared in 1715, yet it is stated that the Commercium Epistolicum (1712) came out while it was at press. There is a cancel in the middle, the paging of which is wrong printed. [This dispute required odd things to be done with the printing: see Com. Epist. next named.] At the end of the English edition is the first appearance of the celebrated *Conti Correspondence*, in which Newton is engaged in person with Leibnitz, on the dispute about the invention of fluxions. But the last letter of Leibnitz is dated April 11, 1716, the work bearing 1715. The fact is that Newton, immediately on hearing of the death of Leibnitz (Nov. 14, 1716), added to Raphson's work this correspondence, with his remarks, which, as long as Leibnitz was alive, he only circulated among his friends. *Quarto.*

Londini, seventeen-twentyfive (in some title-pages seventeen-twentytwo). 'Commercium Epistolicum, &c., una cum Recensione, &c.' The reprint of the Comm. Epist. (seventeen-twelve) and of the account of it in the Phil. Trans. (seventeen-fifteen) with a new preface, alluding to the Conti Correspondence (seventeen-fifteen, sixteen). The English side of the dispute about the invention of fluxions. The paging should not be quoted, but the marginal numbering, the former being disturbed in two places. *Octavo.*

Wittembergæ, no date (marked in ink in our copy seventeen-twentyseven). Weidler, 'De Characteribus Numerorum Vulgaribus.' *Quarto.*

Paris, seventeen-fiftythree. Savérien, 'Dictionnaire Universel de Mathématique,' contains a good deal of historical reference, not the best nor most accurate. *Two volumes quarto.*

Paris, seventeen-eightynine. 'Dictionnaire Encyclopédique des Mathématiques.' This is a part of the immense *Encyclopédie Méthodique:* the introduction by Bossut is not comparable to the History of Mathematics which he afterwards published, as elsewhere noticed. It contains the principal articles on the subject from the Encyclopædia of D'Alembert and Diderot. The little history that there is in the body of the work is for the most part worse than bad. The part of the Encyclopédie which contains the descriptions of games is usually bound with this. *Four volumes quarto* (one of them being plates).

Göttingen, seventeen-ninetysix, ninetyseven, ninetynine, eighteenhundred. Kastner, 'Geschichte der Mathematik,' principally of the sixteenth and seventeenth centuries: the tables of contents at the beginning of each volume are so good that we need not repeat what we have done for Montucla. This work is carefully done, and of good authority. *Four volumes octavo.*

Parma, seventeen-ninetyseven. Cossali, 'Origine, trasporto in Italia, primi progressi in essa dell' Algebra.' The history of algebra in Italy; the *Italian* account, as opposed to the French. *Two volumes quarto.*

Cambridge, eighteen-ten. Woodhouse, 'On Isoperimetrical Problems.' An excellent History of the Calculus of Variations. *Octavo.*

London, eighteen-twelve. Hutton, 'Tracts on Mathematical Subjects.' Tracts xix., xx., and xxi. are the histories of trigonometrical tables and logarithms, which already formed part of the preface to the author's tables of logarithms: Tract xxxiii. is a history of algebra. Hutton is not an historian* of the highest authority, but a valuable guide to sources of information. His library was large and good. *Three volumes octavo.*

Edinburgh, eighteen-sixteen. Playfair, 'A General View of the Progress of Mathematical and Physical Science,' Part i. *Edinburgh*, eighteen-nineteen. Do. do., part ii. In the supplement to the fourth and fifth edition of the Encyclopædia Britannica. *Quarto.*

London, about eighteen-twentysix. Peacock, 'Arithmetic,' in the Encyclopædia Metropolitana. The most complete history of arithmetic. *Quarto.*

Edinburgh, eighteen-thirty. J. Herschel, on the History of Mathematics, being the article 'Mathematics' in Brewster's Edinburgh Encyclopædia, volume thirteen. *Quarto.*

Paris, eighteen-thirtyone. Montucla, 'Histoire de la Quadrature du Cercle.' The first edition, said to be 1754, was Montucla's first essay in history, and, we think, his best. This edition, much augmented, and with an account of the trisection of the angle, includes the useful quadrators, &c., as well as the pretenders. A valuable work. Some things which are in the fourth volume of the second edition of the history are not here. *Octavo.*

London, eighteen-thirtyfour. Peacock, 'Report on the recent progress and present state of certain branches of Analysis.' (Brit. Assoc., vol. ii.)

The volumes of the British Association contain various reports on mathematical physics, but none so closely connected with the present list as the preceding and Mr. Airy's Report on Astronomy presently noticed, which are both excellent. *Octavo.*

Paris, eighteen-thirtyfive. Biot, 'Analyse des ouvrages originaux de Napier relatifs à l'invention des logarithmes." This is an addition to the 'Connaissance des Tems' for 1838, and corrects the mistakes of Delambre on Napier's methods. Biot here speaks highly of Hutton's account. *Octavo.*

Bruxelles, eighteen-thirtyseven. Chasles, 'Aperçu Historique sur l'Origine et le Développement des Méthodes en Géométrie.' An excellent history, particularly on the connexion of the recent improvements with the old systems. *Quarto.*

Paris, eighteen-thirtyeight, do., forty, fortyone. Libri, 'Histoire des Sciences Mathématiques en Italie.' Four volumes published, two more expected; a great addition to Italian history. The citations are numerous. The first volume is a reprint, the original publication having been burnt; only a few copies escaped. If a good index do not accompany the last volume, the value of the work will be materially diminished. *Octavo.*

London, eighteen-thirtynine. De Morgan, 'Progress of the Problem of Evolution,' in the Companion to the Almanac. *Duodecimo.*

* Such an assertion with respect to so respectable a name requires some explanation. Hutton was a most remarkable man; from a labourer in the coalpits at Newcastle he became not only an original mathematician, but well versed in the history and literature of all parts of science. Whenever he writes of an author whom he has read, and whose works were before him at the time, he is trustworthy, though occasionally careless; but he never gained the power of choosing and deciding between authorities. He would follow Sir John Hill on a point of ancient astronomy as soon as Costard.

London, eighteen-seventeen. Colebrooke, 'Algebra, &c., from the Sanscrit, &c.' A translation of the Liliwati, Viga Ganita, &c. *Quarto.*

Bombay, eighteen-sixteen. Taylor, 'Liliwati.' *Quarto.*

London, eighteen-thirteen. Strachey, 'Viga Ganita.' *Quarto.*

All these three works have historical prefaces, that of the first being the most extensive and learned.

London, eighteen-thirtyone. Rosen, 'The Algebra of Mahommed ben Musa.' Arabic and English. Contains historical notes and preface. *Octavo.*

III. *History.—Physics, &c.*

Paris, seventeen-eightyone. Brisson, 'Dictionnaire Raisonné de Physique.' Contains much historical reference, not perhaps the most trustworthy, since many of the articles are taken from the Encyclopédie (Diderot.) *Three volumes quarto* (one of plates).

London, eighteen-seven. Thomas Young, 'A Course of Lectures on Natural Philosophy and the Mechanical Arts.' Lecture xx., on the history of mechanics; Lecture xxx., on the history of hydraulics and pneumatics; Lecture xl., on the history of optics; Lecture xlviii., on the history of astronomy; Lecture lx., on the history of terrestrial physics. See also the bibliographical list presently given. *Two volumes quarto.*

Paris, eighteen-ten. Libes, 'Histoire des Progrès de la Physique.' It seems to us that this work is not as much known or cited as it deserves to be. *Four volumes octavo.*

London, seventeen-seventyfive. Priestley, 'History of Electricity.' Third Edition, *two volumes octavo.*

London, seventeen-ninetytwo. Priestley, 'History and present state of Discoveries relating to Vision, Light, and Colours.' *Quarto.*

London, seventeen-ninetyseven. Beckmann, 'History of Inventions and Discoveries.' A translation by Johnston. *Four volumes octavo.*

Paris, eighteen-three. Biot, 'Essai sur l'Histoire des Sciences pendant la Révolution Française.' Contains also an account of the scientific efforts made to furnish military stores. *Octavo.*

London, eighteen-thirtyseven. Davies, 'History of Magnetical Discovery.' (In the Appendix to a work called the *British Annual.*) *Duodecimo.*

London, eighteen-thirtyseven. Whewell, 'History of the Inductive Sciences from the Earliest to the Present Times.' *Three Volumes Octavo.*

London, eighteen-forty. Lardner, 'The Steam-Engine Explained and Illustrated.' Contains the history of the steam-engine and a memoir of Watt. Seventh edition. *Octavo.*

IV. *History.—Astronomy.*

Parisiis, sixteen-fortyfive. Ismael Bullialdus (Bouillaud), 'Astronomia Philolaica.' The *prolegomena* contain a short historical account of the rise of astronomy; and there is some history (but not much) in the work itself. *Folio.*

Hagæ-Comitum, sixteen-fiftyfive. Borellus, 'De vero Telescopii inventore.' *Quarto.*

Lugduni Batavorum, sixteen-eightyone. Lubienietski, 'Theatrum Cometicum.' The second part is the history of all the comets up to the time, with authorities, or what passed for such. *Folio.*

D

Paris, sixteen-ninetythree. 'Recueil d'Observations faites en plusieurs Voyages.' Contains Cassini, 'De l'Origine et du Progrès de l'Astronomie.' *Folio.*

Vittembergæ, seventeen-fortyone. Weidler, 'Historia Astronomiæ.' Goes down to about 1740; a most excellent work. There is a supplement at the end of the bibliography by the same author, hereinafter noticed. *Quarto.*

London, seventeen-fortysix. Costard, 'Letter to Martin Folkes on the rise of Astronomy among the Ancients.' Costard quotes his authorities as well as cites; his learning in matters of ancient astronomy was not equalled in his day. *Quarto.*

Cantabrigiæ, seventeen-fortyseven. Heathcote, 'Historia Astronomiæ.' A clearly written Latin tract. *Octavo.*

London, seventeen-sixtyseven. Costard, 'History of Astronomy.' Not properly a history of astronomy, but a book on the use of the globes, with the historical remarks of a very learned man. *Quarto.*

Londres et Paris, seventeen-seventyseven. Bailly, 'Lettres sur l'Origine des Sciences.' *Octavo.*

Londres et Paris, seventeen-seventynine. Bailly, 'Lettres sur l'Atlantide de Platon.' *Octavo.*

Good appendages to Bailly's Histories of Astronomy. The root of *his* matter in fewer words.

Berlin, seventeen-seventyseven, seventyseven, seventynine. John Bernoulli,* 'Lettres sur differens sujets.' Contains a great deal of astronomical information, particularly on Observatories. *Three volumes octavo.*

Paris, seventeen-eightyone. Bailly, 'Histoire de l'Astronomie Ancienne.' A romance under the name of a history. [Penny Cyclopædia, Bailly.] Second edition. *Quarto.*

Paris, seventeen-eightyfive. Bailly, 'Histoire de l'Astronomie Moderne, depuis la Fondation de l'Ecole d'Alexandrie jusqu'à l'époque de MDCCXXX.' More of a history, but still founded upon the romance. Second edition. *Three volumes quarto.*

Paris, seventeen-eightyseven. Bailly, 'Traité de l'Astronomie Indienne et Orientale.' An attempt to establish the extreme antiquity of the Hindoo astronomy. *Quarto.*

Paris, seventeen-eightythree. Pingré, 'Cométographie.' The best work on the history of comets, with a complete list of works on the same subject. *Two volumes quarto.*

Paris, eighteen-one (An. ix.). Biot, 'Analyse du Traité de Mécanique Céleste,' and Delambre, 'Des Méthodes pour la Détermination d'un Arc de Meridien.' *Octavo.*

Paris, eighteen-two. Berthoud, 'Histoire de la Mesure du Tems.' *Two volumes quarto.*

Paris, eighteen-six. Dupuis, 'Mémoire explicatif du Zodiaque Chronologique et Mythologique.' *Quarto.*

Paris, eighteen-eleven. Voiron, 'Histoire de l'Astronomie depuis 1781 jusqu'à 1811.' A useful work, which has probably been kept in the background by being called a supplement to Bailly. *Quarto.*

Paris, eighteen-seventeen. Delambre, 'Histoire de l'Astronomie Ancienne.' The first volume takes in all the Greeks except Ptolemy and his commentators, the Chinese, Indians, Chaldeans, &c.; the second gives an

* Not of course the great John, but his grandson, the one who is called John III. in the 'Penny Cyclopædia.'

account of the arithmetic and trigonometry of the Greeks, as shown in Ptolemy and his commentators, and then proceeds to these last-mentioned writers. The account of the Syntaxis is very full. *Two volumes quarto.*

Paris, eighteen-nineteen. Delambre, 'Histoire de l'Astronomie du moyen age.' In three books: the first containing the Arabs, &c.; the second, Sacrobosco and his commentators, Alphonso, Bianchini, Purbach and his commentators, Regiomontanus, Digges, Dee, Stoffler, Ricius, Fernel, Fracastor, Apian, Nonius, Peucer, Gemma-Frisius, Royas, Orontius Fineus, Gauricus, Maurolicus, Jordanus, Stadt, Bressius, Schoner, Vieta, Magini. The third book has the history of gnomonics up to the time of Lahire and Ozanam. *Quarto.*

Paris, eighteen-twentyone. Delambre, 'Histoire de l'Astronomie Moderne.' The first volume contains the reformation of the calendar, Copernicus, Reinhold, Tycho Brahé, Longomontanus, Ursus Dithmarsus, Kepler, Napier, Briggs, Galileo, Riccioli, Scheiner, Marius, Licetus, Nunez, &c. The second volume contains Rheticus, Pitiscus, Adrian Romanus, Torporley, Lansberg, Clavius, Briggs, Gellibrand, Roe, Oughtred, Sherwin, Snellius, Vernier, Metius, Bouillaud, Ward, Reiner, Kircher, Schyrle, Bayer, Descartes, Durret, Morin, Riccioli, Maria Cunitia, Hodierna, Borelli, Gassendi, Mouton, Briggs, Vlacq, Hevelius, Horrocks, Argoli, Roberval, Wing, Street, Levera, De Billy, Tacquet, Duhamel, Lubinietski, Mercator, Greenwood, Huyghens, Gascoyne, Crabtree, Hooke, Buot, Auzont, Picard, Roemer, La Hire, Dominic Cassini. *Two volumes quarto.*

Paris, eighteen-twentyseven. Delambre, 'Histoire de l'Astronomie au dix-huitième siècle.' Posthumous, edited by M. Mathieu; another volume promised on the measure of the earth, which has not yet appeared (1842). It gives Newton, Maclaurin, Pemberton, Whiston, David Gregory, Leadbetter, Flamsteed, Halley, Horrebow, Wurzelbaur, Keill, Lemonnier, Graham, Sisson, Bird, the three Maraldis, James Cassini, Cassini de Thury, Count Cassini, Louville, Delisle, Fouchy, Godin, La Condamine, Bouguer, Maupertuis, Pezenas, Manfredi, Marioni, Ximenes, Briga, Frisi, Bradley, Mayer, Cotes, La Caille, Wargentin, Lalande, Chappe, Maskelyne, Mason, Long, Ferguson, Boscovich, Pingré, Bory, Legentil, Du Séjour, Bailly, Jeaurat, Méchain, Messier. *Quarto.*

To each of these works of Delambre a preliminary discourse is prefixed, giving some general views of the whole, which may be profitably read by any one who has already a little knowledge of the history. Then follow detailed accounts of the works of the several writers. Delambre is the most searching of critics, the most severe of judges, and one of the least national of historians.

Paris, eighteen-seventeen. Gautier, 'Essai Historique sur le Problème des trois Corps.' *Quarto.*

London, eighteen-twentyone. W. Drummond, 'Memoir on the Antiquity of the Zodiacs of Esné and Denderah.' The English Bailly. *Octavo.*

Paris, eighteen-twentyone. 'Précis de l'Histoire de l'Astronomie.' A short and general view: Laplace is not to be trusted when the question of who made a discovery lies between himself and another. *Octavo.*

Calcutta, eighteen-twentythree. Bentley, 'Historical View of the Hindu Astronomy.' The other side to Bailly's theory. *Quarto.*

Altona, eighteen-twentythree. Schumacher's 'Astronomische Abhandlungen.' Contains Olber's list of 125 comets, with elements. *Quarto.*

D 2

Paris, eighteen-twentyfive. Laplace, 'Traité de Mécanique Céleste.' volume the fifth. It contains elegant historical summaries on most of the great points of the solar system. *Quarto.*

London, eighteen-twentyfive. Bentley, 'Historical View of the Hindu Astronomy.' *Octavo.*

Londini, eighteen-twentyeight. Valpy, from Bentley, 'M. Manilii Astronomicon.' The notes are very copious, and contain a quantity of miscellaneous information on ancient astronomy. *Two volumes octavo.*

London, about eighteen-thirtytwo (Lib. U. K.). Rothman, 'History of Astronomy.' *Octavo.*

London, eighteen-thirtythree. Narrien, 'Historical Account of the Rise and Progress of Astronomy.' *Octavo.*

London, eighteen-thirtythree. Airy, 'Report on the Progress of Astronomy during the present Century.' Many references. (Brit. Assoc., vol. i.) *Octavo.*

London, eighteen-thirtynine. Halliwell, 'Notes on Early Almanacs.' In the Companion to the Almanac (reprinted the same year in octavo). *Duodecimo.*

V. *History.—Miscellaneous.*

London, sixteen-fiftysix. Stanley, 'History of Philosophy, in Eight Parts.' A useful work of reference. These eight parts have separate pagings, but in sixteen-sixty the same author published what he called his *third and last volume* in five parts. *Quarto.*

Lipsiæ, sixteen-ninety. Gerard Vossius, 'De Philosophorum Sectis,' with Ryssel's continuation. *Quarto.*

Cantabrigiæ, seventeen-thirtyfive. Johnson, 'Quæstiones Philosophicæ.' A collection of questions for the disputations in the schools, to each of which is attached references to authors who have maintained one side or the other. It therefore becomes useful to a certain extent: were it larger it would be a history of opinions. *Octavo.*

London, seventeen-fiftyeight. House of Commons, Report of the Commons on Weights and Measures. *Folio.* Another in seventeen-fiftynine. *Folio.* Both contain valuable historical information on the history of English weights and measures.

London, eighteen-fourteen, House of Commons. Report of Committee on Weights and Measures (reprinted by the Lords, eighteen-sixteen). Contains the evidence of Playfair and Wollaston. *Folio.*

London, eighteen-nineteen. House of Commons, First Report of the Commissioners of Weights and Measures, appointed by the Prince Regent. Contains an abstract of the statutes on the subject. *Folio.* Their second report, eighteen-twenty, contains the list of weights and measures throughout England. *Folio.* Their third report, eighteen-twentyone, contains their final recommendations, concurred in by House of Commons committee, eighteen-twentyone. *Folio.* The select committees since appointed have done nothing historically important.

London, seventeen-sixtysix. Formey, 'Concise History of Philosophy.' A translation; a compendium which would be useful to those who have no better. *Duodecimo.*

Birmingham, seventeen-eightyeight. Priestley, 'Lectures on History.' *Quarto.*

London, seventeen-ninetyone, ninetyone, ninetysix, eighteen-one, four, seven.

Maseres, 'Scriptores Logarithmici.' Many of the tracts are historically useful: the fourth volume contains 'Dissertation on the Rise and Progress of the Modern Art of Navigation,' by James Wilson, M.D., from the third edition of Robertson's Navigation. *Six volumes quarto.*

Argentorati, eighteen-seven. 'M. Vitruvii Pollionis de Architecturâ.' A good edition for reference, with an excellent index. *Octavo.*

Paris, eighteen-twentyeight. Fourcy, 'Histoire de l'Ecole Polytechnique.' *Octavo.*

Paris, eighteen-twentynine. Tenneman, translated by Cousin, 'Manuel de l'Histoire de la Philosophie.' Abounds in references. *Two volumes octavo.*

London, eighteen-thirtyfive. Malden, 'On the Origin of Universities.' *Duodecimo.*

London, eighteen-thirtysix (Mem. Astron. Soc., vol. ix.). Baily, 'Report on the New Standard Scale.' Contains the history of the standard measures. *Quarto.*

Edinburgh, eighteen-thirtyseven. Milne, 'Treatises on the Law of Mortality and on Annuities,' being the Articles under those heads in the seventh edition of the Encyclopædia Britannica. Contains the history of these subjects; the only one we know of. *Quarto.*

VI. *History.—Royal Society.*

London, sixteen-sixtyseven. Sprat, 'History of the Royal Society.' In those days the history of a Society included what we should now call its Transactions. *Quarto.*

London, seventeen-fiftysix. Birch, 'History of the Royal Society.' *Four volumes quarto.*

London, seventeen-eighty. Hill, 'A Review of the Works of the Royal Society.' A collection of errors, absurdities, and incredibilities, from the Philosophical Transactions; much coloured, no doubt. But it would help any one who was desirous of comparing the disposition of different times to admit singular narratives: the Philosophical Transactions were not always so *dry* as they are now. *Quarto.*

London, seventeen-eightyfour. 'An Appeal to the Royal Society concerning the Measures taken by Sir Joseph Banks, &c.' By a Friend of Dr. Hutton. *Octavo.*

London, seventeen-eightyfour. 'A History of the Instances of Exclusion from the Royal Society. By some Members in the Minority.' *Octavo.*

London, seventeen-eightyfour. 'An Authentic Narrative of the Dissensions and Debates in the Royal Society.' *Octavo.*

London, seventeen-eightyfour. Kippis, 'Observations on the late Contests in the Royal Society.' *Octavo.*

These four tracts have reference to the conduct of Sir Joseph Banks in depriving Dr. Hutton of the office of foreign secretary, and in influencing the election of fellows: very characteristic of times which, it may be hoped, are gone by.

London, eighteen-twelve. Thomson, 'History of the Royal Society' to the end of the eighteenth century. *Quarto.*

London, eighteen-thirty. Babbage, 'On the Decline of Science in England.' Mostly about the Royal Society. *Octavo.*

London, eighteen-thirtyone. Moll, 'On the alleged Decline of Science in England. By a Foreigner.' *Octavo.*

London, eighteen-thirtyone. 'A Statement of Circumstances connected with the late Election for the Presidency of the Royal Society.' *Octavo.*

London, eighteen-thirtysix. Granville, 'The Royal Society in the Nineteenth Century.' A full statistical account. *Octavo.*

London, eighteen-thirtytwo-two-seven. 'Abstracts of the Philosophical Transactions.' Three volumes published as yet: Vol. I., 1800—1814; Vol. II., 1815—1830; Vol. III., 1830—1837. *Three volumes octavo.*

VII. *Biography.—Dictionaries.*

London, seventeen-seventyfive. Granger, 'Biographical History of England' from Egbert to the Revolution. An account of Portraits, with notices attached to them, and a good Index. Second edition. *Four volumes octavo.*

London, eighteen-six. Noble, 'Biographical History, &c.,' being a continuation of Granger to the end of George II. *Three volumes octavo.*

Maestricht, seventeen-seventysix. D'Herbelot, 'Bibliothèque Orientale.' *Folio.*

Cambridge, seventeen-ninetynine. L'Advocat, 'Biographical Dictionary.' A translation. *Four volumes octavo.*

Paris, eighteen-eleven. 'Biographie Universelle.' This celebrated work has been going on down to the present time, the supplements included. Some of the memoirs are excellent; some contain the grossest mistakes. The name of the writer should be referred to, as given at the end of each article. Supplements included, there are now (August, 1842) sixty-nine volumes. *Octavo.*

London, eighteen-fifteen. Hutton, 'Philosophical and Mathematical Dictionary.' The most unevenly executed of all dictionaries: it contains all kinds of historical articles, from those which are themselves authorities, to others which are literal copies of the worst articles of the century previous. Second edition. *Two volumes quarto.*

London, eighteen-thirty. Gorton, 'General Biographical Dictionary.' By far the best of short biographical dictionaries: it usually gives references to further sources of information. The Appendix (196 pages) contains a list of biographical works and authorities. *Two volumes octavo.*

VIII. *Biography.—British Collections.*

Gippeswici (Ipswich), fifteen-fortyeight. Bale, 'Illustrium, &c., Britanniæ Scriptorum Summarium.' A celebrated work. *Quarto.*

Parisiis, sixteen-nineteen. Pits, 'De Rebus Anglicis,' Tomus primus. No second volume. Pits died in 1616. Pits is blamed for plagiarizing from Bale; but as he has omitted the Protestants, whom Bale favoured, and been very full on the Catholics, the two works together are very complete. *Quarto.*

Oxonii, sixteen-seventyfour. 'Historia et Antiquitates Universitatis Oxoniensis.' The Athenæ Oxonienses must be regarded as a continuation of this, though they have some part in common. This edition very much wants an index: it is altogether a bungling piece of editorship. The University bought it of Wood, and instead of publishing it in his English, turned it into Latin, picking up various mistakes by the way. So that, in fact, there is not Wood's authority, absolutely speaking, for any one statement it contains. It is, nevertheless, highly valuable. *Two volumes* (according to the title-page; *two books* they are called in the work itself) *folio.*

Oxonii, seventeen-nine. Leland, 'De Scriptoribus Britannicis.' Said to be a bad edition. *Two volumes octavo.*

London, seventeen-twentyone. Wood, 'Athenæ Oxonienses.' An exact history of all the writers, &c. &c. of Oxford: from 1500 to 1695.

Second edition, with above 500 new lives (Bliss's edition, since published, has more). A great biographical authority. Almost all the science in England, throughout the period in question, came from Oxford, as did that of the centuries preceding. What would Anthony Wood say, if he could be brought back at the present time? *Two volumes folio.*

London, seventeen-forty. Ward, 'Lives of the Professors of Gresham College,' and of the founder. A very valuable collection relative to the history of Gresham College, when it had a history. *Folio.*

London, seventeen-fortyseven to sixtysix. 'Biographia Britannica.' This well-known work, which contains the fullest biographies in our language, contains, of men of science: Alcuin, Barlowe, Barrow, Bassantin, Batecumb, Bede, Bernard, Blagrave, Boyle, Briggs, Brouncker, Chaucer, Cotes, Dee, Digby, Digges, Foster, Gellibrand, Gilbert, Graunt, Greaves, Gregory, Gunter, Harriot, Keil, Maclaurin, Molyneux, Newton, Ogilby, Oughtred, Pell, Tonstall, Wallis, Ward, Wilkins, Wren; and in the Appendix, Collins, Hutchinson, Saunderson, Harrison. *Seven volumes* (so called; that is, six volumes, the sixth in two parts) *folio.*

Londini, seventeen-fortyeight. Tanner, 'Bibliotheca Britannico-Hibernica, sive de Scriptoribus qui in Anglia, Scotia, et Hibernia ad seculum XVII. initium floruerunt....Commentarius.' With a preface by David Wilkins. The names are in alphabetical order. This is so scarce a book that it may almost be considered as a manuscript. There is a copy in the British Museum. *Folio.*

London, seventeen-seventyeight to ninetythree. Kippis. Unfinished new edition of the Biographia Britannica. This edition, so far as it goes, contains, in addition to the former, Berkeley, Bradley, Canton, Cheyne, Collins, Costard; and (vol. iii., p. 143) Fatio de Duillier, Ditton: besides which many of the lives of the preceding edition are considerably augmented. *Five volumes folio.*

London, seventeen-ninetyeight. 'Literary Memoirs of Living Authors.' *Two volumes octavo.* A very meagre and inaccurate performance. The following is very good, and, as far as we have examined, more than usually accurate in the lists of works.

London, eighteen-sixteen. 'A Biographical Dictionary of the Living Authors of Great Britain and Ireland.' *Octavo.*

London, eighteen-twelve (six volumes), thirteen (vol. vii., part i.), fourteen, fifteen, sixteen (vol. vii., part ii.). Nichols, 'Literary Anecdotes of the Eighteenth Century.' The seventh volume contains the index to this multifarious collection; part i. to the first six volumes; part ii. to the eighth and ninth. *Nine volumes octavo.*

London, eighteen-seventeen, seventeen, eighteen, twentytwo, twentyeight, thirtyone. Nichols, 'Illustrations of the Literary History of the Eighteenth Century.' *Six volumes octavo.*

London, eighteen-fourteen. Wilson, 'History of Merchant Taylors' School, part ii.: of its principal Scholars.' Works of this kind (of which there are many) frequently contain points of biography which are not elsewhere: but they must be looked at with great caution, as there is generally a bias in favour of those educated at the place of which the history is written. In the work before us, Titus Oates (who was bred at Merchant Taylors') is a martyr. *Quarto.*

London, eighteen-thirtyseven. De Morgan, 'Notices of English Mathematical and Astronomical Writers between the Norman Conquest and the Year 1600.' In the Companion to the Almanac. *Duodecimo.*

London, eighteen-thirtyseven. Halliwell, 'Brief Sketch of Early English

Scientific Literature.' In numbers XVIII., XX., XXII. of the 'Magazine of Popular Science.' *Octavo.*

The Monthly Notices of the Royal and Astronomical Societies contain, the former in December, the latter in February, accounts of such mathematicians, astronomers, &c., as, being members of the societies, have died within the preceding year. For Englishmen recently deceased, these, and the Gentleman's Magazine, should be the first sources of information consulted. The Societies are generally very imperfect in personal details, and the Magazine in scientific information: both put together generally make up a tolerably complete account.

IX. *Biography.—Single Lives.*

Paris, sixteen-fortynine. De Coste, 'La Vie du R. P. Marin Mersenne.' Trumpery. *Octavo.*

Hagæ-Comitum, sixteen-fiftyfive. Gassendi, 'Vita de Peiresc:' third edition. Same place, date, and author, 'Vita Tychonis Brahei:' second edition: to which are added the lives of Copernicus, Purbach, and Regiomontanus. Same place and author, one year later, 'De Vita et Moribus Epicuri.' We should have been badly off for the lives of the great astronomers mentioned, if it had not been for Gassendi. There are copious indexes; and to the life of Peiresc is appended the French to his Latin proper names—a most necessary addition. *Quarto.*

Carolopoli, sixteen-eightyone. 'Thomæ Hobbes Vita,' auctore R. B. There is, we believe, an English edition. *Octavo.*

Paris, sixteen-ninetyone. Baillet, 'La Vie de M. Descartes.' *Two parts (volumes) quarto.*

London, sixteen-ninetythree. 'The Life of M. Descartes,' translated by S. R. A translation of an abridgment of Baillet. *Octavo.*

London, seventeen-seven. 'The Life of Pythagoras.' A translation from Dacier. *Octavo.*

Paris, seventeen-thirtyseven. 'Vie de Pierre Gassendi.' (Anonymous; attributed to Père Bougerel.) Has good list of works and index. *Duodecimo.*

La Haye, no date. Maty, 'Mémoire sur la Vie et sur les Ecrits de M. Abraham de Moivre, &c.' The most authentic account of de Moivre. *Duodecimo.*

London, seventeen-fortythree. Cheyne, Account of himself and his writings. Second edition. *Octavo.*

London, seventeen-fortynine. Whiston, memoirs of himself. A great quantity of good gossip about the time being, with a good index after p. 662. Two parts, commonly in *two volumes octavo.*

Livorno, seventeen-seventyfive. Frisi, 'Elogio del Galileo.' *Octavo.*

Milano, seventeen-eightyseven. 'Memorie appartenenti alla Vita, &c., del Paolo Frisi.' *Quarto.*

Perth, seventeen-eightyseven. Stewart (Earl of Buchan) and Minto, 'Life and Writings of John Napier.' *Quarto.*

London, seventeen-ninetyone. Emerson, 'Cyclomathesis.' This collection is nothing but Emerson's works, such of them as were octavo, with new title-pages and a memoir of the author (our reason for introducing the collection here) prefixed to the first volume. *Thirteen volumes octavo.*

London (?), seventeen-ninetytwo (?). Hutton, 'Memoirs of the Life and Writings of Thomas Simpson.' This is, we believe, a preface or appendix to the edition of Simpson's Select Exercises, edited by Hutton in 1792. *Octavo.*

London, seventeen-ninetythree. 'Contemplatio Philosophica; a Posthumous Work of the late Brook Taylor, &c. To which is prefixed a Life of the Author by his Grandson, Sir W. Young, Bart.' The only authentic account of Brook Taylor. Some curious correspondence with Conti, Montmort, and others, is added. This work is scarce (see Penny Cyclopædia, TAYLOR, BROOK). *Octavo.*

Amstelodami, seventeen-ninetythree. Mahne, 'Diatribe de Aristoxeno.' *Octavo.*

London, eighteen-six. Turnor, 'Collections for the History of the Town and Soke of Grantham, containing Authentic Memoirs of Sir Isaac Newton.' *Quarto.*

Palermo, eighteen-eight. Scina, 'Elogio di Francesco Maurolico.' Full and good. *Quarto.*

London, eighteen-eight. Kelly, 'Life of John Dollond,' with Appendix. Third edition. *Quarto.*

Bath, eighteen-twelve. Trail, 'Life and Writings of Robert Simson.' *Quarto.*

Padova, eighteen-eleven. Cossali, 'Elogio di Jacopo Stellini.' *Octavo.*

Padova, eighteen-thirteen. Cossali, 'Elogio di Giovanni Poleni.' *Octavo.*

Padova, eighteen-thirteen. Cossali, 'Elogio di Luigi Lagrange.' *Octavo.*

Paris, eighteen-thirteen. Virey and Potel, 'Précis Historique sur la Vie et la Mort de J. L. Lagrange.' A biography of a peculiar kind; both the authors were medical men, and they describe, avowedly, rather the corporeal* than the mental constitution of their subject: what he ailed and what he took for it. *Quarto.*

Modena, eighteen-eighteen. Venturi, 'Memoria intorno alla Vita del Marchese Gherardo Rangone.' *Quarto.*

Paris, eighteen-nineteen. Charles Dupin, 'Essai Historique sur Monge.' A very good account. *Octavo.*

Paris, eighteen-twenty. Sniadecki, 'Discours sur Nicolas Kopernik.' *Octavo.*

Newcastle, eighteen-twentythree. Bruce, 'Memoir of Charles Hutton, LL.D.' *Octavo.*

London, eighteen-thirty. Leybourn, 'Mathematical Repository,' New Series, Vol. V., contains obituaries of Playfair, Hutton, Dalby (mostly autobiography), and Laplace. *Octavo.* A life of Hutton by Dr. O. Gregory, said to be in the Imperial Magazine for March, 1823, is referred to.

Modena, eighteen-twentyfour. Lombardi, 'Notizie sulla Vita, &c. di Paoli Ruffini.' *Quarto.*

Napoli, eighteen-twentysix. Visconti, 'In Morte di Giuseppe Piazzi.' *Octavo.*

Napoli, eighteen-twentysix. Filipponi, 'Elogio di Giuseppe Piazzi.' *Octavo.*

London, eighteen-twentyseven to thirtyone, in various numbers of the Library of Useful Knowledge, were published the Lives of Wren, Newton, Galileo, and Kepler. The life of Wren as a mathematician (and he was one of the first of his time and country) is hardly written at all. The life of Newton is mostly translated from Biot, and serves to give the view of Leibnitz's friends on the dispute relative to fluxions. The list of works at the end is very incorrect (even in the cancel). The lives

* "De telles recherches," say the medical authors, "si elles eussent étés suivies à fond sur les Descartes, les Leibnitz, les Newton, et sur tous les grands précepteurs du genre humain, seraient aujourd'hui bien précieuses." If this be not *esprit de corps*, we do not know what is.

of Galileo and Kepler, by Mr. Drinkwater (now Drinkwater-Bethune), are both of the best kind of biography. *Octavo.*

London, eighteen-thirtyone. Brewster, 'Life of Newton' (Family Library). The largest and most careful life of Newton which was ever published; and we know that the author contemplates one still more extensive, for which he is now collecting materials. *Duodecimo.*

Paris, eighteen-thirtyone. Cousin, 'Notes Biographiques pour faire suite à l'éloge de M. Fourier.' *Quarto.*

Oxford, eighteen-thirtytwo. Rigaud, 'Miscellaneous Works and Correspondence of Bradley.' *Quarto.*

Oxford, eighteen-thirtythree. Supplement to the above, with an account of Harriot's astronomical papers. *Quarto.*

London, eighteen-thirtyfour. Mark Napier, 'Memoirs of John Napier of Merchiston....with a history of the invention of logarithms.' *Quarto.*

London, eighteen-thirtyfive: with a Supplement (*London*, eighteen-thirtyseven). Baily, 'Account of the Rev. John Flamsteed.' *Quarto.*

Oxford, eighteen-thirtyeight. Rigaud, 'On the first publication of Newton's Principia.' *Octavo.*

The five preceding works, coming so close together, make a remarkable epoch in biographical writing.

Boston, eighteen-thirtyeight. Young, 'Discourse on the Life and Character of N. Bowditch.' *Octavo.*

Boston, eighteen-thirtyeight. Pickering, 'Eulogy on N. Bowditch.' *Octavo.*

Salem, eighteen-thirtyeight. White, 'Eulogy on the Life and Character of N. Bowditch.' *Octavo.*

Cambridge, eighteen-thirtyeight. Halliwell, 'A brief Account of Sir Samuel Morland.' *Octavo.*

Paris, eighteen-thirtynine. Arago, 'Eloge Historique de James Watt.' *Quarto.*

Edinburgh, eighteen-thirtynine. 'Life of James Watt....by M. Arago... Historical Account of the Discovery of the Composition of Water, by Lord Brougham; and Eulogium of James Watt, by Lord Jeffrey.' Third Edition. *Duodecimo.*

London, eighteen-forty. Davies, 'Abstract of the Writings of Alexander Anderson' (Appendix to Ladies' Diary for 1840). *Duodecimo.*

London, eighteen-forty. Halliwell, 'The Connexion of Wales with the early Science of England.' An account of Robert Recorde. *Octavo.*

London, eighteen-forty. Vernon Harcourt on the discovery of the composition of Water, with reference to Watt, Priestley, and Cavendish: in the Report of the ninth meeting of the British Association. *Octavo.*

London, eighteen-forty. Halliwell, 'A few Notes on the History of the Discovery of the Composition of Water.' *Octavo.*

Paris, eighteen-fortytwo. Arago, 'Analyse Historique et Critique de la Vie et des Travaux de Sir William Herschel.' An addition to the *Annuaire* for 1842. *Duodecimo.*

X. *Biography.—Epistolary Correspondence.*

Noribergæ, sixteen-one. 'Tychonis Brahei Epistolarum Astronomicarum Libri.' Tycho Brahe's Letters (and his other works) have various titles and dates to the same books. *Quarto.*

Amstelodami, sixteen-eightytwo. 'Renati Descartes Epistolæ.' *Three parts quarto.*

Paris, sixteen-sixtyseven. 'Lettres de M. Descartes.' Second edition. *Two volumes quarto.*

These two do not contain entirely the same letters.

Gedani, sixteen-eightythree. Olhoffius, 'Excerpta ex Literis ad Johannem Hevelium.' *Quarto.*

Oxoniæ, sixteen-ninetynine. Wallis, 'Opera.' The third volume, pp. 615—708, contains a collection of his letters. *Folio.*

Leipsic (?), seventeen-eighteen. Hansch, 'Epistolæ ad Keplerum,' &c. *Folio.*

Amsterdam, seventeen-forty. Des Maizeaux 'Recueil de diverses pièces sur la Philosophie...par Mrs. Leibnitz, Clarke, Newton...' Several letters connected with the dispute on fluxions appeared in the first edition of this work (said to be 1720) for the first time: all of them are, we believe, to be found in the mathematical volume of Leibnitz's collected works. Second Edition. *Duodecimo.*

Lausannæ et Genevæ, seventeen-fortyfive. 'Leibnitzii et J. Bernoullii Commercium Philosophicum.' Epistolary correspondence of John Bernoulli and Leibnitz. Valuable. *Two volumes quarto.*

Hanoveræ et Göttingæ, seventeen-fortyfive. Gruber, 'Commercii Epistolici Leibnitiani, tomi prodromi.' A collection of miscellaneous anecdotes. *Two volumes octavo.*

Hagæ-Comitum, eighteen-thirtythree. Uylenbroeck, 'Christiani Hugenii aliorumque Exercitationes Mathematicæ.' Mostly epistolary correspondence. *Two fasciculi quarto.*

Oxford, eighteen-fortyone. Rigaud, 'Correspondence of Scientific Men during the Seventeenth Century.' (From Lord Macclesfield's collection.) *Two volumes octavo.*

London, eighteen-fortyone. Halliwell, 'A Collection of Letters illustrative of the Progress of Science in England from the Reign of Queen Elizabeth to that of Charles II., (Hist. Soc. of Science.)' *Octavo.*

XI. *Biography.—General Collections.*

Parrhisiis, fifteen-thirty. 'Vita omnium Philosophorum.' More curious than useful. *Octavo.*

Bononiæ, sixteen-fifteen. Blancanus, 'De Mathematicarum Natura Dissertatio, una cum clarorum Mathematicorum Chronologia.' The chronology is good as far as it goes; but the names are few, and only placed in order in their centuries: no years. *Quarto.*

Amstelædami, sixteen-fifty. Gerard Vossius, 'De Quatuor Artibus, &c., et Scientiis Mathematicis, cui subjungitur Chronologia Mathematicorum.' Short accounts of all manner of writers, known and unknown, with a good index for those who know the Christian name of the person they are looking for (it was very common at that time to make indexes in this manner). The higher historians, Montucla, &c., look down on Vossius, Heilbronner, &c.; but any one who wishes to be accurate upon a writer of second-rate reputation (and the history of science is not *all* Galileo and Newton), knows that the latter writers are friends in need. *Quarto.*

Bononiæ, sixteen-fiftyone. Riccioli, 'Almagestum Novum.' Three volumes, the first in two parts, each a folio; the second and third exist, we believe, but are very scarce. The first volume contains a list of astronomers, with short biographies; as far as they go, valuable. The whole work is full of historical information (throughout the first volume, at least, the only one before us), and has a good index. Riccioli was more learned than wise; an excellent authority for a fact, a poor guide in deduction from it. *Folio.*

London, sixteen-seventyfive. Sherburne, 'The Sphere of Marcus Manilius made an English Poem.' This translation of the first book of Manilius is now valuable for the appendix, which contains a list of more than eight hundred names (some of them not mentioned elsewhere), with a short account of each. Also a list of recorded comets and meteors. *Folio.*

Londini, sixteen-ninety. Blount, 'Censura Celebriorum Authorum.' A large and useful collection of notices of writers, with the opinions of others upon them. *Quarto.*

Urbino, seventeen-seven. Baldi, 'Cronica de Matematici.' A wonderfully paltry performance, even for the time at which it was written. It was published nearly a century after the author's death. *Quarto.*

Lipsiæ, seventeen-fortytwo. Heilbronner, 'Historia Matheseos Universæ à mundo condito ad Seculum XVI.' And, to be sure, beginning with Adam, Cain, and Abel, who are the first three mathematicians on the list, there is such a collection of miscellaneous learning as would have been utterly useless, but for the excellence of the indexes. As it is, the work is very useful. *Quarto.*

Lipsiæ, seventeen-fiftynine. Kraus, 'Diogenis Laertii de Vitis, &c.' Gr. Lat. Good indexes. *Octavo.*

London, seventeen-sixtyfour. Ben. Martin, 'Biographia Philosophica.' No authority whatever: useful in the first instance to see the best direction in which to look for other sources. *Octavo.*

La Haye, seventeen-eightyone. De Castres, 'Les Trois Siècles de la Littérature Française, depuis Francois I. jusqu'en 1781.' Not very profound: the alphabetical order is convenient. Fifth edition. *Four volumes duodecimo.*

Paris, seventeen-ninetysix. 'Les Vies des plus illustres Philosophes de l'Antiquité.' A translation of Diogenes Laertius; to which is added, 'Abrégé de l'Histoire de la Vie des Femmes Philosophes,' from the Latin of Ménage. *Two volumes octavo.*

XII. *Biography.—Particular Collections.*

Venetia, sixteen-fortyseven, Ghilini, 'Teatro d'Huomini Letterati.' Useful for Italian biography. *Quarto.*

Lugduni Batavorum, sixteen-fiftyone. Gerard Vossius, 'De Historicis Latinis.' Contains a great deal on the writers of the middle ages. Second edition. *Quarto.*

Leyde, seventeen-fifteen. Teissier, 'Les Eloges des Hommes Savans.' This volume is a collection of the numerous small biographies which are found in De Thou's history; these form the text, and Teissier's additions the notes. This work is very useful. Third edition. *Four volumes duodecimo.*

Paris, seventeen-twentytwo. Baillet, 'Jugemens de Savans sur les principaux ouvrages des Auteurs.' (Par de la Monnoye.) Here and there, in this work, are notices connected with the history of mathematicians. *Seven volumes quarto.*

Paris, seventeen-forty. 'Histoire de l'Académie Royale.' The third volume contains an account of La Loubère.' *Three volumes octavo.*

Paris, seventeen-fortyseven. Mairan, 'Eloges des Academiciens.' Contains Petit, Cardinal Polignac, Boulduc, Halley, Bremond, De Molieres, Hunauld, Fleury, Bignon, Lémery. *Duodecimo.*

A Londres, seventeen-fiftythree. Anonymous, 'Eloges de Trois Philosophes.' Jordan, La Mettrie, and Maupertuis. *Octavo.*

Berlin, seventeen-fiftyseven. Formey, 'Eloges des Académiciens de Berlin.' It contains Vignoles, Lamprecke, Naudé, Jordan, Wagner, Duhan, John Bernoulli, Grischow, Elsner, Dhona, Buddeus, Charles Beausobre, Knobelsdorff, Münchow, Arnim, Vockerodt, Carita, Keith, Althamer, Jerôme Wolfius, Seckendorff, Krosigk, Ancillon, Naudé, Uffembach, Stahl, Duchat, Forneret, Fabricius, Neumann, Isaac Beausobre, La Croze, Reinbeck, Heineccius, Mauclerc, Werenfels, Keysler, Vater, Benzelius, Bohmer, Doppelmaier, Canz, Wetstein, Christian Wolf, Krafft, Marinoni. *Two volumes duodecimo.*

Paris, seventeen-sixtysix. Fontenelle, 'Eloges des Académiciens.' This excellent collection contains Bourdelin, Tauvry, Tuillier, Viviani, De L'Hôpital, James Bernoulli, Amontons, Du Hamel, Regis, Vauban, Gallois, Dodart, Tournefort, Tschirnhaus, Poupart, Chazelles, Guglielmini, Carré, Bourdelin, Berger, Dominic Cassini, Blondin, Pole, Morin, Lémery, Homberg, Malebranche, Sauveur, Parent, Leibnitz, Ozanam, La Hire, La Faye, Fagon, Louvois, Montmort, Rolle, Renau, Dangeau, Billettes, D'Argenson, Couplet, Mery, Varignon, Peter I., Littre, Hartsoeker, De Lisle, Malezieu, Newton, Reyneau, Tallard, Truchet, Bianchini, James P. Maraldi, Valincourt, Verney, Marsigli, Geoffroy, Ruysch, Maisons, Chirac, Louville, Lagny, Ressons, Saurin, Boerhaave, Manfredi, Du Fay. *Two volumes duodecimo.*

Lucca, seventeen-seventyone. 'Elogi degli Uomini illustri Toscani.' A few of them mathematicians. *Four volumes octavo.*

London, seventeen-eightythree. Pringle, 'Six Discourses,' on the delivery of the Copley Medal at the Royal Society to Priestley, Walsh, Maskelyne, Cook, Mudge, Hutton, with Pringle's Life by Kippis prefixed. The discourses are clear and instructive. *Octavo.*

Pisa, seventeen-eightyfour. Fabbroni, 'Elogi d'alcuni illustri Italiani.' Contains Galileo, Giacomelli, Perelli, Leopold de' Medici, Frugoni, Metastasio. *Duodecimo.*

Paris, seventeen-eightyseven. D'Alembert, 'Histoire des Membres de l'Académie Française.' This work contains no lives of mathematicians. *Six volumes duodecimo.* In the sixth volume of D'Alembert's works (by Baslien, *Paris*, eighteen-five, *eighteen volumes octavo*) will be found his éloge of John Bernoulli.

Berlin, seventeen-ninetynine. Condorcet, 'Eloges des Académiciens.' A collection of the *éloges* read by Condorcet during his secretaryship of the Academy, with additional notices of other members. It contains altogether, La Chambre, Roberval, Frenicle, Picard, Mariotte, Duclos, Blondel, Perrault, Huyghens, Charas, Römer, Auzout, Bessé, Borel, Buhot, Carvavi, Coudraye, Cusset, Le Févre, Gayant, Lannion, Marchand, Mignot, Morin, Niquet, Pecquet, Pivert, Pothenot, Richer, Sedileau, Thévenot, Fontaine, La Condamine, Trudaine, Bernard, Jussieu, Bourdelin, Haller, Malouin, Linneus, Joseph Jussieu, D'Arci, Lieutaud, Bucquet, Bertin, Courtanvaux, Maurepas, Tronchin, Pringle, D'Anville, Bordenave, Daniel Bernoulli, Montigni, Margraaf, Duhamel, Vaucanson, Hunter, Euler, Bezont, D'Alembert, Tressan, Wargentin, Macquer, Bergman, Morand, Cassini de Thury, De Milly, Courtivron, Praslin, Guettard, De Gua, Paulmy, Bouvart, Lassone, Luynes, Fouchy, Buffon, Francklin, Camper, Fougeroux, Fourcroy, Turgot, the Chancellor de l'Hôpital, Pascal. *Five volumes duodecimo.*

London, eighteen-seventeen. Leybourn, 'Questions, &c., in the Ladies' Diary,' from 1704, the period of its establishment, to 1816. The

Ladies' Diary was established in 1704, the *Gentlemen's Diary* in 1721, and White's Ephemeris in 1750. When the tax on almanacs was taken off, and as long after as Dr. Olinthus Gregory held the management of them (from 1835 to 1840, both inclusive), they were enlarged, and original communications were given as appendixes. The two former have since been united. The index to Leybourn's collection is a very extensive list of English mathematicians. *Four volumes octavo.*

Paris and *Strasburg*, eighteen-nineteen, nineteen, twentyseven. Cuvier, 'Recueil des Eloges Historiques' (1800-1827). It contains Daubenton, L'Heritier, H. F. Gilbert, Darcet, Priestley, Cels, M. Adanson, Broussonnet, Lassus, Ventenat, C. Bonnet, de Saussure, Fourcroy, Desessarts, Cavendish, Pallas, Parmentier, Rumford, Olivier, Tenon, Werner, Desmarets, Riche, Bruguières, de Sèze, Beauvais, Banks, Duhamel, Haüy, Berthollet, Richard, Thouin, Lacépède, Hallé, Corvisart, Pinel, Fabbroni, Van Spaendonck, Delambre. *Three volumes octavo.*

London, eighteen-twentyseven. Davy, 'Six Discourses delivered before the Royal Society.' Contains historical remarks accompanying delivery of medals to J. Herschel, Sabine, Buckland, Pond, Brinkley, Arago, Barlow, Dalton, Ivory, South: with very imperfect biographical notices of Englefield, W. Herschel, Marcet, Vince, Häuy, Delambre, Berthollet, Hutton, Lambton, Maseres. *Quarto.*

XIII. *Bibliography.—British Libraries, &c.*

London, sixteen-fiftyeight. 'A Catalogue of the most vendible Books in England.' The dates of publication are not given, and the titles are frequently wrong. *Quarto.*

Londini, seventeen-twentyfour. Reading, 'Catalogue of the Library of Sion College.' Both subjects and authors. *Folio.*

Edinburgh, seventeen-fortytwo and seventysix. 'Catalogue of the Faculty of Advocates' Library.' *Two volumes folio.*

Lausannæ et Genevæ, seventeen-fortyfour. Castillon, 'Newtoni Opuscula.' The prefaces are useful for the bibliography of Newton's minor works, which is very confused. There is a short life of Newton. *Three volumes quarto.*

London, eighteen-thirteen. 'Catalogue of the Library of the London Institution.' *Octavo.*

London, eighteen-twentyone. 'Catalogue of Books belonging to the Mathematical Society.' No dates to the books. This society was established in Spitalfields in 1717, where it still exists. *Octavo.*

Edinburgh, eighteen-twentyfour. Watt, 'Bibliotheca Britannica.' *Four volumes quarto.*

London, eighteen-twentyseven. 'Catalogue of the Library of the Royal Observatory at Greenwich.' *Octavo.*

London, eighteen-thirtyeight. 'Catalogue of the Library of the Royal Astronomical Society.' *Octavo.*

London, eighteen-thirtynine. 'Catalogue of the Scientific Books in the Library of the Royal Society.' The best catalogue of scientific works which we know: not as to the extent of it, for the library which it describes is very poor in some departments, nor as to the arrangement, but as to the manner in which the titles are exhibited, and the ease with which any book can be identified. *Octavo.*

London, eighteen-forty. Halliwell, 'A Catalogue of the Misc. MSS. of the Royal Society' (Catalogue of their Letters appended). *Octavo.*

XIV. *Bibliography.—Sale Catalogues.*

Londini, seventeen-fortythree. 'Catalogus Bibliothecæ Harleianæ.' The mathematical parts of this catalogue must be found by the tables of contents at the beginnings of the volumes. *Four volumes octavo.*

London, eighteen-thirtyfive. Kloss, Catalogue of his Library. An immense collection of works, very few later than 1536, was sent over to England to be sold, by Dr. Kloss of Frankfort. This is the sale catalogue: but it is done with superior care, by the owner we suppose. *Octavo.*

Paris, eighteen-thirtynine. Bachelier, 'Catalogue de Livres,' &c. This house had once a monopoly, but there are now other mathematical booksellers at Paris. *Octavo.*

The following Catalogues of the sales of *private individuals'* libraries contain a large number of scientific books: Folkes, 1756; Delambre's;* Dr. Hutton's,† 1816; Dr. Evans's, 1819; Professor Playfair's (Edinburgh), 1820; Degen's (Copenhagen), 1825; Burckhardt's, 1825; Leslie's (Edinburgh), 1833; Miss Lousada's, 1834; Mr. Leybourn's, 1840; Mr. Halliwell's, 1840; Dr. Olinthus Gregory's, 1842.

The most remarkable catalogue of the sale of a *bookseller's* stock which has occurred for many years past is that of Mr. Dickson (better known as of the firm of Davis and Dickson), in three parts, Nov. 1834, Dec. 1834, May 1836. Of shop catalogues, containing mathematical books exclusively, there are Maynard's (six in number, including one on mathematical tracts), Priestley and Weale's (three), Weale's (three), Bohn's, Robinson's, Rodd's, and others. There is no need to particularise the dates of these: all we have here to do is to recommend the historical inquirer to keep every book catalogue which he gets.

XV. *Bibliography, exclusively scientific.*

Francofurti ad Mœnum, sixteen-eightytwo. Lipenius, 'Bibliotheca Realis Philosophica.' These volumes are arranged according to subjects, in alphabetical order, and form a very valuable, though somewhat straggling, record. There are two others on theology, one on medicine, and one on law. *Two volumes folio.*

Amstelodami, sixteen-eightyeight. Beughem, 'Bibliographia Mathematica.' Arranged under nations, with an index at the end: not very extensive, but useful. *Duodecimo.*

Lugduni, sixteen-ninety. Dechâles, 'Cursus seu Mundus Mathematicus.' The first volume contains a sort of bibliographical history of the sciences, which, though imperfect, is often useful. *Four volumes folio.*

Paris, seventeen-thirtynine. Bremond, 'Table des Mémoires Imprimés dans les Transactions Philosophiques,' depuis 1665, jusques en 1735. An excellent triple index, by volumes, by subjects, and by authors. *Quarto.*

Genevæ, seventeen-fortyone. Wolf, 'Elementa Matheseos Universæ.' The above is the date of the fifth volume (all the others are of later date), which contains a bibliographical history imitated from Dechâles, but not so extensive, though very good. Second (or later) edition. *Five volumes quarto.*

Göttingæ, seventeen-eightythree. Eyring, 'Synopsis Historiæ Literariæ.'

* We insert this catalogue, which we have seen more than once; but we cannot at this moment insert the date from actual inspection. Delambre died in 1822.

† This was a splendid collection, and ought to have been bought for the British Museum. Why it was not so bought will be seen in Bruce's 'Life of Hutton.'

A short, well-arranged, and exceedingly useful bibliographical account of all writers down to 1500. *Quarto.*

Wittenbergæ, seventeen-fiftyfive. Weidler, 'Bibliographia Astronomica, accedunt Historiæ Astronomiæ Supplementa.' This suggested his bibliography to Lalande, who has used it as far as it goes, and augmented it greatly. *Octavo.*

Breslau, seventeen-eightyone to eightyseven. 'Einleitung zur Mathematischen Bucherkentniss.' *Three volumes octavo.*

Leipzig, seventeen-ninetyseven and ninetyeight; eighteen-three, four, and five. Murhard, 'Litteratur der Mathematischen Wissenschaften.' A valuable mass of bibliography, very badly arranged and indexed. There is a table of projected contents at the beginning of the first volume, which is not carried out, except for the first volume: each succeeding volume has its own table of contents. But these contents have no paging: this the reader must supply for himself by looking through the volumes, and filling up the deficiency: he may thus render this work usable. The character of the contents answers to the opinion which would be formed of them from the preceding account of the index; generally incomplete, frequently erroneous, titles taken from all manner of sources. Very few English books known to the author. Useful enough for want of better. Vols. I. and II., Arithmetic, Geometry, and Analysis; Vols. III., IV., and V., Mechanics and Optics. *Five volumes octavo.*

Paris, eighteen-three. Lalande, 'Bibliographie Astronomique.' With the history of astronomy from 1781 to 1802. A most useful work, almost perfectly complete. The two indexes (authors and subjects, by M. Cotte) are exceedingly good, so that this is perhaps the most easily used book on bibliography which exists. Lalande was the greatest astronomical gossip in Europe, and his portion of history, and the occasional remarks appended to the titles of the books, contain information which it might be difficult to find elsewhere. *Quarto.*

Göttingæ, eighteen-four and eight. Reuss, 'Repertorium Commentationum à Societatibus Litterariis Editarum.' The first for astronomy, the second for mathematics. Extensive classified indexes. *Quarto.*

London, eighteen-seven. Thomas Young, 'Course of Lectures on Natural Philosophy.' The second volume contains an extensive classed catalogue of books and memoirs, with a good index. Short remarks of the most valuable kind are added, apparently in every case in which Dr. Young had himself examined the work. *Two volumes quarto.*

Paris, eighteen-ten, fourteen, nineteen. Lacroix, 'Traité du Calcul Différentiel et Intégral.' An historical preface, and lists of memoirs at the beginning of each volume. Second edition. *Three volumes quarto.*

Nürnberg, eighteen-twenty. Muller, 'Auserlesene Mathematische Bibliothek.' *Octavo.* And,

Augsburg and Leipzig, no date. Muller, 'Repertorium der Mathematischen Litteratur.' *Three volumes octavo.*

These works have not even a table of heads, and whoever would use them must make one for himself. They seem to be done with more care than Murhard, but are not so extensive.

London, eighteen-fortytwo. Pocock, 'On Life Assurance.' Contains a bibliography of this subject, the only one of which we have heard. *Duodecimo.*

XVI. *Bibliography, not exclusively scientific.*

Paris, sixteen-seventynine. Bullialdus et Puteanus (Bouillaud and Dupuis), 'Catalogus Bibliothecæ Thuanæ.' A catalogue of De Thou's extensive library, excellently well done. *Two volumes octavo.*

Göttingæ, seventeen-forty. Kahl, 'Bibliothecæ Philosophicæ Struvianæ emendatæ, &c.' Full of references. Writers on the history of philosophy are frequently of the greatest use to the inquirer into mathematical history. *Two volumes octavo.*

Neapoli. eighteen hundred. 'Catalogue of the Royal Library at Naples.' *Folio.*

Brussels, eighteen-twentythree. 'Catalogue de la Bibliothèque d'un Amateur.' The second volume contains the scientific books. *Three volumes octavo.*

Bruxelles, eighteen-twentyseven. 'Bibliotheca continens Libros Selectos.' Many mathematical ones. *Octavo.*

XVII. *Chronology.*

Lutetiæ Parisiorum, sixteen-thirty. Petavius, 'Uranologion.' Besides various of the minor Greek astronomers, this work contains extensive additions to the *Doctrina Temporum* of the same author. *Folio.*

London, sixteen-thirtythree. Isaacson, 'Saturni Ephemerides.' A chronological table, the work of a person well read in the old English chronicles. It is useful from its containing the succession of bishops in all the English sees, which is sometimes wanted in investigations relative to the English scientific history of the middle ages. *Folio.*

Francofurti ad Mœnum, sixteen-fifty. Calvisius, 'Opus Chronologicum.' Fourth edition. *Folio.*

Bononiæ, sixteen-seventynine. Riccioli, 'Chronologiæ Reformata.' *Two volumes folio.*

London, sixteen-eightyseven. Helvicus, 'Historical and Chronological Theatre.' An English translation, with continuations. *Sexto.*

Oxonii, seventeen-one. Dodwell, 'De Græcorum Romanorumque Cyclis.' *Quarto.*

Lugduni Batavorum, seventeen-fortyfive. Petavius, 'Rationarium Temporum.' A late edition, with continuations. *Two volumes octavo.*

London, seventeen-fortyfive. Sault, 'Breviarium Chronologicum.' From Strauchius, Beveridge, and Holder. Third edition. *Octavo.*

London, seventeen-sixtytwo. Dufresnoy, 'Chronological Tables.' An English translation, continued to the death of George II. Very good tables. *Two volumes octavo.*

Florentiæ, seventeen-seventy. Lamus, 'Chronologia Virorum, &c.' An alphabetical and chronological list of writers up to the beginning of the sixteenth century. *Octavo.*

Berlin, eighteen-twentyfive and twentysix. Ideler, 'Handbuch der Mathematischen und Technischen Chronologie.' *Two volumes octavo.*

London, eighteen-thirtyfive. Wilson, 'Manual of Universal History.' We mention this work as being more full upon Oriental matters, which are generally neglected in such small works of reference. *Duodecimo.*

London, eighteen-thirtyeight. Nicolas, 'Chronology of History' (Cabinet Cyclopædia). Second edition. *Duodecimo.*

This being the first attempt, as far as we know, to collect a list of works belonging to mathematical history, we must suppose that it will be found very imperfect. Every person who has attended to the subject will be able to make his additions; and probably, in a few years, we may ourselves be in a condition to add a supplement to the present catalogue.

A. DE MORGAN.

University College, London, August 1, 1842.

COMPANION TO THE ALMANAC

FOR

1844.

PART I.

GENERAL INFORMATION ON SUBJECTS OF MATHEMATICS, NATURAL HISTORY AND PHILOSOPHY, CHRONOLOGY, GEOGRAPHY, STATISTICS, &c.

I.—ON ARITHMETICAL COMPUTATION.

For the last two centuries books of arithmetic have borne a very great resemblance to one another in the directions which they have given for performing the rules of calculation. On the ground of correctness there has of course been little to object; and we have nothing to say against the habits of operation which those books strive to form, simply because there is no habit at all, good or bad, to be acquired under their directions. Every learner is left to himself; he is shown the general character of a rule, and he is then tacitly desired to follow his own judgment as to the manner of managing all the details of the rule to the use of which he is to become accustomed. In the present article it will be our endeavour to add to the elementary works on arithmetic a mode by which any one who will take the trouble, and to whom calculation is prolix and irksome, may form habits which will both shorten the work and alleviate the pain. Not that we can facilitate anything: our article is not written for those who desire a lighter load, but for those who are willing to undergo the wholesome exercise which will brace the nerves and strengthen the muscles. There may be difference of opinion upon several of the details of our plan, and, if it be so, we should be well pleased to see other suggestions brought forward. Up to the present time, however, all will admit that there is no work in which the student is shown how to compute, in such a manner that he is not left to himself in the management of the elementary steps.

Why has it happened that so important a branch of the education, not only of the mathematician, but of the merchant and tradesman, of the lawyer and statesman, and indeed of every one who has anything to do with the business of life, has been neglected in all its details? It is not difficult to give the reason why rational, demonstrative arithmetic has fallen before the *practical* tendency (as it is called) of the age; and in a future article we intend to draw attention to the real character of the practical results of this abandonment of thought in a subject on which more men are capable of thinking than on any other. But in the present article we will be content to suppose, with

B

writers in general, that calculation is a topic on which no one need be made to think, but only to work. How do people work? What have been the results of the training in which work, and work only, has been insisted upon? The answer is easy: of all persons who are not computers by profession, there are more who readily pay the income tax than could readily calculate their own shares of it, if their returns of income contained fractions of a pound.

The mode of stating most of the processes of calculation goes back to the beginning of the sixteenth century, at which time arithmetic in its present form was introduced into the rest of Europe from Italy, the inhabitants of which last country had themselves received it from the East some centuries before. The Greeks had indeed arrived at a respectable proficiency in operation before the Alexandrian school lost its character; but their system of numeration was not ours, and their writings were lost to Europe, and did not again become the property of the learned in the West until the Indian system had been fairly established in Italy. The introduction of the Alexandrian writings was not favourable to the spread of arithmetic, since, though it brought in much which was superior to the methods of the middle ages, it established the *sexagesimal* system, or the division of units into fractions by sixtieth parts, which is still retained in the minutes and seconds of an hour and a degree. This mode of division was not favourable to the spread of *decimal* arithmetic. The Romans were prevented by their imperfect arithmetical notation not only from success in calculation, but even from any strong attempt at it. The only Roman whose writing upon arithmetic has come down to us is Boethius (A.D. 600), and, though he had studied in Greece, he does not even produce a system of operations, but confines himself to speculations on the properties of numbers. It is but rarely he mentions or uses any number above a hundred; and his commentators of the beginning of the sixteenth century do not often rise above a thousand, except in dates: up to this period the work of Boethius was a great arithmetical classic.

It is said (by Tonstal for example) that at the beginning of the sixteenth century every country in Europe had treatises of arithmetic in its own language. We have no doubt of the fact, but we do not think it likely that the treatises used the Indian (or Arabic) notation. We have one * before us which, though it explains that notation, does not use it, but adheres to the Roman. In what is perhaps the earliest work of the sixteenth century on arithmetic,† published in Germany, we have

* 'Ain New Geordnet Rechen,' &c., by Jacob (or, as he lets the printer spell him, Jabob) Röbel, Augsburg, 1514. But, by the next year, Jacob had not only taught the printer to spell his name, but himself to *use* the Indian notation, or 'zyffer zale,' as appears in his work on gauging, 'Eyn New Geordnet,' &c., Oppenheym, 1515.

† John Huswirt, 'Enchiridion novus Algorismi,' Cologne, 1501. The author of this production, which is a very respectable one, seems to have feared that his Italian figures would excite calumny, a contingency which he met by putting in the title-page a request that all opposers would either hold their tongues or go to the devil; as follows:—

"more Ytalorum," a proof how much the Italians had advanced by help of their eastern importations.

It was not till towards the middle of the century that England received any decided impression from the advancing system. The treatise 'De Arte Supputandi,' published by Bishop Tonstal in 1522, gives as little as possible of what is operative, and confines itself to explanations, particularly on the subject of fractions. Robert Recorde's 'Grounde of Artes,' published in 1540, is the first work in which there is anything like a digested system of rules. From this period to the end of the century the power of computation was growing, both at home and abroad. Two circumstances gave it a great impulse: the increasing demand for trigonometrical tables, and the desire on the part of mathematicians to obtain the definite quadrature of the circle. In 1533 there was nothing in print more extensive than Regiomontanus's table of sines. In 1596 the 'Opus Palatinum' was published, containing a complete trigonometrical canon, to every ten seconds, and to ten places of figures. Still the want of a direct system of decimal fractions was an unseen obstruction, which was not generally removed till the beginning of the seventeenth century. At this epoch the invention of logarithms called out new energy in the direction of computation, and brought on the time when almost every mathematician was a good calculator, and when many of the most eminent among them performed feats of mere arithmetic which their successors are more disposed to admire than to imitate.

It is worth noting that the logarithmic branch of arithmetic, which has, up to a certain point, rendered operations so easy that expertness in the higher sorts of unassisted computation is no longer completely indispensable to the mathematical student, was for a long time the means of stimulating the power which it subsequently checked. The original methods of finding logarithms were, fortunately for computation, so exceedingly laborious as to require exertions far surpassing any then on record. The efforts of Vieta, Van Ceulen, &c., in the quadrature of the circle, sink into absolute insignificance when compared with those required in almost any page of the tables of Napier or Briggs. Euclid Speidell * ('Logarithmotechnia,' 1688) says he had heard that Briggs's logarithms took eight persons a whole year; and Pardies, in his 'Geometry' (first published, we believe, at Paris, in 1671), speaks of twenty persons having been employed more

"Invide ne latres; linguam compesce furentem,
Nec nimium rabidis garrulus esto labris.
Aut pete tartareas (superis incognitus) umbras
Et Phlegetonteos labere adusque lacus;
Atque illic potius lites et jurgia misce,
Et vivis pacem concubitare sine."

* The son of the John Speidell who was known as one of the earliest calculators of logarithms. The assertion seems to refer to Briggs's *first chiliad*, published in 1617, in which there could have been little more than a year for computation: it can hardly be meant for the twenty chiliads published in 1624. Ferguson (Select Exercises) quotes this as though it took eight persons a whole year to calculate the logarithm of 2 only, which is not what Speidell meant, though one word of bad grammar makes him seem to have meant it at first sight.

B 2

than twenty years, though it is impossible to conjecture to what tables he refers.

We have said that the higher class of mathematicians, at the end of the seventeenth century, became excellent computers; and this was particularly the case in England, of which Wallis, Newton, Halley, the Gregorys, and De Moivre are splendid examples. Since the beginning of that century we are fully justified in saying that Britain has borne away the palm of computation. This may seem a singular opinion, after the opinion we have expressed of the general state of the science of arithmetic in this country. But it must be remembered that we were then speaking of the present time, and that we are now taking the average of a couple of centuries; also that we were then alluding to the knowledge of the community in general, and that we are now speaking of the professed mathematicians. Before results of extreme exactness had become quite familiar, and while the doctrine of series and its consequences were novelties, there was a gratifying sense of power in bringing out the results of the new methods, which made it a pleasing exercise to go to a higher degree of approximation than could be advantageously applied. Newton says, in one of his letters to Oldenburgh, that he was at one time too much attached to such things, and that he should be ashamed to say to what number of figures he was in the habit of carrying his results.

During the last century, however, a schism has taken place among the mathematicians of our country: those not educated in the universities have preserved the art of computation, while those who come from the universities have well nigh abandoned it entirely. Let any one compare the questions proposed in the Lady's or Gentleman's Diary,* the classical works of the non-academic mathematical student in England, with those given to candidates for honours at Cambridge, and he will see that the problems which are to be answered numerically are at least three times as large a per-centage of the whole in the former as in the latter. We regret this prevailing fashion of the university of Cambridge: we think that a little systematic attention to the acquisition of arithmetical expertness would very much increase the amount of power acquired by the students. We think also that the demand for such a qualification would act most advantageously upon the preparatory education.†

The growth of the power of computation on the Continent, though considerable, did not keep pace with that of the same in England. We might give many instances of the truth of our assertion. In 1696, De Lagny, a well-known writer on algebra, and a member of the Academy of Sciences, said that the most skilful

* After this was written we saw in a Sunday newspaper, among the columns of answers to correspondents, the information that the Lady's and Gentleman's Diaries were the media of communication of the most eminent mathematicians.

† When an examiner actually does give a question which requires the use of logarithms, it is the custom of the University (books of logarithms not being allowed to be brought by the competitors) for the proposer of the question to pick out the necessary logarithms, and to print them as appendages to the question. This is a make-believe mode of solving questions which a future generation will smile at.

computer could not, in less than a month, find within a unit the cube root of 696536483318640035073641037. We would have given something to have been present if De Lagny had ever made this assertion to his contemporary, Abraham Sharp. In the present day, however, both in our universities at home and everywhere abroad, no disposition to encourage computation exists among those who attend to the higher branches of mathematics, and the elementary works are very deficient in numerical examples. Their writers flounder through what work they cannot avoid as they best may, leaving their algebra as little as they can, and getting back again as quick as possible. A good mathematician is or is not a decent computer, just as he is or is not tall or short, by the accidents of nature and circumstances.

After these remarks on the history of the subject, we now proceed to our main question, the method of forming readiness in arithmetical computation. We give up the scientific question; indeed, as far as mere manual dexterity is concerned, it matters little or nothing* whether the student is or is not versed in the principles of arithmetic. Speaking then of an art, and an art only, we say that those who abandoned the science of arithmetic did not take up the art, but proceeded to teach the latter on a plan which, though it would hardly have suited the former, yet was more fit for it than for the latter. General directions, unaccompanied by attention to detail, may do well enough in a process which is to be filled up demonstratively, but cannot form good habits, nor avoid bad ones, where nothing but habit is to be the guide. If Euclid should forget, as he does, to specify, for instance, that "of four magnitudes of the same kind, the two greater will, together, compose a greater magnitude than the two lesser," nobody will imagine much harm to arise from leaving the student to see and acknowledge this self-evident maxim in his own way, when it is wanted. But the lights of science are not our guides in this matter, which is one of mechanical operation merely: the drill-sergeant and the dancing-master are, as models, worth a hundred of Euclid and even of Cocker. If those who have merely to teach figure-work would abandon the demi-semi-scientific manner of the latter as completely as they have done the scientific manner of the former, and take a lesson from those who train the human body, whether for the ball-room or the field of battle, they would, we think, learn something worth their knowing. The principles of the mode of proceeding adopted by the trainers whom we would imitate, seem to be as follows,— First, all bad habits are to be abandoned at once: not one moment's compromise with them is permitted, however uneasy the abandonment may make the learner. Secondly, all that is to be newly acquired, in the positive sense, is broken up into its smallest elementary portions, of which one only is taught at a time. When a raw countryman first enlists, he is not allowed to retain his old habits of holding himself until they can be con-

* With respect to the mere rules of operation: the application of them is another thing.

veniently got rid of: he is not permitted to stoop until he has made it easy to himself to turn out his toes, nor told to dispense with the latter until he has learnt to step in time; but he is at once put into the position which he is finally to acquire, in every particular, and the eye of a veteran is always upon him, to see that he loses simultaneously all the habits of his former life. It matters nothing that at first the recruit is neither so steady, nor so active, nor so much at his ease as before; the incessant effort to abandon all that is bad soon ceases to be an effort, and a more useful form of steadiness, activity, and ease is the result. But the young arithmetician is merely presented with a rule, and left to manage its details as he pleases; he is treated as the young soldier would be if he were told nothing but that marching is the operation by which an army moves from one part of a country to another, and left to do it his own way, with instruction as to the manner in which a regiment performs evolutions, but none as to his mode of bearing himself. Again, when the recruit, having some little portion of new habits formed, begins to acquire the positive part of his business, the method is entirely changed. The various motions which he is required to learn are subdivided as much as it can be done, and each movement is practised by itself until reasonable perfection is acquired, nor are two ever put together until each has been learnt separately. No one has ever seen the process of drilling without becoming fatigued with the apparently never-ending reiteration of some one simple exercise. Indeed, whenever, save only in teaching arithmetic, there is a mechanical part which can be distinguished from what requires thought, genius, or taste,* it seems to be the rule of good teaching to separate that part into its simplest elements, and to dwell on each element separately. But in arithmetic, the fundamental rules are made a confused mixture of different operations, which never are made the subjects of separate exercise.

The processes which we should recommend, as preparatory to everything in the form of combined operations, are as follows:—

A. Supposing the learner to be able to count with sufficient rapidity, backwards and forwards, by single units—the only exercise which is now methodically learnt and properly practised in the first instance—he should then learn to count backwards and forwards by twos, by threes, by fours, up to tens at least, perhaps to twelves; beginning with different numbers. As in 3, 7, 11, 15, 19, &c.; 61, 57, 53, 49, 45, &c.; 2, 6, 10, 14, &c., &c. No reiteration should be allowed; it should not be "3 and 4 make 7," "7 and 4 make 11", &c., but simply 3, 7, 11, &c. If there

* A good singing master makes his pupils practise scales until the hearer is wearied; and the best singers, in every stage of their career, have continual recourse to this alphabetic process. The story of the manner in which Porpora taught the celebrated Caffarelli is well known to musicians. He kept him for five years at the scales, and a few passages written on a single sheet of paper. One year was then given to articulation, &c. Caffarelli thought he had only learnt the merest elements, and wondered how many years it would take to finish the course; but, at the end of the sixth year, Porpora said to him, "You have now nothing more to learn from me—you are the first singer in Italy, if not in the world!" Of course Porpora only professed to teach the mechanical part of the art, not taste, nor knowledge of music.

be difficulty, let the pupil be allowed to take his own time, but let him be prevented from repeating any single word, except one which expresses a result. This exercise is the preparation for addition; and continual repetition of the same addend will prevent the formation of those habits of error which grow up in the common modes of training. Some persons form tendencies to add wrongly in particular numbers, which they never completely get rid of: here the addition of every number is an exercise by itself. The use of addends exceeding ten may be desirable as a preparation for various abbreviatory processes, but is not essential in the fundamental rules of arithmetic.

It is also of importance to learn to add easily the smaller number, generally a single digit, to the larger, as 9 and 43, 7 and 85, &c. This should be made as easy by practice as 43 and 9, 85 and 7, &c.

B. The next exercise is the formation of the defect of a lesser number from a greater, when that defect does not exceed nine. The manner in which it should be acquired is by giving the lesser number, and the units only of the greater, the learner having to supply for himself the tens which should be in the greater, so that the defect may not exceed nine. Thus, having 56, and seeing 4, the exercise consists in learning immediately to supply both the 8 in "56 and 8 make 64," and also the 6 tens which, as it will turn out, are always to be carried. To perform this exercise by itself, write down any line of figures, as

7126605832417.

Make examples by taking the two first figures (71) for the lesser, and the next (2) for the units of the greater: then the second and third (12) and the fourth (6); and so on. The process then is to make out, as rapidly as possible,—71 and 1 are 72; 12 and 4 are 16; 26 and 0 are 26; 66 and 4 are 70; 60 and 5 are 65; 5 and 3 are 8, &c.

C. The multiplication table is now to be learnt up to nine times nine at least, but not in the common way. Of all the drawbacks upon rapidity of computation, none is so great as the common habit of reproducing in regular form the assertion "8 times 7 are 56" every time that 8 and 7 are seen, and multiplication is known to be coming. The exercise we now speak of consists in stating instantly the product of two digits, as soon as they are seen. Take a line of figures as before, and learn to repeat rapidly the product of every pair, without naming either of the pair.

72698437718.

The following products are to be caught instantly, 14, 12, 54, 72, 32, 12, &c.; not "7 times 2 are 14," "2 times 6 are 12," &c. One advantage of this process will be that the learner will become equally habituated to the products, whether the greater factor be seen first or the lesser. Some persons know 6 times 8, but are puzzled with 8 times 6 until they have inverted the factors, particularly those who have learnt the products from the contracted table which lies in a triangle instead of a square.

D. The next thing to be acquired is the formation of a product increased by a given digit, or a given digit by a product, instantly, without repetition of the factors and addend. Instead of "4 times 8 are 32 and 3 are 35," 4, 8, and 3 should suggest $4\times8+3$ or $4+8\times3$, as required. If rows of figures be taken again, and if the exercise be repeated on each three figures consecutively, slowly at first, if necessary, but keeping strictly to the rule of allowing no additional words to be either articulated or thought of, it will not be found very difficult to make the results come as readily as those of the simple multiplication table. Thus, taking

62987401328,

the object is to arrive rapidly at 21, 26, 79, &c., or $6\times2+9$, $2\times9+8$, $9\times8+7$, &c., and also at 24, 74, 65, &c., or $6+2\times9$, $2+9\times8$, $9+8\times7$, &c.

E. The next process is to catch the result of the preceding process, and to add it to another figure, naming the first result only and none of its constituents. Thus, taking the last line of figures, the exercise consists in learning to form rapidly "21 and 8, 29;" "26 and 7, 33;" "79 and 4, 83," and so on; arising from

$$(6\times2+9)+8;\ (2\times9+8)+7;\ \&c.$$

the reason of the distinction between the two addends being that the first of them will be a carried figure, and therefore retained in the memory, while the other is on the paper. In the same way it should be done when the carried figure is put first, which, in the preceding row, would require the rapid formation of 32, 81, 69, &c., or

$$(6+2\times9)+8;\ (2+9\times8)+7;\ \&c.$$

F. The next of these exercises resembles that in (B.), only that the smaller number is formed as in (D.). A product increased by a digit is to be taken from a number of which the unit's place is before the operator, while the ten's is to be supplied as wanted to make the defect not exceed nine. Thus, out of 7861 is to be instantly supplied, "62 and 9 make 71," or $7\times8+6$ is to be made up to the next number that ends with 1.

Thus, from the row of figures 79812563 is to be rapidly formed "71 and 0, 71;" "73 and 9, 82;" "10 and 5, 15;" and so on.

G. The last process is the inversion of (E.), namely, finding the quotient and remainder of tens and units divided by a single digit. But this should be practised without repeating as in "8 in 59, 7 times and 3 over;" it should be at most "8 in 59, 7 and 3." A row of figures may be used for practice, as in the preceding cases.

As soon as these seven rules become as familiar as counting, so soon, and no sooner, is the drudgery of computation annihilated. These are the steps by which the calculator walks; and let his journey be in what direction it may, no single pace can be anything but one or another of the preceding. Before proceeding to apply these *ultimate processes* * to the principal rules of

* It is commonly said that the elementary processes of arithmetic are addition, subtraction, multiplication, and division. This we deny; the rules given under these names are extensive systems of combinations of the elementary processes. Even the seven rules in the text are not all elementary.

arithmetic, we have one or two remarks to make:—1. The mode of acquiring rapidity is by practising rapidity. That every one must walk before he runs is true; but it does not therefore follow that walking will teach running. In the school teaching of arithmetic it is one of the greatest defects that no cognizance is taken of the rate of working, and no attempts made to increase it. Hence, nothing but correctness being aimed at, rapid calculation is seldom learnt: and to those who are obliged to work slowly, arithmetic is never anything but drudgery.

2. We cannot offer any decided opinion on the question whether the use of the voice should be allowed in computation. Some persons have so accustomed themselves to repeat what they are doing that they cannot give it up: with others the actual articulate sound has subsided into a kind of buzzing accompaniment; others, again, can do without any kind of vocal assistance. It is a point to be ascertained whether there are not some persons with whom the ear is so much more correct than the eye that they receive sensible assistance from repeating the figures they are to use: we suspect this to be the case with many. But we should recommend every one to give the silent system a fair trial before coming to any decision: the matter cannot be one of indifference; the use of the voice, if not a decided assistance, must be a hindrance.

3. Among the sources of retardation, if not of liability to error, is the difficulty of bearing in mind a particular number, as a multiplier, for example. In some of the more prolix kinds of calculation, the use of Horner's method, for instance, it is rather difficult to do this well. In the first instance, if not permanently, the computer might, as we have found by experience, receive considerable assistance by using the left hand in such a manner as to make a signal indicative of the multiplier or divisor in use. After a little time and practice, the position of the left hand may be made to suggest the figure wanted in such a manner that the name of that figure will hardly want either mental or oral enunciation. Our plan is as follows:—the number 1 is denoted by placing the fore-finger only on the paper; 2, by the first and second fingers, and so on up to 4; 5 is denoted by the thumb, 6 by the thumb and fore-finger, and so on up to 9, which is signified by the thumb and the four fingers.

We now proceed to the ordinary rules of arithmetic:—

Numeration.—The learner is usually taught to lay his pen upon the successive figures, repeating units, tens, hundreds, &c. Now, in reality, the eye is afterwards to detect the meaning of a figure, not by its place, but by the number of figures which come before it on the right. It may seem of little consequence whether we say that the seventh place is that of millions, or that when six figures are cut off all that precede are millions; but it is of considerable importance when the learner comes to decimal fractions, if not before. Let the primary denominations, 1, 10, 100, &c. be carefully associated, not with the number of *places* they contain, but with the number of *ciphers*. Thus ten is a

B 3

one-cipher number, a hundred a two-cipher number, a million a six-cipher number, and so on. It is not true that, for instance, only the fourth place is that of thousands; all that remain when three figures are cut off are thousands; and the student should form the first answer to such a question as "how many thousands are there in 452739," instantly and mechanically, by cutting off three places: answer 452. If any one should think that four will (as is usual) be more usefully connected in the mind with thousand than three, let him run over the following assertions, and propose his substitutes; a thousand has *three* ciphers; when *three* places are cut off, the rest are thousands; any digit is multiplied a thousand fold by removing it *three* places to the left; one thousand is the *third* power of ten; to divide or multiply by a thousand, remove the decimal point *three* places to the left or right; the logarithm of a thousand is *three*, &c.

Addition.—All statement of partial processes should be strictly checked, the only enunciation, mental or oral, being that of the results obtained by adding the digits one after another.

In the following example, all that should come upon the attention, whether spoken or not, is the following set of words. When the accented syllables are thought of or spoken, the figure or cipher should be written down at the same instant.

76483	five, eight, ten, nineteen, twenty-seven, thirty′;
9198	nine, ten, eighteen, twenty-two, twenty-four, thirty-three, forty-one′;
60729	eleven, thirteen, nineteen, twenty, twenty-seven, twenty-eight, thirty-two′;
4142	
6683	twelve, seventeen, twenty-three, twenty-seven, thirty-six, forty-two′;
5210	twelve, eighteen, twenty′-five.
89765	
252210	

The common mode of verifying the process by striking off the top line and adding again, is as absurd a mode as could be given. Whatever tendency to error there may be, by wrong association, it exists exactly in the same form in both additions. The best method is to go over the question again, adding the other way, thus,—three, eleven, twenty, twenty-two, twenty-five, thirty, &c. Those who add up one column and down the next (which is a very convenient way, as it removes all chance of slipping a column) should adopt one uniform mode; as for instance, beginning upwards at the first trial, and downwards at the second.

When the columns are very long, it is useful to preserve a record of the figure carried, and the following is convenient enough; if the sum of a column be 136, set down 6 in the column, with the 13 to be carried, written thus—6_{13}

Subtraction.—For the last two hundred years books on arithmetic have directed that a subtraction should be accompanied by such a monologue as the following (we take Cocker's own instance, in his own words, in the subtraction of 153827 from 437503), "Say 7 from 3 I cannot, but 7 from 13 and there remains 6; 1 that I borrowed and 2 is 3 from 0 I cannot, but 3 from 10 and there remains 7; 1 that I borrowed and 8 is 9 from 5 I cannot," &c. &c. All this useless phraseology, both in this rule and others, has descended to us from the early time which we have described, when what is now a moderately difficult question for a schoolboy of a year's standing was the highest attempt of an elementary writer, wishing to show off all the resources of his art. It is worth while to show this extreme case, and at the same time to point out that the same redundancy of expression prevails, in a less degree, in almost every rule which is practised.

Subtraction may be performed in two ways, either by a mental subtraction, properly so called, or by an effort to recover the number which added to the less gives the greater, as in the exercise (B.). Of these we recommend the second decidedly, not only as having advantage over the other in the simple subtraction, but as being a much better preparation for a subsequent process, which we mention under division. We take the preceding instance both ways.

```
437503   First by actual subtraction. Seven from thirteen, six′; three from
153827     ten, seven′; nine from fifteen, six′; four from seven, three′; five
------     from thirteen, eight′; two from four, two′.
283676
```

Next, by recovery of additions. It is here unnecessary to name the higher number; all that will be required is the following, the accented digit being written down when named or thought of. Seven and six′; three and seven′; nine and six′; four and three′; five and eight′; two and two′.

The sum of several numbers may easily be subtracted from another, as in the following instance, using the process acquired in (B.); again,

```
From 64197  Eight, fourteen, twenty-three, thirty-one, and six′, thirty-seven;
     ⎧3618  four, eleven, sixteen, seventeen, and two′, nineteen;
Take ⎨ 559  five, seven, twelve, eighteen, and three′, twenty-one;
     ⎪9276  six, fifteen, eighteen, and six′, twenty-four;
     ⎩4418  two and four′.
     -----
     46326
```

Multiplication.—The exercises (C.) and (D.) will, when properly learnt, much diminish the proverbial vexation of this rule. It would be a good thing if the first instances chosen had the same figure repeated in the multiplier, and were done both ways. Thus, 277385 might be multiplied by 77777, and 77777 by 277385.

The following is all that need be repeated, aloud or mentally, in a question of multiplication:

```
     72146  Twenty-four′; eigh′teen; five′; eight′; twenty′-eight′; thirty′;
      9854  twenty-three′; seven′; ten′; thirty′-six′; forty-eight′; thirty-
----------  six′; eleven′; seven′teen; fifty′-seven′; fifty-four′; forty-one′;
    288584  thir′teen; nine′teen; sixty-four′. The addition need not be
   360730   here repeated.
  577168
 649314
----------
710926084
```

Division.—The common mode of performing this operation is not the most convenient. Those who have been accustomed to the performance of each multiplication and subtraction at one step instead of two find that the risk of error is not increased, and the tediousness of the operation very materially diminished. The simple division by one figure ought to require in the way of statement, oral or mental, only the digit of the result. Thus in dividing 617483 by 9, nothing should be repeated but 6,8,6,0,9, 2 over. This will become familiar by practice; but at the same time, since the process is what we may call a clause of an operation by itself, and not one which is frequently repeated as part of a chain of processes, it is not of the same importance to avoid useless labour which it is in simple multiplication.

We have discussed all the usual elements of the longer process, and need not put them together. Previously to attempting the shorter mode, just mentioned, in which the multiplications and subtractions are combined, the exercise (F.) should be practised. Learn to dwell emphatically on the tens introduced (on which the carrying figure will depend) almost in the instant of writing down the accented figure: the power of dwelling in thought upon one figure at the moment of writing down another is very necessary to be acquired.

	Repeat mentally	Remember without saying it to carry
4 7 6 3	34 and 9′ are 43	4
2 7 0 8	14 „ 4′ „ 18	1
8 8 10 2	74 „ 8′ „ 92	9
2 2 2 9	6 „ 3′ „ 9	0

The exercise may be practised upon sets of consecutive figures, like the others; and when the process is tolerably familiar, that of division will be as in the following instance:—

76428) 19164277038 (250749
387867
572770
377743
720318
32466

As soon as 2 is ascertained to be the first figure of the quotient and 191642 is separated for operation, the first remainder is found at once,* thus:—16 and 6′, 2|2; 6 and 8′, 1|4; 9 and 7′, 1|6; 13 and 8′, 2|1; 16 and 3′, 19. The first remainder is then 38786 to which 7 is brought down, and 5 is ascertained to be the new quotient figure. The process for the second remainder is then 40 and 7′, 4|7; 14 and 2′, 1|6; 21 and 7′, 2|8; 32 and 5′, 3|7; 38 and 0 (not written) 38. The second remainder 5727 has 70 brought down, 0 being placed in the quotient. The next quotient figure is 7, and we proceed thus:—56 and 4′, 6|0; 20 and 7′, 2|7; 30 and 7′, 3|7; 45 and 7′, 5|2; 54 and 3′, 57. The third remainder, with 3 brought down, is 377743, and the quotient figure being 4, we have 32 and 1′, 3|3; 11 and 3′, 1|4; 17 and

* The carried figure is separated from the other by a line.

0′, 1|7; 25 and 2′, 2|7; 30 and 7′, 37. The fourth remainder is then 72031, and from 720318 the quotient figure is 9: accordingly, 72 and 6′, 7|8; 25 and 6′, 3|1; 39 and 4′, 4|3; 58 and 2′, 6|0; 69 and 3′, 72.

In the trial by which the quotient figure is found it will be desirable to make the first trial with the left-hand figure of the divisor increased by 1, if the next figure be 5 or upwards. Thus if the divisor be 873, &c., and the remainder 6142, &c., it is not the *eights* in 61 which should be first thought of, but the *nines*.

The operations of the extraction of the square-root and that of Horner's* method are much simplified by the ready use of the preceding modes. But before the last-mentioned process is attempted it will be advisable to practise the making of four figures suggest the product of the two first increased by the third and fourth: thus 8798 should give 65 and 8, 7|3′. As an instance, we give De Lagny's *month's work*, already noticed, carrying it on to a few decimal places, which he would perhaps have estimated at a fortnight more.

```
                886437165·39328 Answer.
 0          0             696,536,483,318,640,035,073,641,037
 8          64            184536
16          19200          15064483
            21184           1030027318
24 8        2323200            87606774640
25 6        2339076            16890934933
            23549880 0           389883585
26 46       23560513 6           154152392
26 52       23571148 8 00         12713651
            23571946 5 6 9          927088
26 58 4     23572744 3 4|7          219894
26 58 8     23572930 4 9|7            7736
               3116 6|4|7              664
26 59 23       3119 3|0                193
26 59 26       3121|9|6                  5
|26|59|29      3123|5
               3125|1
```

These are all the figures which a computer of a very moderate degree of expertness need put down: the student who starts in the manner described in the first article cited in the note will soon see that the steps here omitted to be written down are supplied by the rules above described.

For the computer who is afterwards to use logarithms, there is a process which, though it is not essential in ordinary arithmetic, is not to be neglected. The arithmetical complement of a number is its defect from the simple unit of the next higher denomination. Thus 7, 84, 293, 1630, are the arithmetical complements of 3, 16, 707, and 8370, being the numbers which, with the former, severally make 10, 100, 1000, and 10000. The arithmetical complement is most readily formed from the left, by subtracting every digit from nine, except only the last *significant* digit, which is to be taken from ten. Thus in forming the arith-

* This method is described in the 'Penny Cyclopædia,' "Involution and Evolution;" and its history is given in the 'Companion to the Almanac' for 1839.

metical complement of 7204136500, or 2795863500, repeat nothing but 7 and 2′; 2 and 7′; 0 and 9′; 4 and 5′; 1 and 8′; 3 and 6′; 6 and 3′; 5 and 5′; placing the two ciphers after the new digits thus obtained. It is desirable that the student should learn to take out the arithmetical complement to any number written in a table, without taking down and writing out the number itself. Again, if the sum of certain numbers be to be subtracted from the sum of others, the whole may be done as in addition, by using the defect from nine or ten (according to the position of the digit) and subtracting a unit in the first column in which any one of the subtrahends has no figure written: as in the following example, in which the subtractive units are entered in their proper places as soon as the question is arranged. Throughout the subtractive part of the process it will be observed that the digits of the arithmetical complement are substituted for those of the number itself.

From the sum of 7249616, 8381374, and 2978, it is required to take that of 3806125, 4176838, and 23852.

$$\begin{array}{r} 7249616 \\ 8381374 \\ 2978 \\ \bar{1}\ 6193875 \\ \bar{1}\ 5823162 \\ \bar{1}76148 \\ \hline 7627153 \end{array}$$

This process is not one which requires much exercise for its own sake, but it is, with its verification by the ordinary mode, a useful assurance to the young computer, that he understands the use of the arithmetical complement.

A person who is thoroughly conversant with the preceding processes cannot fail to be a good computer. He may need instruction in the manner in which these elements are combined in one or another question, but that is soon given, and is seldom difficult to retain or to practise. The real difficulty of computation is the performance of these operations, time after time, with correctness and rapidity; and failure in this respect is the principal source of error and prolixity: while the feeling of difficulty and insecurity is the main reason why many persons, whose minds are capable of and attracted by, mathematical reasoning, never conquer their aversion to arithmetical application. If boys, instead of being neglected in their *habits* of computation, were drilled to that which would ultimately come most easily to them; and if, instead of their being led through a maze of rules with a very imperfect power of correctly performing the most simple operations, their attempts at complicated combinations were arrested until they had firm ground under their feet in the more simple ones, the rapidity and certainty of their progress towards the end of their course would compensate a hundredfold its apparent slowness at the beginning.

These elementary rules, namely, addition, subtraction, multiplication, and division, with the extraction of the square-root, and those of the third and fourth root by Horner's method, furnish an ample field in which to learn computation. Some readers will be surprised that we do not name many more rules; but it is to be remembered that skill in computation is one thing, and skill in the application of computation another. To use these rules in application, principles must be understood: we cannot even recommend the mere computer, as long as he is nothing more, to try his power upon questions involving fractions. Whatever he may have to do in this latter subject, or in commercial arithmetic, he must, to gain the encouragement arising from success, have some ideas on what he is doing which mere power of computation will never give.

For procuring examples, again, in mere computation, we know of no work which will be of so much assistance to him as the tables of squares, cubes, &c., recently republished from Barlow's Tables.* We shall briefly point out how examples are to be obtained in each of the rules above mentioned, with answers.

Addition.—Take the numbers to be added consecutively out of the column of cubes; those who desire to practise many additions at a time may take any page of cubes as it stands, which will save writing: for example, page 5 beginning with 201^3 or 8120601, and ending with 250^3, or 15625000. When the addition has been made, it may be verified as follows: Multiply the first root by its preceding, and the last one by its following number. Multiply the sum and difference of the results; and the fourth part of the product is the result of the addition. Thus 200×201 is 40200, and 250×251 is 62750, the sum and difference of which are 102950 and 22550. The product of these is 2321522500, the fourth part of which is 580380625, which ought to agree with the addition.

Subtraction.—An easy way of producing subtractions with verification, is to subtract a given number successively from ten times that number; the proof of which is that the subtractive number itself becomes one of the remainders. It is a useful exercise to allow the subtractive number to remain in its first place, so that its distance from the other number of operation is gradually increased: as in the following example, in which 740896313 is successively subtracted from 7408963130, then from the remainder, and so on.

```
 740896313
7408963130
6668066817
5927170504
5186274191
4445377878
3704481565
2963585252
2222688939
1481792626
 740896313
```

* Barlow's tables of squares, cubes, square roots, cube roots, reciprocals, of all integer numbers up to 10,000 (stereotype edition). London. Taylor and Walton, 1840.

Multiplication and Division.—Since the square is the number multiplied by itself, and the cube is the square multiplied by the number, the table sets out twenty thousand questions of multiplication, and thirty thousand of division, with their answers. This is independent of all the approximative questions which are solved by the other parts of the table.

Square, Cube, and Fourth Roots.—Of the two first kinds there are ten thousand questions of each sort, with exact answers, and as many with approximate answers. Of fourth roots there are one hundred examples, by taking numbers which have exact square roots. But with regard to these rules it should be observed that the use of the table is only for verification at the outset, as far as its decimals go. Horner's method should be practised as far as twelve places of decimals at least.

Students who have examples with given answers get into the habit of looking at the answer before their own work is finished, to see if they are *right so far*. Any one who does this may rely upon it that he will never succeed when he comes to throw away his book: he will never learn that confidence in himself which fair play would give in time. Nothing in the way of temperament is more injurious to the formation of an easy habit of correct computation than fear: it is much better to go on boldly, and not to suppose the possibility of a mistake, until the time for verification arrives, than to learn the practice of suspecting mistakes and of never going over many steps without some verification. We speak merely of questions done of exercise: in matters of actual application every one must judge for himself how far it will be right for *him* to go without applying a test; but those who find they are not safe for a single step without reiteration should improve themselves by mere exercise questions, done more boldly than would be judicious if the result were of importance. Take a question in which mistake, if any, must appear at the end of the process: the multiplication of two numbers, for instance, followed by the division of the product by one of the factors, and before beginning the division rub out the steps of the multiplication. Proceed freely and carelessly, not thinking of the result until it arrives. If the answer be wrong, as assuredly it will be at first, do not attempt to find out the error, but destroy the work and try again. Do not be discouraged by failure, remembering that though incorrectness in one of the elementary processes may alter every figure of the result, yet for one such error there must have been dozens of correct steps. In this way learn to work without fear of consequences, attending only to the current step of the moment, as if this were a thing *per se*, to be made correct, even if all the rest were wrong.

All that precedes has reference to *written* computation. As far as we know of systems of *mental* arithmetic, there is considerable difference of opinion about the modes which should be adopted, and the extent to which they should be carried. The subject involves two questions; first, in what manner, and to what extent, is it desirable to teach young persons to calculate mentally; next, how much can those who have been educated in the usual way attempt with good prospect of utility and chance of success.

In more than one place of education, a degree of attention has been paid to mental arithmetic, which the testimony of competent eye-witnesses assures us has been followed by extraordinary results. We ourselves are not much surprised at these results, knowing that in childhood and early youth the faculty of *remembering* and combining numbers is frequently very active, and generally, as to the former, more active than it is in the same individuals grown up. In computation, the paper represents memory and nothing else: grant a sufficient power of retaining results, and there is nothing in the successive formation of them which requires writing. When, in the printed accounts of the schools to which we allude, we find the teachers themselves expressing their amazement at the manner in which they are outstripped by the boys, which obliges them to have recourse to a ready reckoner or some such help, we do not find them express themselves as if they were aware that the principal difference between them and the pupils is in *power of retention*. This must fall off in after-life, and with it the extraordinary appearance of readiness in mental arithmetic: at least, unless it be constantly kept up to an extent which the ordinary wants of life do not require, and for which there will therefore be no inducement. The world is now aware that the exercise of memory on one matter is not the same thing as the exercise of memory on another; and that it does not appear that one sort of exercise is equivalent to another. These premises are the heads of an argument which would be conclusive against the acquirement of such an amount of mental readiness as that attained in the schools alluded to, unless it could be shown that the time expended in the pursuit is not so great as to be a serious hindrance to other learning. But, at the same time, a greater degree of training than is usually given is most desirable, because it is afterwards wanted in the business of life.

There are two sorts of mental arithmetic. The first consists in the actual performance of rules in a manner different from that which is usual on the paper, and more fitted to the mode in which computation and mental retention are to go together. It should never be forgotten that the retention is the thing to be principally considered. The second consists in the substitution of easier rules in particular cases, rules framed according to the particular numbers to be used. In this second case the easier rule is as much the proper rule for the paper process as for the mental one; but, as capable of being mentally performed, which the common rule is not, it gets the character of belonging to mental arithmetic. As an example of the first sort, take the addition of 85 and 47: in doing this mentally, the 80 and 40 should be added first, and 120 retained; the 5 and 7 should be then added, and 120 and 12 put together. Or else 85 and 40 should be first added in thought, giving 125, and 7 then added, giving 132. This process put down on the paper would require more writing than the common one. As an example of the second kind take the rule for calculating the price of a pound from that of a hundred-weight;—or two-pence and one-seventh for every pound sterling in the price of a hundred-weight. Thus 14*l.* per cwt. is instantly 28+2, or 30 pence per lb.; 18*l.* 10*s.* per cwt. is 37+2$\frac{4}{7}$, or 3*s.* 3$\frac{4}{7}$*d.* per lb. This process is as much of an abbreviation on the paper as it is in the mind.

The first sort of mental arithmetic has been but little studied; and it would seem as if what there is of it in the schools referred to were the result of the pupil's own thought applied to a clear perception of the meaning of what he is about. In all probability, the method which is most easy to one mind will not be so to another. That young pupils can, when instructed in an intelligent manner, frame processes for themselves many teachers know: the following is the evidence of Mr. Wood, in his 'Account of the Edinburgh Sessional School,' though it would have been more satisfactory if the time previously given to the subject had been stated: "They will multiply such a line of figures as

768592816548793876‌4

by 7, 8, or any other figure, in less than the sixth part of a minute. From such a line they will *subtract* another of the same length in the ordinary way, in about seven seconds; and if allowed to perform the operation from left to right, while the question is under dictation (though it should be dictated with a rapidity which would not permit ourselves to take down merely the original figures), they will present the whole operation, both question and answer, in scarcely one second from the time of announcing the last figure. In *addition* they will sum up seven lines of eight figures each, in the ordinary way, in less than one-third of a minute; and if allowed to perform the operation while the question is dictating, in about three seconds. All other calculations they perform with proportional celerity. These modes of working during dictation (*when allowed*) are suggestions of their own, in their zeal to surpass each other, and not taught by the master."

We turn to the second sort of mental arithmetic; that in which the processes are only mental as being more easy, and therefore being better on paper than the usual ones, as well as possible without it. Of all the aids to abbreviated computation there is nothing like the ready knowledge of decimal fractions; and of all the advantages of decimal fractions, none is so conspicuous in commercial arithmetic as the use of the decimal parts of a pound, described in a paper in the number of this work for 1840. It will be worth while to recapitulate and extend this mode of operation, which sets the advantage of a decimal coinage in a strong light: showing that it is better, as our coinage stands, to learn a rule by which to go out of it into decimals, and working the question in decimals, to return the answer into ordinary denominations, than to remain in those ordinary denominations, and work the question by them.

For the first three places of decimals, the rule is as follows:—

Allow 1 in the first place for every *pair* of shillings, and fill the second and third places with one for every farthing above the shillings, and one more for sixpence, together with 50 for the odd shilling if there be one. Thus 4*s.* 4½*d.* is ·218*l.*; 4*s.* 7¾*d.* is ·232*l.*; 5*s.* 4½*d.* is ·268*l.*; 5*s.* 7¾*d.* is ·282*l.*; 17*s.* 11¾*d.* is ·898*l.*

For the fourth and fifth place of decimals, fill them with 4 for every farthing above *the last sixpence*, and 1 more for every 24 in the result (or, if found more easy, for every three halfpence in the farthings used). Thus for 4½*d.* the fourth and fifth places are (18×4 or) 72+3 or 75; for 7¾*d.*, 28+1 or 29; for 8*d.*, 33; for 11¾*d.*, 95.

For all decimal places after the fifth let the number of farthings *above the last three halfpence* be a numerator, let 6 be a denominator, and use the figures in this fraction reduced to a decimal. These can be formed in the head, and will soon be remembered, namely:

$\frac{1}{6}$ = ·166666 &c.	$\frac{3}{6}$ = ·5	$\frac{5}{6}$ = ·833333 &c.
$\frac{2}{6}$ = ·333333 &c.	$\frac{4}{6}$ = ·666666 &c.	

It is not often perhaps that more than five decimals are wanted, and three will most frequently be sufficient; but those rules are valuable by which, in case of need, approximation can be extended. A very little practice would enable any one to use the preceding rules so as to write down at once, to any extent required, any decimal of a pound; instead of 176*l.* 13*s.* 10¾*d.*, for instance, to put down £176·6947916666 &c. Considering the very great facilities afforded by this rule, which are more than those who have not tried it have any idea of, it would be worth while to save trouble yet further by learning the multiples of 4 up to 23×4 or 92, and those of 1½*d.* up to one shilling.

The inverse rule, to turn a decimal of a pound into money, is as follows:—A pair of shillings for every unit in the first place; a shilling for 50 (if so much) in the second and third places; a farthing for every unit left in the second and third places, after abatement of 1, if what is left be 25 or upwards. Thus ·477*l.* is 9*s.* 6½*d.*, and ·217*l.* is 4*s.* 4¼*d.*

The labour of questions in commercial arithmetic is much more than halved by the use of the preceding rule, and the risk of error reduced in proportion. It is not easy to make those believe this who have grown up in the use of the ordinary methods. We shall point out a few of the advantages attending the transformation in rendering questions capable of being solved mentally with ease and dispatch.

The two great modes of estimating fractions of a sum, namely, the number of shillings, &c. in the pound, and the per-centage which the part is of the whole, can be reduced to one another, or either mode compared with the other, almost instantaneously. Suppose, for example, it is required to know how much per cent. of his debts he can pay whose assets are 8*s.* 7¾*d.* in the pound. The regular book-arithmetician would proceed as follows:

As £1	:	8*s*. 7¾*d*.	::	100
20		12		415
20		103		960)41500(43$\frac{11}{48}$ per cent.
12		4		384
240		415		310
4				288
960				22

The one who is used to shorten his rules will say that 8*s*. is 40 per cent., and 8*s*. 6*d*. 42½ per cent., and will perhaps rest satisfied with that approximation. But the method we advocate enables us to write down at once ·432 as the fraction which 8*s*. 7¾*d*. is of a pound, and thence to name 43$\frac{3}{16}$ per cent. as the proportion required. If still greater precision be demanded, the advantage is still greater; for ·43229 gives ·43*l*. 4*s*. 7*d*. per cent.

With regard to most rules of abbreviation there is this to be said, that however convenient they may be, each in its place, they will be so seldom required that they will be forgotten: so that, when an occasion for the use of any one of them arises, one person will have finished the question in the ordinary mode, while another is recovering the abbreviated rule. It is only with respect to processes which are sure to be often wanted that such rules are therefore worth their trouble: each kind of business has its own cases, which need only be learned by those who are to attend to that business. In the books of arithmetic there is the fault of presuming that every person is to learn every variety of commercial rule *by rote;* compound interest and the determination of the price of a mixture from that of the materials are supposed to be equally necessary to all, though the subdivision of business makes it as certain as any thing which can be predicated of two rules of arithmetic that no one who requires one of the above rules will require the other. A mistake of the same kind is very likely to be committed in teaching rules of abbreviation.

When a process very frequently occurs it is the custom to have recourse to tables, such as those contained in the ready reckoners, tables of interest, &c. These may be divided into two kinds; tables in lieu of computation, and tables in aid of computation. On both it may be said that a person who does not understand their construction, and could not dispense with them, rarely makes a good and sure use of them. It is unwise to dispense with any mode of calculation, so as to make tables absolutely necessary. No one can tell, in any question, that a table will not present some little circumstance which will confuse any one who does not understand its construction. Besides, it is to be remembered that tables are subject to error, both of calculation and printing, and that in point of fact a perfect table is a rare thing. Those who use common tables of interest should not grudge the trouble of making some simple verification out of the table itself; such, for instance, as finding the interest on the half sum as well as on the sum, and comparing the results.

The time is not yet come for a very extensive use of tables in mercantile affairs; nor can it come until a decimal coinage is established. When the present coinage has become matter of history, it will not be the least amusing anecdote connected with it, that the great financiers of the nation were so much afraid of the fractions of their own money, that they preferred to lose eight pounds out of three hundred in the collection of a tax, that it might be *exactly* seven pence in the pound, rather than let it be three per cent. It is obvious enough that the income tax was meant for the nearest approximation to three per cent. which would give an exact number of pence in the pound, and it was supposed, for example, that in calculating the tax on 587*l.* the first of the following calculations being the easier would be substituted for the second or more difficult:

```
      587                     587
        7                       3
     ----                    ----
 12)4109                    17,61
 --------                      20
 20) 342-5                   ----
 --------                   12,20
   £17 2 5                     12
                             ----
                             2,40
                                4
                             ----
                             1,60
                             ----
                 Answer £17 12 2¼
```

But, in truth, had the rule for procuring the decimal parts of a pound been well known, it would have been easier to do the second than the first, as follows:

```
 587
   3
----
1761   17·610 = £17 12 2¼
```

If any person, not much accustomed to computation, and feeling its difficulty, should endeavour to mend his habits by the practice of the foregoing recommendations, we warn him that he will have his period of difficulty, during which he will not be able to see that he gains anything. He has to acquire what he probably never aimed at before, quick and ready habits of doing a few simple things. Without resolute determination he will do nothing; if he feel that he is not of the nature of those who face difficulties and conquer them, he had better not proceed to the trial. There is no use in disguising the fact that persons of ordinary memories never become good computers without hard work and steady perseverance; but, on the other hand, it must be allowed that long practice is not necessary. As far as mere computation goes, there is no reason why excellence should not be attained in six months. Almost any person may become such an arithmetician that his figures are as steady as the laws of the Medes and Persians, years and years before an active student in algebra will attain as great a chance of uniform correctness in the use of the signs + and −. And as a quaint Hindoo writer sets off thus, "He who distinctly and severally knows addition and the rest of the twenty logistics, and the eight determinations, including measurement by shadow, is a mathematician;" so we will end by saying:—He who can easily, rapidly, and accurately add, subtract, multiply, and divide, is a computer.

A. DE MORGAN.

University College, Sept. 5, 1843.

COMPANION TO THE ALMANAC

FOR

1845.

PART I.

GENERAL INFORMATION ON SUBJECTS OF MATHEMATICS, NATURAL HISTORY AND PHILOSOPHY, CHRONOLOGY, GEOGRAPHY, STATISTICS, &c.

I.—ON THE ECCLESIASTICAL CALENDAR.

THE year 1845, to the Almanac of which the present number* of the Companion belongs, is (like 1818) one of those remarkable years in which, to all appearance, the Calendar is wrong: and in which the British version of it certainly contradicts itself. If, say the instructions given in the prayer-books to find Easter, the full moon that comes next after the 21st of March fall upon a Sunday, Easter Sunday shall not be that Sunday, but the one after it. Now in 1845, the full moon that comes next after the 21st of March is on Sunday the 23rd, at some minutes past eight in the evening: *and yet that same Sunday the 23rd is Easter Sunday, though according to the rules laid down it should be Sunday the 30th.*

In looking at the explanations of the Calendar which are accessible to readers in general, we do not find one which combines a description of what is astronomical with a proper indication of what is matter of convention: and we need hardly tell our readers that the definition laid down in the Act of Parliament is adopted in every book published in Britain, and is the one inserted in the prayer-books of the Established Church. When astronomers write about the Calendar, they blame its complexity, and what they call its astronomical errors; and they very frequently mistake its construction: when theologians, who are not astronomers, do the same, they treat the Calendar with a degree of respect, as an astronomical production, which it does not deserve; or else, if informed of its departures from astronomical correctness, they treat those departures as errors to be deplored and corrected.

* Seeing that there was every prospect of a revival of the discussion of 1818, we sent the main conclusions of this paper to the 'Athenæum,' in which Journal they were published July 13th, 1844.

B

In the Roman Catholic Church, the Calendar rests upon the papal authority, not upon the moon. That Gregory XIII. chose to continue and amend the use which had always been made of the moon is merely a fact connected with the Calendar. It is also a fact connected with the Calendar that the learned Jesuit to whom its preparation was intrusted had seven letters in his name: but neither of these facts is of more consequence than the other with respect to the authority of the Calendar, in the Church of which we are speaking. In this country the same Calendar now rests on an Act of Parliament, in which Act the moon is stated to be the basis of the mode of finding Easter: here again the rule is determined by law, and the fact of the moon being employed does not add to the authority of the statute. Both the Roman Church and the British Parliament have a right to impose upon those who acknowledge their several ecclesiastical powers any way of using the moon to find Easter, or an Easter found without the moon, at their pleasure: both admit this principle, the Roman Church avowedly, the British Parliament indirectly. Nevertheless many Catholics and many Protestants have an undefined reverence for the moon's share in the determination of Easter, and feel as if every error, as they call it, of the Calendar, were a profanation of a sacred festival.

The Roman Church published a full and correct, if not a clear, account of the reasons which induced it to construct the particular Calendar on which we are writing, and of the mode of proceeding: the British Parliament added nothing to the extracts which it gave from the Roman *rules*, except the following sentence of *explanation*:—"Easter Day, on which the rest depend, is always the first Sunday after the full moon which happens upon or next after the twenty-first day of March: and if the full moon happens upon a Sunday, Easter Day is the Sunday after." (24 Geo. II. cap. 23, A.D. 1751, Statutes at Large, vol. vii. pp. 329, 345.) But the Parliament either forgot, or did not know,* that the bright, round, or horned body which shines in the heavens, which people call the moon, and which certainly is the moon, neither is, nor ever was, nor (since 1582) was intended to be, nor was asserted to be, the *moon of the Calendar.* On the contrary, all reasonable trouble was taken to make it clear that the "moon of the Gregorian Calendar" was a creation of the church, which, though allowed to maintain certain relations with the moon of the heavens, was in every particular contrived

* It might be more correct to say that it forgot in the Appendix what it had hinted at in the body of the Act. In the preamble to the enactment about Easter, the following words occur:—"And whereas a Calendar, and also certain tables and rules for fixing the true time of the celebration of the said Feast of Easter, and the finding the times of the full moon on which the same dependeth, so as the same shall agree, *as nearly as may be,* with the Decree of the said General Council [*i. e.* the Nicene], and also with the practice of foreign countries, have been prepared and are hereunto annexed." This is not quite accurate; it should have been *as near as may be* with the decree of the Council, and *exactly* with the practice of foreign Catholic countries.

to suit the convenience of those who invented it. Again, the British Parliament, in common with most astronomers, and many modern writers on the Calendar, was not aware that it was not the *full moon* even of the fictitious moon, from which Easter was determined, but the *fourteenth day* from the day of the new moon *inclusive:* which was always one day, sometimes two days, before the full moon. There is no pretext for this mistake in Antegregorian or even in Antenicene history: the decree of Pope Victor, and the description of the Nicene Calendar by St. Ambrose, are both perfectly explicit, for the 14th: and those who would persist in adherence to the Judaizing mode of keeping Easter were not called *Plenilunarians,* but *Quartadecimans.* We shall make all this plain enough in the sequel: in the mean while we remark that the discrepancy on which this article is written arose from the Parliament taking the papal Calendar with the papal moon omitted, and a nominal superintendence assigned to the moon of the heavens. We now proceed to some account of the history of Easter, and of the several Calendars.

The convenience of subdividing the solar year by means of lunations, each lunation comprising the interval between the appearance of two new moons, caused very early attempts to combine the sun's and moon's motions in the formation of the Calendar. As a general rule, in fine climates, the moon is first seen as a very thin streak of light, just after sunset on the first or second day following the conjunction of the sun and moon. Our day of new moon is the day on which (by reasoning and observation combined, for no one ever *saw* the sun and moon in astronomical conjunction, except when the conjunction was so close as to cause an eclipse of the sun) we know the conjunction to take place: but the new moon of the times which preceded the application of reasoning, was the first appearance of the moon on the eastern side of the sun, and was generally at least a day later than the conjunction. On both systems the day of appearance may be called (though it is not now so called) the *first* day of the moon; the day of conjunction being the last day of the old moon. Now the interval between two conjunctions is 29 days and a half very nearly, that is, the excess above 29 days varies from a quarter to three quarters of a day, but is about half a day (more correctly 12 hours, 44 minutes, 3 seconds) on the average. The average interval from new to full moon is the half of this, or 14 days, 18 hours, 22 minutes, $1\frac{1}{2}$ seconds; in round numbers, 15 days. Not counting then the day of the conjunction, the full moon will generally happen on the *fourteenth or fifteenth day of the moon,* though there will be occasional exceptions. To try this, we put down the days of new and full moon for a couple of years, 1843 and 1844, with the intervals, not counting the day of new moon.

Year 1843.				Year 1844.			
New.		Full.		New.		Full.	
Jan. 30^{12}		Feb. 14_{8}	15			Jan. 5_{5}	15
Mar. 1^{6}	30	Mar. 16^{6}	15	Jan. 19_{6}	29	Feb. 4^{9}	16
Mar. 30_{11}	29	April 14_{2}	15	Feb. 18^{9}	30	Mar. 4_{9}	15
April 29_{4}	30	May 13_{10}	14	Mar. 19^{0}	30	April 3^{7}	15
May 29^{7}	30	June 12^{7}	14	April 17_{4}	29	May 2_{8}	15
June 27_{7}	29	July 11_{5}	14	May 17^{9}	30	May 31_{11}	14
July 27^{5}	30	Aug. 10^{5}	14	June 16^{0}	30	June 30^{6}	14
Aug. 25_{3}	29	Sept. 8_{7}	14	July 15_{2}	29	July 29_{2}	14
Sept. 23_{11}	29	Oct. 8^{11}	15	Aug. 14^{3}	30	Aug. 28^{0}	14
Oct. 23^{8}	30	Nov. 7^{5}	15	Sept. 12_{1}	29	Sept. 26_{1}	14
Nov. 21_{6}	29	Dec. 7^{0}	16	Oct. 11_{11}	29	Oct. 26^{5}	15
Dec. 21_{5}	30			Nov. 10^{10}	30	Nov. 24_{11}	14
				Dec. 9_{8}	29	Dec. 24_{7}	15

This table is to be read thus:—1843, a new moon Jan. 30th, at 12 at noon; the following full moon Feb. 14th, at 8 in the evening; on the 15th day after that of the new moon exclusive: the next new moon on March 1st, at 6 in the morning, on the 30th day of the preceding moon; followed by a full moon on March 16th, at 6 in the morning, on the 15th day after that of the new moon exclusive; and so on. It sometimes happens that the full moon falls really on the 16th day of the moon, but when this happens, the moment of full moon is in the morning, so that the preceding night seems to be the brightest night of the moon.

Now from the earliest times,* the 14th night of the moon has been used where we might have expected that of the full moon; in fact, from the time of Moses downwards, the technical phrase of the Calendar has not been "the full moon," as set down in our Act of Parliament, but "the 14th day of the moon." It may naturally be supposed that, among rude nations, the time at which the moonlight reached its brightest would rather be anticipated, since the matter on which prediction would be required would be rather "bright moonshine" than "full moon." Also, any loss of the moon on the first evening, such as might

* Riccioli (*Chronol. Reform.* l. i., c. xii.) gives ample authority for the assertion that, just before the Christian era, the Jews reckoned from the appearance of the moon, not from the astronomical conjunction. Persons were appointed to watch on the day on which the new moon was to be expected: if by any accident it was not seen, a rule was applied to determine, for that one month, when it should begin. Long after the adoption of the astronomical new moon it was still customary to watch for the first night of the planet, and to pronounce certain words of blessing. There is then little doubt that the 14th of the moon meant the full moon, possibly the day after, among the Jews.

happen from haze or cloud, would make the night of brightest moonlight the 14th or 13th of the visible moon, instead of the 15th or 14th. Be this as it may, the 14th day of the visible moon is held * to have determined the Jewish feast of the Passover. "In the first *month*, on the fourteenth day of the month at even, ye shall eat unleavened bread, until the one and twentieth day of the month at even." (Exodus, xii. 18.) Again, "In the fourteenth *day* of the first month at even *is* the LORD's passover." (Leviticus, xxiii. 5.) This feast of the first month was intended to be held as near as might be to the vernal equinox, and the first month of the old Jewish year is generally held to have begun with the new moon belonging to the full moon which came next after the equinox, which new moon therefore sometimes preceded and sometimes followed the equinox. There has been discussion upon this point, some thinking that the first month always began with the new moon which came next after the vernal equinox. The weight of opinion, however, is all the other way, and the point is of no consequence for our purpose, since it is certain that the Christian substitute for the Passover, the festival of Easter, has always been regulated by the first 14th of the moon which follows the equinox, or at least the intention has been so to regulate it.

The 14th day of the moon, reckoning from its first visibility, is the evening of full moon or that preceding it; and the Passover was therefore most probably a feast of the full moon. But astronomers, when they came to have some knowledge of the moon's period, and some idea of the mode of finding the conjunction of the sun and moon, naturally transferred their reckoning to the moment of conjunction, and ceased to consider the moment of first visibility as the commencement of the lunation. The 14th day of the moon, counting the day of conjunction as the first, was adopted as the day by which the Christian festival was settled, and has long been adopted even by the Jews.

There is not much information as to the manner in which the Christians of the first three centuries kept Easter, except that the proper mode was much disputed; that there was one division as to whether it should be kept on the day of the old Passover

* Many writers suppose that the Jews brought with them from Egypt an astronomical system and a Calendar deserving of the name. This may or may not be true; but there is certainly no direct evidence for it. Nay, so little do we know of what the Jewish year actually was in the time of Moses, that Archbishop Usher positively, and Petavius doubtingly (to say nothing of Kepler), contend for its having been entirely solar, and it is impossible to answer them out of the Pentateuch: of course it is hardly necessary to say that a host of later authorities maintain that this year was luni-solar, and depended for its beginning upon the moon: in fact, that it was the same as the Jewish year after the captivity. If Usher and Petavius be right, the *original* Passover was no lunar feast at all, being fixed merely to the 14th day of a month: and this seems to contain the answer to them; it is very difficult to imagine any time at which so conspicuous a celebration as the Passover should, either by mistake or intention, be transferred in dependence from the sun to the moon. Nevertheless in 1 Kings, vi. 38, it is said that Solomon finished the temple "in the month Bul, which is the eighth month." Here are two words for month, the first derived from the moon, the second from a word which signifies *innovation* (new style, possibly). But in the work from which we cite these etymologies (Lindo's Jewish Calendar) it is simply stated that this text proves the Jewish months to have been always lunar.

or on the first day of the following week, another as to what mode of constructing the lunar calendar should be adopted. Leaving out the Montanists, who are said to have kept Easter on a fixed day, the dispute seems to resolve itself entirely into the schism of the Eastern and Western Christians; and Eusebius says that the parties were nearly equally balanced. No doubt there were some subdivisions of opinion: but the great mass of the Eastern Christians celebrated Easter on the 14th day of the moon, and of the Western on the Sunday following: the two sides pleading two different apostolic traditions. There is also some probability that the Easterns used the lunar cycle of 84 years, which the Jews are known to have used, and are supposed to have learned during the captivity: while there is every reason to suppose that the Westerns calculated their new moons by the aid of the cycle of nineteen years introduced by Meton. That the Western churches used the Sunday after the 14th day of the moon is certain from the letters of Popes Pius and Victor on the subject in the second century. The schism has an historical existence from the middle of that century, and probably there never had been any agreement on the subject. Uniformity of practice was introduced by the Nicene Council, in a manner which will require some description.

The Council of Nice (A.D. 325) issued the following announcement in their epistle to the church of Alexandria, preserved by Socrates and Theodoret:—"We also send you the good news concerning the unanimous consent of all in reference to the celebration of the most solemn feast of Easter, for this difference also has been made up by the assistance of your prayers: so that all the brethren in the East, who formerly celebrated this festival at the same time as the Jews, will in future *conform to the Romans* and to us, and *to all who have of old* observed our manner of celebrating Easter." Here, it will be observed, no rule is fixed, nor pretended to be fixed; all that is told is that the Eastern Christians shall or will in future conform to the practice of the Western ones. There is not a word about the moon, nor about any rule for determining Easter.

Writers, both Catholic and Protestant, have endeavoured to infer that the Council laid down the strict use of the cycle of nineteen years, and all that constituted the rule afterwards established. Both parties had motives which gave them this bias. The Catholic writers, naturally enough, disliked doubtful or inferential authority, and preferred the positive decree of the most celebrated of all the Councils, to the silent usage of centuries; the Protestants, who consider the Nicene Council as having preceded papal encroachment in point of time, are glad to refer to it not only for the old but for the new Calendar, instead of acknowledging the latter to spring from Pope Gregory and the Roman church. In England, in particular, the same feeling which long kept out the Jesuits' bark operated much longer against the Jesuits' Calendar. It is this which renders it necessary for us to state minutely what was actually done by the

Nicene Council; for we think it likely enough that there would be found some to defend the faulty definition which has given rise to this article, upon the supposition of its being a return to the genuine decree of the Council. Though, had it been so, the rule should have been to fit a calendar to the moon of the heavens itself, and to the full moon, or else to have predicted Easter from time to time by astronomical observation, as is done by the German* Protestant States. It was decidedly clumsy to take the Pope's Calendar, and disguise it by stating it to depend on a moon which contradicts it. The man who stole the saucepan and put on a new lid that the owner might not be able to swear to it, took care the new lid should fit.

The first mention of the Nicene decree in the next generation is by St. Ambrose. Clavius (with whom, as we shall see, we have more to do than with any other writer on the subject) cites a passage which can hardly have been written before A.D. 375, and which mentions that persons skilled in calculation were collected, who put the cycle of nineteen years into a form fit for use. But his expressions about the beginning of the first month are so inaccurate, that Clavius is obliged to explain them in a sense very different from their literal meaning: and a writer to whom it is no object to make an authority of St. Ambrose, may doubt if that bishop perfectly understood himself. Clavius proceeds to inform us, but without citing the passage (though he is generally most careful in citing), that St. Ambrose states that the cycle of nineteen years was arranged for the purpose by Eusebius of Cæsarea † (the historian, and one of the bishops of the Council) and some Alexandrian astronomers; "quod idem," says he, "*planius* Beda profitetur," &c., and he then goes on to cite Bede (A.D. 700) for the share of Eusebius in the matter, and also for the information that before the time of the Council it was usual for the Bishop of Alexandria to promulgate the proper time of keeping Easter, year by year, through all the churches. But Eusebius himself, much as he says of the Council in his Life of Constantine, and of the dispute about Easter in his History, never mentions this reference of the matter to himself and others; and without consulting ‡ St. Ambrose, we feel tolerably sure that any explicit passage on the subject would have been triumphantly cited by Clavius. This adroit covering of the epistle of the Council by that of Ambrose, and of the words of Ambrose by those of Bede,

* An account of the Lutheran reformation of the Calendar, with which we have here nothing to do, may be seen in Montucla's *Histoire des Mathématiques*, vol. iv., p. 322. One of the consequences of the *explanation* given by the British Parliament, on which this article is written, is, that all the readers of Montucla are left to suppose that our Calendar resembles that of the Germans. Lalande, who put together Montucla's fourth volume, conjectures that the English must sometimes differ from other Protestants, because probably they do not use either the same meridian or the same lunar tables as the Germans.

† Riccioli says that it was the Bishop of Alexandria, as being the spiritual head of the city of astronomers. He does not cite his authority; we had it once, but have lost it.

‡ Since this was written, we have looked carefully through the epistle of St. Ambrose in question, and find it to be exactly as we supposed. The epistle is No. 83 of the collection, headed 'Ad Episcopos per Æmiliam constitutos.'

is so well managed, that Delambre rose from the work of Clavius with the impression (which he has distinctly stated, 'Hist. de l'Astron. Mod.' vol. i., p. 3) that the epistle of the Council is not* in existence, but that some writers have given the spirit of it without citing the expressions. On looking attentively at the work, we see that this inaccuracy is very pardonable: the long and heavy argument by which Clavius intends to extract from the synodical epistle more than can be found in it, looks very like an attempt to restore its accurate meaning from versions of its spirit; and this argument has succeeded, down to the present day, in securing quiet acquiescence in the belief that the Council actually established a rule.

Had the Nicene fathers really called the astronomers of Alexandria together, and desired them to form the Metonic cycle into the sort of Calendar which was in use for a thousand years before the time of Pope Gregory, it is difficult to imagine that the simple cycle arising out of this rule should have borne the name of a priest of the sixth century. The first question which an ancient astronomer would ask, in arranging a chronological reckoning, would be, What is its cycle? After what period does it begin to recommence? Supposing the Metonic reckoning to be accurate, nothing is more easy than to see that, with leap-year *every* four years, a period of 19 × 28, or 532 years, will bring round the Easters in an order which will be repeated in the next 532 years, and so on. But this cycle bears the name of the Scythian Dionysius, surnamed Exiguus, an abbot at Rome about A.D. 530. From all this it seems clear to us that the Nicene Council ordained nothing but that the Eastern churches should assimilate their practice to the Western ones; that the latter used the Metonic cycle to find the Easter moon in the same way as to find any other, the Sunday after the 14th day of the moon being the well established rule; that the particular adaptation of this cycle to the peculiar rule which forms the principal part of the Antegregorian Calendar, did not take place till afterwards; and that the reference to Eusebius is not supported by reasonable testimony.

So far we have been content to make out our point from the writings or citations of Clavius himself. But when we come to look into other history, we find that we have to charge our author with something very much resembling unfair suppression. Granting that the Council fully established a unanimous observance of the Sunday after the 14th of the moon, we shall see that not only did they not succeed in framing a lunar cycle, but that the Church itself never had an undisturbed rule till the sixth century. First comes Theophilus of Alexandria, A.D. 380, with a cycle of 437 years; after him, Cyril of Alexandria, A.D. 412, with one of 95 years, which obtained great celebrity. Next, Victorinus of Aquitaine, the real author of the Dionysian cycle of 532 years, was actually employed by Pope Hilarius to correct the

* It is most formally given both by Socrates and Theodoret, and there are not ten words of difference between the two copies.

Calendar, in the year 463. The authority for this account of Victorinus is his contemporary Gennadius, who mentions as his predecessors in the art of cycle-making, Hippolytus (Antenicene), Eusebius, Theophilus, and Prosper. Dionysius Exiguus seems to have done no more than accommodate the cycle of Victorinus * (or Victorius, as he is often called) to his new mode of reckoning; he being the person who first abandoned the era of Diocletian, and reckoned from the supposed year of the birth of Christ. Had all parties had their due, Pope Hilarius would have gone halves with the Nicene Council in the credit of having settled the question of Easter; their joint work lasted a thousand years.

The Dionysian cycle entailed a gradually increasing error, both as to the time of the year at which the feast should be kept, and its coincidence with the moon. This error began to be fully recognised about the beginning of the sixteenth century, and after various attempts † had been made to excite attention to the subject, it was taken up in earnest by Pope Gregory XIII. in the year 1577. In this year a circular was forwarded from Rome to all the Catholic Princes and Universities of Europe, stating that a work had been presented to the Pope by the brother of Aloysius Lilius, deceased, containing a mode of treating the subject which was neither incommodious nor difficult. The circular goes on to desire that the attention of the learned should be directed towards this method, with a view to its improvement or the substitution of a better one; and then, after describing the conditions of the question, and the failure of the existing Calendar, it gives the heads of the plan of Lilius. What suggestions were received cannot now be known, without searching the records of the Vatican. The next step was the publication of a bull, dated ‡ March 1st, 1582, abolishing the old Calendar, giving a description of the new one, amply sufficient for use, and referring, for the grounds on which the new Calendar was adopted, to a work which was to be published. This work was intrusted, as well as the previous superintendence of the whole reformation, to the Jesuit Clavius,§ well known to mathematicians as the commentator on Euclid and other ancient geometers. Clavius was a laborious and voluminous writer, excessively prolix and learned, one of those dreadful exhausters of their subjects with whom the mathematical world abounded

* The writings of Victorinus, with all others of note relating to the Calendars, both Antenicene and Antegregorian, are found in the 'Doctrina Temporum' of the Jesuit Ægidius Bucherius, Antwerp, 1634. With regard to the mistake about his name, it is very probable that a lurking confusion between him and Victor, the Pope of the second century, who is so often mentioned on the matter, has given rise to Victorius instead of Victorinus.

† Such as that of Stöffler in his 'Calendarium Romanum,' Oppenheim, 1518, a very respectable work; that of Gauricus, mentioned by Delambre (*Hist. Astron. Moyenne*), very much the contrary; and that of Pitatus, which we shall presently have occasion to notice.

‡ This is the attested date of publication; but the document itself bears date "sexto Calendas Martii" (Feb. 24th), 1581.

§ Others are known to have been joined with Clavius; but it is clear enough that he was both the head and hand of the commission. His real name was Christopher Schlüssel.

before the sciences began to move with the rapidity of the seventeenth century, who thought shame of themselves if their comment ever fell short of twenty times the length of the text—in whose hands Latin, the most concise of all languages, became the most wordy of all. But he was sound and accurate, knew well what he was about, was free of all superstition about times and days, and looked upon the festivals in their true light, namely, as religious ceremonies, the real objects of which are independent of the planets. His work, 'Romani Calendarii a Gregorio XIII., Pontifice Maximo restituti Explicatio,' appeared in 1603, at Rome, and was reprinted in the collection of his works published in five volumes at Mayence in 1612, the year of his death, at the age of 75. The reprint, which we have now before us, contains more than 600 folio pages, of which about 400 are tables; but even with this deduction it well bears out what we have said about the prolixity of Clavius.

It was a question much agitated at the time, as it has been since, whether it was desirable to make Easter a moveable feast at all. Why would it not have done to name a fixed Sunday at any convenient period after the 21st of March? This is the question which Delambre and all the astronomical critics of Clavius never fail to ask. Clavius has answered the question himself by simple reference to ancient custom, which should not be altered without pressing reason. So far as Easter *Sunday* is concerned this is a sufficient answer; but when it is remembered that the whole of Lent depends upon Easter, which interferes, particularly in Catholic countries, with the business of life, it may be very much doubted whether the historical recollections connected with this one mode of keeping Easter are worth the advantage which certainly would be gained by making the intervals of the great feasts the same in all years. Clavius insists particularly upon the fact that his church is under no obligation* whatever to make Easter moveable, and rests his defence of the system simply upon the expediency of adhering to antiquity. When we see one liberty taken with the moon to insure regularity, another to prevent Easter from being kept at the same time as the passover, others to procure easy rules of calculation,

* It is rather amusing, all things considered, to compare the perfect freedom of Clavius, and the various modes in which his proceedings demonstrate that his Church is the master and not the slave of the festival, with the tone of Protestant writers, and still more of Protestant speakers in ordinary life. But Clavius had ecclesiastical authority at his back; and it is still more amusing to see how nice a line is drawn for those who are not in the same strong position. In 1705, Michel Touraine, a French priest, published at Paris his criticism on the Calendar, dedicated to a Cardinal Archbishop, approved by the censure, privileged by the king. His criticisms are strong and unqualified, showing, however, that he was not aware of the intentional liberties which Clavius had taken: showing also that a very considerable latitude of discussion was allowed. But he was unlucky enough to suppose that a day had been slipped, from which, says he, "the days which we celebrate, thinking them Sundays, are in reality Mondays." This was too much: the publication seems to have been arrested (for the approbation and the *privilège* are dated 1697 and 1699), and the book finally appeared, in 1705, with an additional *avertissement*, stating that the Sundays are Sundays and the Mondays Mondays, really and truly, the contrary being only stated to draw more attention to the error by which the days of the month given to Mondays are those which should have been given to Sundays.

we may well ask why, seeing that the real paschal full moon is sometimes abandoned for convenience, it was not thought the best plan to name the first or second Sunday after the 21st of March, and thus to simplify the relations of the festivals of religion to ordinary life, and their use in chronological reckoning. But it may be very much doubted whether it would be advisable to adopt such a plan now, in any one country. If the Parliament of this country were to fix Easter Sunday, which it could only do for the Protestant portion of the empire, the odds are that the confusion which would arise from the different sects keeping Easter at different times, would counterbalance the advantage of the more simple reckoning.

Confining ourselves particularly to the determination of Easter Sunday, we shall now describe the Calendar, reserving to the end the description of the old Calendar. The thing to be settled is the mode of determining the Sunday which follows the 14th day of the moon which follows (or arrives on the same day as) the entry of the sun into the vernal equinox. The first point is the manner in which the determination of the vernal equinox is saved by making an arrangement which keeps it always on or about the same day of the year. This part of the Calendar is the one which is best known, and has least difficulty.

The time which elapses between two entries of the sun into the vernal equinox is 365 days, 5 hours, 48 minutes, 50 seconds. The Calendar of Julius Cæsar takes this as 365 days, 6 hours, and accordingly allows an additional, or 366th, day, to every fourth year. This makes the average year to be 365¼ days, which is too long by 11 minutes, 10 seconds. This will be found to make 400 years too long by about 3 days; so that the error may be nearly corrected by reducing 3 leap-years in 400 years to common years. The mode of doing this was by neglecting* to make leap years of the *anni domini* which end with 00, unless the preceding figures are divisible by 4. Thus 1600 is leap-year,

* This artifice, which is the most elegant part of the new Calendar, was not due either to Lilius or Clavius. Stöffler (above mentioned) had proposed the omission of one day in 134 years, but Pitatus of Verona (who had proposed a new Calendar to Paul III. in 1537), in a petition presented to the Council of Trent in 1552, proposed that 1600, 1700, 1800, ... 2000, 2100, 2200, ..., should not be leap-years. (Compendium Petri Pitati Veronensis, &c., Veronæ, 1560.) As Pitatus is thus probably the first proposer of the plan actually adopted, with a slight variation, and as we never saw his claim stated, we extract his own words from the work just cited, fol. 4, prop. 2, premising that he wanted to destroy nominal days enough to bring the equinox to the 25th of March, where he supposed it to be at the death of Christ: and so he may be held to have originated the worst as well as the best part of the reformation:—"Atque exinde primus, secundus, ac tertius centesimus quisque Christi D. annus, qui secus bissextilis esse debuisset, communis modo pertranseat, quartusque dehinc post tercentenos (veluti in annis quatuor communes tres intercalaris quisque subsequitur) bissextilis haud secus quotusquisque centesimus quartus communes tercentenos tres ex ordine subsequatur. Sic nanque a quingentesimo post mille Christi D. transactos (qui bissextilis dudum effluxit) inchoando, subsequentes ad quartum usque millesimum noningentesimum, tres scilicet 1600, 1700, 1800, communes efficiantur anni: millesimus vero noningentesimus intercalaris ex ordine permaneat. Itidem 2000, 2100, 2200, Christi D. anni communes æque fiant, atque 2300 haud secus quam 1900 atque 2700 (servata in reliquis de quarto in quartum usque ad centesimum quenque prisca intercalandi series) bissextilis æque pertranseat, ut firma stabilisque æquinoctii verni sedes 25 communi atque 24 Martii die bissextili anno Romano Calendario tandem statuatur."

but 1700, 1800, 1900, are not leap-years; 2000 is leap-year, but 2100, 2200, 2300 are not leap-years. In the Julian Calendar every year is leap-year which is divisible by 4; in the Gregorian Calendar the exception is, that the years which close centuries (or end with 00) are not leap-years unless they are divisible by 400. This correction leaves the year still a little too long, by about a day in 3600 years. Delambre proposed, and it was adopted in the French Revolutionary Calendar, that the years 3600, 7200, 10800 should not be leap-years: but the Gregorian Calendar did not pretend to legislate for a day in periods of such length. Indeed Clavius is expressly of opinion that there would be no harm in the equinox being occasionally as far away from the 21st as the 11th, if it would only come back again of itself. So far the Gregorian reformation was well managed: we now come to its great blot, as we think. It happened that at the time of the Nicene Council the equinox fell upon the 21st of March. Now too long a year made the equinox gradually recede, insomuch that by the year 1582 the equinox was supposed to be on the 11th of March. There was no need to make the equinox proceed forwards again to the 21st, because it was so at the time of the Nicene Council: if anything, it should have been carried still further forward, so as to stand at the same day as it did at the Crucifixion. Considering how closely the events which Easter celebrates coincided in point of time with the Jewish passover, we can treat with all respect the feelings of those who like to determine the festival in the same manner as the passover was determined, even though the inconvenience of moveable feasts may result. But no Christian association is at all connected with one day or another of the month which was named after the God Mars; the events of the life of Christ were connected with the Jewish, not the Roman, year; nor was the accident of a general council interfering when the equinox happened to be at the 21st of a Roman month an event worth remembering. But it was thought otherwise: and the stupid expedient was resorted to of destroying ten nominal days, which has created more confusion and more chronological error than all the anomalies of the old Calendar put together. The day which should have been the 5th of October, 1582, was ordered to be called the 15th, so that the 5th, 6th, &c. and 14th of October were never allowed to exist. All the confusion arising from difference of style, as it was called, embarrassed the dealings of Catholic and Protestant countries for many years. This was merely a mistake; it did not happen to strike Clavius and his advisers that the Protestants were not likely to receive so startling an amendment of the Calendar at the hands of Rome. There is much reason to suppose that this violent change placed as great a difficulty in the way of Protestant governments acceding to the new Calendar as religious feeling. When, in England, in the eighteenth century, it was at last introduced, the mob pursued the minister in his carriage, clamouring for the days by which, as they supposed, their lives had been shortened:

and the illness and death of the astronomer Bradley, who had assisted the Government with his advice, were attributed to a judgment from Heaven.

A slight mistake was committed in making this change of style. The equinox vibrates between one day and the next, and is brought back again by the leap-year. Since the common year is about 6 hours too short, suppose that the equinox should fall, say at 2 in the afternoon of the 21st of March, the year being leap-year: next year it will fall at 8 in the evening, next at 2 in the morning of the 22nd, next at 8 in the morning of the 22nd. In the next year it would fall at 2 in the afternoon of the 22nd, but that year is leap-year again, the additional day brings everything after the 29th of February a nominal day earlier, and the equinox is again at 2 in the afternoon of the 21st. The incorrectness of the 6 hours' alteration makes 11 minutes odd of difference, which gradually accumulates until it is set nearly right at the end of the century by the Gregorian correction. Now it was intended that the equinox should have vibrated in this way between the 20th and 21st of March; but an error of a day in the Alfonsine tables, which were consulted, made the equinox vibrate between the 21st and 22nd, as it now does. So that, even if the moon of the Calendar were the moon of the heavens, here is a source of occasional error: for if a 14th of the moon were to fall on the 21st, at a time when the equinox is on the 22nd, that 14th would be held to be the paschal 14th, and Easter Day would be the next Sunday, according to the Calendar, though the 14th really fell before the equinox. This would make a whole month of "error" at once. The *calendar-equinox*, nevertheless, is to be held to fall on the 21st of March, and the next question is, how to settle beforehand what days will be Sundays in any one year. The old method of the *dominical letter* was of course retained in the new Calendar, as follows:—Take the seven letters A, B, C, D, E, F, G, and let A belong to the 1st of January, B to the 2nd, and so on, as in the next page.

If the 2nd of January be a Sunday, its letter being B, all the other days of a common year are Sundays which have also the letter B. All the Sundays of a common year have the same letter, and this letter is called the *dominical letter* * of the year. In leap-year the 29th of February has no letter; but it has a day of the week: accordingly there are two dominical letters, the first for January and February, the second for all the rest of the year. For example, the year 1844 begins with the dominical letter G; its first Sunday is the 7th of January; the 25th of February therefore, from the table, is Sunday, the 28th is Wed-

* It is sometimes directed to remember the letters attached to the first of the several months A, D, D, G, &c., by the following lines:—

At Dover Dwell George Brown, Esquire,
Good Christopher Finch, And David Friar.

Here is a more ancient and a graver specimen:—

Astra Dabit Dominus, Gratisque Beabit Egenos,
Gratia Christicolæ Feret Aurea Dona Fideli.

	Jan.	Feb.	Mar.	April	May	June	July	Aug.	Sept.	Oct.	Nov.	Dec.	
1	A	D	D	G	B	E	G	C	F	A	D	F	1
2	B	E	E	A	C	F	A	D	G	B	E	G	2
3	C	F	F	B	D	G	B	E	A	C	F	A	3
4	D	G	G	C	E	A	C	F	B	D	G	B	4
5	E	A	A	D	F	B	D	G	C	E	A	C	5
6	F	B	B	E	G	C	E	A	D	F	B	D	6
7	G	C	C	F	A	D	F	B	E	G	C	E	7
8	A	D	D	G	B	E	G	C	F	A	D	F	8
9	B	E	E	A	C	F	A	D	G	B	E	G	9
10	C	F	F	B	D	G	B	E	A	C	F	A	10
11	D	G	G	C	E	A	C	F	B	D	G	B	11
12	E	A	A	D	F	B	D	G	C	E	A	C	12
13	F	B	B	E	G	C	E	A	D	F	B	D	13
14	G	C	C	F	A	D	F	B	E	G	C	E	14
15	A	D	D	G	B	E	G	C	F	A	D	F	15
16	B	E	E	A	C	F	A	D	G	B	E	G	16
17	C	F	F	B	D	G	B	E	A	C	F	A	17
18	D	G	G	C	E	A	C	F	B	D	G	B	18
19	E	A	A	D	F	B	D	G	C	E	A	C	19
20	F	B	B	E	G	C	E	A	D	F	B	D	20
21	G	C	C	F	A	D	F	B	E	G	C	E	21
22	A	D	D	G	B	E	G	C	F	A	D	F	22
23	B	E	E	A	C	F	A	D	G	B	E	G	23
24	C	F	F	B	D	G	B	E	A	C	F	A	24
25	D	G	G	C	E	A	C	F	B	D	G	B	25
26	E	A	A	D	F	B	D	G	C	E	A	C	26
27	F	B	B	E	G	C	E	A	D	F	B	D	27
28	G	C	C	F	A	D	F	B	E	G	C	E	28
29	A		D	G	B	E	G	C	F	A	D	F	29
30	B		E	A	C	F	A	D	G	B	E	G	30
31	C		F		D		B	E		C		A	31

nesday, the 29th is Thursday; consequently the 3rd of March (F) is Sunday, and F denotes Sunday for all the rest of the year. Its dominical letters are therefore said to be G F; and in all cases the dominical letter of the last ten months of leap-year is one behind that of the first two months. Again, a common year has its first and last day of the same name; consequently the Sundays of the next year have a dominical letter which is one behind that of the preceding year: for the new year begins with A as well as the old one, and if A be the *Sunday* letter of a common year, it is the ***Monday*** letter of the next, or G is the Sunday letter. We can easily go on then with the formation of Sunday letters, if we know that 1844 has G F to start with; thus, 1844 G F, 1845 E, 1846 D, 1847 C, 1848 B A, 1849 G, 1850 F, 1851 E, 1852 D C, &c. And since Easter always falls after the 29th of February, the rule is, recede one for a common year, two for leap-year. We have now to put this rule into an arithmetical form. Number the dominical letters thus:—

A	B	C	D	E	F	G
6	5	4	3	2	1	0

It is clear from what precedes, that so far as the part of the year in which Easter lies is concerned, the number of the letter increases by 1 in passing into every common year, and by 2 in passing into a leap-year, striking off 7 as fast as it arises, since we have to pass from 6 to 0 again. The dominical letter of 1583 was F in the old style and B in the new. But if the *old* Calendar had continued from A.D. 1, the cycle of dominical letters would have been repeated every 28 years just as in the preceding 28, consequently, if B had been the letter for 1583, it would also have been that of A.D. 15, whence, reckoning backwards, E would have been the letter of A.D. 1, and 2 its number in our list. We do this merely for the convenience of starting from A.D. 1, and we have supposed none of the peculiarities of the new Calendar to enter until after 1600, except that we have taken such a letter for A.D. 1 as will, by A.D. 1583, bring us to the proper letter of the new Calendar. If then we begin with A.D. 1 and Sunday number 2, every addition of 1 to the A.D. would add 1 to the Sunday number, and every complete series of four would add one more for the leap-year. From this set the sevens are to be rejected as fast as they arise; but it will be convenient to defer this rejection to the end of the process. Set down then first 1; secondly, the year required (remember that A.D. 1 begins not with 1, but with 1 more than 1); thirdly, the quotient of the year required divided by 4, neglecting the remainder: the rejection of the sevens, or the remainder after division by seven, would give the Sunday number, but for the provisions about the rejection of leap-years at the end of the centuries. If we reject 1 from the preceding amount for every century complete after 1600, that is, if we subtract 16 from the two first figures of the year and subtract the remainder, we shall reject too much by a unit for every complete four centuries, because 2000, 2400, 2800, *are* leap-years; consequently, after rejecting one for every unit above 16 in the two first figures, we must add one for every unit in the quotient of that excess divided by 4. Hence the following rule, which puts us in possession of the dominical letter for any year, or the *second* dominical letter, if it be leap-year:—

I. Add 1 to the given year.
II. Take the quotient of the given year divided by 4 (neglecting the remainder).

III. Take 16 from the centurial figures of the given year, if that can be done.
IV. Take the quotient of III divided by 4 (neglecting the remainder).
V. From the sum of I, II, and IV, subtract III.
VI. Find the remainder of V divided by 7: this is the number of the dominical letter in the system
A B C D E F G
6 5 4 3 2 1 0.
VII. Subtract VI from 7: this is the number of the dominical letter in the (for subsequent purposes) more convenient system
A B C D E F G
1 2 3 4 5 6 7.

EXAMPLE 1.—What day of the week is May 15th, A.D. 4779:—

I = 4780 III = 47 − 16 = 31
II = 1194
IV = 7
―――
5981
31
―――
7) 5950 = V
―――
850 rem. 0 = VI

Dominical letter* G.

Now May 13 has the letter G, and is Sunday, whence May 15, 4779, is Tuesday.

EXAMPLE 2.—What is the first Thursday in February, 1844:—

I = 1845 III = 18 − 16 = 2
II = 461
IV = 0
―――
2306
2
―――
7) 2304 = V
―――
329 rem. 1 = VI

Second dominical letter F.
First letter G.

Before March, G is Sunday, and D is Thursday, the first D in February is opposite the first of the month, the day required.

We now come to the determination of the full moons of any year. Those who constructed the Nicene system of finding Easter, either took the mean moon of the astronomers, or else really imagined that the moon's motion was uniform. The first is the more likely of the two, if we are to suppose that the arrangement was actually referred to others; since the school of astronomy at Alexandria, a city whose inhabitants had more reason than those of any other to take interest in the result of the Council, was of higher reputation than any other, and it is reported that astronomers of this city were among those to whom the Calendar was referred. This mean moon of the astronomers†

* This agrees with Clavius, who has given in a table the results of the Calendar for every year from 1600 up to 5000! All the examples in this paper are to be understood as confirmed by this table.

† The Jews of the middle ages used the mean moon, and did not in the least mind

is a fictitious body, which moves with the average motion of the real moon; if started in its proper place, and supposed to move in the ecliptic with the average motion of the real moon, the latter would be in the long run as much to the east as to the west of the former. If there were really a mean moon in the heavens, the number of new moons which precede the new mean moons, would be in the long run the same as the number of those which follow. A mean new moon may be as much as nine or ten hours in advance of, or behind, the corresponding real new moon. Clavius saw very distinctly that it would be impossible to foretell Easter for any very long period with never-failing accuracy from the moon of the heavens. The reason is as follows:—There must be some uncertainty about the prediction of the place of the moon, and the times of new and full moon: they cannot be foretold even in our day within half a second; in the time of Clavius they could not be foretold within a good many minutes. Suppose that there are five minutes of uncertainty, and suppose it happens that the paschal full moon is predicted by astronomical tables to take place at a minute to 12 on Saturday evening, meaning within five minutes one way or the other. It is clear that no one can tell whether that full moon is to take place on Saturday evening or Sunday morning; yet upon this it will depend whether the Sunday or the Sunday week is to be the day of the festival. In these doubtful cases, then, the moon of the tables must be the one adopted, whether it shall be found to agree with the moon of the heavens or not. Clavius and his advisers seem to have thought, very reasonably, that since this departure from the real moon must sometimes take place, it would matter little if it were allowed to take place somewhat oftener than absolutely need be, if any considerable simplification of calculation could thereby be gained.

We reject the real moon, and put the mean moon in its place: but we are not yet come to the ecclesiastical moon. There are some paragraphs of Clavius which are worth translating (cap. xviii. of his work above cited); it being remembered that our Easter is by statute the Easter of Clavius, with a wrong explanation added by our legislators. "Who, except a few who think they are very sharp-sighted in this matter, is so blind as not to see that the 14th of the moon and the full moon are not* the same things in the Church of God? . . . Although the Church, in finding the new moon, and from it the 14th day, *uses neither the true nor the mean motion of the moon*, but measures only according to the order of a cycle; it is nevertheless undeniable that the mean full moons found from astronomical tables are of

the fact of the moon being occasionally visible and horned on the day of their Calendar new moon. But, beginning their day at six o'clock in the evening, they did not allow any day to be that of the new moon, unless the mean moon came in the first 18 hours, that is, before noon. If it happened in the last 6 hours of the day, they referred it to the next day.

* Clavius in another place gives a good list of authorities, all for the *fourteenth*: all understand the *fifteenth* to be the full moon, and several say that Easter must not be kept *before* the full moon: our statute says it must be kept *after*.

the greatest use in determining the cycle which is to be preferred . . . . the new moons of which cycle, in order to the due celebration of Easter, should be so arranged that the 14th days of those moons, reckoning from the day of new moon *inclusive*, should not fall two or more days before the mean full moon, but only one day, or else on the very day itself, or not long after. And even thus far the Church need not take very great pains . . . . for it is sufficient that all should reckon by the 14th day of the moon in the cycle, even though sometimes it *should be more than one day before or after* the mean full moon. . . . . We have taken pains that in our cycle the new moons should *follow* the real new moons, so that the 14th of the moon should fall either the day before the mean full moon, or on that day, or not long after; and this was done on purpose, for if the new moon of the cycle fell on the same day as the mean new moon of the astronomers, it might chance that we should celebrate Easter on the same day as the Jews or the Quartadeciman heretics, which would be absurd, or else before them, which would be still more absurd."

Here we have before us the whole principles of the Calendar according to Clavius. The Jews and Quartadecimans celebrate their Passover and their heretical Easter on the 14th day of the real moon, the day before * the real full moon: take care that the Easter Sunday of the Church comes after, not before or on, their festival day. This point gained, the sooner Easter comes the better. So that, according to the real principles of this Calendar, and in spite of our legislators, there is not the least objection, but the contrary, to Easter Sunday being on the day of a full moon: for the Jewish and Quartadeciman 14th is sure to be over by that day. It is desirable, he distinctly intimates, not to celebrate the feast *before* the real full moon, because the chances of coming in contact with Jews and heretics are then very great: but he does not say a word against the very day of the full moon; nay, we are even left to infer that the oftener the very day of the full moon, being Sunday, can be made Easter Sunday, the better does the cycle fulfil the intentions of its framers.

It was not so childish a thing to wish to keep Easter on a

* Clavius assumes the 15th from the day of new moon (our 14th in page 4) to be the real full moon. If any should think that possibly the 14th day of the moon, so often mentioned, meant from the day of the new moon exclusive (which no one can do who has actually consulted authorities), he should remember that this would have been such a violation of the ordinary idiom of the Latin language, that it would have been most expressly mentioned. The Calendar of Julius Cæsar was running into confusion very fast (till corrected by Augustus), because the priests mistook *once in four years* for *every fourth year:* now every fourth number was construed by them as meaning 1, 4, 7, 10, &c., always counting the first number; so that they gave a bissextile, as we should say, every third year. Again, in their Calendar, they had the Kalends of each month, the day before the Kalends, and then not the second day before the Kalends, as we should say, but the *third* day. To the question, what is the first day before the 10th, the Latin idiom should answer—the 10th itself. Thus Macrobius says (Dr. Smith's Dictionary of Greek and Roman Antiquities, art. *Calendar, Roman*), that bissextiles take place every *fifth* year, "quinto quoque incipiente anno:" and there is a passage of Livy in which many commentators have not recognised the cycle of *nineteen* years, because it is said to begin again *vicensimo anno.*

different day from Jews and Quartadecimans as may at first sight appear. To the Jews, the passover was an obligatory ordinance, and the 14th of the moon was as much a part of the ordinance as the unleavened bread. The Quartadecimans were a Judaizing sect, who wanted to make the passover itself an ordinance of Christianity. Against this the ecclesiastical authority always raised its voice from the very earliest times and in the most express words: and it was to provide a constant proof that Easter is not the feast of the passover that the day was changed. If the new Calendar had been adopted in England at the time of its first publication, and if such pains had been taken with it as are taken in our day with much more difficult subjects, to make it possible for a mere arithmetician to understand it, and to appreciate the extent of the liberties which had been taken with the real 14th of the moon, it would have been impossible that so much superstition * on the subject of Easter should have existed in 1752.

In order to secure this first great point, it is necessary, Clavius says, that the new moons of the cycle should come after the new moons of the heaven (true or mean; it hardly becomes worth while to distinguish them when Clavius is talking of a whole day as no great matter). Suppose, says he, for instance, that we make the new moon of our cycle coincide with that of the heavens, which happens, say at noon on the 8th of March. Then the 14th day, according to us, is the 21st, and if the 22nd should be Sunday, we should keep it as Easter Day. But the mean full moon comes on the morning of the 23rd, and the real full moon will often come on that same day, so that Easter will thus sometimes be kept before the real full moon, and the Church and the heretics may sometimes use the same day. To avoid this he has to contrive that his new moons shall follow

* Some of those who read this article will, perhaps, remember the following story, which was told, as of his own knowledge when a boy, by an eminent and excellent follower of science, not long deceased. A worthy couple in a country town, scandalized at the change of style in 1752, continued for many years to attempt the observance of Good Friday on the old day. To this end they walked seriously and in full dress to the church door, at which the gentleman rapped with his stick: on finding no admittance, they walked as seriously back again, and read the service at home. But on the new and spurious Good Friday, they took pains to make such a festival at their house, as would convince the neighbours that their Lent was either ended or in abeyance. But there must have been some years of comfort, for between 1752 and 1800 there were 18 years in which old and new Easter Day coincided. This still happens occasionally, and will do so, though less and less frequently, till A.D. 2698, when it will happen for the last time. There are more singular manifestations still to be found. We know there was in England (but we cannot lay our hands upon authority for it at this moment), on the part of some persons, a belief that the cattle always went down on their knees in their stalls, at the moment when Christmas Day began: and it was averred that, when the change of style was introduced, the cattle would not acknowledge it, but kept the commencement of the old Christmas Day in the manner above mentioned. But these were Protestant cattle, the sirloins of whose ancestors had smoked upon boards at which confusion (and sometimes worse) to the Pope and all his adherents was a standing toast. Among the Catholics, on the other hand, the truly learned and laborious Jesuit Riccioli, in whom every student of the history of astronomy finds a most useful friend, informs us with the utmost gravity, that the blood of St. Januarius, which liquefied on the 19th of September, and a supernatural rod which always budded on the morning of Christmas-day, both acknowledged the change of style the instant it was made; thus affording a testimony to the new Calendar, of which we cannot anywhere find that Clavius has thought it right to avail himself.

those of the heavens, and they do sometimes follow by two or even three days. It would have been as easy to have made the 15th or 16th of the moon the regulating day; but Clavius was as much bound to the *fourteenth* as any Jew or Quartadeciman of them all, by the old definition of Easter. The Jews (the Quartadeciman Christians were extinct by the time of Clavius, at least in the countries which were or had been under the papal government) kept the Passover on the day before the real full moon; the Catholic Church would not keep Easter on the same day, even when its own distinct rules happened to give that day, and deliberately pushed on the beginnings of its months, in order that, without ceasing to depend nominally on the *fourteenth day*, it might never bring its Easter to the Passover. It is evident that, by pushing on the day of the first new moon from which we start, we equally push forward the beginnings of all succeeding full moons, so that this curious mode of making a quintadeciman or sextadeciman church under quartadeciman colours, will not give any trouble in the explanation.

Denoting by I the day of the new moon, or the first day, by II the day after the new moon, or the second day, we have seen that I of the new moon is either XXX or XXXI of the old moon, because it is either the 29th or 30th day, not including the day of the new moon. The first step of the Gregorian lunar Calendar is to suppose that the months shall be alternately of 30 and 29 days, or that I, the new moon, shall be alternately XXXI and XXX of the old moon. The mean lunation being very nearly 29½ days, the Calendar moon will always be kept very near the mean moon by this supposition, for a very long period. The error will be that the Calendar moon will be sometimes in advance* of the mean moon, sometimes behind it, but gradually falling less in advance and more behind. Let the mean and Calendar moon set off together at a new moon, and reckon onwards mean and Calendar lunations as follows:—

	MEAN.				CALENDAR.
1st new moon . .	0^d	0^h	0^m	0^s	0^d
	29	12	44	3	30
2nd " . .	29	12	44	3	30
	29	12	44	3	29
3rd " . .	59	1	28	6	59
	29	12	44	3	30
4th " . . .	88	14	12	9	89
	29	12	44	3	29
5th " . .	118	2	56	12	118

At the fifth new moon the Calendar moon is nearly 3 hours behind the mean moon, at the third only 1½ hours: at the fourth, the Calendar is not 10 hours in advance of the mean, at

* It is often convenient to speak of time in the same manner as of distance measured from a point. Thus, of two events, the later is the most advanced, and the earlier comes behind it: ten o'clock in the evening is more advanced than six o'clock.

the second more than 11. The reason is obvious enough; the Calendar loses 44m. 3s. in each lunation, on the average, or about a day in 33 lunations. Part of this loss is made up as follows:—the 29th of February is actually passed over in the Calendar lunation and not counted, while it is counted in the mean lunation: so that in fact one lunation in 4 years is a day longer than it ought to be, and the Calendar new moon is thrown forward a day. But this is not enough, for a day in 33 lunations is about 7 days in 19 years. The rest is nearly made up by allowing all but one of the additional or *embolismic* lunations, as they were called, and to which we shall immediately come, to have 30 days, and not alternately 30 and 29.

Twelve lunations of 30 and 29 days alternately, make 354 days. If then January 1 be a Calendar new moon, it follows that December 21 will be the same; or if January 1 be marked i, December 21 will be marked i, and the next January 1 will be the twelfth day of the moon, or xii. The *epact* of a year means the day of the Calendar moon on which the 1st of January falls; accordingly *eleven* must be added to the epact of one year to get the epact of the next, upon the supposition of twelve lunations in a year. Now suppose a year which we will call year 1, begins with the last day of the Calendar moon; then year 2 begins with xi, year 3 with xxii, and year 4 with xxxiii, or the third year requires another lunation. Let this *embolismic* lunation (so called) be inserted, and let it be 30 days, then year 4 begins with iii of the Calendar moon. Again, year 5 begins with xiv, year 6 with xxv, year 7 would begin with xxxvi, but for another embolismic lunation of 30 days, which makes it begin with vi. Proceed in this way and we shall find that, if the first epact be xxx, as supposed, the epacts of the successive years will be as follows, embolismic years being marked with an asterisk:—

Year.	Epact.	Year.	Epact.	Year.	Epact.	Year.	Epact.
1	xxx	9*	xxviii	17*	xxvi		
2	xi	10	ix	18	vii		
3*	xxii	11*	xx	19	xviii	19*	xviii
4	iii	12	i				
5	xiv	13	xii	20	xxix	20	xxx
6*	xxv	14*	xxiii				
7	vi	15	iv				
8	xvii	16	xv				

Now let 19 be an embolismic year with a month of 29 days, and the year 20 will then begin with the last day of the moon, and its epact should be xxx as at first. We shall hereafter see the artifice by which the last day was always marked xxx.

Here is the cycle of 19 years, which in the old Calendar was supposed to be exact: let us now see what its lunation is in the long run, and in the Julian Calendar. We know that 19 years have sometimes 4 and sometimes 5 leap-years; but taking them at an average we have, for 19 years of 365¼ days each,

6939¾ days. We have in this 12 times 19, or 228, common lunations, and 7 embolismic ones; 235 in all. Now 235 mean lunations are 6939 days, 16 hours, 31 minutes; so that this mode of throwing the 29th of February into its lunation without further mention, and making all the embolismic months but one to consist of 30 days, makes the cycle of 19 years exact to the mean moon within the difference of 16h. 31m. and 18h., or 1 h. 29 m.; and the cycle is too long by this quantity. But this is an average quantity: for the cycle with 5 leap-years has 6940 days, that with 4 only 6939. And the mean quantity being right, or very nearly so, we see it is the principle of the Calendar not to mind the occasional error of a day which may arise from the deferment or proximity of the intercalations. When an embolismic month of 30 days, which ought to be 29, happens in leap-year, there are two days which are not counted in the moon's Calendar age; there are other years which will have every day counted. There are also occasionally months of 31 days.

It appears that our average cycle is too long; its tendency then will be to throw the Calendar new moons too forward. At 1½ hours per cycle, this would amount to a day in 16 cycles, or 304 years; according to the data assumed by Clavius, 8 days in 2500 years was thought more correct, and on this supposition the new Calendar was framed. At the time of the reformation of the Calendar the real new moons usually fell on the 5th day before, from the day of the Calendar new moon inclusive; and it was usual to pass to the probable day of real full moon by going from the Calendar day backward with the syllables *Nova luna hîc*, or *In coelis est hîc:* thus, supposing the Calendar full moon to be on the 17th, the real one would fall on the 13th—

hîc	-na	Lu	-va	No
13	14	15	16	17

It was the device of Lilius* to put down numbers opposite to the several days of the year, beginning from xxx (which Clavius marked with an asterisk *) and reckoning backwards, as in the following page.

If the epacts were written backwards in this way, and if they went up to xxix and xxx (or *) alternately, then, assuming the month in which the year begins to have 30 days, the next one 29, and so on, the epact *of the year* would always be found opposite to the day of Calendar new moon. But Clavius prefers to carry the days of every month up to 30, and in the *hollow*† months he contrives to reduce this number virtually to 29, by writing two of the numbers, xxv and xxiv, in the same line.

* Delambre, in a note added to the table of contents (at the word *epact*) of the *Hist. de l'Astron. Mod.*, informs us of a statement of Calandrelli, that the idea of the epacts belongs to Giov. Tolosani, whose work, known to exist in 1535, was printed at Venice in 1575. As epacts were certainly known and used long enough before either, we are to suppose that it is this particular disposition of the epacts which is alluded to: but we cannot be certain.

† Months of thirty days were called *full;* those of twenty-nine, *hollow.*

January.			February.			March.			April.			May.			June.		
1	*	A	1	xxix	D	1	*	D	1	xxix	G	1	xxviii	B	1	xxvii	E
2	xxix	B	2	xxviii	E	2	xxix	E	2	xxviii	A	2	xxvii	C	2	{25 / xxvi}	F
3	xxviii	C	3	xxvii	F	3	xxviii	F	3	xxvii	B	3	xxvi	D	3	{xxiv / xxv}	G
4	xxvii	D	4	{25 / xxvi}	G	4	xxvii	G	4	{25 / xxvi}	C	4	xxv	E	4	xxiii	A
5	xxvi	E	5	{xxv / xxiv}	A	5	xxvi	A	5	{xxv / xxiv}	D	5	xxiv	F	5	xxii	B
6	xxv	F	6	xxiii	B	6	xxv	B	6	xxiii	E	6	xxiii	G	6	xxi	C
7	xxiv	G	7	xxii	C	7	xxiv	C	7	xxii	F	7	xxii	A	7	xx	D
8	xxiii	A	8	xxi	D	8	xxiii	D	8	xxi	G	8	xxi	B	8	xix	E
9	xxii	B	9	xx	E	9	xxii	E	9	xx	A	9	xx	C	9	xviii	F
10	xxi	C	10	xix	F	10	xxi	F	10	xix	B	10	xix	D	10	xvii	G
11	xx	D	11	xviii	G	11	xx	G	11	xviii	C	11	xviii	E	11	xvi	A
12	xix	E	12	xvii	A	12	xix	A	12	xvii	D	12	xvii	F	12	xv	B
13	xviii	F	13	xvi	B	13	xviii	B	13	xvi	E	13	xvi	G	13	xiv	C
14	xvii	G	14	xv	C	14	xvii	C	14	xv	F	14	xv	A	14	xiii	D
15	xvi	A	15	xiv	D	15	xvi	D	15	xiv	G	15	xiv	B	15	xii	E
16	xv	B	16	xiii	E	16	xv	E	16	xiii	A	16	xiii	C	16	xi	F
17	xiv	C	17	xii	F	17	xiv	F	17	xii	B	17	xii	D	17	x	G
18	xiii	D	18	xi	G	18	xiii	G	18	xi	C	18	xi	E	18	ix	A
19	xii	E	19	x	A	19	xii	A	19	x	D	19	x	F	19	viii	B
20	xi	F	20	ix	B	20	xi	B	20	ix	E	20	ix	G	20	vii	C
21	x	G	21	viii	C	21	x	C	21	viii	F	21	viii	A	21	vi	D
22	ix	A	22	vii	D	22	ix	D	22	vii	G	22	vii	B	22	v	E
23	viii	B	23	vi	E	23	viii	E	23	vi	A	23	vi	C	23	iv	F
24	vii	C	24	v	F	24	vii	F	24	v	B	24	v	D	24	iii	G
25	vi	D	25	iv	G	25	vi	G	25	iv	C	25	iv	E	25	ii	A
26	v	E	26	iii	A	26	v	A	26	iii	D	26	iii	F	26	i	B
27	iv	F	27	ii	B	27	iv	B	27	ii	E	27	ii	G	27	*	C
28	iii	G	28	i	C	28	iii	C	28	i	F	28	i	A	28	xxix	D
29	ii	A				29	ii	D	29	*	G	29	*	B	29	xxviii	E
30	i	B				30	i	E	30	xxix	A	30	xxix	C	30	xxvii	F
31	*	C				31	*	F				31	xxviii	D			

July.			August.			September.			October.			November.			December.		
1	xxvi	G	1	{xxv / xxiv}	C	1	xxiii	F	1	xxii	A	1	xxi	D	1	xx	F
2	xxv	A	2	xxiii	D	2	xxii	G	2	xxi	B	2	xx	E	2	xix	G
3	xxiv	B	3	xxii	E	3	xxi	A	3	xx	C	3	xix	F	3	xviii	A
4	xxiii	C	4	xxi	F	4	xx	B	4	xix	D	4	xviii	G	4	xvii	B
5	xxii	D	5	xx	G	5	xix	C	5	xviii	E	5	xvii	A	5	xvi	C
6	xxi	E	6	xix	A	6	xviii	D	6	xvii	F	6	xvi	B	6	xv	D
7	xx	F	7	xviii	B	7	xvii	E	7	xvi	G	7	xv	C	7	xiv	E
8	xix	G	8	xvii	C	8	xvi	F	8	xv	A	8	xiv	D	8	xiii	F
9	xviii	A	9	xvi	D	9	xv	G	9	xiv	B	9	xiii	E	9	xii	G
10	xvii	B	10	xv	E	10	xiv	A	10	xiii	C	10	xii	F	10	xi	A
11	xvi	C	11	xiv	F	11	xiii	B	11	xii	D	11	xi	G	11	x	B
12	xv	D	12	xiii	G	12	xii	C	12	xi	E	12	x	A	12	ix	C
13	xiv	E	13	xii	A	13	xi	D	13	x	F	13	ix	B	13	viii	D
14	xiii	F	14	xi	B	14	x	E	14	ix	G	14	viii	C	14	vii	E
15	xii	G	15	x	C	15	ix	F	15	viii	A	15	vii	D	15	vi	F
16	xi	A	16	ix	D	16	viii	G	16	vii	B	16	vi	E	16	v	G
17	x	B	17	viii	E	17	vii	A	17	vi	C	17	v	F	17	iv	A
18	ix	C	18	vii	F	18	vi	B	18	v	D	18	iv	G	18	iii	B
19	viii	D	19	vi	G	19	v	C	19	iv	E	19	iii	A	19	ii	C
20	vii	E	20	v	A	20	iv	D	20	iii	F	20	ii	B	20	i	D
21	vi	F	21	iv	B	21	iii	E	21	ii	G	21	i	C	21	*	E
22	v	G	22	iii	C	22	ii	F	22	i	A	22	*	D	22	xxix	F
23	iv	A	23	ii	D	23	i	G	23	*	B	23	xxix	E	23	xxviii	G
24	iii	B	24	i	E	24	*	A	24	xxix	C	24	xxviii	F	24	xxvii	A
25	ii	C	25	*	F	25	xxix	B	25	xxviii	D	25	xxvii	G	25	xxvi	B
26	i	D	26	xxix	G	26	xxviii	C	26	xxvii	E	26	{25 / xxvi}	A	26	xxv	C
27	*	E	27	xxviii	A	27	xxvii	D	27	xxvi	F	27	{xxv / xxiv}	B	27	xxiv	D
28	xxix	F	28	xxvii	B	28	{25 / xxvi}	E	28	xxv	G	28	xxiii	C	28	xxiii	E
29	xxviii	G	29	xxvi	C	29	{xxv / xxiv}	F	29	xxiv	A	29	xxii	D	29	xxii	F
30	xxvii	A	30	xxv	D	30	xxiii	G	30	xxiii	B	30	xxi	E	30	xxi	G
31	{25 / xxvi}	B	31	xxiv	E				31	xxii	C				31	{19† / xx}	A

† The 19 is for the last year of the cycle.

Certain days, therefore, have two epact-numbers, xxv.xxiv, written after them. We shall presently describe this further, and also the 25.xxvi which occurs.

Suppose the epact of the year to be vii, January 1 is the 7th of the moon, January 2 is the 8th, &c., and the moon of January is to have 30 days. Accordingly, Jan. 24th is the first day of the February Calendar lunation (remember that the month in which a lunation *ends* is that to which it is always made to belong). But the trouble of counting is saved by our seeing vii, the epact of the year, opposite to January 24. Similarly, the first day of the March lunation is February 22, and so on.

In the same cycle there may arise two out of the nineteen years which have xxv and xxiv for their epacts. Both of these will have, for anything yet explained to the contrary, some of their Calendar new moons on the same days, on account of the coalescence of xxv and xxiv in the alternate months. Now it will never happen as to mean lunations, and rarely as to real ones, that in the same cycle there should be the lunation of a given month beginning on the same day in two different years of the cycle; and such a thing never happened in the unreformed Calendar. Clavius thought it desirable to imitate this in the new Calendar; and he observed, that by taking the preceding day whenever the epact was xxv, and the year of the cycle after the 11th, he could avoid the reiteration, and thus make the desired resemblance. Take the previous table of epacts, and write those of the first eleven and last eight years of the cycle in two lines, as they are when year 1 has xxx or *.

xxx	xi	xxii	iii	xiv	xxv	vi	xvii	xxviii	ix	xx
i	xii	xxiii	iv	xv	xxvi	vii	xviii			

If the first epact be i, we have but to add 1 to every one of these; if it be vii, we have but to add 7, always striking off 30 when it can be done, and we shall thus get all the successions of epacts that can possibly arise in any cycle of 19 years. Now it is clear that in the first line there are no two contiguous numbers, nor in the second: while all the numbers of the second line are, as far as they go, contiguous to those of the first. The same thing will happen when all are increased by the same number, and it follows that when xxiv and xxv come together in the same cycle, xxiv must be in the first line and xxv in the second; and xxvi will not then be in the second. If then for xxv, when xxiv is in the first line, we use xxvi, or the day opposite xxvi for the day of the Calendar new moon, it will prevent a repetition of those which are used when xxiv is the epact: which amounts to the following rule;—Whenever the epact should be xxv, the year of the cycle being upwards of 11, say that the epact is 26. This is not an astronomical correction, but a mere conventional mode of reconciling the choice which Clavius made of the mode of writing the epacts with an essential peculiarity of the old cycle of 19 years which that mode of writing would have otherwise destroyed. If Clavius did this merely to avoid offending his contemporaries, we cannot enough admire the solemn astronomical face which he put upon his goodhumoured attention to their preconceptions: if he himself thought it necessary, we must set him down as the grandest folio trifler that ever lived. The greatest mischief that could have arisen from omitting this 25.XXVI would have been that Easter Sunday would sometimes have fallen on the same day of the month twice in 19 years: to avoid this, he does not scruple to make the moon wrong by a day more than it otherwise would have been. Several of those who have described the Calendar since the time of Clavius either did not know or would not tell the real reason of this arbitrary correction, but have tried to give astronomical reasons for it.

We can now proceed to find the mode of ascertaining the epact: in the following list of circumstances necessary to be taken into account, we include some recapitulation of what has gone before:—

1. Since the use of the epact is to find Easter, it will be convenient to take into account, from the beginning of the year, anything which happens between that and the equinox; such, for instance, as the Gregorian omission of a leap-year.

2. It has been seen that we want *all* the leap-years to make the proper lunar intercalations and to keep the cycle of 19 years from gradually getting more and more from the truth. Now the Gregorian solar year dispenses with three leap-years in 400 years; and every one of these, from the moment it happens, makes the age of the mean moon a day less than it otherwise would have been, on any given subsequent day of any month. We must therefore, every time this Gregorian omission takes place, lessen by one the epact of the Calendar moons, at least for new moons happening in and after March. There ought to be two epacts for those years from which the Gregorian correction throws out the 29th of February, one for January and February, the other for the rest of the year: but, just as in finding the dominical letter there is no use in any but the second one of a leap-year, for finding Easter, so, in those years which the Gregorian Calendar makes common, there is no use in any but the second epact, and this second epact is always put down as that belonging to the year.

3. The correction of 8 days in 2500 years must be added to the epacts, as already noticed, or the Calendar moon of the 1st of January must be made a day older than the other rules direct, as near as may be at the rate of one day in 300 years, seven times following, and then one day in 400 years. Clavius chose A.D. 1800 (considered as the end of a set of 400) as his first year of this correction,* which would accordingly be made

* Hence it is that the table in the Act of Parliament holds good from 1700 to 1899, through *two* hundred years, for though the loss of a leap-year in 1800 diminishes the epact by 1, yet the correction for the moon increases it by 1 at the same time.

C

as follows: (+1) meaning, add one to the epact otherwise determined:—

A.D. 1800	(+1)	3300	
300		300	
2100	(+1)	3600	(+1)
300		300	
2400	(+1)	3900	(+1)
300		☞ 400	
2700	(+1)	4300	(+1)
300		300	
3000	(+1)	4600	(+1)
300		&c.	
3300	(+1)		

4. But for these corrections, the epacts would be repeated in every cycle of 19 years, the same as in the preceding cycle: as would have been the case in the old Calendar had not this very circumstance rendered the use of epacts unnecessary.

5. The year A.D. 1 was always considered as year 2 of the cycle of 19 years, whence the remainder of one more than the year divided by 19 will give the year of the cycle, 19 being the year when the remainder is nothing. It only remains to ask what epact is to be taken as that of year 1 of the cycle in the seventeenth century, or for all years of the Reformed Calendar preceding 1700, in which for the first time a 29th of February is not allowed. Clavius chose the epact 1 for the first year of the cycle, meaning thereby to put his calendar full moon a day in advance of the mean full moon, for reasons already explained. Consequently his next epact, for year 2, is 1 and 11; for year 3, 1 and twice eleven, and so on; rejecting 30 as fast as it arises. Hence the starting rule for the epact is, find the remainder of the year of the cycle increased by ten times the next less* number, divided by 30. It will easily be seen that, for instance, 1 and 4 times 11, is 5 and ten times 4, and so on. We have now to find the effect of the corrections. Every leap-year lost is, as we have seen, a diminution of 1 to the epact: we must therefore subtract the III (page 15) which we made use of in finding the dominical letter, and because we have then subtracted too much we must add the IV of the same rule.

Now for the correction of 300 years. At 1800 we add 1 to the epact, and then for every 300 years we must add one to the epact, except for the eighth 300 years, for which we must write 400 years, and add one for that 400 years, after which we begin with 300 years again. For anything short of A.D. 4200 this gives as a rule, add the quotient (neglecting the remainder) of the

* This is because 1 increased by 11 taken one time less than the year of the cycle, is the same thing as the year of the cycle increased by ten times the next less number.

$$1+11(x-1) = x+10(x-1).$$

number of centuries less 15 divided by 3: but when we come to 4200, the correction waits till 4300. And 4300 is treated as 4200 would have been, but for the deferment of the correction, and so on; therefore it is, from 4200 use one less than the number of centuries. This does until 6700, when there is another deferment of the correction for a century, and two less than the number of centuries should be used. Now 42, 67, &c., are successive twenty-fives added to 17; and instead of the number of centuries, we should take 17 from the number of centuries, divide by 25, keep the quotient only, and diminish the number of centuries by that quotient; then subtract 15, divide by 3, and add the quotient to the epact. We shall now put down all the rules together, both for the dominical letter and the epact:—

I. Add one to the given year.

II. Take the quotient of the given year divided by 4, neglecting the remainder.

III. Take 16 from the centurial figures of the given year, if it can be done.

IV. Take the quotient of III. divided by 4, neglecting the remainder.

V. From the sum of I., II., and IV., subtract III.

VI. Find the remainder of V. divided by 7.

VII. Subtract VI. from 7; this is the number of the dominical letter

A	B	C	D	E	F	G
1	2	3	4	5	6	7

VIII. Divide I. by 19, the remainder (or 19, if no remainder) is the *golden number.**

IX. From the centurial figures of the year subtract 17, divide by 25, and keep the quotient.

X. Subtract IX. and 15 from the centurial figures, divide by 3, and keep the quotient.

XI. To VIII. add ten times the next less number, divide by 30, and keep the remainder.

XII. To XI. add X. and IV., and take away III., throwing out thirties, if any. If this give 24, change it into 25. If 25, change it into 26, whenever the golden number is greater than 11. If 0, change it into 30. Thus we have the epact.

When the Epact is 23, or less.	*When the Epact is greater than 23.*
XIII. Subtract XII., the epact, from 45.	XIII. Subtract XII., the epact, from 75.
XIV. Subtract the epact from 27, divide by 7, and keep the remainder.	XIV. Subtract the epact from 57, divide by 7, and keep the remainder.

XV. To XIII. add VII., the dominical number, (and 7 besides, if XIV. be greater than VII.,) and subtract XIV., the result is the day of March, or if more than 31, subtract 31, and the result is the day of April on which Easter Sunday falls.

Up to XII. we have explained the rules; the change of 25 into 26 has also been explained, and 25 is preferable † to 24 in the construction of the subsequent rules (both being given in XXV. XXIV.). It now remains to explain these rules after XII.

* The name given to the number of the year, in the cycle of 19 years, from its being written or printed in gold letters.

† Let it be remembered that we have nothing to do with any lunation except that which follows March 21, in which XXIV., when it occurs, enters only with XXV.

Having the epact and the dominical letter, *and the epact-almanac*, the rest of the process is easy enough. For example the epact is VII. and the dominical letter G, required Easter Sunday. Look at the equinox day, March 21, and we see VII. opposite to March 24: this is the paschal new moon (or the one belonging to the paschal full moon) of the calendar: count on 14, from the VII. inclusive, and we come to April 6; this is the calendar paschal fourteenth. The next Sunday (G) is April 8, this is Easter Sunday. What we have now to do is to contrive an arithmetical calculation independently of the epact-almanac. Two cases present themselves, for the paschal lunation may be either that which ends in April or in May. If the epact be 23, the calendar fourteenth of the moon is on the equinox day, and the new moon on the 8th of March: this is the earliest. But if the epact be 24, the fourteenth is on the day before the equinox, and the paschal fourteenth is not till April 18. If the epact be 23 or less, the paschal new moon is in March; if more than 23, in April. Also, for an arithmetical rule, we may call April 1 the 32nd of March, April 2 the 33rd, and so on. With this understanding, the day of the month of the calendar new moon is found by subtracting the epact from 31 or from 61, according as the epact does not or does exceed 23. Again, the day of the month of the calendar *fourteenth* is thirteen days from that of the new moon *exclusive*, so that it is found by subtracting the epact from 44 or 74, according as the epact does not or does exceed 23. Hence the first possible day for Easter Sunday is found by subtracting the epact from 45 or 75, because it must be the day after the fourteenth, at least. Hence the first possible day is found in rule XIII.

But the number of the calendar letter* is thus found: March 1 has D (whose number is 4), March 2 has number 5, and so on: add 3 to the day of March (extended into April, it may be, by going beyond 31), and you have the number of the calendar letter by throwing out the sevens, if any. For the number of the calendar letter of the first possible day of Easter subtract the epact from 48 or 78, and throw out the sevens; or rather, from 27 or 57, because throwing out three sevens still leaves enough behind to take the epacts from: this gives us (XIV.). Now to pass from this number to the next actual dominical number (VII.) for the year, we must pass on, from the earliest possible day, as many days as there are units in (VII.) more than in (XIV.), or if (VII.) be not so great as (XIV.) we must pass on as many days as there are units in the excess of 7 more than (VII.) above (XIV.). Thus, if the dominical number were 5, and that of the first possible day of Easter were 2, the real Easter Sunday must be 3 days after the first possible day. But if the dominical number were 4, and that of the first possible Easter Day were 6, we must pass into the next week (1 2 3 4 5 6′ 7 1 2 3 4′ 5 6 7) to arrive at Easter Day, and the number of the day from 6′ to 4′ is found

* Every day of the year has its letter, but it would cause confusion if we were to call it always a *dominical* letter; thus E is not dominical except when Jan. 5 is Sunday.

by subtracting 6 from 7 more than 4 (giving the fifth day). Hence the reason of rule (xv.) is apparent. We shall now apply this to a couple of Examples, taken at hazard, and then to the year 1845.

EXAMPLE 1.—Required Easter Sunday of A.D. 4610:—

```
 I. = 4611      III = 30        19) 4611 (242
II. = 1152                            81
IV. =    7                             51
     -----                      rem.  13 = VIII. Golden Number.
      5770
III.    30                         46              13
     -----                         17             120
  7) 5740 = V.                     --            ----
      820 rem. 0 = VI.        25) 29 (1 = IX.  30) 133
VII. =  7 Domin. Lr. G.            4                 4 rem. 13 = XI.
 XI.    13                          3) 30
  X.    10                             --
 IV.     7                             10 = X.
     -----
        30                         75              57
III.    30                         30              30
     -----                         --              --
         0 Epact 30 = XII.         45 = XIII.   7) 27 rem. 6 = XIV.
45 and 7 less 6 is 46
                   31
                   --
      April 15, Easter Sunday.
```

EXAMPLE 2.—Required Easter Sunday of A.D. 1854:—

```
 I. = 1855      III. = 2        19) 1855 (97
II. =  463                            145
IV. =    0                      rem.   12 = VIII. Golden Number.
     -----
      2318                         18              12 = VIII.
III.     2                         17             110
     -----                         --            ----
  7) 2316 = V.                25)   1 (0 = IX.  30) 122
      330 rem. 6 = VI.              1                 4 rem. 2 = XI.
VII. =  1 Domin. Lr. A.
                                   3) 3
 XI.     2                            --
  X.     1                            1 = X.
 IV.     0
        --                         45              27
         3                          1               1
III.     2                         --              --
        --                         44 = XIII.   7) 26 rem. 5 = XIV.
Epact    1 = XII.
44 and (7 and 1) less 5 = 47
                          31
                          --
            April 16, Easter Sunday.
```

Now to show that March 23 is really the canonical Easter Day of the year 1845, we have the following:—

I. = 1846	III. = 2	19) 1846 (97	
II. = 461		136	
IV. = 0		rem. 3 = VIII. Golden Number.	
2307		IX. = 0	X. = 1, as in last Example.
III. 2			
		3 = VIII.	
7) 2305		20	
329 rem. 2 = VI.		30) 23	
VII. = 5 Dom. Lr. E.		0 rem. 23 = XI.	
XI. 23		45	27
X. 1		22	22
IV. 0		23 = XIII.	7) 5 (rem. 5 = XIV.
24			
III. 2			
Epact 22 = XII.			

23 and 5 less 5 = 23, the day of March for Easter Sunday.

Clavius relied more on tables than on rules. Gauss is the first, as far as we know, who translated Clavius into algebra; but his method still makes use of a table. Delambre made the complete transition to formulæ of calculation independent of tables, and it is his rule,* broken up into short arithmetical directions, which we have given, after having compared it with Clavius.

Clavius asserts that, generally speaking, he has put the new moon forward about a day, or made the epacts too small by 1, for reasons already explained. Increasing his epacts by 1, he compares the resulting new moons with the tables of his time, in a manner which it may be curious to imitate for the present time. The following table contains the new moons of the years 1839, 1840, 1841, and 1842, from the modern almanacs, and also from the epact-almanac before given, the first column being the truth, and the second arising from Clavius's epacts (which are 15, 26 7, and 18), each increased by unity.

It thus appears that for four years running the calendar epact increased by 1 gives either the day of the full moon or the day after, with two days after in three cases only, and the day before in two cases. If an epact-almanac were made for full moons, and list of *best epacts* attached to it, it would be quite sufficient to determine what most persons want, namely, the mere settlement of the question whether it is good moonlight or not on any particular evening. Such a table for fifty years might be made on a card. This comparison serves to confirm Delambre's remark that he found the rules which Clavius had given more correct, astronomically considered, than Clavius himself appeared to think they were likely to be.

There is, in this mode of determining Easter, no mention made of the meridian of the place, and yet it is very obvious that if either the real or mean moon were the object of considera-

* Delambre, *Astronomie*, vol. iii. p. 686, and *Histoire de l'Astronomie Moderne*, vol. i. p. 1.

	1839.		1840.		1841.		1842.	
January . .	15	15	4	4	22	23	11	12
February . .	14	13	3	3	21	21	10	10
March . . .	15	15	4	4	23	23	12	12
April . . .	13	13	2	3	21	21	10	10
May . . .	13	13	{2, 31	2	20	21	10	10
June . . .	11	11	29	{1, 30	19	19	8	8
July . . .	10	11	28	30	18	19	8	8
August . .	9	9	27	28	16	17	6	6
September .	7	8	25	27	15	16	4	5
October . .	7	7	25	26	14	15	4	4
November .	6	6	24	25	13	14	2	3
December .	6	5	23	24	12	14	{2, 31	2

tion, different parts of the world must keep the feast occasionally on different days. Suppose the mean moon to be taken, and suppose that the mean new moon happens at Rome at ten o'clock on Sunday evening. In places three hours eastward of Rome in longitude, it is one o'clock on Monday morning; accordingly, Sunday is the first day of the moon at Rome, and Monday in the place three hours east. Then the fourteenth day of the moon is Saturday at Rome, and Sunday three hours eastward; or at the latter place the festival falls a whole week later than at Rome. This inconvenience is obviated by a calendar moon, contrived under prescribed rules: and the simple mode of settling such a question is this:—If the calendar make Easter fall on the 25th of March, then the 25th of March at any place is the day on which it is to be kept. There is only left one way in which a difference might really arise, and certainly was once in the way to have arisen, but whether it did so or not, and how it was settled, we have no information. When a person goes round the world, keeping his reckoning of days in the usual way, by alternations of light and darkness, it is very well known that he will gain a day in his reckoning upon those who stay behind, if he travel eastward, or that he will lose a day if he go westward. And if two persons, setting out with the same reckoning, meet, having made between them the round of the world, one east and the other west, then if they meet on a day which the eastern traveller calls Tuesday, the western one will call it Monday; and both will be right, according to the times they have severally seen day followed by night. It was noticed long ago that the Portuguese and Spaniards in the East were different in their names of the same day, the Portuguese being a name in advance of the Spaniards: the fact being that the Portuguese had gone eastward by India, the Spaniards westward by America. Clavius does not provide for this case.

We now proceed to describe the ancient calendar; and after making a due allowance of pity for the generations which kept their festival by a calendar moon not nearly so like the real one as that which Clavius furnished to their posterity, we may congratulate ourselves that they had so simple a calendar that Easter gives no trouble in chronology. Had their rule been as complicated as that of Clavius, it is most likely that ages in which the art of printing did not exist would have fallen into many varieties of usage. After describing the calendar of Clavius, we have only to say that they took the cycle of nineteen years to contain exactly 235 lunations, and that the paschal 14th of the moon was the same in every year 1 of every cycle, the same in every year 2, and so on: for example, in every seventh year of the cycle of nineteen years the paschal 14th was on the 30th of March. A very simple table then will supply the place of all rules, as soon as the golden number and the dominical letter are known.

Mar. 21	C.	xvi.	Mar. 30	E.	vii.	Apr. 8	G.		Apr. 17	B.	xix.
22	D.	v.	31	F.		9	A.	xvii.	18	C.	viii.
23	E.		Apr. 1	G.	xv.	10	B.	vi.	19	D.	
24	F.	xiii.	2	A.	iv.	11	C.		20	E.	
25	G.	ii.	3	B.		12	D.	xiv.	21	F.	
26	A.		4	C.	xii.	13	E.	iii.	22	G.	
27	B.	x.	5	D.	i.	14.	F.		23	A.	
28	C.		6	E.		15	G.	xi.	24	B.	
29	D.	xviii.	7	F.	ix.	16	A.		25	C.	

The dominical letter of A.D. 1 was B in the old style, whence, by principles already explained, it appears that if we add 4 to the number of the year, and add the quotient of the number of the year divided by 4, the remainder after division by 7, subtracted from 7, is the dominical number. Thus, for 1639, old style, we have—

```
 1639
    4
  409
 ----
7)2052
 ----
  293 rem. 1.
```

Whence 6 is the dominical number and F the letter, for 1639, *old style*.

Now 1639 and 1, divided by 19, gives the remainder 6, which is the golden number: vi. in the table is opposite to April 10, which is the paschal 14th, the next Sunday letter (F) is opposite to April 14, which is therefore Easter Sunday.

If the formulæ for the new calendar be properly cut down and altered, the following is the rule, in which, however, an epact is introduced, a thing which was unnecessary. We keep the numberings of our former rule, so far as the same steps are found, and thus we have several numbers omitted.

I. Set down the given year.
II. Take the quotient of the given year divided by 4, neglecting the remainder.
V. Take 4 more than the sum of I. and II.
VI. Find the remainder of V. divided by 7.

VII. Subtract VI. from 7; this is the number of the dominical letter,
A B C D E F G
1 2 3 4 5 6 7.

VIII. Divide one more than the given year by 19, the remainder (or 19 if no remainder) is the golden number.

XII. Divide 3 less than 11 times VIII. by 30; the remainder (or 30 if there be no remainder) is the epact.

When the Epact is 23, *or less.*	*When the Epact is greater than* 23.
XIII. Subtract XII., the epact, from 45.	XIII. Subtract XII., the epact, from 75.
XIV. Subtract the epact from 27, divide by 7, and keep the remainder.	XIV. Subtract the epact from 57, divide by 7, and keep the remainder.

XV. To XIII. add VII., the dominical number, (and 7 besides if XIV. be greater than VII.,) and subtract XIV., the result is the day of March, or if more than 31, subtract 31, and the result is the day of April on which Easter Sunday (old style) falls.

The difference of the styles is the difference of names given to the same day in the reformed and unreformed calendars. The reformation suppresses ten days to begin with, and so puts the reckoning ten days forward: moreover, it puts it one day forward for every leap-year which it turns into a common year, as soon as February 28 of the reduced year is past. Thus from 1582, October 5, of old style, Octoble 15 of new style, up to 1700, February 28, old style, March 10, new style, the addition is 10 to the day of the month. But in old style there is such a day as February 29, 1700, which there is not in the new style, hence from the last day named in both styles up to 1800, February 28, old style, March 11, new style, the difference of styles is eleven days, and so on. At present it is 12 days.

It is not necessary for us to enter into any description of the vast machinery of tables by which, instead of by methods and formulæ, Clavius described the structure of his new calendar. We will now state, in recapitulation, the points of this paper which particularly concern the apparent discrepancy on which it is written.

1. The law which regulates Easter in Great Britain declares that whenever the full moon on or next after March 21 falls on a Sunday, that Sunday is not Easter Sunday, but the next: it also prescribes rules for determining Easter.

2. In defiance of the precept, though in accordance with the rules, the Easter Sunday of 1845 is on the very day of the full moon next following March 21.

3. One part of the reason of this is that the British Legislature misunderstood the definition of Easter, used in the rules which they adopted, thinking that it depended upon the *full moon*, whereas it depends upon the *fourteenth day* of the moon, the day of new moon being counted as the *first*. Now full moon never happens before the *fifteenth* day of this reckoning.

4. The other part of the reason of this discrepancy is that the

c 3

legislature supposed the moon of the calendar to be the same as the moon of the heavens, which neither is nor was intended to be the case: the moon of the calendar being not only made to vary from the moon of the heavens for convenience of calculation, but also to prevent Easter Day from falling on the day of the Jewish Passover.

5. These two errors very often compensate one another, for though the fourteenth day is very often a day behind the calendar full moon, yet the calendar moon is also very often a day before the real moon, so that the fourteenth day of the calendar moon is frequently the day of the real full moon. But they do not always do so; and it should never be matter of surprise if Easter fall on the Sunday of the full moon, whether real or calendar.

6. It is not correct to say that Easter is made to fall wrongly in 1845: it falls where the legislators, who correctly copied the rule of the Roman Church, intended it should fall, though they did not correctly give the explanation of the rule they intended to use.

The last time that Easter Sunday fell on the day of the full moon was in 1818, in which year both the festival and the full moon were on the 22nd of March, the earliest possible day. It excited some stir that the definition of Easter, as contained in the Act, should so palpably be violated, and an Oxford* clergyman publicly protested against the observance of Easter on the, as he thought it, wrong day. More than one writer discussed the matter on the supposition that the Parliamentary definition was correct, and also that the extreme of astronomical correctness had been always sought after and considered essential to the due observance of the day. No person who had ever examined the volume of Clavius, *the only authority† on the subject*, appears to have taken any part in the discussion. It seems even to have been supposed that the proceedings of the courts of law might possibly be called in question, since an error in Easter would occasion a corresponding error in the commencement of Easter term. A lawyer would no doubt answer that a positive enacted rule is law, even though the grounds of that rule were incorrectly stated, or though there were no grounds at all. But it is desirable that those who like discussion upon this and similar subjects should not be allowed, in mere ignorance of existing facts, and without any opportunity of knowing what they are doing, to agitate for the reconsideration of what with all its defects is a *fixed rule*,‡ the thing most wanted.

The advantages of the present system are as follows:—

1. There is a fixed rule which prevails throughout the Roman, English, and Scottish churches, and from which the remaining Protestant churches vary but little.

* 'Investigation of the cause of Easter, 1818, being appointed to be celebrated on a wrong day....By a member of the University of Oxford.' Second Edition. London, 1818.

† Of course any one has a right to his opinion as to what *ought to be*: we mean that Clavius is the only authority as to that which *is*.

‡ Clavius asserts that his church cares much more for peace and concord than for the equinox or the moon; that a certain amount of agreement with the Jewish passover is kept up merely for the better understanding of the origin of the feast; but that any day, fixed or not, if observed by all, would be better than dispute or trouble.

2. The general desire of the Christian world, namely, to make Easter an anniversary of the last days of Christ, is substantially satisfied, since it always must come close upon the full moon which comes next after the vernal equinox. No one can know how Easter is kept without attending to the chronological connexion of the death of Christ with the passover, and of the resurrection with the first day of the week following.

3. All necessary warning against the mere observance of days for the sake of the days is given by the very nature of the rule which determines Easter, when known. There is no answer to any manifestation of superstitious feeling on the subject which can be so good as a reference to Clavius putting the moon backwards or forwards a day to suit convenience of calculation.

The disadvantages of any alteration of the rule will be as follows:—

1. The advantages stated in the first and third reasons preceding are destroyed, and the contrary disadvantages introduced.

2. Unless astronomical tables could be rendered absolutely perfect, there must be, as Clavius remarks, the substitution of a fictitious for the real moon.

3. Any change must introduce an inconvenient schism, since it is certain that all Roman Catholics must adhere to the present system. It is hardly to be supposed that the papal see will acquiesce in any alteration.

4. An astronomical Easter is impossible, unless the festival be sometimes kept on one day on the east of *a variable meridian,* and on another day on the west; the difference being a week. It might happen, for instance, that those on one side of the meridian of London should have to keep Easter a Sunday after those on the other side: nay, astronomical tables are exact enough to make it possible that a true astronomical Easter, according to a definition drawn from the real moon, should be observed on one Sunday in St. Paul's, and on another in Westminster Abbey; and as astronomy advances, it is perfectly conceivable that the true astronomical Easter should be one Sunday or another in St. Paul's only, according as it is to be solemnized at one end or the other of the building.

As we are satisfied that there are persons who really have a lurking religious veneration for the ceremonial part of Easter, and for the apparently astronomical definition from which it is drawn, we will demonstrate the assertion about Westminster Abbey and St. Paul's.

The difference of longitude of the two cathedrals is about seven seconds, say six to make sure of the argument; that is, the clock of St. Paul's, the more eastward of the two, ought to be more than six seconds faster than that of the Abbey. Hence Sunday morning begins at St. Paul's six seconds before it begins at Westminster Abbey. Now suppose Easter regulated strictly by the paschal full moon, as implied in the Act of Parliament, and suppose that, on a Saturday evening (at the Abbey) the paschal full moon happens at three seconds before midnight. Then at St. Paul's it will happen three seconds after midnight, on Sunday morning. That is, the Sunday just named is the next after the paschal full moon, at the Abbey, and is Easter Sunday. But at St. Paul's the paschal full moon falls on the Sunday, and Easter Sunday is the next Sunday.

But it will be said this is trifling with the subject; nobody means to stand out about a few seconds. We answer, that whoever gives up a few seconds gives up the principle on which the discussion to which we have alluded was raised, and adopts that of Clavius, namely, that perfect astronomical accuracy must *at some point* give way to convenience. Again, in the time of Clavius, from the less amount of accuracy then existing, there was as little disposition to stand out about a day as there now is about six seconds; the time will come when more will be thought, astronomically, of the tenth part of a second than now of six seconds. If it were granted that the astronomical definition should be used, without minding four hours, still Easter cannot be always kept on the same Sunday in Calcutta and London, or in Montreal and London; carry the love of astronomical truth so far as not to reject ten minutes, and Exeter and London cannot always keep Easter on the same Sunday.

5. It can only happen very rarely that Easter is a perfect anniversary of the events which it commemorates. The Passover (fourteenth of the moon) took place on Thursday evening, the Crucifixion on Friday, the Resurrection on Sunday. The observance of the Friday and Sunday is properly anniversary, but it only happens now and then that the fourteenth of the moon is on Thursday. Since, then, in the nature of things, the moon's appearance can but seldom lead to a true recurrence of the chronological character of the circumstances commemorated, it matters little that the connexion of the moon with Easter, arbitrary as it must be in some respects, should be a little more arbitrary still.

6. Every alteration of the calendar is an additional trouble and risk of error in questions of history; the Gregorian reformation has done much in this way, another attempt would go near to render the chronology of the country in which it was made an unfathomable mystery.

There is but one reformation of the *British* calendar which we should wish to see. It is not desirable that a statute should exist which contains a complete misunderstanding of its own provisions, however little the legal force of those provisions may be thereby affected. A short Act of Parliament, repealing the words about the full moon in 24 Geo. II. cap. 23, and substituting a definition which should not lead to mistake, would be of service; it being remembered that the erroneous words are not merely buried in the statute-book, but are directed to be attached to all the prayer-books used in the service of the Established Church.

A. De Morgan.

University College, London,
July 11, 1844.

COMPANION TO THE ALMANAC

FOR

1846.

PART I.

GENERAL INFORMATION ON SUBJECTS OF MATHEMATICS, NATURAL PHILOSOPHY AND HISTORY, CHRONOLOGY, GEOGRAPHY, STATISTICS, &c.

I.—ON THE EARLIEST PRINTED ALMANACS.

THE attention which our paper of last year *On the Ecclesiastical Calendar* made it necessary to give to some astronomical writings of the century preceding the reformation of the calendar, led us to the subject of almanacs, and thence to the examination of those of Regiomontanus and Stöffler. Happening to find a third of the same period, printed by Zainer of Ulm, which has been little mentioned, and never described, we thought, considering that the same may nearly be said of the other two, that a description of these early astronomical almanacs would not be out of place in a work like this. We propose to prefix to our account some notice of one or two proposed reformations of the calendar, with some matters supplemental to the contents of our last paper, above referred to.

The history of science is almost entirely the history of books and manuscripts: but few historians have been able to command access to all the works they mention. For a long time it was not thought necessary to describe the titles of books with accuracy, and even the lists of professed bibliographers were drawn up with mere paraphrases of titles. Subsequent lists were partly taken from these; and it has happened very frequently that a work has been set down twice in the same catalogue—once from the maker's inspection of the book itself, and again as a different work from somebody else's wrong title-page. The historians of mathematics and astronomy have made use of these incorrect lists in many cases, and those of most note in the last century never learned * any very accurate mode of proceeding. In the meanwhile, however, the division of labour

* In the first edition of Montucla's History of Mathematics he spoke of books with all possible negligence, though he had been a most diligent user of large libraries; and Lalande was just as bad. The Abbé Rives, a noted bibliographer, fell foul of both, and treated them as the interests of the subject required. In his second edition, Montucla mended his habits materially, and wished his critic had lived to see the *vernis de bibliographe* which he had given himself. Lalande also was amenable, and lived to publish the *Bibliographie Astronomique*, far the most useful piece of scientific bibliography which exists, precisely because it is the work of a person who had knowledge both of the books and their contents.

B

had introduced the class of students who are called *bibliographers*, men whose profession it is to investigate the dates, places, titles, colophons, authorship, paper, print, &c. &c. of books, their several degrees of rarity, the libraries in which they are to be found, and all the physical history, as we may call it, of a volume. If one of these most useful inquirers should happen to know anything about the contents of a work, it is so much above his bargain: he is not bound to it by the terms of admission to his company. It is to be wished that it were otherwise, for there is much confusion arising from those who know the books not knowing the contents, and *vice versâ*. Some instances will appear in the course of this article. We now proceed to our subject.

As we expected, no sooner had the almanacs for the present year fairly made their appearance than it was discovered that Easter Sunday was made to fall on the day of the full moon. This was immediately pointed out to the newspapers by their correspondents: the blame was laid on the supposed negligence of the almanac-makers, who came in for a sound rating; they were "sapient chronologists," and several things besides. Before, however, there was time for them to answer, the attention of editors was directed to our last article, and the extracts which were made from it set the part of the question at rest which refers to the correctness or incorrectness of the day named. Some discussion was raised in various newspapers and magazines as to points connected with the history of the calendar; and there the matter ended. According to the account which the 'Times' newspaper gave of the several answers to the attack on the almanac-makers, which it was not thought necessary to publish in full, one appealed to the table in the prayer-books, and asserted that the almanac-makers had no power over that table: this was a right answer. Another thought that the astronomers must be wrong by a day in their actual full moon—a thing next to impossible. It was also asserted that the mean moon, or moon freed from the various periodical irregularities of her motion, was the means of determining Easter-day. This was no answer, for it happens that the mean full moon of March, 1845, falls on the Sunday, as well as the real full moon.

The answer given to the explainer who thought that the almanacs had erred by a day, was a reference to the fact of *all* the almanacs published agreeing with one another. But in all probability if any one of them were wrong (otherwise than by mere misprint) all would be wrong: for it may well be supposed that they all take their materials from the Government 'Nautical Almanac,' which is published four years in advance.

The erroneous definition of Easter was, we find, in the prayer-books from the time of Charles II., which must very frequently have produced discussion at a time when the new moon of the calendar was always four or five days after that of the heavens. But even in 1640, John Booker, the almanac-maker, was summoned before Charles I. to explain why his Easter did not accord with the moon. In 1664 a petition was presented to the King in Council against the sapient chronologists, and the sapient chronologists—being Dr. John Pell, George Wharton, John Booker, Vincent Wing, and John Gadbury—were obliged to appear and vindicate their methods and almanacs. In that same year Booker published a *Tractatus Paschalis* by the King's command, and Pell is said to have written an *Exercitation on Easter:* we have never met with the former of these works; the latter is 'Easter not Mistimed: A Letter written out of the Countrey to a Friend in London concerning Easter Day,' London, 1664, 4to. It was written by Pell to Theodore Haak. In 1735 and 1737 there was another stir on the subject; and a pamphlet was written by one Henry Wilson in the former year, explaining the actual state of the calendar.

Throughout the English writers there runs the definition of Easter by the full moon, and not by the fourteenth day. We give as much account of this as we have been able to collect.

The last time the Liturgy underwent any revision was at the hands of the Convocation which met immediately after the accession of Charles II. (May—December, 1661). The prayer-books of the English Protestant Church previous to this time do not contain any description of Easter, but only the table for finding it: thus those printed by R. Barker, 1639, and by John Bill, 1661, both King's printers, do not contain any *explanation* connected with the calendar, except that the ecclesiastical year is made to begin on the 25th of March, because the world is supposed to have been created at that time, and the conception of the Virgin Mary to have then taken place. The first prayer-book which appeared with all the alterations* and additions made by the Convocation was the folio of 1662, in which for the first time the explanation of Easter appears as it has stood ever since: the commencement of the year is also noted, but the reasons are omitted. Dr. Pell quotes from it all the modern explanation, except the part which postpones Easter Sunday for a week when the full moon is on a Sunday. From some statements which appear in works of no great authority it might be supposed that this latter portion was not added till 1664. We must of course suppose that the explanation of Easter was introduced by the Convocation itself, and was not an addition made by the King's printer, which last we at one time suspected. Not being able to find any full account of the proceedings of the Convocation, we cannot quote anything of first-rate authority on this point; but the following is conclusive enough. Baxter and the other non-conforming divines, when the Act of Uniformity was about to pass (see Calamy's 'Life of Baxter') compelling "unfeigned assent and consent to all and

* The distinction between the old and new editions which first catches the eye is the difference between the translations of the Bible. The authorised version of James I. was not introduced into the prayer-books till 1662. Thus the old copies begin the morning service with "At what time soever a man doth repent him of his sinne from the bottome of his heart," &c. References for the alterations may be found in Neal's 'History of the Puritans.'

B 2

everything contained and prescribed in and by" the book of common prayer as it now stands, argued that the definition of Easter was inconsistent with the tables, that they were required to subscribe to the full moon of the heavens, and also to rules which were acknowledged to depend on a different and fictitious moon. To this it was not replied that the preface was not a constituent part of the book, nor that it contained explanations inserted at the discretion of editors; but the word *moon* was used in the shifting sense which we have seen Clavius applying to it, and the non-conformists were stated as wanting to have the moon of the reformed calendar applied to the unreformed calendar. At least we judge this to have been the reply from the answer of later date, cited by Calamy's editor ('Abridgment of Calamy's Life of Baxter,' London, 1713-17, 2 vols. 8vo., Index "Easter"). It is clear enough, however, that Baxter and his side looked upon the calendar rules and tables as matters of subscription, and considered the palpable contradiction as not the least of the hardships contained in the terms offered to them.

The use, however, of the full moon instead of the fourteenth day of the moon is not peculiar to the *Protestants* in England. The quarto missal published in the second of Queen Mary (1555) has the following explanation (cited by Pell):—

> Post veris Æquinoctium Quære plenilunium
> Et dominica proxima sacrum celebra Pascha
> Non verius invenies si mille legas codices.

But the full moon or *plenilunium* here mentioned is not the full moon of the heavens; for the old calendar table is given to find it from, with directions to find out the full moon of the heavens by *in cœlis est hic*, as explained in our last paper. What we have here chiefly to notice is, that there is no objection to the day of the calendar full moon being Easter-day. Dr. Pell mentions a Prayer-Book printed at Edinburgh in 1637, and a church Bible at London in 1640, in which the Golden Numbers were altered in position four days, so as to give the real full moons: but they were restored to their old places in 1662. Pell's pamphlet has the *imprimatur* of the chaplain of the Archbishop of Canterbury, with the unusual addition *et imprimi forte necesse est*, &c. The insertion of the explanation had opened people's eyes to the state of the lunar cycle. Sir George Wharton, the astrologer (who had promised* Charles I. from the stars all manner of success when he set out from Oxford to try his fortune at Naseby), objected to the explanation altogether, as proper for the Gregorian Calendar (which he obviously supposed to be founded on the real moon, without alteration), but not for the Julian. Of all the writings on this subject, we have not met with any one

* He tried also to help towards the fulfilment of his promise at the head of a troop of horse, raised at his own expense. But the troop was dispersed, and the captain severely wounded. After the Restoration he was made Treasurer of the Ordnance, and afterwards a baronet; which, however, did not prevent his publishing almanacs. Lilly, his great opponent, was the parliament astrologer, and had a pension of 100*l*. from the council of state. But though Lilly prophesied more correctly than Wharton, the art of the former was treated with contempt by the royalists; and Butler is supposed to have ridiculed him in particular under the name of Sidrophel.

in England previous to the change of style, the author of which must either have read Clavius himself, or have copied from some one else who had done so, except a reply to a question proposed to the 'Weekly Oracle' for 1737.

We suspect that the explanation of the mistake is as follows: The fifteenth day of every calendar moon is the full moon (of the calendar); if Easter Sunday is to be the Sunday following the fourteenth of the moon *exclusive*, it is the Sunday following the fifteenth of the moon or the (calendar) full moon *inclusive*. Either of the following rules would be good, and both had been used:—

1. Easter-day is the Sunday next after the fourteenth day of the Moon which happens on or next after the 21st of March. 2. But if the fourteenth be on a Sunday, Easter-day is the following Sunday.	Easter-day is the Sunday on or next after the full moon (fifteenth day of the moon), which happens next after the 21st of March.

Perhaps some person who did not clearly understand the difference, transferred to the second a provision resembling the second clause of the first, and so produced the explanation as it now stands. Wallis (*Phil. Trans.*, No. 240) states the rule as on the right above, and sets down the additional clause as a blunder. But we do not find any writer who understood this, except one Gamaliel Smethurst, who published (without date, but before 1752, probably in 1750, at Manchester) a useful and unpretending little work called 'Tables of Time.' We must, perhaps, except also the writer of a paper on the calendar in Holliday's 'Miscellanea * Curiosa Mathematica,' for 1747, though his expressions are ambiguous.

In the reign of Queen Elizabeth, John Dee proposed a mode by which to make the English reckoning coincide with the Gregorian. Digges, Savile, and Chambers were appointed to report upon this plan to the minister Burleigh. Greaves and Wallis at different times gave their opinions upon this plan; and the report, with their opinions, is in the Philosophical Transactions. But what somewhat surprises us is, that neither Dee nor his reporters, nor Greaves, nor Wallis, nor the advisers of the Imperial Diet of Regensberg, at which the reformation of the German calendar was enacted, though all their prepossessions are against the Roman Church, and they all find fault more or less—not one of them seems to have known that the moon of the heavens was intentionally departed from. They all speak as if they thought the moon of the Gregorian Calendar was meant to be identical at least with the mean moon: and this though Cla-

* This was a quarto periodical, somewhat on the plan of the 'Ladies' Diary,' published variously, yearly, half-yearly, and quarterly, from 1745 to 1754. A part of the number for 1755 was printed, but never published. Holliday's translation of Stirling's '*Methodus Differentialis*' was intended for this work: and about a third of his translation of Taylor's '*Methodus Incrementorum*' had been actually published in it when the work was discontinued. We are not aware that the rest of this translation ever appeared.

vius has not only stated the reverse in the plainest words, as we have seen, but has actually shown (in the manner we have imitated in our last article) by tables of instances, that epacts which are *not* those of the Gregorian Calendar agree better with the heavens than those *which are.* We suppose that none of them read Clavius, and we cannot much wonder at it. And hence it has arisen, that when the English style was altered, the Gregorian moon was taken to be the real one, and the explanation was prefixed. It is well, perhaps, that it was so, for if it had been known how the case stood, the astronomical falsehood might have put some difficulty in the way of the arrangement. Many persons knew that the Gregorian Calendar was imperfect; but we cannot find that any one was aware of the *intentional* error, or of its cause.

There are two other writers in the Philosophical Transactions between 1700 and 1720, Messrs. Thornton and Jackman, who both discover from the tables that, as they phrase it, the full moon is the fourteenth: that is, they find out that nothing will make the explanation and the tables agree except reading the fourteenth of the moon as the full moon. Never suspecting the explanation to be in error, they presume therefore that the two things are identical. That is to say, there were communications to the Royal Society, running over a period of thirty years, beginning from Wallis, which showed that there was error or confusion somewhere: and yet not a single reference appears to have been made to original writers to ascertain what was meant by the rule.

This is a sweeping assertion, and will be looked at with proper distrust. It can only be verified by those who have read the various books and papers referred to, after making some study of the work of Clavius: * we rest assured that it will be so verified, and say no more on a point which is not without its importance as illustrating the years 1818 and 1845, in which no maker of almanacs, no astronomer, no clergyman, was found able and willing to state the reason of the contradiction which occurred in the Calendar.

At the time of the change of style in 1751 a table was published by Lord Macclesfield, who, with Lord Chesterfield (both, it so happened, pupils of the celebrated De Moivre), was the great promoter of the measure, drawn up, we suppose, by his astronomical† adviser, whoever that might have been. We find

* In our last article, page 7, we noted that Clavius has made use of a passage of St. Ambrose in a manner which, to say the least, looks very unfair. He certainly does this: but we have since found a passage of St. Jerome, cited by Tillemont, which, had it been in the place of the passage of Ambrose above noted, would have exactly answered his purpose. It is possible enough that he may have substituted one citation for the other by inadvertence; in which case his argument is honest, after the correction is made, though, in our opinion, not more sound than before; for we cannot admit Jerome as authority for a reference to Eusebius which that bishop himself, much as he speaks of the Nicene Easter, and of his own *theological* work on the subject, never alludes to in any way; and on which the professed historians of the Council are equally silent.

† A note in the Gentleman's Magazine for 1751 mentions, "Peter Daval, Esq., of the Middle Temple, Secretary to the Royal Society, who drew the bill (and prepared

it in almanacs as "composed by the Rt. Hon. George E/ of Macclesfield, showing by the Golden Numbers on what days of the month the paschal limits or *full moons* have happened, or will hereafter happen," &c.; but it was first published in the Philosophical Transactions for 1750 (No. 494, p. 417), at the end of a letter from the above-named nobleman on the subject of the calendar generally. This table gives for the Gregorian reckonings a set of tables resembling that which answered every purpose connected with Easter in the old style, and which is given in our last article, page 32. A new table is required whenever either a Julian leap-year is abandoned, or a correction is made for the error of the lunar cycle; but not when both of these come together, for the effect of one correction destroys that of the other. Lord Macclesfield, or Bradley, whichever it was, had closely attended to the details given by Clavius, so far as calculation is concerned: a superficial acquaintance with them would most certainly have led to the neglect of the arbitrary alteration which was made to prevent the calendar fourteenth from occurring on the same day twice in nineteen years, as explained in our last article, page 24. No astronomer, as such, could possibly have explained this; but it is properly represented in the tables we are speaking of, in which its effect is instantly perceptible.* It is then strange that the full moon of the heavens should take its place at the head of these tables, instead of the fourteenth day of the calendar moon. For it is to be remembered that tables of this form are not given by Clavius at all; so that they could not have been simply copied from his work: they must have been deduced from it. It is certainly possible either that they should have been deduced by mere classification from the great table of results ranging from A.D. 1600 to 5000, or that they may have been taken from some intermediate work. But it is not fair to impute so slovenly a proceeding as the first to either of the probable framers; and as to the second supposition, we have seen many works on the Gregorian Calendar, and never found any such table in any one of

most of the tables), under direction of the Earl of Chesterfield, the first former of the design. And the whole was carefully examined and approved by M. Folkes, Esq., President of the Royal Society, and Dr. Bradley, his Majesty's astronomer at Greenwich, who composed the three tables at the end of this bill." We suppose we may judge of the general temper of the public with regard to this question from the above-named periodical, which was then at the head of its order. Some opposition appears to have been made to the specific change, on the ground that the Gregorian Calendar has errors, but what they are is not specified. One correspondent affirms of it, as a fault, that Easter is sometimes kept at the same time as the Jews keep their Passover; an imputation which would doubtless have raised Clavius from his grave, if it had not been that the utter absence of all reference to his name and writings, which prevailed everywhere, showed that there was nobody either to mean it or to take it as directed against them. The indefatigable veteran, William Whiston, then in his eighty-fourth year, came forward once more to speak a word for the authority of the Apostolical Constitutions, in a letter addressed to the editor, March 8, 1751. His deduction from the Constitutions is that Easter must be kept after the vernal equinox, and that the full moon must fall *in* Passion Week. Daval, above mentioned, wrote a tract in vindication of the new calendar, which we have never seen; it was published in 1761.

* See some specimens of these tables in the Supplement to the 'Penny Cyclopædia,' article Easter.

them, though its superior facility and clearness would almost certainly have perpetuated it.

To this it may be added, that Lord Macclesfield's letter above referred to, which must be looked upon almost as the official forerunner of the change, is perfectly silent as to the authority for its statements on the Gregorian methods. So total an ignorance of the authorised writings on the Roman calendar, accompanied with so correct a transference of its rules (we are not speaking of *explanations*) into the Act of Parliament for the change of style, puzzled us exceedingly, accustomed as we are to find it impossible to be very correct except from original sources. But it was casually brought to our notice that Charles Walmesley,* who was well known as a mathematician, and had just (1750) been brought into the Royal Society, was said to have been one of those who were consulted by the framers of the bill. Combining the characters of a priest and an astronomer, he had probably made himself acquainted with the details of the reformed calendar. In the short obituary which is given in the *Gentleman's Magazine*, he is mentioned as the last survivor of the mathematicians who were consulted on the change of style. But not a trace of mention of his name can we find at the time: for which it is not difficult to conjecture the reason.

A perpetual almanac, in its easiest form, would be a work consisting of thirty-five common almanacs, one for each time of happening of Easter Sunday, from March 22 to April 25, both inclusive, and numbered 1, 2, 3, &c. A preliminary table should give for every *Annus Domini* from 1 to 2000, the number of the almanac belonging to that year. It would not be necessary to call the reader's attention to old style or new style: the author would make the proper arrangement of his numbering, and the reader would turn at once to the almanac for the year he wants. For an English work indeed, it would be necessary to have two numberings from 1582 to 1752, one for each style. And if the thirty-five Calendars were made for common years, all that would be necessary for leap-years would be to take January and February from the *next* Calendar: thus, when the Calendar for the year is No. 25, the January and February, if it be leap-year, must be read from No. 26. Such a Calendar has been published by M. Francœur, of the Academy of Sciences, under the title 'Théorie du Calendrier et Collection de tous les Calendriers des Années Passées et Futures,' Paris, Roret, 1842, 12mo. In reading history, it must be convenient to be able to turn to the actual almanac for the year, and not to have to make out everything moveable by a table of Easters. Of course this work gives nothing but new style after 1582: an English reader wishing to

* Born 1721, died 1797. At the time we mention he was a Catholic priest of the Benedictine order, and six years afterwards was made vicar apostolic of the western district. The statement relative to the Calendar occurs also in the *Annual Register* for 1797, where it is certainly taken from the *Gentleman's Magazine*. We are told that it also occurs in the *Catholic Directory* for 1802, and in the memoir of Walmesley, prefixed to the Dublin reprint (1805) of his work 'Pastorini's History of the Church:' of these we have not been able to get sight.

have his own Calendar up to 1752, must subtract 532 from the year, and use the number opposite to the remainder. Thus, opposite to 1698 in the book is No. 9, which is for new style; subtract 532, and there remains 1166, the number opposite to which is 34: accordingly, for English history, the number 34 is to be used; for French, &c. history, the number 9.

Among the opposing statements which our article on Easter called forth, we saw hardly any which we could not be very well contented to leave unanswered, feeling sure that those who inquired into the subject, with original* authorities in their hands, could come to no other conclusions than we had done. After exposing (page 7) the manner in which Clavius had used Bede of the eighth century to establish the doings of the Nicene Council in the fourth, we saw in newspapers and magazines references to the same writer for a similar purpose. Leaving to those who feel better satisfied with an authority of three or four centuries' distance from the event referred to than with contemporary testimony the fullest permission to draw their inferences, on the simple condition of not being obliged to join them, we proceed to notice one statement, namely, that the fourteenth day of the moon was chosen as the means of determining Easter, because it was supposed to be the day of the full moon. We answer this because it is the most natural error for a person unacquainted with Greek astronomy to fall into. This assertion was meant, we presume, to lead to the conclusion that the British Parliament, in naming the day of the full moon, had followed the spirit of the Nicene directions in departing from the letter, and had only done what the bishops of the Council would have done, if they had not unfortunately been wrong a whole day in their description of the full moon. We have seen that the Council did not enter into the astronomical part of the matter at all, leaving that to the ordinary sources of information. But, supposing them to have consulted the astronomers, we shall now show how many words it takes to answer an assertion which could not have proceeded from any one acquainted with the science of the period.

It is matter of astronomical notoriety that the length of the lunation was never wrongly known by so much as an hour from the time of the Chaldeans, that Hipparchus settled its value within a second correctly, that Ptolemy followed Hipparchus, and that Pto-

* Many have been misled by Wheatley's 'Rational Illustrations of the Book of Common Prayer,' which they call the *History of the Prayer-Book*—a name which its excellent author never meant to give it. He (or else the editor who adapted the chapter on the calendar, for Wheatley died before the change of style) has evidently followed some erroneous leader in the following point. He gives four most distinct and formal canons, directing the manner in which Easter is to depend upon the moon, and fixes them on the Nicene Council, upon the authority of Eusebius, in his Life of Constantine, book iii., chapter 18. Not one word is there in this chapter about the moon, or its connexion with Easter, or any specific direction. The chapter cited is a part of Constantine's letter to the churches, and dwells upon nothing but the importance of all Christians keeping Easter at one time, and of that time not being the Jewish time: for anything there stated to the contrary, Easter-day might have depended on the weather. The reader will find a very full account of all that the Council did in Tillemont's history of it. The canons, be they genuine or spurious, attributed to the Council, are there given, and there is not a word about Easter in them; nor in any other record of its proceedings, except only in the synodical epistle, as quoted in our last year's article.

B 3

lemy was the great authority in all such matters at the time of the Nicene Council. The decline of Greek learning was so rapidly accelerated by the capture of Alexandria, that in thinking of a monk of the eighth century, such as Bede, we do not look to him for any information about those who took their astronomy from Ptolemy; we rather inquire into the probability of his having seen any Greek mathematical writing at all. But however this may be, it is certain that Greek acuteness had discovered that the middle of a period of 29½ days is in the *fifteenth* day, and nearer to the beginning of the sixteenth than to that of the fifteenth; they put the full moon on the fifteenth day, and would rather have put it on the sixteenth than on the fourteenth. They knew that the irregularities of the moon's motion do sometimes actually push the full moon into the sixteenth day, but never draw it back into the fourteenth. If we were to cite the phrase "the fifteenth day" used for the full moon, those of whose argument we are now speaking might assert, perhaps, that the day after the full moon was meant: if we were to cite instances of this phrase, "the full moon, that is, the fifteenth day," it might then be said that the writer, by thus explaining himself, intended to signify that *he* differed from the common usage in the day on which he put the full moon. We might get instances enough of both; but we prefer to bring forward one in which a comparatively popular writer, who shows little knowledge of Hipparchus, leaves it to be inferred that the mode of describing the moon's phases common in his time could not place the full moon earlier than the middle of the fifteenth day. Geminus, who wrote in the first century, says,* "The moon is horned towards the beginning of the month, halved about the *eighth* day, gibbous about the twelfth, and full about the middle of the month: after the middle she is again gibbous, and is halved about the *twenty-third*, and horned towards the end of the month." The half moons accordingly are on the eighth and twenty-third: if, as we have some right to suppose, we take these to be intended for average epochs, and put the middle of the eighth and twenty-third days, then Geminus implies that the new moon, half way between them, is at the very end of the fifteenth day. And if, to make it tell as much as possible against us, we take his words as meaning that the half moons do not occur before the beginnings of the eighth and twenty-third days, then we have him implying that the new moon does not arrive before the middle of the fifteenth day. Thus much we have given in compliment to the supposition that those who could determine the mean lunation within a fraction of a second were unable to divide it correctly into two equal parts, after using it more than 700 years.

There are treatises on the calendar, or *de computo ecclesiastico*, attributed to Bede, to Roger Bacon, and to John Sacrobosco (Holywood or Halifax), all Englishmen by birth. We agree with Dr. Peacock (*Encyclop. Metropol.*, Arithmetic, p. 441) in

* Ἐισαγωγη ἐις τα φαινομενα (Petavius, *Uranologion*, Paris, 1630, folio).

thinking, that the separate treatises attributed to the two former are almost certainly spurious:* but this opinion does not affect what there is on the subject in Bede's Ecclesiastical History, or in Roger Bacon's '*Opus Majus.*' The latter is, in regard to the calendar, what he was as to many other things—a reformer: the earliest remaining on record of those who pointed out the defects which were at last held decisive in favour of alteration. It may be worth while to give a short account of his views on this subject, as contained in pp. 169–180 of the '*Opus Majus*' above mentioned, addressed to Pope Clement IV. in 1267.† After noting that the cause of the errors was ignorance and negligence, and stating that they were known and written on both by astronomers and computers of ordinary almanacs, he proceeds to assert that the length of the year implied in the Julian calendar (365¼ days) is too much by about one day in 130 years. This he afterwards calls one day in 125 years. For these determinations of the length of the year he gives no authority but the general opinion‡ of astronomers, "estimatur a sapientibus." Now it is very remarkable that neither Ptolemy nor the Arab writers give any value for the length of the year so correct as either of the above; and even the Alphonsine Tables, which were constructed about fifteen years before Bacon wrote, and which are nearer the truth than anything before them, are still not so near the truth as this. The first idea which strikes the mind is that Bacon might have been aware of the result of the Alphonsine Tables, and might have represented it in round numbers. But the truth is, that according to those tables the error of the Julian year amounts to a day in 134⅛ years, so that he would have chosen 135 in preference to either 130 or 125, if he had followed them. Bacon then proceeds to point out the effect of the error upon the day of the equinox, which he says has at last become so great that not only the astronomer but any person whatsoever may detect it by merely looking at the sun's shadow from a wall. He recommends the taking off one day in 125 years, as making the best approximation which existing knowledge would furnish. He then proceeds to consider the mode of finding Easter, and shows with his usual clearness (for he is one of the most distinct writers of any time) that the cycle of nineteen years is wrong, and that in every 304 years the new moon of the calendar will advance a day upon that of the heavens. Hence he infers that the new moon should not be sought from

* It seems pretty certain that it was a habit of the manuscript ages to write out portions of a treatise, a chapter for instance, and to put on the proper heading, with an acknowledgment of the author: thus making the appearance of a separate treatise by that author. And it is probable that this practice was followed when the extract was not literal, when it was only abstract or epitome, and even when comments or other amplifications were added.

† The date is very certain, for Bacon describes the golden number and dominical letter of the year in which it was written: and there is but this one year of Clement's pontificate to which they can apply.

‡ The reputation of the Alphonsine tables seems to have ultimately prevailed against this opinion. In the manuscript Oxford Almanac, written about 1347, described by Mr. Harris in the 'Memoirs of the Royal Astronomical Society,' vol. xv., now in the press, the length of the year is that used in the Alphonsine tables.

cycles of this kind. His recommendation is to predict Easter from time to time astronomically, or else to adopt, for ecclesiastical use at least, a luni-solar year resembling that of the Saracens. Should the former mode be preferred, he recommends the adoption of what he calls the Hebrew tables, stating that he considers the Jews to be, and always to have been, the most accurate astronomers. What he means by the Hebrew tables cannot now be ascertained: nor have we ever been able to find, in any work professing to describe the Jewish calendar of the middle ages, a length of the year more correct than that of Hipparchus. Bacon then proceeds to discuss the question, whether the directions of the Nicene Council be not against the plan he proposes: and as Bede is here his authority, he comes to the same conclusions as Clavius about the adoption of the cycle of nineteen years by that Council. He is as defective as those of his time in the power of criticising authorities: it is quite enough that some one has written it. For instance, Bede brings forward a miraculous attestation of the truth of the rule for finding Easter, which he states to have occurred in the time of Pope Leo, about two centuries and a half before him. There is probably now not a priest of the Roman Church in Great Britain who would receive, assuredly not one who would maintain, a miracle upon the sole authority of a writer of 250 years after the asserted event. But Bacon says, that since Bede has brought forward a miracle, it must be admitted that there was then no error in the rule. This disposition to receive authorities without discrimination neither belongs to Bacon's faith nor age in particular, but was universal and continued for centuries. Tycho Brahé, Lubienietski, and Riccioli, in the seventeenth century, the followers respectively of Luther, Socinus, and Loyola, received a comet upon evidence beginning from a much later period than the asserted phenomenon, as readily as did Roger Bacon a miracle.

Bacon concludes with a strong address to the Pope in favour of a reformation of the calendar. All men of science, he says, are aware of its numerous and scandalous errors, and laugh at the ignorance of the prelates who maintain it. Saracens, Jews, and Greeks, whether residing among Christians, or in their own countries, are also aware of the same errors existing in the Christian calendar, and abhor the folly of those who are misled by them. Should the reformation, says he, take place in your time, it will be one of the greatest, best, and most glorious achievements that ever was attempted in the Church. It will be seen that Bacon was really in possession of all the knowledge necessary for the attempt, and that his calendar, had he been attended to, would have been better than ours, astronomically speaking, inasmuch as he would not have shifted his moon to keep out of the way of Jews or Quartadecimans.

Our means of recovering the authorities of which Bacon made use in his astronomical data are very scanty; but the number of manuscripts yet remaining, written by English astronomers of his and the preceding day, is by no means small. There is one predecessor whom he calls "the famous Gerlandus, whom all computists and astronomers follow in all that relates to the calendar." This Gerlandus must have lived in Bacon's younger days, for Tanner cites from his '*Opus Minus*' a dispute about whether a Latin word was to be spelled *Oricalcum* or *Auricalcum*, upon which point "Magister Johannes de Gerlandia vituperavit omnes, sicut ego ex ore ejus audivi," says Bacon. There is one John Garland, and a writer on the calendar too, who is stated to have lived about A.D. 1040, and to have been driven abroad by fear of the Danes. He is mentioned by Roger Yonge, another writer on the same subject, who lived in 1150. But there is another John Garland, who was alive in 1245, and who also wrote on the calendar: several of his works were printed in the sixteenth century. That two persons of the same name should have left treatises on the calendar is by no means extraordinary, since nothing was more common than to write on this subject. There are manuscripts of the work of one or the other* yet remaining, and, considering that Bacon does not pretend to be himself an astronomical observer, and that he refers to Gerlandus as the great authority of his age, it would be worth inquiry whether these manuscripts contain the value of the length of the year which Bacon gives, as above noted.

Nearly contemporary with Roger Bacon was the celebrated Sacrobosco, whose name is supposed to have been John Holiwood, or Halifax. His work *De anni ratione, seu Computus Ecclesiasticus*, though not so famous† as his treatise on the Sphere (which was the common book of astronomy till long after the invention of printing), was still widely known, and was finally printed several times. Sacrobosco disposes of the errors in the calendar very summarily: he points out that the new moon is three or four days wrong, but adds that, as a general Council has prohibited any alteration, the moderns have been obliged hitherto to tolerate errors of this kind. After showing that the embolismic months are interpolated in the most convenient way, he makes use of Bede's authority in a more reasonable manner than did Clavius, namely, to establish the maxim that of two evils the least should be chosen, thus:—

> Si tibi concurrant duo turpia, *dilige* neutrum;
> Sed quod turpe minus *elige* Beda docet.

On the length of the year he contents‡ himself with saying that it is a little less than 365 days 6 hours, but that the difference is too small to be accurately measured.

* Heilbronner mentions one in the Vatican Library. See Pits and Tanner for information on the Garlands generally.

† The authors whose names appear most frequently in Lalande's *Bibliographie Astronomique* are Sacrobosco, Regiomontanus, and Ptolemy, but the first much the most frequently. The work on the Sphere was printed, with or without commentary, eighty times at least, according to Lalande: and it is probable that his list is far from complete.

‡ Cardinal Cusa, in his writing on the calendar presently mentioned, more than once asserts that Sacrobosco took the excess of the Julian year to be one day in 288 years.

About the time of the invention of printing two distinguished advocates of reform in the calendar appeared: Nicholas of Cusa (the Latin form of the name of some village in the diocese of Treves), commonly called Cardinal Cusa, who died in 1464; and John Muller of Königsberg, in Franconia, thence called Regiomontanus, who died in 1476. The treatise of the former, entitled *Reparatio Calendarii*, which was certainly written before 1437, was published with the author's other works, in three volumes, Paris, 1514, folio. In his historical account he is more confiding than Bacon. He quotes the opinion of Gamaliel, the teacher of St. Paul, on the authority of Sacrobosco, which is even a better joke than taking the acts of the Nicene Council from Bede. His exposition of the defects of the Calendar is true and clear: his proposed reformation consists in destroying a number of nominal days, as was afterwards done, to set the equinox right, and altering the golden number so as to make the ordinary rules bring the new moons right. He makes no attempt to prevent a recurrence of the same necessity at some future period.

Regiomontanus, the first astronomer of his day, attacked the existing Calendar in the almanac of which we shall presently give a full description. He describes the cause of the difficulty very briefly, and seems to propose as a remedy the determination of Easter by astronomical prediction from time to time. He gives for fifty-five years, from 1477 to 1531, every instance in which the *usus ecclesiæ* was about to differ from the *decreta patrum*—the difference being always either one, or four, or five weeks. Regiomontanus is the next after Bacon who declared that the *usus ecclesiæ* was not in accordance with the *decreta patrum:* he does not attempt the usual historical falsehood of fixing the existing method upon Eusebius and the other Nicene bishops, but refers it (and truly, as we showed in our last number) to the Abbot Dionysius, "whose pascal calendar," says he, "we are using to this day." He mentions the diffusion of the Scriptures among the people in their native tongues as giving rise to many doubts and questions upon the subject; and alludes to this diffusion as having taken place in Italy, and specifically as having caused a reference to Cardinal Bessarion when he was legate at Venice. He further says that the Jews very frequently raised objections to the existing Christian mode of determining a feast which was asserted to be connected with the Passover; and so far had the dissatisfaction gone, that some priests of Bremen* (*sacerdotes premenses*) had undertaken to correct the error themselves, and had kept Easter a month before all the rest of the Church—for doing which the adjective *premenses* had become a standing joke against them with a double meaning.

Between clocks and watches, lighted roads, and public announcements, we are now so well provided with information,

* At least Bremen is called *Premis* and *Premen* in contemporary cosmographies: but it may be that these priests were of Parma, and that *Parmenses* was altered into *Premenses* for the joke's sake.

that an almanac is not a matter of the first necessity to most persons. It was otherwise in the fifteenth century; and if we consider how much comfort must have depended upon being able to arrange business with reference to the numerous holidays and the moonlight, we shall see that the mere list of saints' days and moons must have been a matter of consequence. The boys of the thirteenth century learnt a string of useful jargon, which we find in verse at the end of Sacrobosco's *Computus*, to enable them to remember the order of these days. It was called *Cisio Janus* from its two first words, which must have been as familiar to the urchins as *As in præsenti* is now to those who suck their Latin from the Eton grammar. The two first lines were—

Cisio, Janus, Epi, sibi vendicat, Oc, Feli, Marc, An,
Prisca, Fab, Hag, Vincenti, Paulus, nobile lumen;

Meaning that the month of January contained the Circumcision, the Epiphany, its octave, Felix, Marcellus, Antonius, Prisca, Fabian, Agnes, Vincent, and Paul. Another of these delightful inspirations of the Muse is—

Pe, Steph, Steph, Protus, Sin, Don, Cyr, Ro, Lau, Tybur, Hyp, Eus.

There are many manuscript almanacs of the fifteenth century yet remaining. They may be divided into what we now call *astrological* (a word which was then frequently used in its original sense), and those which were simply astronomical. The majority of the astronomers in the middle ages believed in the prognosticating power of the stars: even Roger Bacon inclines to the supposition that they have a physical influence on the body, and through it on the acts of the mind. The Church of Rome always, collectively, set itself against this absurdity. Not that its popes and cardinals were by any means universally free from belief in it; but those who believed considered it as magic, while those who disbelieved thought it of course an organized fraud. So that, upon the whole, divination by the stars met with little public encouragement, and its professors were obliged to write about it under modified phrases. We have never found, in ancient astrologers, such impudent pretensions as those which have been published in London in our own day. If the members of the Stationers' Company, who still continue to publish an astrological almanac, could see this trash, and could know that there are shops in London which sell nothing else, and that there are persons who make a trade of imposing on the ignorant, and who doubtless quote their authority, we suppose they would not allow any consideration of profit to induce them to lend their names to the continuance of so vile an imposition. But if they will persist, we should recommend them to insert the qualification contained in the last sentence of the following quotation; and most particularly we should recommend them never again to do what they have done in *Moore's Almanac* for this present year, namely, insinuate* that

* We quote our ground for this charge, from Moore's Almanac for 1845. The Italics, when mixed with Roman, are our own:—

"ASTROLOGICAL

unbelief in astrology is infidelity, and denial of the providence of God. This is an oversight, arising from an unskilful attempt at imitation of the old almanacs; it must show the Stationers' Company that if they will play with edge tools, and call it sport, they ought to be very careful how they use them.

The following is the preface to a prophecy for the years 1554 and 1555, by Antonio de Montulmo, who must have lived a century earlier at least, for he is mentioned as one of the writers from whom Regiomontanus intended to print some extracts. It was translated in 1554 by F. van Brunswike, and at first we supposed that it might be his own composition, with the name of Montulmo used as a bait; but on further examination we believe it to be really a translation, with some adaptation:—

"For asmuche, as God whiche is the begynnynge of all Creatures, (good gentle reader) hath made man to his own similitude and likenes, whereby he is almoste cōparable to the holy Aungels of God, who is it then that wyll denye that man maye by the disposition of the starres prognosticate thynges to come, as *Berosus Chaldeus* writeth, that holy Noe did unto the Armenians and Scithians after the uniuersall deluge or floude, wherfore thei called hym the partaker of the mind of God. Moreouer, S. Hierome writeth, that what so euer can be compreheded with humane understanding is ruled by the wheles of the sunne. Wherefore in asmuche as bothe the diuine and the prince of Philosophers Aristotle in his boke *de generatione et corruptione* confesseth that *propter accessum et recessum solis in curculo obliquo fiūt generationes et corruptiones ī mūdo sublunari:* It foloweth that knowynge the disposition of the celestiall bodies, by the whiche all elementate bodies be ruled, that we men may by the science of Astronomie (whiche science *Petrus Cameracensis** calleth worthyly naturall theologie) prognosticatyng as is aforsayd thyngs to come, as mutations of the ayre, pestilences, and al other infirmities, warre penurie, peace, or plentie of victuals, whyche depende of the heauens as seconde causes of God, to suche effectes, appoynted wherfore (good gentill reader) the premisses beyng wel pensed, I desyre the not to take all thynges that shall bee prognosticated in this yeares prognostication and the next yeres prognostication as contigentes of fatall necessitie, but that it shall chaunce, if god doo not forbyd, thus I bydde thee well to fare."

"ASTROLOGICAL PREDICTIONS.

Judicium Astrologicum, pro anno 1845.

VOX CŒLORUM, VOX DEI: The Voice of the Heavens is the Voice of God. He speaketh in all the Changes of the Seasons and of the Times.

Courteous Reader,—In this my annual production, I have a long time sounded the *above important truth* in your ears, and I trust *not in vain.* It is, *however,* to be lamented that there is a *great deal of infidelity* upon the face of the earth, and even no small portion thereof cleaves to the skirts of Britannia.......That wonder-working Hand, which placed each mighty orb......is clearly manifest in our earth in the changes of seasons....... Let those who are disposed to *deny the existence of Divine Providence* reflect on these words of holy writ, Not a sparrow falleth......"

Such an inuendo would have been too bad for the sixteenth century.

* This was Cardinal D'Ailly, who drew confirmations of Christianity from the horoscope of Jesus Christ.

This is a fair specimen of the reasons by which astrology was supported in the fifteenth and sixteenth centuries, as well as of the scientific English of the middle of the sixteenth century, before either the words or the orthography were settled. Did the Roman Church remain undecided on the calendar question for nearly three centuries, during which the admitted errors became larger and larger, from suspicion of the consequences of giving authority to *astrologers*, as most astronomers then were? Was there any fear that, if the intended reformation should involve the necessity of permanent astronomical assistance, as proposed by Bacon, astrologers would take advantage of their being thus called in to disseminate their principles? It is possible, we think, that such an apprehension may have had its effect: the more so as those who treated the pretended science with contempt had seen and read of popes, cardinals, and bishops who were addicted to it. However this may be, the Roman Church has never had due credit among Protestants for its uniform resistance to this absurdity, in ages so ignorant that it might have added to its power by assuming a monopoly of it, and when many, perhaps most, of those who could afford it never engaged in any affair of consequence without having a lucky hour named for the commencement. And here again, there was no creed nor country which was a protection against the belief: the followers of Luther were as much given to it as those of the Church they had left. Rheticus, the colleague of Melancthon, tried to establish that the changes of monarchies depended upon the position of the perigee of the sun's orbit; and Melancthon* himself wrote a laudatory preface to Schoner's work on judicial astrology. Kepler, a Catholic, though an enemy of ordinary astrology, had one of his own exactly like it. We dismiss this part of the subject not without a feeling of shame at the existence of a public body in our own country which turns a penny by the publication of pretended prophecies, and which, as to the pretext that it is only a matter of amusement, might know, and ought to know, that it is keeping up a real belief in the old nonsense. A company, in a Protestant

* Those who observe that Bayle has let Melancthon off for his astrology with two words and a reference, would hardly imagine the lamentable extent to which this weakness was his master. On his death-bed, he half predicted that he should die on the day on which a conjunction of Mars and Saturn would take place: and on one occasion, being seen bathed in tears, and showing every sign of the greatest grief, it was found on inquiry that on looking at the stars in his evening's walk, he saw that a dreadful war was going to burst upon Germany. The first story is from his most accredited biographer, Melch. Adam; the second from a less respectable source, the 'Jocorum atque Seriorum Centuriæ' of Otho Melander, whose grandfather, Dion. Melander, was the person who asked Melancthon what was the matter with him. This same Otho relates a severe rebuke which Melancthon received from Luther, for saying that persons born when Libra was in the ascendant must be miserable through life: and a ridiculous attempt to cast the nativity of an infant child of his (Otho's) grandfather, whom he supposed to be of the male sex, and to whom he promised learning, honours, and religious contests. Being told that the child was a girl, he was out of countenance for a time, and at last said—then she will rule her husband. But on second thoughts he cast a new scheme, in which he condemned her to death at seven years old, and staked the validity of his art upon the prediction. The girl lived, however, till the age of fourteen. Otho was of the reformed church, and was therefore free from one bias against Melancthon, at least.

country, and in the nineteenth century, profits by a superstition which was condemned in the Councils of the dark ages: and serves Mammon in the very point in which those whom they charge with superstition served God.

The earliest almanacs published are free from *predictive* astrology, though of two at least of the three we intend to describe the authors had a belief in it. Regiomontanus was one of those who could assign the end of the world by the heavens, and Stöffler is famous for the deluge which he is *said* to have predicted, and was an inveterate consulter of the stars.

In the earliest times, as now, a great distinction is to be drawn between the *Calendar* and *Almanac*, and the astronomical *Ephemeris*. In the first, nothing but the immediate wants of civil society are consulted, as in the common almanacs of our own day: in the second, the places of the moon and planets are given, with various helps to the astronomical calculator. In the present article we describe nothing but Calendars, the lineal progenitors of our present race of almanacs. A great deal of confusion has been made by neglecting the distinction between these two different classes of works, in the various accounts of early astronomical books. For instance, it is said sometimes that the almanacs of Regiomontanus were so valuable that they sold for ten crowns of gold, or a great deal more than their own weight. But this statement is intended to apply to the *Ephemerides*, and any one, who looks at the immense quantity* of printing in the latter, will see that, for the fifteenth century, this was no very remarkable price for 832 closely-printed pages of tables, with manuscript additions, requiring the utmost care, inserted in red ink.

Regiomontanus set up a printing-press at Nuremberg, at the expense of his friend Bernard Walther, who combined the characters of a merchant and an astronomer. The earliest book known to have been printed at Nuremberg has the date 1470, but it is not from this press. The calendar or almanac which we are going to describe was printed by Regiomontanus, and it is the earliest *printed* almanac which has ever been produced,† or even mentioned. In fact, the earliest astronomical book which has been found at all is the poem of Manilius, '*Ex officinâ Johannis de Regiomonte habitantis in Nuremberga*,' 4to. It has no date, but the German bibliographers make it out to be of 1472 or 1473, probably the former. The doubt about the date introduces a claim on the part of another work, The Sphere of Sacrobosco, in 4to., to which there is a date, 1472, and Verses by Andrew

* The expense of this work must have occasioned manuscripts to be made from it. Mr. Halliwell (in the 'Companion' for 1839, p. 57) mentions the following manuscript:—"The Effemerides of John of Mounte Riol, a German Prince of Astronomers." This can have been no other than Regiomontanus.

† We say this with a knowledge of the *almanac* for 1457, from the press of Gutenberg himself, found at Mayence, which was a single sheet, and seems to have been nothing more than a calendar, in the lowest sense of the word. If, however, this sheet should turn out to contain astronomical predictions of the sun, moon, and their eclipses, then it must take precedence of that which we are speaking of in the text, as an almanac, properly speaking.

Gallus, importing that though he gets his name from France, he is a citizen of Ferrara; thus indicating the place at which the book was printed.

When ordinary printing was well established, it was necessary, for mathematical works, to overcome the difficulties of tables and of diagrams. The first was managed by Regiomontanus, both in his Calendar and in his Ephemerides; the second by Ehrard Ratdolt, in his *Editio Princeps* of Euclid, Venice, 1482. But in printing almanacs, it was also absolutely necessary to use both black and red ink,* so as to imitate the manuscripts. Our forefathers were a jealous race, and would not have kept a saint's day if it had been announced in black. In the *Ephemerides* Regiomontanus does not attempt to overcome the difficulty, and everything is black except the signs of the zodiac, which are inserted by hand. And in the *Almanac*, though he contrives it in the pages of the Calendar, where there is not much red ink, he does not attempt it in the page of new and full moons presently described. Here, accordingly, the parts which are in red are written with a pen. Both begin with the year 1475: and as it is not to be supposed that accounts of past years would be published, it is concluded that 1474 is the proper date of both: Panzer and others say 1473 or 1474. The Almanac was so soon in vogue that it was reprinted at Venice in 1476 by Ehrard Ratdolt, and again in 1482 with some additions. German editions are said to have appeared early, De Murr (*Notit. trium cod. autograph. Joh. Regiomont*, Nuremberg, 1801, 4to.) may seem to speak of one in his possession to which he gives as early a date as 1473-4. But unfortunately he is so vague that we hardly know whether by *Calendarium Germanicum* he means calendar calculated for the latitude of Germany, or calendar written in the German language. And here we may give a caution which may save some young bibliographer from being puzzled now and then.

Before regular title-pages commenced (and there are few of the fifteenth century) the contents, date, and place of a book are either not described at all, or are to be gathered from a short announcement at the end, or a few words at the beginning, of the matter of the book itself. As in the instance above, in which we can only collect the book to have been printed at Ferrara by giving Andrew Gallus credit for sense enough to see that his announcement would mislead the world if it were not so. Inferential descriptions of books ought always to be mentioned as such; but bibliographers frequently construct title-pages out of the announcements just alluded to. With very great respect and liking for their labours, we cannot but know that they want

* There is remaining a constitution of Charlemagne, confining the saints' days to those which were then usually *marked in red*, and which were Christmas, St. Stephen, St. John the Evangelist, Innocents, the Octave of our Lord (Circumcision), the Epiphany and its Octave, Purification, eight days at Easter, the Great Litanies (rogation-days), Ascension, Pentecost, St. John the Baptist, St. Peter, St. Paul, St. Martin, St. Andrew; and the Assumption is left for further directions. But in the Gregorian calendar there are nearly fifty of these days.

a little watching from those* who are concerned with the *insides* of books. We find, for example, in Lalande, the following announcement from the catalogue of the Ratisbon Library:—"1476. *Venet. in fol.* Joann. ITALI Aureus liber seu Gemma, kalendaria solis, lunæ omniumque temporum notitiam demonstrans." Now what are we to infer from this? Was there a John of Italy, a lost astronomer, who published almanacs directly after Regiomontanus? Or are we to suspect the public-spirited printer Ratdolt of an attempt to turn Regiomontanus into an Italian in the earliest work which issued from his press, that he might propitiate the national vanity of his adopted at the expense of his real country? We must have left this point unsettled if we had not happened to find in the Preface to the '*Planisphærium Stellatum*'† of Bartsch (Kepler's son-in-law), that he (Bartsch) had seen at Nuremberg a reprint of the Calendar of Regiomontanus made at Venice by Ratdolt and two others, with an announcement in verse, which he gives at length. We have put in Italics the words out of which the catalogue which misled Lalande formed its title-page.

Aureus hic *liber* est: non est pretiosior ulla
Gemma Calendario: quod docet istud opus.
Aureus hic numerus: *lunæ; solisq;* labores
Monstrantur facilè, cunctaq; signa poli
Quotq; sub hoc libro terræ per longê sequuntur
Tempora quisque dies, mensis et annus erit,
Scitur in instanti quæcunq; sit hora diei.
Hunc emat Astrologus qui velit esse citò.
Hoc *Johannes* opus Regio de Monte probatum
Composuit: tota notus in *Italia*
Quod *Veneta* impressum fuit in tellure per illos
Inferius quorum nomina picta loca.

1476.

Bernardus Pictor de Augusta,
Petrus Leslein de Langencen,
Erhardus Ratdolt de Augusta.

We shall now proceed to the contents of this almanac. It has 64 pages, and would be called quarto, but the printing is conducted in a manner different from that of modern times. It is printed in the manner of a folio, one double leaf, or four pages

* And it is just as true that laborious students have frequently led their readers into sad mistakes from not knowing how to describe books properly. Any person who mentions a book, descriptively, ought to give as much of the title-page as will be sure to distinguish the book, and then its date, place, and form, with the name of the author (correctly spelt, if convenient). It may be of little use at present, in the case of a modern work, but the time may come when attention to such minutiæ will preserve the work which shows it from oblivion.

The printers have been for a long time falling into the habit of constructing inferential title-pages by marks of quotation, which authors have allowed them to insert until the practice has become almost a nuisance. We see 'Bible' continually and 'Scriptures,' though if there must be quotation, it should be 'Holy Bible' and 'Sacred Scriptures.' Ordinary abbreviations ought not to be cited as actual titles. It may be permitted to speak of Gibbon's Roman history, but not of his 'Roman History,' for there is no such work.

† This work is also described by Panzer, from whose collection much might be added to scientific bibliography; that is, something to its extent, and a great deal to its accuracy, in all that relates to the early printing.

at a time, and several double leaves are then put together each inside the next, as in a quire of letter-paper, four in a lot, in one case five. Thus in the first set (we cannot call it sheet) pages 1, 2, 15, 16, are printed together, then 3, 4, 13, 14, &c. The part of a common page covered by the printer is 4¼ by 5¾ inches. There is neither paging, signature, title, date, nor place: at the end of page 62 (63 and 64 being diagrams) are the words DVCTV IOANNIS DE MONTEREGIO, which is the only clue to the authorship. Some circumstances connected with the copy we are examining tell us that nothing could have preceded what we have called page 1: and Brunet describes the work from his copy as consisting of 32 folios (64 pages). The type is that woodcut-looking Roman letter preserving an approach to the characteristics of the black letter, by which the practised eye recognises the fifteenth century, or the very beginning of the sixteenth. The numerals are those facsimiles of the numerals used in manuscripts which were totally abandoned before the end of the fifteenth century, except perhaps in reprints. The following woodcut gives a rough imitation of their form, and further examples may be seen in Leslie's 'Philosophy of Arithmetic.'

1234567890

There are but few contractions. By those which there are we detected a little circumstance, the mention of which may possibly save some one from charging an opponent with misquotation. In the copy in the British Museum (King's Library, *Mullerus J. Kalendarium Latinum*) we found, before the table of moveable feasts, a paragraph pasted on an empty space, in addition to the rest of the page; which, thinking it might not be in the other copy to which we had access, we transcribed with attention to the contractions and the endings of the lines. On making the comparison, we found the same addition pasted in the other copy, but printed in a different manner, and with another set of contractions and different endings. Probably a large portion of the impression had been forwarded from Nuremberg to some other place before the omission was discovered, and it was judged more convenient to send instructions for supplying it than to forward the correcting slips themselves.

Page 1 being vacant, 2 and 3 are devoted to January, 4 and 5 to February, and so on; 24 and 25 to December. Each month occupies an opening of the book. Into this double page is to be condensed (with a little assistance from one or two slight tables) the almanac of the month for three complete metonic cycles, or 57 years. On the left hand page is the succession of days of the month, and of dominical letters, with the days of the Roman calendar. This last is the one actually used by Regiomontanus, for he makes his astronomical corrections for leap-year depend on an additional day being inserted in the Roman manner, by doubling the sixth day before that of the Kalends of March inclusive; in conformity with the ecclesiastical usage of his time. Then follows the description of the saints' days in a broad column, which notes also the *claves*, or *key-days*, a peculiarity of the ancient almanac, which we shall explain.

Opposite to January 7 we see *Clavis* lxx., which must be translated *the key* of *Septuagesima Sunday*, which is the ninth Sunday before Easter-day. If, instead of counting nine weeks, we choose to select any day in March, and call it *clavis paschæ*, or the key of Easter, then if we call the same day nine weeks before it the key of Septuagesima, the two keys will be thus related. If Easter-day, for example, should fall ten days after its key, Septuagesima Sunday would also fall ten days after *its* key. The key number for the year depended simply upon the golden number: thus when the golden number was 3 the key number was 34, or Easter, &c. all fell at 34 days' distance from their several keys. The key-day chosen for Easter was the 11th of March, the day of the equinox in old style, as it then was. A more awkward contrivance can hardly be imagined, since it would have saved all counting to have written down the golden numbers in their proper places for every feast or fast depending on Easter; and then the Sunday would have been found by looking out the golden number and looking forward to the next dominical letter.

After the column of saints' days comes a column for the place of the sun, containing certain degrees and minutes for each day. At the end of the almanac is a table of the sun, containing a number for the beginning of each year, which, added to the number for any one day in the monthly table, gives the sun's longitude at noon of that day. Next follow two columns for the moon's longitude; one for the mean, the other for the true. At the end of the almanac is a little instrument, consisting of a circle graduated, and printed on the page, with two other graduated circles attached to the page by a thread through the centre of all three, and therefore capable of revolving: there is also a string which revolves about the same centre as an index. This machine is set for any given year by the help of two small tables, in a manner which we cannot undertake to describe in any reasonable space: and when set, a use of the two numbers of degrees and minutes, written in the monthly page for any day, gives the degrees of the mean and true place of the moon for that day of the month in the year for which the machine was set.

The left-hand page of each month is devoted to the new and full moons which take place in that month, for all the 57 years of the three Metonic cycles; that is, from 1475 to 1531, both inclusive. Many of our readers are not aware that the distribution of moonlight throughout the year is not altogether the capricious phenomenon which it seems to be. All persons know that the almanac of any one year is no guide to the next year as to the moon, though it may do well enough as to the sun. But though one year is far from a repetition of the preceding, any eight years is nearly a repetition of the preceding eight years, and any nineteen years is much more nearly a repetition of the preceding nineteen. If it be full moon, for instance, on January 23, 1845, we may safely assert that in 1853 the full moon of January will fall on the 24th or 25th, one or two days in advance. Persons in the country must sometimes be puzzled to settle a plan with reference to the moonlight two or three months before the time, at that unlucky period * of the year when next year's almanac is wanted, but has not appeared. How many persons of those who keep their old almanacs are aware that, in the matter of moonlight, this year is always eight years ago with an advance of either one or two days? For example, in 1837, the days of full moon are January 21, February 20, March 22, April 20, May 20, June 18, July 17, August 16, September 14, October 13, November 12, December 12. In 1845, the days of full moon are January 23, February 22, March 23, April 22, May 21, June 19, July 19, August 17, September 15, October 15, November 14, December 13. But in nineteen years the accordance is much more closely established, to the day generally, and if not, to the day before or the day after. Thus the days of full moon † in January and February 1856, nineteen years from 1837, are January 21 and February 20, as in 1837. Accordingly, eight years in the first instance, and afterwards nineteen years, have been celebrated astronomical periods, and in the almanac before us an arrangement in cycles of nineteen years is made of the new and full moons, of which the following is a specimen. The left-hand page of the month of March begins thus: the figures which are in Roman numerals being those which are *written* in red ink in the original work, and which are the Golden Numbers:—

	1475.		1494.		1513.	
	CON.	OPPO.	CON.	OPPO.	CON.	OPPO.
1	viii 23 46	iv 2 25	viii 8 58			iv 8 30
2		xii 21 15		xii 5 12	viii 1 11	xii 18 44
3			xvi 18 31		xvi 2 7	i 23 9
4	xvi 6 48			i 15 19	v 20 58	
5	v 18 17	i 1 1	v 11 18			ix 18 14
6		ix 10 48		ix 6 25		
7	xiii 1 45	xvii 17 41	xiii 2 41	xvii 18 15	xiii 0 33	xvii 17 50
8	ii 10 13		ii 9 40		ii 10 17	
9	x 21 46	vi 3 31	x 17 9	vi 1 13	x 17 41	vi 2 10
10		xiv 7 24		xiv 16 18		xiv 9 42
11	&c.		&c.		&c.	

This is, for the month of March, the table for the three periods of nineteen years severally beginning with 1475, 1494, and 1513, CON. and OPPO. denoting conjunction and opposition, or new moon and full moon. For example: to find the full moon of March 1479; the Golden Number for this year is 17, and looking down

* Every almanac ought to give the changes of the moon for at least the three first months of the year following that for which it is made. The British Almanac for this year, and from henceforward, will contain an abbreviated almanac for the year following the current year: thus 1846 will contain an epitome of 1847, &c.

† In a work on the globes, recently published, by the author of this article, will be found tables by the use of which any person who can look into a table and do a few simple additions, can find for himself, within about a quarter of an hour, and always within twenty-five minutes, the time of new or full moon for any month of any year from B.C. 1000 to A.D. 2000.

the first column of oppositions till we come to xvii, we find it opposite the 7th of the month, with 17 41 after it. This means that the full moon is 17h. 41m. after noon of the 7th, or at 41 minutes past five on the morning of the eighth. We leave to the reader to detect, in the preceding sample, specimens of the properties of the cycles of eight and nineteen years.

After the monthly pages comes a table of the longitudes and latitudes of about sixty of the principal places in Europe, the longitudes in hours and minutes, the latitudes in degrees only. In these tables, when given in almanacs, it was always customary to reckon longitudes from the meridian of the place for which the almanac was made; accordingly, the longitude of Nuremberg is marked 0h. 0m. A degree of error both in latitude and longitude is not uncommon in this table. Then follow descriptions of the eclipses supposed to be visible at Nuremberg, giving the time of the middle of the eclipse, its semi-duration, and the number of *points,* or, as we should now say, digits, eclipsed: there is also a diagram of the magnitude of the greatest phase of the eclipse. After the eclipses comes an account of the golden number and other matters connected with the moveable feasts, and a table of these feasts for all the combinations of the golden number with the dominical letter. Before the reformation of the Calendar, if, for instance, the golden number were 5, and the dominical letter E, Easter Sunday could be nothing but March 23, and the other feasts accordingly. A table like that in page 32 of our last article, with all the principal feasts, &c. annexed for each case, is given. Then follows description of the mode of using the almanac, with the supplementary tables already noticed. And then comes one page of astrology, namely, the influences of the signs of the zodiac on the human body. For example, "Capricornus is cold, dry, and earthy; the knees belong to it; it is of no use in letting blood." Regiomontanus proves himself to have been an astrologer, and not merely deferring to public opinion, by announcing his intention to prepare a larger work on the subject.

Tables of the length of the day are then given, or rather of the half day, in the form of double entry, for every three degrees of the sun's place in the zodiac, and for every degree of latitude from 36° to 55°. There is then a description of the mode of constructing the horizontal sun-dial and other instruments which are now totally superseded: Delambre (*Hist. Astron. Moyenne*) has given an account of them. At the end come the remarks on the Calendar to which we have adverted.

The great celebrity of Regiomontanus brought his almanac into universal use, and caused many reprints of it to be made. It is natural that those who afterwards made almanacs should appear to have followed Regiomontanus very closely; since the latter himself in a great measure followed the plan of the manuscript almanacs of the day. Nevertheless, as early as 1478, another almanac was published, which seems to be known to the astronomical world, only by one simple announcement in

Lalande's Bibliography, as follows: "1478, *Ulmæ in fol.* Calendarium per Jo. ZAINER." This J. Zainer was the first introducer of the art of printing into Ulm; and Panzer (who mentions several of his productions, and this one among the rest) states that several of his predecessors consider that his types were cut and not cast, in which opinion he (Panzer) does not agree. We cannot pretend to decide this point; we only know that the printing is a black letter of the very roughest character, and that the numerals, though they are more like the modern ones than those of Regiomontanus, are as uncouthly intelligible as possible. It is called folio, but the mode of printing is of the sort already described: there are 36 pages, the black part of the common page being 7½ by 4½ inches. There is no title-page, title, paging, or signature: it begins *Usum hujus opusculi breviter exponemus.* At the end of the introduction are the words *Impressum Ulme per Johannem Zainer Anno dominice incarnationis,* 1478. With respect to this introduction there is something to say.

In the library of the Royal Society, in that of the Museum, and in various catalogues, are copies or notices of a supposed work of Regiomontanus which is not an almanac, but a description of an almanac, and it always begins with *Usum hujus opusculi breviter exponemus.* In the Royal Society's catalogue it is called 'Canon in Ephemerides, a fragment of an unknown work by Regiomontanus.' Lalande (page 27) cites a work of Regiomontanus not mentioned, he says, by Panzer, and having *Usum Ephemeridis cujuslibet breviter exponemus.* This work of Zainer, we see, begins in the same way. Now, on comparing this *Usum hujus*, as we may call it, in Zainer, with the detached fragment, we find they are substantially the same, with such alterations as the plan of the former requires. And we find, moreover, that though there is no formal beginning to the explanation in the calendar of Regiomontanus which we have been describing, yet that the contents of Zainer's introduction are for the most part to be found word for word in the explanations of this Calendar. We have not examined any manuscript calendars for this particular purpose: but we strongly suspect that it had been for some time customary to word the explanations in the different calendars after one ancient model, which both Regiomontanus and Zainer followed. At any rate, no one must hastily assume that two calendars are from the same hands, because the explanatory parts are in the same words. We think it not very improbable that this introduction was separately printed before table printing was accomplished, that it might have manuscript calendars fitted to it.

Zainer's calendar was not reprinted, that we can find: nor is this extraordinary; for though there is a slight table of latitudes (without longitudes) at the commencement, everything is confined to the latitude and longitude of Ulm. There is one page to each month containing the modern and Roman calendars, the saints' days (but without any red printing), one column

c

for the sun and two for the moon, as in Regiomontanus: also a column for the moon's ascending node, and the ascensions (in the old sense) of the various degrees of the zodiac for the latitude of Ulm. There is a radical table for the sun's place, and for that of the moon's node, to be used with the columns in the almanac. There is one radical table for the moon's place, and a circular zodiac resembling that of Regiomontanus, with the two moveable circles removed. The moon's age for the given day, as deduced from the table of new moons, supplies the place of those moveable circles. There are four complete Metonic cycles of new and full moons, from 1477 to 1552, arranged by themselves, a table of moveable feasts, and a table for the length of the day, in the latitude of Ulm only. There is more of astronomical pretension about the eclipses than in Regiomontanus: for not only is the middle of the eclipse given to seconds, and the magnitude to sixtieth parts of points or digits, but in the diagrams of the greatest phase of the eclipse the direction as well as the magnitude of the phase is shown.

It is extraordinary, but certainly true, that Riccioli, the most learned astronomical writer of the seventeenth century (whose *Almagestum Novum* was published in 1651), knew nothing either of the Ephemerides or Calendar of Regiomontanus: of course he knew nothing of Zainer. In his Life of Regiomontanus he does not enumerate either in the list of works: and in his list of ephemerides he gives no writer previous to 1506 by name; saying only that he has an ephemeris from 1484 to 1506, of which the author and place are both unknown. The account of Regiomontanus given by Melchior Adam in his collection of biographies, describes the ephemerides, but does not mention the calendar: this account was obviously unknown to Riccioli. The most extensive biography we have was published three years after Riccioli's work. Of the Ephemerides, Gassendi, its author, says that he never saw more than one mutilated copy, except in an Augsburgh reprint of the last fifteen years of the series. He had the Calendar which we have described before him, but he is unfortunate in his account of it. Seeing 1475, 1494, 1513, at the heads of the columns, he concludes that it is made for these years only: which shows that he must have looked rather inattentively for a biographer. But Delambre has made precisely the same mistake (*Astron. Moyenne*, p. 323) with the book before him: he also confounds the *Ephemerides* with the *Calendar*, never having seen the former, though he mentions the ephemerides separately afterwards. This confusion could not be avoided; for Gassendi, Weidler, and all the predecessors of Delambre do not distinguish between the two works in a very clear manner. Lalande himself (p. 11) shows that he was not aware of the distinct character of the two. The history of science in the fifteenth and sixteenth centuries abounds with misunderstandings of the same kind, from the difficulty of procuring exact accounts of books. The early writers are exceedingly careless about their descriptions of books, for which they have a very good excuse, namely, that their readers were presumed to be, for the most part, tolerably familiar with the works cited. But in our day the historian must describe books which not one in a hundred of his readers will ever see.

Previously to our examination of Stöffler's almanac, we shall say something about himself, and about the curious story to which we have alluded, namely, that he announced that the world would be almost destroyed by a deluge in 1524. That such an occurrence was announced, that it created the greatest fright, that those who were near the sea or a river left their homes, and so on, is beyond all doubt: as also that learned treatises were written against the prophecy by Augustine Niphus, Paul Middleburg, Bishop of Fossombrone (Paulus Forosemproniensis), and others. Until we came to write this article, we had always supposed, on more than a dozen good authorities, that Stöffler did announce such a deluge; and this, although we could not find any allusion to it in his almanacs appended to the *Calendarium Romanum*, among which is one for the year 1524. We saw in his treatise on the Calendar great learning, and (astrology excepted) no small share of sense, and the same in his commentary on Proclus. But the voice of history spoke very plainly against him, and it added an account of his death, which seemed to point him out still more decidedly as an unlucky prophet. The stars told him that on a certain day he would be killed by something falling on his head; accordingly he determined not to quit his house on that day, and invited friends to dine with him. But an argument arose at dinner which obliged him to go for a book to settle it; the book-shelf was badly secured, and tumbled with all its contents upon him, to the entire fulfilment of his destiny. This story, very fairly attested,* joined to the known fact of his being an astrologer, seemed to give tolerable corroboration to the great mass of authority which pointed him out as

* We may give a little idea of the confusion into which biographical history may be thrown by the successive misstatements of writers, each taking his predecessor's errors with one or two more of his own. Sherburne, a very respectable compiler of short biographies, informs us that the story in the text about Stöffler's death is false; for that Joachim Rudolph Camerarius states, in his *second* century (or set of a hundred) of verifications of astrology, that he (Camerarius) had it from Andrew Ruttell, a pupil of Stöffler's, that the latter died of the plague, February 16, 1531; the time of his death being explained by saying that the sun and Venus were approaching to a quadrature. We do not know of any *second* century of such things published by Joachim Rudolph Camerarius (who, by the way, must not be confounded with either Joachim the father, of that name, or Joachim the son, or Rudolph the botanist—all men of greater note; and a pretty batch of mistakes might occur from confounding the four); but we have the *first*, and, as far as we know, the only one. And we find that the common story about Stöffler's death is not only given, but proved from his horoscope; for Jupiter was in conjunction with the sun at his birth, in an airy sign, which is a sufficient reason for anything tumbling upon him from above. And as to Mr. Ruttell, we find *his* horoscope too, and also that his death took place fourteen years before Stöffler's, in 1517, by an arrow in the lower part of the abdomen, which he might very well have known would have happened, for the position of Saturn at his birth distinctly announced the weapon which would kill him, and that of Libra the part it would wound.

Now the simple truth is, that Melchior Adam, whose collections of German biographies, made at the beginning of the seventeenth century, are always cited as high authority, says that Stöffler died of the plague on the day above mentioned. And the astrological story is the *ben trovato* of Seth Calvisius, who gives it as hearsay in his '*Opus Chronologicum*.'

C 2

the predicter of the deluge. But on examination we found reason to suppose that the prophecy could not have come from him. Bayle, beyond whom we cannot trace the attribution of the prophecy to Stöffler, gives everything but just the account of the place in which the prophecy is to be found; all the biographers whom we have seen, and who are anterior to him, as Melch. Adam, Riccioli, &c., do not mention it at all. The writers against this deluge do not mention Stöffler in their title-pages, but only a certain prophecy: though it was the custom of that day always to mention in the title the name of any one against whom the book was written. Nor do the numerous quotations which Bayle has made from them refer to Stöffler in the smallest degree. The secondary accounts give the most contradictory collateral circumstances: some say the prophet acknowledged at last the vanity of his science, others that he persisted all the more after his failure; some say that the character of his ephemerides was destroyed, others that they were more in vogue than ever: some attack his edition of Proclus, which they had never seen, on presumption; the author of such a prophecy *ought* to have written in a silly manner. The author of the article 'Stöffler' in the Biographie Universelle, states that he had before him a copy of the ephemerides, printed in 1522, and that at the head of the one for 1524 the prophecy is given; he does not say that he had seen it there, but makes the assertion in one part of his article, and the statement about the book being before him in another. This made us a little suspicious, for we have always observed that persons who write with the original authority before them are apt to give the *very words*, and these we can find nowhere. The ephemerides, as they are called, of Stöffler, were published at various times, and in various editions, the earliest being in 1499; the title is *Almanach nova plurimis annis venturis inservientia*. The only edition within our reach is that in the library of the Royal Society, Venice, 1513, and in this we could not find any prophecy: there is not even a separate ephemeris for the different years, but the book consists of rules and tables for constructing astrological schemes. These may serve for an ephemeris, but do not constitute a set of them; for the results of the several years are not collected together. It remains then for any one who repeats the charge against Stöffler to point out where the prophecy is to be found. The truth seems to be as follows: from his ephemeris, which the astrologers would naturally prefer to all others, both as the work of a decided believer, and as intentionally adapted for their pursuit, the users of it would find out that in February 1524, three planets would be in conjunction in Pisces, a watery sign. This, as one may easily see, would be enough to drown a globe fifty times as big as ours, if there be anything in astrology. This result would be noised abroad as having been derived from *Stöffler's ephemeris;* and thus he would get the credit of the prophecy. If such a conjunction were to happen in our day, our modern Dariots would appeal to the Nautical Almanac for the fact, and the Superintendent, or perhaps even the Lords of the Admiralty, would get credit for the astrological consequence. We all know that the astronomers who gave the orbits of the comets for 1832, were cited as the authorities for one of them being about to strike the earth; and similarly, John Booker's *Bloody Almanacs* of the time of Charles I., the thunder of which is drawn from the interpretation of the Revelations by Napier, have been sometimes attributed to the inventor of logarithms.

John Stöffler (born 1452, died 1531) was professor of mathematics at Tübingen. His '*Calendarium Romanum*,' which contains the almanacs we intend to describe, was published at Oppenheim in 1518, folio. The title is torn out of the copy we have before us: Lalande gives it thus—'Calendarium Romanum magnum, Cæsariæ majestati dicatum.' It has 250 pages, exclusive of indexes, &c., and the large folio page, the print of which covers $8\frac{7}{8}$ by $5\frac{1}{4}$ inches, with 52 lines, contains a great quantity of matter: the brevity of the Latin being aided by the numerous contractions. There is so much evidence of exactness in the paging, the indexes, the list of errata, the division of each page into lettered paragraphs, &c., that the first impression the reader of books of that date receives of the author is very favourable. This impression may require a deduction for his astrology, and for his incapacity to use authorities properly, two great faults of his time. In the latter failing he is very conspicuous, as in the following instance:—Moses, "the Egyptian, the Jew, the lawgiver, of all men the most intimate with God," has said that the moon is the planet of the night. Though holding the authority in high reverence, Stöffler seems to think that the assertion is one of those which must not rest on any one authority, however good. He looks about for confirmation, therefore, and finds it in Hali Aben Ragel, one of Alphonso's astronomers, Ptolemy, and the Peripatetics; all of whom assert most positively that the moon is the planet of the night; so he goes on, like an overcautious general, leaving a strong garrison in every place which he takes, however unimportant.

The introduction to the almanacs occupies 148 pages, most of which is spent on discussions about the history of the calendar and its reformation. As to the latter, Stöffler would have a leap-year omitted once in every 134 years, and a table drawn up without any attempt at a cycle, which should give Easter for a thousand years to come. He sees that the mode of fixing the equinox would destroy the ordinary use of the cycle of nineteen years, and the method afterwards proposed did not suggest itself to him. On the history of the calendar he has written with great clearness and length, both as to the times preceding and following the Nicene Council, and even now we find his account very useful. To the introduction is appended a long list of places with their longitudes to hours and minutes, and their latitudes to degrees.

These almanacs begin at 1518, the year of publication, and extend through three periods of nineteen years to 1574 inclusive though the parts which are subsidiary to the monthly pages are extended several years further. The difficulty of printing in black and red ink is completely conquered. The two pages for each month resemble those of Regiomontanus, except that there are in the right-hand page three columns for the moon's place instead of two. The additional column is used instead of that which depends upon the age of the moon in Zainer's almanac, whose subsidiary diagram, or lunar zodiac, Stöffler takes, instead of that of Regiomontanus. Then follow eclipses, tables of moveable feasts, of risings and settings, and radical tables, as in Regiomontanus, but more extensive. It seems however that then, as now, arithmetical calculation was rather shunned even by mathematicians. "Complures sunt," says Stöffler, "nostri temporis homines, qui sese mathematicos vocari volunt, primam tamen mathematicæ disciplinam, Arithmeticam, utpote antiquissimam, cæterarum principem, radicem, matrem, atque nutricis vicem gerentem, fugiunt, abhorrent, et detestantur!" For such persons he has computed what he calls his great table of the moon, which gives her degree of the zodiac for every day of every year from 1518 to 1579, both inclusive. The diagrams of instruments close the book, as in Regiomontanus.

In the monthly pages, at the side of the almanac, are drawings of the twelve signs of the zodiac, one to each month. There is only one of these which deserves a remark; but it is a curious one. Luther had begun (1517) to raise his opposition to the indulgences the year before these almanacs were published, and it was while they were being printed (which we collect from themselves must have been from March to August, 1518) that the Pope issued his first summons to Luther, and the Elector of Saxony made his first interference in favour of the latter. In drawings of the zodiac, the Twins are usually represented in some brotherly relation to one another; but Stöffler has given a significant hint about the state of the times in his picture of them. One of the Twins has his hand upon the throat of his brother, and has forced him to the ground; the suffering party holds in his hand a large key, the emblem of the Church. And yet Stöffler was so good a Catholic that he refuses to admit places in Greece into his list of latitudes and longitudes, that he may not (such is his expressed reason) be suspected of hostility to the orthodox faith.

At the bottom of the almanac-pages are pictures descriptive of the occupations of the agriculturist in the several months of the year, with a couplet to each, as follows:—

January.	Carnes torreo Janus en trementes Et lætus comedo biboque ad ignem.
February.	Incedo glaciem Februs securi Nec non mitia culta stercorizo.
March.	Et sum Martius qui puto gementes Vites, nec minus arbores comosas.
April.	Aprilis patulæ nucis sub umbra Post convivia dormio libenter.
May.	Maius nunc equito per arva lætus Atque hac glorior aucupatione.*

* The reader must judge for himself whether this is a misspelling for *occupatione*, or an allusion to fowling.

June. Aestivo meto Junius calore
Has lætas segetes diu cupitas.
July. Granum Julius arridis flagello
Et spicis quatio coquente sole.
August. Augustus vegetos cados coarcto
Quam possum bene circulis papyro.
September. Nigris impleo dolium racemis,
September bene vina concoquantur.
October. October bove semiuo juvante
Ut tellus ferat omnibus legumen.
November. Pingues ditibus anseres November
Vendo et ligna seco favente luna.
December. Lætus vivere nunc volo December
Occido quoniam suem triumphans.

The following is the sense of these couplets :—

January. I roast the quivering flesh, and eat and drink joyfully at the fire.—*February.* I go over the icy ground with my axe [to cut timber] and manure the fields.—*March.* I am he who prunes the vines and the trees [groaning vines and leafy trees, which are hardly March epithets].—*April.* I sleep at ease after the feast, under the shade of the spreading nut.—*May.* I ride gladly over the fields, and take pride in the occupation.—*June.* In the summer's heat I reap the glad harvest so long expected.—*July.* I beat the grain with the flail, and turn it with the pitchfork while the sun dries it.—*August.* I do my best to make the casks sound with hoops of paper [to press the grapes in].—*September.* I fill the cask with black grapes that the wine may be well made.—*October.* I sow, with the help of the oxen, that the earth may bear fruit for all.—*November.* I sell the fat geese to the rich, and cut down wood at the proper time of the moon.—*December.* Now I will live a jolly life, for I kill the sow in triumph.

June is represented as the harvest month, but from what is said of July it appears that the corn was cut down before it was ripe, and then dried in the sun. This could have been done on no other principle than that half a loaf is better than no bread, to put the crop as soon as possible in a state to be conveyed away on the approach of an enemy. It is also to be remembered that, at the time when these descriptions were written, the equinox was ten and eleven nominal days earlier than we make it to be; so that their 1st of June was as far advanced in summer as our 11th. With all this, however, we cannot suppose that anything but rye is spoken of: and in like manner our English notion (poetical notion) of the harvest being in autumn, in September, may be a relic of the time when barley was much more nearly the staple of food than it is now.

As to the astronomical source of these almanacs, they are all three taken from the Alphonsine tables which appeared in 1252. There was then no other well-established modern guide. The results of these tables differ very little indeed from those which might have been deduced from Ptolemy. There is little to choose between any of the astronomical predictions which followed those of Regiomontanus, until we come past the time of Tycho Brahé, who died in 1601.

A. De Morgan.

University College, August 5, 1845.

III.—RECURRENCE OF ECLIPSES AND FULL MOONS.

We request the attention of the reader to an astronomical matter which, so far as the 'British Almanac' is concerned, has a particular relation to the years 1846 and 1847.

This work made its first appearance as an almanac for 1828. At the beginning of 1846, eighteen years had elapsed since its commencement, and nineteen years at the beginning of 1847. Those who have kept up the set from the commencement may now begin to verify by their almanacs the astronomical recurrence of eclipses and new and full moons: on the periods of which we add a few words.

[Continued next page]

Look at the eclipses for 1828 and 1846, and also for 1829 and 1847. They stand as follows:—

	1828.			1846.
Sun eclipsed	April 14	. .	Sun eclipsed	April 25
" "	Oct. 9	. .	" "	Oct. 20
	1829.			1847.
Moon eclipsed	Mar. 20	. .	Moon eclipsed	Mar. 31
Sun "	April 3	. .	Sun "	April 15
Moon "	Sept. 13	. .	Moon "	Sept. 24
Sun "	Sept. 28	. .	Sun "	Oct. 9

So it would seem as if whatever eclipse there may be, there is the like in eighteen years and eleven* days. But, first, remember that, in the periods above given there are only four leap years; in other periods of equal length there are five. Thus if there had been an eclipse in January 1828, there would have been five leap years (or five intercalary days) between it and the similar eclipse of 1846. In this case the period would have seemed to be of eighteen years and *ten* days. The truth is, that in a great majority of cases (not all) the eclipses do recur after periods of eighteen years, ten or eleven days, and one-third of a day, on the average. The short popular explanation of this is as follows.

Take an eclipse of the sun, which happens whenever the moon comes into the same part of the heavens as the sun. This can only be when there is a new moon at the *moon's node*,† that is, when the moon is in the plane of the ecliptic. Suppose such a thing to happen very accurately, that is, let the sun's and moon's centres be both in a line drawn through the centre of the earth: so that the new moon takes place when the moon is exactly at her node. There would then be a total eclipse of the longest duration. Now suppose the motions of the bodies to go on, the several eclipses of the sun taking place whenever new moon happens so near to the moon's node that any part of the moon's body hides any part of the sun. A long series of eclipses of various degrees would, it might be supposed, occur, until in process of time the first eclipse was again brought about: that is to say, until the new moon took place, when the moon was again exactly at her node. Such would be the case, but the period would be of enormous length. There is however a short period of 223 moons, or 18 years, ten or eleven days (according to the leap years), and one third, nearly, in which the re-establishment of the circumstances is very nearly, but not quite, effected. If the eclipse be perfectly central at the beginning of such a period, there will be an eclipse, but not of so decided a character, at the beginning of the next period. Let one period more elapse, and there will still be an eclipse, but less in amount than even the last, and so on, until the eclipse which marks the beginning of the period becomes a mere contact, after which the commencements of the subsequent periods

* From April 3 to April 15 is twelve days; but it is to be remembered that from late on the third to early on the fifteenth may be little more than eleven days.

† See the Article on the Moon's Orbit in the Companion for 1834.

will show no eclipse at all. But while corresponding eclipses of successive periods are thus becoming of smaller and smaller amount, and, as it were, gradually going out, others are coming in, and gradually increasing in effect, as follows. Suppose an orrery, with a revolving earth and moon, properly placed, made to revolve with considerable rapidity. A person whose eye was accurate enough to enable him to state each eclipse as it happens would see in the successive periods a great many moments at which there were narrow misses of eclipses, the moon coming very nearly between the earth and sun. If he watched any one miss of an eclipse, as we may call it, and waited for the corresponding circumstances of the next period, he might, and often would, see a narrower miss than the former one. At last there would be no miss at all, or a slight eclipse would take place, which would grow greater and greater in amount in still further periods. Between eclipses which go out, and those which come in, the number is kept up; being about 70 in all, 29 of the moon, and 41 of the sun, in each period. The lunar eclipses may be described in the same way, with the full moon instead of the new.

Accordingly, it is not to be expected that every period will altogether agree with the next in its eclipses, still less that it will agree with distant periods. The reader who has the complete set of the almanacs may ascertain, as they arrive, the eclipses which go out and those which come in.

This period is mentioned by Ptolemy, Pliny, and others: and Geminus attributes the knowledge of it to the Chaldeans. Some have supposed it to be the same as a period known to have been employed by the Chaldeans, which was called the *Saros:* this is doubtful; but the period of eighteen years has been often called the Saros in modern times.

The period of nineteen years, which was introduced by Meton among the Greeks, and which, under the name of the Metonic cycle, and the cycle of the golden number, has been described in the articles on Easter (see the *Companion* for 1844), is a perfectly distinct thing from the preceding, though sometimes confounded with it. It is founded upon this, that 235 moons last 6939 days, 16¾ hours, very nearly. Now 19 years with five leap years is 6940 days, and with four, of course, 6939. The consequence is, that after 19 years, the new and full moons begin to fall on the same days of the months as in the preceding nineteen years, or at most one day before or behind. This will be seen by putting down a few of the new and full moons for 1828 and 1847:—

	1828.	1847.
Full Moon	Jan. 2	Jan. 1
New ,,	Jan. 17	Jan. 17
Full ,,	Feb. 1	Jan. 31
New ,,	Feb. 15	Feb. 15
Full ,,	Mar. 1	Mar. 2
New ,,	Mar. 15	Mar. 16
Full ,,	Mar. 31	Mar. 31
&c.	&c.	&c.

A. De Morgan.

COMPANION TO THE ALMANAC.

FOR

1848.

PART I.

GENERAL INFORMATION ON SUBJECTS OF MATHEMATICS, NATURAL PHILOSOPHY AND HISTORY, CHRONOLOGY, GEOGRAPHY, STATISTICS, &c.

I.—ON DECIMAL COINAGE.

SEVEN years ago (in the "Companion" for 1841) we endeavoured to call attention to the advantages of decimal coinage, and the ease with which they might be secured. The article in question had a particular reference to the introduction of logarithmic calculation into commerce, which, though one of the benefits in question, is perhaps, of all, the one which is least appreciated by persons in general. Since that time an opinion on the subject has gradually grown, and it has been, in the main, for the change. There are those with whom it is a positive predilection, arising out of a strong sense of the conveniences of the proposed system. There are others, and many more in number, who believe that the balance of advantage is with the proposal, on the whole, and who would rather assent than dissent; but more from a consciousness that the thing must be, and will be, than from any such strength of conviction in favour of the change as would lead them to face opposition for the sake of it. There are others who have not any clear idea upon the positive benefits to be derived, but who are swayed by a general impression that a decimal system of counting ought to be accompanied by a decimal system of measuring. We have met with such a number of men engaged in business (almost every one we have spoken to on the subject) who express no fear of the change, but rather a desire for it, that we are satisfied the usual difficulty, attachment to established usage, and apprehension of the consequences of disturbance, does not exist to any serious extent. The proceedings in Parliament during the last Session, on the occasion of Dr. Bowring's motion, seem to confirm us in what we have stated. There was not opposition enough to make a debate, worthy of the name; and the Chancellor of the Exchequer, in yielding the first step, the introduction of a two-shilling coin, rested his non-acquiescence in the whole extent of the motion on the want of a strong public interest in favour of the question, and the slow growth of belief adverse to existing usages. He said, as plain as a Chancellor of the Exchequer could speak, Force me; and here I am, ready to be forced. In perfect conviction that we are within a little of the change, we try to keep the subject before our readers, and to introduce it, in the most elementary form, to those who may have been frightened by the logarithms in our former paper.

When the decimal coinage is contemplated, the questions we naturally ask are—1. What is the new state of things, as compared with the old? 2. What effect will be produced on the ordinary usages of life? 3. How are those who cannot easily adapt themselves to the change, to manage to speak the old language to themselves, and the new to the rest of the world? 4. How, when the new state of things is established, are those who must do it to return to the old system as occasion arises? 5. What amount of legislation will be required? 6. What do we gain upon a fair balance of conveniences and inconveniences? We shall take these questions in order.

1. What is the new state of things, as compared with the old?

If the change be made in that simplest way in which there is a tolerably general agreement that it must be made, if at all, there will be no necessity to withdraw from circulation any one coin now in use, nor to change any name. It may perhaps be thought advisable to abolish the half-crown and five-shilling piece; but this is not essential. The great alteration will be in what is technically called the *money of account*. By this phrase is meant the various denominations of money which are set down in calculations. Thus when, during the war, the Spanish dollar was circulating at 4*s*. 9*d*. (and, by the way, no difficulty was ever made about reckoning these fractional coins), it never became money of account, never interfered with the £ *s*. *d*. of the books.

Our present style of account is in pounds, shillings, pence, and farthings; but as the farthing is rarely used, it will be convenient not to mention it as old money: in this article it shall be entirely a coin of the new account. This new account will be, say in pounds, royals, groats, farthings, 10 of each going to the one before it; so that 10 farthings are a groat, 10 groats a royal, 10 royals a pound. The pound is precisely the same thing in both systems; the royal is two of our actual shillings; the groat is 5 new halfpence, or 10 new farthings. The new halfpenny differs from the old one in that 25 go to the shilling, instead of 24. If, after the change is completed, any one should forget or refuse to distinguish the new and old halfpenny, the difference that it would make is as follows:—The receiver would lose four per cent. on the odd copper of the money which passes between them, and the payer would gain it. Thus, suppose 6*s.* 4½*d.* was due the day before the change, and is to be paid the day after it. Six shillings of the old account is three royals of the new, absolutely; 9 new halfpence (supposed to be paid instead of 9 old ones) are 9-25ths of a shilling, or half-royal, instead of 9-24ths: the difference is three quarters of a new farthing.

When such writing is seen as £45,768, it would mean 45 pounds, 7 royals, 6 groats, 8 farthings. To find the tenth part of this we should write down £4,5768, meaning 4 pounds, 5 royals, 7 groats, 6 farthings, and 8-tenths of a farthing. To multiply it by 26, or to divide by 12, we should proceed by common arithmetic, as follows:—

```
 45,768            12)45,768
     26               ------
-------                3,814
 274608
 91536
-------
1189,968
```

And thus we should get 1189 pounds, 9 royals, 6 groats, 8 farthings, and 3 pounds, 8 royals, 1 groat, 4 farthings.

2. What effect will be produced on the ordinary usages of life?

Hardly any need be produced. The royal and groat would soon pass under their names, because they would come out as new coins. But they would appear, and be thought of, as *two shillings* and as *twopence-halfpenny*. We are well used to 2½ by means of the half-crown; and the succession of groats, 2½*d.* 5*d.* 7½*d.* 10*d.* &c., would be all the easier. The great pinch, if it really be one, lies in the odd halfpenny which would go to the shilling. Imagine silver so scarce and copper so plenty, that any one who gives you change for a shilling gives a halfpenny for the accommodation. The change is always a halfpenny more than now, and a farthing on the sixpence. The price of the shilling has risen to twelvepence-halfpenny. We are, in truth, well accustomed to deal, and suddenly, with much more difficult changes in the price of articles; but an alteration in the money startles us more than any other.

In all probability, the reckoning by groats would go as far as the (two shilling) royal, just as we now often reckon by pence as far as eighteenpence, at least.

3. How are those who cannot adapt themselves easily to the change, to manage to speak the old language to themselves, and the new to the rest of the world?

This, when the pressure comes, will be more easily done than any one supposes who is not much of an arithmetician, and looks at an article of this kind from curiosity, without any particular desire to master difficulties. Suppose, for example, he has 4*l.* 13*s.* 10¼*d.*, got out in the old way. There can be no difficulty about the shillings: six royals and five groats give the 13 shillings. Accordingly, £4,65 represents 4*l.* 13*s.* A royal for each pair of shillings; 5 groats for the odd one, if any. Turn the rest into farthings (41), and here are the remaining groats and farthings (new), provided only that one be added (making it 42) when the overplus above shillings is sixpence or more. Thus, 4*l.* 13*s.* 10¼*d.* is £4,692, or 4 pounds, 6 royals, 9 groats, 2 farthings.

This rule gives a little too small a result in new money, except when the sum is an exact number of sixpences. But the error never amounts to a farthing. Properly, 4*l.* 13*s.* 10¼*d.* is £4,692708333, being above the preceding account of it by 7-tenths of a farthing, 0 hundredths, 8 thousandths, 3 ten thousandths, &c. But this will be of no consequence.

The three places of figures that follow the pounds in the new system may then be explained thus:—

First.	Second.	Third.
This column takes 0 for any thing under two shillings, and 1 for every two shillings.	This column takes 5 for the *odd shilling*, if there be one. Also 1 for every ten farthings in the pence and farthings of the old reckoning.	This column takes the units left of the farthings in the pence and farthings, with an additional unit if they amount to sixpence.
Royals.	*Groats.*	*Farthings.*

If the difficulty were a real one, it might be said that this way out of it would be harder than itself. But it will be more imaginary than real; and the preceding process is, in fact, the readiest way, both for those who feel it and those who do not, for turning the old money into new. Those who at first *think* that they cannot reckon in the new money, will *find* that they can; but they may gain assurance in the meanwhile by seeing an option. The following are some examples, the figures after the comma being royals, groats, and farthings.

s.	*d.*	*r.*	*g.*	*f.*	*s.*	*d.*	*r.*	*g.*	*f.*	*s.*	*d.*	*r.*	*g.*	*f.*	*s.*	*d.*	*r.*	*g.*	*f.*
0	4¼	,0	1	7	1	0	,0	5	0	2	5¼	,1	2	1	7	0¾	,3	5	3
0	6	,0	2	5	1	1½	,0	5	6	2	8½	,1	3	5	13	11	,6	9	5
0	7	,0	2	9	1	2¾	,0	6	1	3	0½	,1	5	2	16	1½	,8	0	6
0	9¾	,0	4	0	1	6	,0	7	5	3	9	,1	8	7	17	11¾	,8	9	8
					1	10½	,0	9	3	4	6	,2	2	5	18	2¼	,9	0	9

The error of this transition may be easily remembered, as follows:—The new account, in the above rule, falls short of accuracy by such a fraction of a new farthing as the copper in the old account is of sixpence. Thus, when we say that 15*s.* 10¾*d.* is ,794, the copper being 4¾*d.* which is the same fraction of sixpence as 19 is of 24, the new account is too small by 19-24ths of a farthing. Take 4 for every farthing in the copper, and an additional unit for every six of them, and we have the two next places, the tenths and hundredths of a farthing. Thus, 4 times 19 being 76, which with 3 for the three sixes in 19 is 79, we have ,79479 for 15*s.* 10¾*d.*, or 7 royals, 9 groats, 4 farthings, 7 tenths of a farthing, and 9 hundredths of a farthing. Thus, 3*s.* 7¼*d.* is ,18020. This is a superfluous degree of accuracy for nearly all purposes; but power gives confidence, even when it is not wanted.

4. How, when the new state of things is established, are those who must do it to return to the old system, as occasion arises?

By an inversion of the last rule. Allow two shillings for each royal, one shilling for five groats, and an old farthing for every new farthing left, abating 1 for 25 or more. Thus, ,692 is 13 shillings (six pairs and one) and 42 new farthings, 42 being left out of 92 when 50 is taken out for the odd shilling. Say 41 *old* farthings, or 10¼*d.* Accordingly, ,692 is 13*s.* 10¼*d.* Thus, ,083 is 1*s.* 10½*d.*; and ,644 is 12*s.* 10¾*d.*; and ,218 is 4*s.* 4½*d.*; and so on.

These rules are actually in use among accountants who are acquainted with decimal fractions, though not to the extent which they ought to be. It is easier to turn the old money into imaginary new money, to work with the latter, and then return the answer into the old system, than to practise the ordinary methods. It is true that, instead of calling the places which follow the pounds by the names of royals, groats, and farthings, they are called *decimals of a pound*, which they are; the royal being the tenth, the groat the hundredth, and the farthing the thousandth. Thus, if we had the following question:—16*l*. 10*s*. 2½*d*. yields 23*l*. 1*s*. 11¼: how much does 146*l*. 3*s*. 2¾*d*. yield? The ordinary mode of doing it by those who are not expert at the rule of *practice*, the way which is taught in schools, has not a figure less than the following:—

```
 £  s.  d.      £  s.  d.          £   s.  d.
16 10  2½      23  1  11¼        146   3  2¾
20             20                 20
-----          -----             -----
330            461               2923
 12             12                 12
-----          -----             ------
3962           5543              35078
   4              4                  4
-----          -----             ------
15850          22173             140315
                                  22173
                                 ------
                                 420945
                                982205
                               140315
                              280630
                             280630
                             ----------
                   15850)3111204495(196290
                         15850          49072  2
                         -----           4089  4
                         152620           204  9
                         142650          £204  9s. 4½d.
                         ------            + ½ a farthing.
                           99704
                           95100
                           -----
                            46044
                            31700
                            ------
                            143449
                            142650
                            ------
                              7995
```

In order to compare this with the result of the new coinage, we are to remember that our present system, which turns the student away from ordinary decimal integer arithmetic before he is master of it, prevents him from practising useful abbreviations. The system which would soon be taught would give the following mode of working, in which the trouble is increased by taking the tenths and hundredths of farthings. If we had begun with the new system, instead of representing the old, these fractions of farthings would have been avoided.

```
16,51041        23,09687        146,16145
                                  7869032
                                ---------
                                292322900
                                 43848435
                                  1315453
                                    87697
                                    11693
                                     1023
                                ---------
                       1651041)337587201(204,469
                                  737900 £204 9s. 4¾d.
                                   77484
                                   11442
                                    1536
                                      50
```

We shall presently resume this subject.

5. What amount of legislation will be required?

By legislation we mean actual imposition of rule, whether by the Crown or by Parliament. The issue of two new coins, the royal and the groat, and the fixing the royal at 100 farthings, or the shilling at 25 halfpence, is all that is required. It is not necessary to enact that accounts shall be kept in the new coins, except as to law proceedings. The bankers, and particularly the Bank of England, have a slight inducement to make the change. At present, nothing below a penny is carried to the credit of any one; that penny would be changed into five farthings, or the half groat. Accordingly, in all such accounts, the last column would be filled with noughts or fives.

6. What do we gain by the change upon a fair balance of convenience and inconvenience?

The inconveniences are soon enumerated. We have the difficulties common to all changes, which, as we have seen, are in this case slight, and the inconvenience of historical reference. This last is not much, as we have also seen: even in referring to old accounts, the nearest new farthing may, with very little practice, be brought out in the head from any sum in the old ledger. Both these difficulties become less and less from day to day, and pay their own expenses in a limited time, leaving what all admit to be an immense facilitation to be enjoyed in perpetuity.

The great advantages of the measure are threefold. First, all the operations of arithmetic are reduced to those of the decimal system. In the example which we have given, intended to show that, *even now*, decimal reckoning is more advantageous than that by pounds shillings and pence, we have not arrived at the full benefit of the new system. Take the same question, with new farthings, omitting their fractions, and we have, assuming the shortest mode of working, the following:—

If 16*l.* 5*r.* 1*gr.* 1*f.* yield 23*l.* 0*r.* 9*gr.* 7*f.*, how much will 146*l.* 1*r.* 6*gr.* 2*f.* yield?

```
16,511        23,097              146,162
                                    79032
                                  -------
                                  2923240
                                   438486
                                    13154
                                     1023
                                  -------
                          16511)3375903(204,464
                                    73703  £204 4r. 6 gr. 4f.
                                     7659
                                     1055
                                       64
```

We do not attend to the possible objection that we have introduced contracted methods of calculation (explained in part in the article on *computation* in the "Companion" for 1844), and have thereby magnified the advantage gained by the new method. For the only reason why those abridgments, which save about half the trouble of multiplication and division, are not now in common use, is simply that the paramount necessity of practising our present money system prevents young calculators from having time to learn them, or opportunity to use them. The want of them is a consequence of our present system; the introduction of them would follow the new one.

There are two modes of comparison common in business, one by how much per cent., the other by how much in the pound. To reduce one to the other with accuracy is a problem (in common usage) of the rule of three: generally speaking, a man of business is satisfied with a guess. Thus, 13*s.* 9½*d.* in the pound give 65 per cent. for the 13*s.*, 2½ for the 6*d.*, and 1¼ for the 3*d.*: about 69 per cent. The rules of reduction given under the third and fourth heads are worth learning, were it only for this one question of reducing poundage to per centage, and the converse. Thus, 13*s.* 9½*d.* gives ,68958, which is £68,958 per cent., or 68*l.* 19*s.* 2*d.* in 100*l.*: all done by the head. Now, let us have it *secundem artem*:—

```
£1         13s. 9½d.          £100
             12
           ------
            165
              2
           ------
          2)33100
           ------
         12)16550
          20)1379  2
           ------
             68   19           £68 19s. 2d.
```

Conversely, suppose we want to know how much 43*l.* 17*s.* $4\frac{3}{4}$*d.* per cent. is in the pound. Let us first take the grand process.

```
£100          £43  17s. 4¾d.          £1
               20
             ----
              877
               12
             ----
            10528
                4
             ----
          4)421,15
             ----
         12)105  1
              8  9          Ans.  8s. 9¼d.
```

But we might at once write—43*l.* 17*s.* $4\frac{3}{4}$*d.* as £43,869 and ,43869 (use ,438) gives 8*s.* $9\frac{1}{4}$*d.*, the answer.

In the new system, the distinction of per centage and poundage would soon disappear altogether. Thus, 62,314 per cent., or £62 3*r.* 1*gr.* 4*f.* in 100*l.* would be identical with ,62314 in the pound, or 6 royals, 2 groats, 3 farthings, 1 tenth, 4 hundredths in the pound.

We might swell this paper with detail of the advantages which a decimal system would give in mere calculation; but they are never disputed, though not always fully seen.

Secondly, the introduction of decimal coinage would prepare the way for the change to decimal weights and measures. The difficulties of this last alteration are immeasurably greater than those of the coinage, which the Government holds in its own hands. Nothing but a strong demand on the part of the public would bring it about; and it is not likely that such a demand would ever be created except by experience of the advantages of the coinage plan. The members of the last Commission upon weights and measures (who reported in December, 1841) were afraid of proposing as much of the decimal system as they approved; but they saw the step which was in the power of the Government, and recommended a decimal system of money in the following words:—

" The first point which has called for our special notice, is the general question of decimal scale. In introducing this subject, we beg leave to invite the attention of the Government to the advantage and the facility of establishing in this country a decimal system of coinage. In our opinion no single change which it is in the power of a Government to effect in our monetary " [metrical ?] " system, would be felt by all classes as equally beneficial with this, when the temporary inconveniences attending the change had passed away. The facility consists in the ease of interposing between the sovereign (or pound) and the shilling, a new coin equivalent to two shillings (to be called by a distinct name); of considering the farthing (which now passes as the 960th part of a pound) as the 1000th part of that unit; of establishing a coin of value equal to the 100th part of a pound; and of circulating, besides these principal members of a decimal coinage, other coins of values bearing a simple relation to them, including coins of the same value as the present shilling and sixpence. We do not feel ourselves at liberty to enter further into this subject; but we have felt it imperative on us to advert to it, because no circumstance whatever would contribute so much to decimal scale, in weights and measures, in these respects in which it is really useful, as the establishment of a decimal coinage."

Though thus sensible of the advantage of decimals in matters of coin, the Commissioners leaned to the binary system in many cases of weight and measure for buying and selling, apparently led by the great simplicity of this division, and its fitness, as things now stand, for the uninstructed classes. And in truth, if it were contemplated that education should remain in its present state, no such thing ought to be attempted as a violent alteration of ordinary measure or weight. But having the coinage by which to teach a decimal system, and having at last some chance of a considerable extension of the means of education, of which common decimal arithmetic is to form a prominent part, there is much reason, we think, to look forward to a wish, on the part of those whose fathers are now the ignorant portion of the community, for the complete adoption of a decimal system.

The Commissioners appear to us not to have sufficiently weighed, while recommending decimal money in any and every case, the fact that there is an extraordinary point of simplicity in a complete decimal mode of adjustment, which is wanting when, with decimal coinage only, a binary system of weights and measures is adopted. Suppose we had, for instance, a pound of 10 ounces, an ounce of 10 drachms, a drachm of 10 grains; and also the coinage of pounds, royals, groats, and farthings. For every farthing in the price of a grain, there is a groat in the price of a drachm, a royal in the price of an ounce, a pound in the price of a pound of weight. Thus £63,792 per hundred weight (of 100 pounds) is, ,63792 per pound ,063792 per ounce, and so on. This connection would be invaluable; the price of one denomination would be changed into that of another by simple removal of a comma from one place to another.

We repeat, however, that education must promote the demand for a complete decimal system: but that the application of the principle to coinage only must first promote education. And this brings us to the third of the advantages which we assert to be the necessary consequence of this simple alteration of the money. Has any one of our readers ever taken the pains to form an idea how much of the time actually spent in education in Great Britain and Ireland is spent in overcoming the disadvantage of our present system of coinage? We say coinage, because by far the greater part of practice in commercial arithmetic is devoted to pounds, shillings, and pence. We believe that five per cent. is under the mark, taking in all classes: we believe that in purely commercial schools it is a great deal more; but that in all together, from Oxford and Cambridge down to the lowest village school, more than five per cent., more than one twentieth of the whole time passed in every kind of learning and practising, is lost by the having two systems of arithmetic to learn, the common decimal, and the monetary. We put down arithmetic, looking at the mass of places in which only reading, writing, and ciphering, are taught, as more than 20 per cent. (we cannot say how much more) of the whole: and we estimate the dead loss of time which arises out of our monetary system, as one quarter at least of that 20 per cent. We speak of time only: were we to compare what is done, as to efficiency, as to sound result produced, we should say much more. But suppose it five per cent., and we think we could maintain this statistically; or go lower, suppose it three per cent.; and what is the result? Three hours out of every hundred spent in education are employed in mere consequence of a system which is in itself, and without reference to the trouble of learning it, a positive disadvantage. It would be well to abolish this system even though it saved nothing in teaching, and, besides this, we turn every 97 hours of useful school work into 100. Add to this, the relief given by the abolition of the worst part of the drudgery of learning computation, which lies in this, that there is a lower deep beneath the lowest. As soon as the unfortunate schoolboy has mastered the four rules, there is a weary recommencement of his toil.

In all the earlier rules of arithmetic, there is nothing which appears to apply. The most ordinary questions of every day life seem to be beyond the power of addition, subtraction, multiplication and division, of ordinary numbers. And so they are: because the money, weights, and measures are all numbered on other systems. If the change were made in the money only, an immense power of application would be immediately given. The addition and subtraction of common units would give that of money: by common multiplication and division the total price of integers, or the price of one from the total price, would be found. Thus 138 at 3*r.* 2*gr.* 7*f.* a piece, or ,327*l.* would be determined by the ordinary multiplication of 138 and ,327 to be 45,126*l.* or 45*l.* 1*r.* 2*gr.* 6*f.* At first, many would be slow to see, perhaps to believe, the closeness of the connexion which exists between ordinary numeration and the decimal system applied to any thing concrete. It is in fact more than connexion; it is identity. A great part of the apparent difference arises as follows. Though it be the principle of our numeration, and of every other, to make collections, and to *monadize* those collections, that is, to make units of them, to consider each collection as a new kind of *one*, yet the want of language prevents us from a clear view of the process. We ought to have a word for *one* ten; expressive of the distinction (which is not one of value but of form) between it and ten *ones*. Such distinctions we have in weights and measures, but not in ordinary numeration. Twelve pence are one shilling, which is not merely the twelve pence, but a new formation out of the collection, *another thing*. To have this as clearly in common numeration, we must invent some names. If we call 10 a parcel, 100 a basket, 1000 a sack, then 4683 read as 4 sacks, 6 baskets, 8 parcels, and 3, presents the view the want of which may make many think that there is really some essential difference between abstract numeration and calculation in concrete measurement.

Some persons have proposed that, when a decimal system is completely introduced, the subdivisions should have common names, expressive only of their derivation: that a pound of weight, a pound of money, a mile, an acre, &c. should each have its *dime*, or tenth part, its *cent*, or hundredth part, its *mill*, or thousandth part, &c. Accordingly, when the coinage is altered, they would have what we have called royal, groat, and farthing, to be at once the dime, cent, and mill, of a pound. To this we see serious objection. It is desirable beyond a doubt, to have these general names, expressive of the relations of tenth, hundredth, thousandth: but that the same name should at once express a piece of money, a weight, a length, a surface, a measure of capacity, &c. would lead to a hopeless and ludicrous confusion. Perhaps the ingenuity of swindlers would make it serve many purposes of fraud.

There is one inconvenience connected with the matter of the new coinage. The groat of ten farthings must be a distinct coin. It cannot be silver, for it would be too small; and a very adulterated silver would give rise to much spurious coin.

At the same time, twopence-halfpenny in one mass of copper, is thought to be too big. We have heard this held to be a sufficient objection to the whole scheme, but only by those who cannot tolerate any plan except one which is supremely advantageous in every respect. Those who remember that the copper coinage is not really of its value will not perhaps see much difficulty. The penny is not worth the 240th part of the sovereign: it keeps its value because Government is bound to receive 240 pence for one sovereign. At the same time it is so near its value that it is not worth the swindlers' while to make a spurious coinage. If a coiner of base money were to turn copper into halfpence, he would get a profit, but not enough. Surely then the groat might be a copper coin, a little more under value than the penny; say double of the penny in weight. As long as the common fourpenny and sixpenny pieces remain, nobody would be obliged to carry too much copper. The whole of this matter however, is for the consideration of the Mint officers, who will find a satisfactory arrangement as soon as the country has declared in favour of the plan.

We have now had for a long time the advantage of a uniform coinage. With the exception of the brief recourse to the introduction of Spanish dollars above noted, there is no one, we believe, who remembers having to reckon any thing except pounds, shillings, and pence. We are therefore apt to imagine that the difficulties of departure from an existing system are much greater than in truth they really are. If we want to know what can be done, or rather what can be borne with, let us look back to the Restoration. There were then in circulation, of *silver* coins, differing in weight, two halfpence, a three-farthings, three different pence, a three-halfpence, three twopences, two three-pences, four four-pences, two sixpences, two shillings, two half-crowns, two crowns. The gold coinage was advanced in value by proclamation in 1660. There were *fifty-nine* coins of gold, ranging the country at *thirty-seven* different values, most of them with odd pence in them, as the sovereign of Elizabeth at 11*s.* 11*d.* And all of them were advanced at once, into other odd pence: thus the largest of all, the old double rose noble, was ordered up from £1 16*s.* 4*d.* to £1 18*s.* 8*d.*, while the least, the quarter angel, passed from 2*s.* 9*d.* to 2*s.* 11*d.* What would a banker's clerk of our day, who verifies his counting in a few seconds by throwing a heap of sovereigns into a scale, say to having the correct value of a thousand pounds worth of this medley to ascertain? A great deal of confusion can be tolerated for the sake of a currency: compare what is asked for in the present case with what was done at a stroke by the restored monarch. We want a shilling of twelvepence halfpenny, one alteration; a coin of twopence halfpenny and another of two shillings, two additions. All the trouble which would be caused (for a time) to ordinary tradespeople would probably not be more than would be given to them if, for the same time, the silver halfpenny and silver three-farthings of Elizabeth were introduced into our currency in the state at which they were at the Restoration, and they had to make the distinction as often as the coins were offered.

The greatest inconvenience which would be found would probably be in mercantile accounts, &c. during the period in which the two systems must work together. One house with another, there would be half a year of some difficulty, or rather of some precaution against difficulty. While the reference to both systems lasts, a small card should be at hand, giving the value of every old farthing, in new ones, up to two shillings. Even after the new coinage has become law, it will be perfectly practicable for people to take their own time of falling into it, the transitions being so easily made, that accounts kept on the old system may have their results presented in the new one, as wanted. Some persons will find it convenient to begin by making columns for shillings, pence, and farthings (old) to be written in new denominations. In this manner, a person of much less practice than a banker's clerk could write in the new denominations as fast as another usually reads (for checking) in the old ones. The pounds of course are unaltered. Hearing the shillings, he could as easily write down the pairs (with 0 if even and 5 if odd), as write down the shillings themselves. Thus 80 for 16*s*., 65 for 13*s*., &c. Hearing the pence, he would write down four times as many in the next column; as 24 for 6*d*., 32 for 8*d*., &c. The farthings, if any, he would put down, as read, in the farthings column with one more if the pence were as many as 6*d*. A little practice would enable any person used to business (and we are speaking of no others at this moment) to take down the second column here written on hearing the first.

£.	*s.*	*d.*		£.	*r.*	*gr.*	*f.*
166	13	7	$\frac{3}{4}$	166	65	28	4
28	0	4	$\frac{1}{2}$	28	0	16	2
66	1	10		66	5	40	1
2	12	8	$\frac{1}{4}$	2	60	32	2
48	3	6		48	15	24	1
12	19	9	$\frac{1}{2}$	12	95	36	3

In the second reckoning the figures which are separated by

a single line belong finally to one column. Thus the first sum is £166,682, the two first columns after the double line giving 678, and the 4 turning 78 into 82.

Again, there are those who will find it makes a material difference to *them* whether they treat the transformation as we have previously done, or in the following way. We have described it by *places: one* in the first place for each *pair* of shillings, &c. But some will find it more convenient to take each part in new farthings at once. And then the rule will be;—100 for each pair of shillings, 50 for the odd one, and one more for each farthing left, with one additional at sixpence. Many, by this rule, would read 13*s*. 10½*d*. into 650+43=,693 much more easily than by the separate filling up of the places. This is the plan that we should recommend, if we were to assume that what is best for ourselves is best for others also (an assumption which no person must rashly make in matters of calculation). But we observe that the other plan has a great majority of those who have described the rule in its favour. Beginning in this way with the shillings, and learning first to read shillings alone, rapidly and correctly, as 1*s*. into 50, 2*s*. into 100, &c., we think many very moderate arithmeticians would soon *read off* into new farthings at once, as we have seen many young persons do with an hour or two of practice.

Indeed, any one who knows the above secret has no personal object in desiring a decimal coinage, which would abolish an advantage now existing in his favour. The little process of changing into decimals, or new farthings, the turning 13*s*. 10½*d*. into ,693, and the like, is all that stands between him and the full advantage of a decimal coinage. But this process will never become the property of the people at large. Ask one of those who have mastered the process, what is the price of one when a thousand cost £972? and he answers at once 19*s*. 5½*d*. What do a thousand cost at 5¾*d*. a piece? Answer at once, roughly, £28. If more accuracy be required, the extensions of the rule can be given, in a manner not necessary to describe here.

Perhaps there is no case in which the existing system looks so feeble, as in the attempt to divide one concrete by another. How often does £4 17*s*. 6½*d*. contain 17*s*. 9¾*d*.? Watch the books of arithmetic, and it will be found that they are shy of proposing such questions in the proper place, under the head of division. They wait until the prosy operation can be invested with the dignity of the rule of three, as in—how many at 17*s*. 9¾*d*. a piece can be bought for 4*l*. 17*s*. 6½*d*.? There is a class of books in which as many questions as possible are made to take the form of the rule of three, which involve the use of coins, weights, &c. When the question is "at 3 a penny, how many for twopence" the instructions given are to say "as 1 is to 2, so is 3 to $3 \times 2 \div 1$, or 6." This tendency is a consequence of the complexity of the metrical system. When one division requires two previous multiplications, the technical form of the rule of three is convenient for making an arrangement of the preliminaries.

There is impolicy in putting forward any argument as the most important when it is not likely to take that character with those who are to be convinced. For all who do not like to be convinced are quite ready to give infallibility to those who try to argue with them, in every point on which such concession will help their own view. Accordingly, when a reason is given as being the most forcible which exists, those with whom it is of small effect will, without giving due weight to the arguments which remain, allow the relative place of that reason to remain undisputed, and lower all the others accordingly. Protesting against this, and without prejudice to the absolute force of the grounds of convenience which we have shown, we are inclined to maintain that the alleviation of the drudgery of arithmetical education which would follow the establishment of a decimal coinage, is an argument of greater force still. If that drudgery were ever so successful, if we were a nation of expert calculators, we should still be of this opinion. But is it so? are we famous for the results of our schools as to arithmetic? We answer confidently in the negative. There is no lack of skilful calculators in business: the man who can pay (and not very high either) can command any quantity of good computation. But ask these ready numerists how much of their skill they acquired at school: and note the answers. In truth, common sense tells us that if primary education, such as all the middle classes get, itself gave power, we should find that power out of daily business, as well as in it, and we should find that men in business are good calculators upon matters which do not enter into their routine. But it is not so: the man of business is rarely good for anything but his immediate concerns, and of other persons, not one in twenty is above ridicule. Nor will it be otherwise until the coinage, the great subject matter of calculation, is assimilated in its processes to the ordinary arithmetic of simple number.

But the time of contending for the change is near at hand: and to give it a fair chance of success, extraneous matter must be avoided. It generally happens, when a definite alteration is proposed, that those who wish for other alterations seize the opportunity of trying to make riders of their own plans. Some would adopt a substitute for the great coin of account, the pound: a change which would carry confusion into all the operations of commerce, and would never be listened to by the banking and other monetary interests.

This, and twenty other schemes, may be argued each upon its own merits: but those who are sincerely desirous of aiding in the useful task of simplifying calculation from one end of the empire to the other, should remember that the way to divide, and the what to divide, are two different things: to unite against the plan all who oppose the decimal system, with the fifty times stronger body which sets itself against adopting a new fundamental coin, would be its ruin. It is also much to be wished, for the success of the whole scheme, that the favourers of the complete decimal system, in weights and measures as well as coinage, would forbear to press the two former until the latter is carried. They surely must feel that the monetary change would not retard that in the weights and measures; if so, the former must be favourable to the latter; for we can hardly suppose that experience of a decimal coinage would leave opinion as to further progress on the same route undecided either one way or the other. Moreover, the whole decimal scheme has no chance whatever of immediate success; that of the coinage alone has: the probabilities in favour of its speedy adoption are very much greater than they were seven years ago. The events of that period, the tone of the publications which have made mention of the subject, show that there is little more than inertia to arouse, hardly any opposition to contend with. Things standing thus, a resolute and united attempt at the practicable would soon force a result.

It is worth notice that the commissioners to whom we have already alluded, all men of science, did not appear to admit the expediency of an entire decimal system, at any period. In this we differ from them entirely: and we think it fortunate that the various half measures which they proffered were not adopted. But, though differing, we take the state of their opinion as a sufficient index of what would be likely to happen, if the whole decimal system were forced upon the attention of Parliament. The same commissioners strongly urged the change in the coinage, admitted the ease with which it could be done, enforced its desirableness as a preliminary to all they thought could be usefully adopted from the complete system, and showed the firmness of their opinion by travelling out of the matter intrusted to their deliberation in order to offer this advice.

A. De Morgan.

University College, October 14, 1847.

COMPANION TO THE ALMANAC

FOR

1849.

PART I.

GENERAL INFORMATION ON SUBJECTS OF MATHEMATICS, NATURAL PHILOSOPHY AND HISTORY, CHRONOLOGY, GEOGRAPHY, STATISTICS, &c.

I.—SHORT SUPPLEMENTARY REMARKS ON THE FIRST SIX BOOKS OF EUCLID'S ELEMENTS.

THE following short annotations are intended as suggestions to teachers, or to learners who are proceeding without a teacher: many other remarks might be added. Discussion is avoided, and matters of opinion are given, for brevity, in the short and peremptory form of memoranda, which may occasionally lead to some faults of language. Simson's Euclid is used, as being the basis of most modern English editions. The letters are the same in all editions of Simson, and in many others, particularly in Williamson's *real translation* of Euclid. The order of the English alphabet is substituted for that of the Greek; thus, Α Β Γ Δ Ε Ζ Η Θ, &c. in Euclid are A B C D E F G H, &c. in Simson and others. All the annotations here given are strictly of that character; they propose no alteration in the fundamental part of any one of Euclid's methods. The reader may consult with these comments the various geometrical words in the Penny Cyclopædia; or *Eucleides* in Dr. Smith's Biographical Dictionary, on matters connected with editions.

BOOK I. *Definitions.* Of these iii, vi, xiii, are obvious statements, but not definitions of words; viii, xxvi, xxxi to xxxiv, are never subsequently used; xviii, if semicircle have its etymological meaning, as seems the intention, is a theorem, which ought to be III. 1. The remaining definitions are of two kinds: first, those which do not explain their terms, but demand a notion already existing in the student's mind; they are, i, ii, iv, v, vii, ix: secondly, purely verbal definitions; they are, x, xi, xii, xiv to xvii, xix to xxx, and xxxv. Insist on angle as a *magnitude:* on the comparison of angles as to greater, equal, or less, by superposition; on the rights of angles equal to and greater than two right angles. The angle made by a straight line with its own continuation is a definite angular magnitude;

and its half is the best definition of a right angle. It is to be regretted that there is no single phrase for "two right angles."

Postulates and *Axioms:* in Euclid, *postulates* and *common notions.* All *geometrical* demands are *postulates* in Euclid; *his* axioms or common notions are in every instance notions common to all kinds of magnitude as well as space magnitudes. Restore this: that is, let the postulates be, Simson's postulates and axioms x, xi, xii; but instead of xi, substitute "if two right lines coincide in two points, they coincide when produced," as more self-evident. From this it is seen that the doubles of all right angles are equal, and thence that all right angles are equal; and this should come between I. 12 and I. 13, as a proof of the theorem "all right angles are equal." For xii substitute "two lines which cut one another are not *both* parallel to any third line," from which, after I. 28, prove Simson's axiom xii as a theorem. Remark that the distinction of postulate and axiom, as *problem* and *theorem*, could not have been Euclid's notion, for he does not recognise the last distinction; both are with him simply propositions. The expressed six postulates of Euclid are not the only ones which occur; others are tacitly adopted, as will presently appear. Nothing should be tacitly assumed by those who will not assume * without express statement, that "two straight lines cannot inclose a space."

I. 1. The following postulates are demanded: "if two figures which have one or more points in common have each a point which is not in the other, the boundaries of those figures must cut," and "every point is within or without a circle, according as its distance from the centre is more or less than the radius." With less, the intersection of the circles cannot be proved. I. 2, 3. Insist here upon the restrictions imposed by the first three postulates, which do not allow a circle to be drawn with a compass-carried distance; suppose the compasses to close of themselves the moment they cease to touch the paper. These two propositions extend the power of construction to what it would have been if *all* the usual power of the compasses had been assumed: they are mysterious to all who do not see that postulate iii does not ask for *every use of the compasses*. I. 4. This postulate is assumed, "any figure may be removed from place to place without alteration of form, and a plane figure may be turned round on the plane." But for this right to turn, I. 4 would not prove I. 5. I. 5. Euclid's construction is necessary only for the part relat-

* Simson, defending I. 22, for not proving the intersection of the circles, says, "Who is so dull as not to perceive," &c. He admits then tacit postulates, which any one could not fail to see; but who is so dull as not to perceive that two straight lines cannot inclose a space?

ing to external angles; for the rest the triangle might be turned round, set down on the plane, and I. 4 applied to the original and removed forms. I. 6. This is never wanted till after it might be readily proved again with I. 19. I. 7 would be made more easy to beginners if they were first familiarized, as a common notion, with "if two magnitudes be equal, any magnitude greater than the one is greater than any magnitude less than the other." I. 9. Of this I. 11, which is "to bisect the angle made by a straight line and its continuation" should be a particular case; the constructions are the same. I. 12. These postulates are assumed:—"if one point be taken on each side of an indefinitely extended straight line, any line which joins the two must cut the straight line;" and "if one point of a straight line be inside a figure, the straight line must, if sufficiently produced, cut the figure in two points at least." I. 13. Before this, introduce "all right angles are equal," as above noted. I. 13 and I. 14 are mere consequences of Euclid's refusal to consider the double right angle as a magnitude, in any other point of view than as the sum of its halves. Of O A, O B, O C, he never proves that A O B and B O C are together equal to the two halves of A O C, except when O A and O C are in the same straight line. I. 18, I. 19. The latter of these, and I. 6, is a purely logical consequence of I. 18 and I. 5, following independently of the meaning* of the terms. After I. 21, introduce the following, "The perpendicular is the shortest straight line that can be drawn from a given point to a given line; and of others, that which is nearer to the perpendicular is less than the more remote, and the converse; and not more than two equal straight lines can be drawn from the point to the line, one on each side of the perpendicular." I. 22. The postulates assumed in I. 1 are used again. I. 24. The proof is incomplete. To make it perfect, the following proposition (easily proved from that following I. 21), should be made to precede; "every straight line drawn from the vertex of a triangle to the base, is less than the greater of the two sides, or than either, if they be equal." I. 25. This (and I. 8, had it not been proved before) are connected with I. 4 and I. 24 in the manner noticed under I. 19. After I. 26 there is an omission which is supplied in the sixth book. This proposition ought to be introduced, "If two triangles have a pair of angles equal, the sides about another pair of angles equal each to each (the opposite of the first to the opposite, and the adjacent to

* If A B C be three propositions, of which one, and one only must be true; and if P Q R be of the same kind; and if A always give P; B, Q; and C, R; then P always gives A; Q, B; and R, C. This is the connecting logical proposition of I. 6 and I. 19 with I. 5 and I. 18.

the adjacent), and the third pair of angles both acute, both obtuse, or one right, the triangles are equal in all respects." I. 27 is a logical equivalent of what is already in I. 16. Before I. 29 introduce proof of Simson's axiom xii from the more simple postulate, as already noted. I. 30 is a logical equivalent of that more simple postulate. The following is much wanted, "If two lines be parallel to two other lines, each to each, the first pair make the same angles as the second;" and this also, "The perpendiculars to two lines make the same angles with one another as the lines themselves." Introduce the definitions of a parallelogram and of a rectangle. Note the alteration of meaning of the word equal at I. 35, which is thenceforward used, as to areas, for equality of area, with or without identity of form. The postulate, "an area taken from an area, leaves the same area from whatever part it may be taken," should be mentioned by a writer who specifies that "two straight lines cannot inclose a space;" it is particularly important as the key to equality of non-rectilinear areas which could not be cut into coincidence geometrically. I. 43. The first proposition in which appear equal areas with unconnected bases; and the foundation of the whole comparison of such areas. Note the manner in which the postulate in I. 35 is used. I. 46. Define the square as a rectangle with equal adjacent sides; extend this proposition to the construction of a rectangle with given sides. Introduce the phraseology of the "rectangle *under* two given lines." Introduce the propositions, "the squares on equal lines are equal," and "equal squares must be on equal straight lines." I. 47. Note the mode of demonstration, namely, that each square is equal to the rectangle under the hypothenuse and the adjacent segment. I. 48. An appearance of avoiding indirect demonstration by drawing the triangles on different sides of the base, and appealing to I. 8, because drawing them on the same side would make the appeal to I. 7 (on which, however, I. 8 is founded).

Book II. Introduce (to avoid confusion) the arithmetical connexion of the rectangle and multiplication, for rectangles of commensurable sides. Prove the existence of incommensurable magnitudes (by the usual example of the side and diagonal of a square), and point out the insufficiency of the arithmetical theory. If this be first done, so as to fix the notion that the propositions of the second book are not arithmetical, the resemblance to algebraical form may be increased with safety, and to the great abbreviation of the demonstrations. As follows: introduce this definition, "When a straight line, or the same produced, is cut in a point, whether internally or externally, the distances of the point of section from the ends of the line are called segments of the line, internal or external according as the point of section is internal or external." The line is, therefore, the sum of its internal, the difference of its external, segments. Show separately that the difference of internal, or the sum of external, segments, is double of the distance of the points of section and bisection. II. 1. Annex to this the following extension, "If two straight lines be each of them divided into any number of parts, the rectangle under the two lines is made up of all the rectangles under all the parts of the one, taken separately with all the parts of the other." Treat II. 2 and II. 3 as corollaries of the first part of II. 1; worthy of statement, but not needing new demonstration. Demonstrate II. 4, directly from the new part of II. 1, without proving that the smaller squares are *about the diagonal* of the larger one; state it as "the square *on** a line is greater than the sum of the squares on its internal segments by twice the rectangle under those segments." With it, or after it, join II. 7, as "the square on a line is less than the sum of the squares on its external segments by twice the rectangle under those segments." II. 5 and II. 6. Make these corollaries or included cases of "the rectangle under the sum and difference of two lines is equal to the difference of the squares on those lines," which can be easily proved from direct construction of the rectangle and completion of the figure. Then from II. 4 and II. 7, deduce "the square on the sum and on the difference of two lines, are together double of the sum of the squares on those lines," and "the square on the sum exceeds the square on the difference by four times the rectangle under the lines." This last is II. 8; and II. 9 and II. 10 are cases of the first, proved by Euclid in a most ingenious indirect manner. This much of the book is general; the rest is special application. II. 11. An excellent opportunity for comparison of geometry and arithmetic; note the manner in which Euclid gives the most direct construction of the solution of the quadratic equation. II. 12 and II. 13. Take these together, observing the complete analogy of the corresponding steps, a thing impossible while II. 4 and II. 7 remain as left by Euclid. II. 14. Prove the property of the circle first, and apply it afterwards. When the connexion of number and magnitude, and the distinction, are properly attended to as preliminaries, the more the analogies of geometry and algebra are made visible the better; geometry takes no harm and algebra receives some illustration.

Book III. *Definitions.* Simson has noted that i. is a theorem: strange that he should not here have inserted and *proved*

* Use the phrase "the square *on* a line" as helping to establish a distinction (much wanted) between the figure in geometry and the arithmetical square *of* a number.

def. xviii of the first. Defer the definitions of contact, omit that of the angle of a segment (vii), and explain that in "similar segments" the word *similar* is an anticipation, and that similarity* *of form* is meant. This definition is a theorem: or would be if "similar" had taken its final meaning. III. 1. Prove the fundamental theorem "the line which bisects a chord perpendicularly must contain the centre," and then make III. 1, III. 25, and IV. 5 immediate corollaries of it. III. 2. Make this an immediate consequence of the new proposition preceding I. 24, and the circle postulate assumed in I. 1. Introduce and prove that "a straight line cannot cut a circle in more than two points." III. 3. Put this after III. 1, as a converse. III. 5. This equally refers to cutting and touching, and is, in fact, "circles which have the same centre, and one point of circumference common to both, coincide altogether." III. 6. See the last remark. III. 7. There is an unproved assumption; namely, that D E B is greater than D E C, when the hypothesis is only that D F B is greater than D F C. That B cannot fall *within* the triangle F E C, is assumed. Prove it by help of the new proposition preceding I. 24. III. 8. How much Euclid is depending upon the eye for his conclusions is obvious, by the introduction of the terms convex and concave without definition. A line drawn through D, meeting the circle twice, is said to have its intersection nearest to D on the concave side, that furthest from D on the convex. These are two unproved assumptions, namely, that K falls within the triangle D L M, and E without the triangle D F M. Let D E meet M F and M L in Y and Z; then M Y and M Z being each less than the radius, show from the proposition introduced after I. 21, that only one of the points K and E can lie on one side of Y or of Z, and that neither can lie between. III. 9. A logical equivalent of part of III. 7; when it is proved that every *non*-central point is *not* a point from which three equal straight lines can be drawn, then III. 9 is also proved. III. 10. It can be proved directly from IV. 5 placed as above recommended. III. 11, 12, 13. The contact of circles is in a state of confusion in Euclid: there is a positive assumption that the circle which touches internally is entirely within, and that the circle which touches externally is entirely without, the other circle. Leave the *visual* notions of *contact* till they can be otherwise established, and proceed as follows. Circles cut in two points at most:

* Students have sometimes a slight confusion from the common meaning of the word *similar*, which implies likeness of a smaller degree than that expressed by *same*. It would not be contradiction, nor unintelligible, to say, "those colours are not the same, but they are similar." But in mathematics similarity is absolute sameness, and the sameness of the technical term of geometry is sameness of form.

distinguish the case in which there are two *coincidences* (whether contacts or intersections) from that in which there is only one. Then proceed by the following steps, easily proved:— A. If two circles have two coincidences, both cannot be on the line (or line produced) joining the centres. B. Then neither: since, for any coincidence on one side of that line, another can be constructed on the other side. C. By the same construction, when there is only one coincidence, it *is* on the line joining the centres. D. When there is only one coincidence, the one circle is entirely within, or entirely without, the other. E. When there are two coincidences, the circles intersect (shown by help of the postulate assumed in I. 1). III. 17. Note, when the means of proof arrive (which might very well have preceded) the easier construction by drawing a circle on A E. After III. 19, introduce "If two circles have but one coincidence, they have the same tangent at the point of coincidence." III. 20. Extend the proposition to the case in which the angle at the centre is equal to and greater than two right angles: make III. 31 an immediate consequence. III. 21. Bring all the cases under one by means of the extension in the last. III. 22. Make this depend upon the sum of the two central angles being four right angles. III. 25 has preceded, as above noted. III. 26, 27, 28, 29. Put these together, and augment, under the general enunciation, "In equal circles either of the five pairs, arcs, chords, angles at the centre, angles at the circumference, or sectors, being equal, the other four pairs are equal." III. 31. After the extension of III. 20, this is the merest corollary. III. 34. The construction might be made more simple, though Euclid's reason for taking another is not difficult to find. III. 35, 36. After the definition of internal and external segments, as noted in book II, these propositions are one, under the enunciation, "if two chords cut one another, the rectangles under their segments are equal." The want of proper coupling of propositions in book II. prevented this being made apparent. III. 37. The same defect as in I. 48. If D B be not a tangent, produced it cuts, and then the square on D B is shown at once to be the rectangle under D B and a longer line, which is absurd.

Book IV. This book is of no direct use in what follows, and might well be omitted for a time by a learner who wishes to proceed to proportion. The idea of dividing the revolution into equal parts should be made prominent; and the description of regular polygons treated as incidental. The general design is somewhat concealed by the first propositions relating to *any* triangle: it happens that it is as easy to divide the circle into *three* parts which are in the ratio of the angles of a triangle, as into three equal parts. IV. 3 and 4. Euclid insists that the figure shall be circumscribed, without this limitation, the problem is of a four-fold answer. IV. 5 has been anticipated as before mentioned. IV. 10 is better enunciated as "To divide a right angle into five equal parts," or at least, this effect of the problem should be insisted on. The method of IV. 11 is not so natural as making a direct use of the angle obtained in the last. IV. 12, 13, 14. These propositions supply the place of the following: "Having given a regular polygon of any number of sides inscribed in a circle, to describe the same about the circle: and having given the polygon, to inscribe and circumscribe a circle." The method should be applied generally. Add to this book the following proposition: "Having given a number, no matter how great, to inscribe a regular polygon in a circle which shall have a number of sides greater than that number."

Book V. In development and maintenance of many of the following remarks, the reader may consult the articles *Ratio* and *Proportion* in the *Penny Cyclopædia; Ratio, Composition of*, in the Supplement to that work; and a tract *On the connexion of Number and Magnitude*, by the author of this article. Note that the fifth book is on *magnitude* of any kind, and not on space only: provided that the magnitude be one of those of which we have the power to form multiples, and to compare the multiples of different specimens, with the power of selecting the greater, or affirming equality, if it exist. V. def. iii. *Ratio* is best founded on the notion of *relative magnitude:* but Euclid, who adheres to the etymological meaning of words to the utmost of his power, found the word λόγος established. This means *communication*, and the word which is translated *quantity* ought to have been rendered by *quantuplicity*, referring to the number of times one magnitude is contained in the other. We cannot describe magnitude in language without quantuplicitative reference to other magnitude. The peculiar manner in which the Greeks made a substantive, alone, takes the sense for which the moderns would form a compound,* is seen in ellipse, hyperbola, anomaly, &c. We imagine Euclid's predecessors to have designated the notion on which our *power of communication* depends by the simple name of *communication;* and this rough definition refers to *quantuplicity*, much as that of a straight line refers to *evenly lying*. V. 4. There is nothing about less or greater in Euclid: magnitudes have a ratio when their multiples can exceed one another,

* Euclid himself makes a slight restriction, which confirms our view. The term ἄλογος, "which cannot be expressed," is used by him (book x. definition vii.) not as merely expressing the relation of incommensurability, but that of being incommensurable with the *standard line* which, throughout the book, is the unit of rational (ῥηταί) lines.

meaning when the relation of greater or less can be predicated of their multiples with respect to one another. As this follows when it can be predicated of the quantities themselves, this definition amounts to saying that the magnitudes are of the same species. V. 5. Give a clear view of the existence of incommensurables. Dwell on the fact of a notion of proportion existing in the mind previously to entering on geometry, as evinced by sameness of *relative* magnitude, so far as the eye can judge of it, being demanded by the eye in all front pictures, models, &c. Remember that common sense requires that we should satisfy this notion of proportion, not invent a new one for the occasions of geometry. Point out the manner in which, if all magnitudes were commensurable, finite arithmetical ideas and language would be sufficient for reasoning on proportion; and also that arithmetic itself may be made a science of *ratios*, all abstract numbers (fractions included) being capable of being considered as expressions of ratios. The existence of incommensurables shows that ordinary arithmetic cannot be made the science of *all* ratios. Prove that any magnitude being named, however small, two incommensurable magnitudes may be made commensurable by adding or subtracting to or from either a magnitude less than the magnitude named. Define the *scale of relation* of two magnitudes of the same kind, namely, a list of the multiples of both *ad infinitum*, all arranged in order of magnitude, so that any multiple of either being assigned, the scale of relation points out between which multiples of the other it lies. Point out how this scale must detect commensurability, if it exist; and if it do not exist, how it gives the means of making any required degree of approach to arithmetical expression of one magnitude in terms of the other; by enabling us to assign arithmetical fractions of either, as near to each other as we please, between which the other lies. Further, that when the relative scales of two pair of magnitudes are looked at together, difference of relative magnitude is detected by any one instance in which a multiple of the first is in a place (or interval) among the multiples of the second different from that in which the same multiple of the third is found among the multiples of the fourth. Hence pass to Euclid's *indirect** de-

* Or *negative* definition. That of parallels is another instance: the positive notion is intersection, which *may* be affirmed by examination of one pair of points, one in each line, when it happens that these points coincide. But parallelism is the absence of all intersection, and cannot be affirmed after the examination of any number, how great soever, of pairs of points. The consequence of incommensurability is that, while *dis*proportion (which is here made the positive term) *may* be affirmed upon a settlement of one item of the distributions in the relative scale, proportion can only be reasoned on as the absence of all disproportion.

finition of proportion, consisting in the absence of every positive test of *dis*proportion, as seen in "four magnitudes are proportional when *every* multiple of the first is found in the same position among the multiples of the second, in which the same multiple of the third is found among the multiples of the fourth." *Position* is a somewhat more definite term than interval: thus three different possible positions of 13 A, are, that it should be equal to 20 B, that it should be between 20 B and 21 B, and that it should be equal to 21 B. V. def. vii. The first is to the second in a greater (or less) ratio than the third to the fourth, when a multiple of the first takes a more (or less) advanced position among the multiples of the second, than the same multiple of the third takes among those of the fourth. Proof should be given that the same pair of magnitudes can never offer both tests to another pair; that is, the test of greater ratio from one set of multiples, and that of less ratio from another. V. viii quite unmeaning after vi: perhaps its original place was after iii, as the same rough introduction to proportion, which iii was to ratio. V. def. ix, defines nothing. V. def. x, xi. Omit these until compound ratio has been considered. Impress the notation following; capital letters signify magnitudes, *not numbers* representing them. The following theorems are very useful, and may be introduced here, or as wanted. (1.) If four magnitudes be not in proportion, a multiple of the first may be found which shall lie among the multiples of the second, in a position removed by as many intervals as we please from that in which the same multiple of the third lies among the multiples of the fourth. (2.) If the multiples of the first and third, *from and after any given multiples whatsoever*, be similarly distributed among the multiples of the second and fourth, the four magnitudes are proportional. (3.) If, for any multiple which may be named, a higher multiple of the first can be found, which is in the same position among the multiples of the second, in which the same multiple of the third is among the multiples of the fourth, the four magnitudes are proportional. (4.) Division can be performed upon two *magnitudes* in just the same manner as upon two numbers, and as many places as we please of (perhaps) an interminably significant decimal fraction can be found, from which an approximate expression of the ratio of the two magnitudes may be formed. (5.) If a pair of magnitudes give the same interminable decimal as another pair, the four magnitudes are proportional by Euclid's definition, and the converse.

We cannot enter into the manner in which the demonstrations of the fifth book may be rendered clear by the very simple* artifice of using 2 A, 3 A, &c., as signifying the mul-

* We have before now heard this called an *algebraical* method. But it

tiples of A. But rigorous reasoning does not require adherence to the rude forms which the arithmetic of the Greeks contained. If there be any who do not consider the fifth book an *application of arithmetic to geometry*, we have no ground in common with them.

V. 1, 2, 3, 5, 6. These are simple propositions of concrete arithmetic, covered in language which makes them unintelligible to modern ears. The first, for instance, states no more than that *ten* acres and *ten* roods make *ten* times as much as one acre and one rood. V. 4. The corollary in Simson is only necessary to those who will not admit M into the list M, 2M, 3M, &c.; the exclusion is grammatical,† and nothing else. V. (A). Needless to those who believe *once* A to be a proper component of the list of multiples, in spite of *multus*‡ meaning *many*. V. (B). A very proper insertion. V. 7. On this and various other propositions it should be remarked that their purpose is not only to fill up the links of the chain of demonstration, but to make trial of the definition of proportion, and to certify that this complicated test produces the evident results of the simple and previous notion.

Of the simplification which, without abandoning any one principle of Euclid, may be introduced into the demonstrations, the following is a sample. It is required (V. 18) to show that if A be to B as P to Q, then A+B is to B as P+Q to Q. Whatever multiple of A+B we choose to examine, take the same multiple of A, say 17A, and let it lie between some two multiples of B, say 23B and 24B: then 17P lies between 23Q and 24Q. Add 17B to all the first, and 17Q to all the second; then 17(A+B) lies between 40B and 41B, and 17(P+Q) between 40Q and 41Q; and in the same manner for any other multiples.

The following is the sketch of a proof that, B being any magnitude, and P and Q two magnitudes of the same kind, there does exist a magnitude A, which is to B in the same ratio as P to Q. The right to reason upon any aliquot part of any magnitude is assumed; though, in truth, aliquot parts obtained by continual bisection would suffice: and it is taken as previously proved, that the tests of greater and of less ratio are never both presented in any one scale of relation as compared with another. (1.) If M be to B in a greater ratio than P to Q, so is every magnitude greater than M, and so are

is essential to algebra, in the sense intended, that the letters should signify not magnitudes, but numerical representations of them.

† The ancient Jews might sometimes have needed corollaries of completion for 15A: they could not designate 15 as 10 + 5, since the letters signifying 10 and 5 made a name of the Deity.

‡ Is *twice* A a multiple of A, etymologically?

some less magnitudes; and if M be to B in a less ratio than P to Q, so is every magnitude less than M, and so are *some greater magnitudes.* Part of this is in every system: the rest is proved thus. If M be to B in a greater ratio than P to Q, say, for instance, we find that 15 M lies between 22 B and 23 B, while 15 P lies before 22 Q. Let 15 M exceed 22B by Z, then if N be less than M by any thing less than the 15th part of Z, 15 N is between 22 B and 23B : or N, less than M, is in a greater ratio to B than P to Q. And similarly for the other case. (2.) M can certainly be taken so small as to be in a less ratio to B than P to Q, and so large as to be in a greater; and since we can never pass from the greater ratio back again to the smaller by increasing M, it follows that while we pass from the first designated value to the second, we come upon an intermediate magnitude A, such that every smaller is in a less ratio than P to Q, and every greater in a greater ratio. Now A cannot be in a less ratio to B than P to Q, for then some greater magnitudes would also be in a less ratio; nor in a greater ratio, for then some less magnitudes would be in a greater ratio: therefore A is in the same ratio to B as P to Q. The previously proved proposition above mentioned shows the three alternatives to be the only ones.

The mention of compound ratio only occurs once (in VI. 23, if we regard the definition at the beginning of the book as spurious), and then only as a phrase. A is *said* to have to C the ratio compounded of the ratios of A to B and B to C. We see much to regret in the phrase, and something more, not being of continual application in geometry. Treat ratio as an engine of operation. Let that of A to B suggest the power of altering any magnitude in that ratio. This notion of altering magnitude in a ratio involves the idea of something which is certainly possible, and capable of being pictured as done by an act of the mind, even by those who are not aware that the operation is identical with that which takes place in multiplication by an integer or fractional number; and those who have recourse to ideas of mechanical operation, and lose the notion, deprive geometry of much of its life, and of all its distinct power of representation. Every alteration of a magnitude is alteration in some ratio, two or more successive alterations are jointly equivalent to but one, and the ratio of the initial magnitude to the terminal one is as properly said to be the compound ratio of alteration, as 13 to be the compound addend in lieu of 8 and 5, or 28 the compound multiplier for 7 and 4. Composition is used here, as elsewhere, for the process of detecting the single alteration which produces the joint effect of two or more. The composition of the ratios of P to Q, R to S, T to U, is performed by assuming A, altering it in the first ratio into B, altering B in the second ratio into C, and C in the third ratio into D. The joint effect turns A into D, and the ratio of A to D is the compounded ratio. The propositions added by Simson at the end of the fifth book will be good exercises in the comparison of the language of composition of ratio with that of multiplication.

Duplicate ratio ought to be defined as the ratio arising from the composition of two equal ratios; triplicate of three equal ratios, and so on. Euclid gives only the process, and neglects to give the idea, in which the force of the sixth book much consists. Any one who, after being used to the usual arithmetical way of stating the following, learns to appreciate the language of composition of ratios, will find a vast accession of clearness. "If the sides of a triangle AB, BC, CA, be cut by a straight line in C′, A′, B′, the ratio compounded of the ratios of AC′ to C′B, BA′ to A′C, and CB′ to B′A is a ratio of equality." And the treatment of this proposition after the manner of VI. 2, will add as much light to the proof.

Book VI. VI. 1. Much instruction is necessary to enable learners to take the proposition properly: as the first instance in which proportion is applied, it should be illustrated on every notion. Rectangles alone had better be used, and the other cases deduced: and the composition of the sides of the rectangles should follow.

The rationale of this proposition is best understood, after comparison with the last of the book, by the following enunciation—'If there be two species of magnitudes, *named* A and B, and if every A be accompanied by one B and one only, and if every B be accompanied by one A and one only; and if, according as one A is greater than, equal to, or less than another, the B of the first is greater than, equal to, or less than that of the second; and if, when one A is a multiple of another, the B of the first is the same multiple of the second—then any two As have always the same ratio to one another as their corresponding Bs.' VI. 2. Prove it also by formation of the relative scales, one being formed from the other by parallels. VI. 3 and A. Put the two propositions together, as in the pairs of the second book: introduce the notion of harmonic division. A development ought to be made of the sixth book, involving the fundamental modern use of harmonic division, anharmonic ratio, &c., the forms not being algebraic. VI. 4. Define similar (in form) figures: dwell on the possibility of creating a complete definition by ratios only, or by angles only, if *diagonals as well as sides* be introduced. Point out how Euclid, by throwing diagonals out of consideration, takes his definition from a mixture of the two homogeneous ones. Abandon the peculiar mode of construction by which Euclid proves two cases at once: make an angle coincide with its equal, and suppose this process repeated three times, one for each angle. VI. 5. Attend to the remark on I. 48. VI. 7. This is the extension of the proposition supplied after I. 26. VI. 8 is out of place; it should follow VI. 12. Observe (and prove presently) that the property in I. 47 is a consequence of the right-angled triangle being capable of division into two of the same form: that if any other triangle could (but it cannot, which prove) be divided in the same way, the property of I. 47 would result. VI. 9, 10, 11, 12. Note that the first is a particular case of the second, and the third of the fourth: remark the manner in which Euclid never notes either analogy, or genus and species, when any difference of nomenclature interposes the smallest necessity for adaptation or extension of language. Note the command obtained, so far as straight lines are concerned (and ultimately rectilinear areas) over operations which supply the place of multiplication, division, and their combination, the rule of three. VI. 13. Connect this, after the manner just mentioned, with the extraction of the square root. VI. 14. Owing to the disjointed manner in which Euclid treats compound ratio, this proposition is strangely out of place. It is a particular case of VI. 23, being that in which the ratio of the sides, compounded, gives a ratio of equality. The proper definition of four magnitudes being reciprocally proportional, is that the ratio compounded of their ratios is that of equality. To this proposition should be added the other converse 'equal parallelograms, having sides reciprocally proportional, have their angles equal, each to each.' VI. 15. The immediate consequence of the last. The converse wanting is 'equal triangles, with the sides about a pair of angles proportional, have those angles equal or supplemental.' System requires that *all* propositions relative to the area of a triangle should follow from that of its parallelogram. Any independence is not real, since the first establishment of the area of the triangle must depend on the parallelogram. VI. 16. A mere case of VI. 14, illustrating the remark made on VI. 9, &c. VI. 17. A mere case of VI. 16. VI. 18. A theorem spoiled; its converse is in VI. 20. Put VI. 18 and part of VI. 20 into one in this manner:—'Pairs of similar triangles, similarly put together, give similar figures; and every pair of similar figures is composed of pairs of similar triangles, similarly put together.' Then make the *problem* of VI. 18 an application of the first part. VI. 19. Join to this the remaining part of VI. 20. Prove it from the proper definition of duplicate ratio, namely, the ratio compounded of two equal ratios: this is done instantly if VI. 23 be in its proper place, which will give the proposition at once for the case of similar parallelograms. The method of Euclid is an elegant application of the *operation* requisite to compound equal ratios, by which the conception of the process is lost sight of. VI. 21. Has Euclid assumed in many places that two figures which agree in any particulars with a third, agree in those particulars with one another? If so, this proposition is needless. To show that he has done so, remark that *all reasoning* assumes that any two things (figures or not) agree with one another in every particular in which they agree with a third. Nevertheless, this proposition, and V. 11, are of a class which might be retained, not as proving their *results*, but as proving that the *terms employed in them* are so far not self-contradictory. VI. 22. In this proposition there is a most palpable failure: it is assumed that similar and equal figures must have their homologous sides equal, each to each. In most of the manuscripts of Euclid, and in many of those of highest authority, a *lemma* follows this proposition, which supplies the defect. That Euclid would ever appeal to *what follows* is wholly incredible: this lemma, then, is not his; but the defect is. Dr. Simson omits the lemma without a word; and preserves the defect. The best way of remedying it is to insert the following proposition, which immediately proves the second (or defective) case of the theorem. 'It is impossible that two *different* ratios can have the same duplicate ratio.' For, if possible, let the ratio of A to B be duplicate both of A to X and A to Y: from which A is to X as X to B, and A is to Y as Y to B. Let X be greater than Y: then A has to X a less ratio than A to Y: that is, X has to B a less ratio than Y to B; or X is less than Y. But X is also greater than Y, which is absurd, &c. VI. 23. See what has been said throughout on compound ratio. VI. 24 and 26. Put these consecutively: how could VI. 25 have got between them? VI. 25. Let the student see this in the following way:—'To make a figure which shall have the form of one given figure, and the size of another,' as well as in Euclid's language. It will be instructive to compare the arithmetical process with the geometrical. VI. 27, 28, 29. The use of these propositions to a learner, is historical: they show how Euclid, in the tenth book, proceeded in cases of which the arithmetical counterparts would require the solution of quadratic equations. VI. 30. A process derived from II. 11; little more than the latter under a new name. VI. 31. Take notice that I. 47 has never been employed in the real preliminaries of this demonstration; so that I. 47 is really demonstrated again. VI. 32 is useless; but the following would often be useful:—'If two similar triangles be placed with their bases parallel, and the equal angles at the bases towards the same parts, the other sides are parallel, each to each: or one pair of sides are in the same straight line, and the other pair are parallel.' VI. 33. Here the angle breaks prison.

Euclid's definition of proportion requires the consideration of every *multiple* of the magnitudes in question; accordingly, not only are angles greater than *two* right angles employed, but greater than *two million*. Dr. Simson makes no note in reference to this subject. VI. B, C, D. These introductions of Simson's are sometimes omitted, very injudiciously we think.

If the preceding suggestions were all adopted to the fullest extent, no material alteration would be made in the fabric of the elements. The greatest apparent alteration here proposed, the extension of the angle, is, as just seen, ultimately made by Euclid himself. The doctrine of compound ratio, which, as far as the first six books are concerned, is only a new phrase occurring in one proposition (if we dismiss the definition at the beginning of the sixth book as spurious), has a process which is really employed on several occasions. The small defects which we have noted in several of the demonstrations are such as can be amended by introductions which strictly accord in character with the rest. The mode of making, not the fifth book more arithmetical, but the existence of the arithmetic of the fifth book more discernible, by the application of modern phraseology and symbols, preserves Euclid's definitions and modes of proceeding. Logic, and language critically considered with reference to its connexion with thought, were subjects which had no formal existence when Euclid began to write: the necessity of extending terms, and the desirableness of not confusing logical equivalences and distinct propositions will, we suppose, be conceded. If the study of Euclid have been almost abandoned on the Continent, and have declined in England, it is because his more ardent admirers have insisted on regarding the accidents of his position as laws of the science.

There never has been, and till we see it we never shall believe there can be, a system of geometry worthy of the name, which has any material departures (we do not speak of *corrections*, or *extensions*, or *developments*) from the plan laid down by Euclid. If there be one worthy of consideration, it is the commencing with a strict theory of proportion. But it may very well be doubted whether any complete treatment of the fifth book could be made intelligible to students *of our day* before they have had some familiarity with demonstration as applied to particular species of magnitude. We say of our day, because it is impossible to foresee what the advance of education may do. It is perfectly conceivable that the rapid advance of demonstrative arithmetic, as a study preliminary to that of geometry, may ultimately render the change desirable.

A. De Morgan.

University College, London,
Oct. 30, 1848.

COMPANION TO THE ALMANAC

FOR

1850.

PART I.

GENERAL INFORMATION ON SUBJECTS OF MATHEMATICS, NATURAL PHILOSOPHY AND HISTORY, CHRONOLOGY, GEOGRAPHY, STATISTICS, &c.

I.—ON ANCIENT AND MODERN USAGE IN RECKONING.

THE year 1850, to which the present number of our work belongs, is a year one designation of which depends upon the settlement of a controversy. About the end of last century, there was a fierce contest upon its proper termination: some said that the nineteenth century began with January 1, 1800, others with January 1, 1801. The former would say that 1850 is the *first* year of the *second* half of the century, the latter that it is the *last* year of the *first* half. In taking up this subject, we do not look at the century question (which is of no consequence whatever, and is easily settled in favour of the second-named interpretation) so much as at the mode in which it arose, and at the genus of which it is a species. We make it a peg on which to hang a short dissertation on the distinction of *ancient and modern reckoning*, which the lawyer generally ignores, the scholar often disregards, and the mathematician almost always denies the existence of upon demonstration.

Those who are accustomed to settle the meaning of ancient phrases by self-examination, will find some strange conclusions arrived at by us; but nothing, we believe, which may not be justified by even a moderate examination of old writers.

Language and counting both came before the logical discussion of either. It is not allowable to argue that something is or was, because it ought to be or ought to have been. That two negatives make an affirmative, ought to be: if *no* man have done *nothing*, the man who has done nothing does not exist, and *every* man has done *something*. But in Greek, and in uneducated English, it is unquestionable that "no man has done nothing" is only an emphatic way of saying that no man has done *anything;* and it would be absurd to reason that it could not have been so, because it should not.

B

The manner in which any common reckoning of time is made would, we might suppose, be a matter admitting of neither dispute nor ambiguity, and of little, if any, change. It is the object of this paper to point out that such is not the case—that language is to this day but ill adapted to express precise meaning—that serious and not sufficiently marked changes have taken place in the modes of reckoning—and that the confusion which these changes have made continues.

At the same time, there is nothing on which we are so positive, each for himself, as upon what is and what is not, right in the matter of reckoning time. Every one has, or thinks he has, a permanent meaning attached to the phrases in common use; which meaning no small number think these phrases must of necessity bear: others, aware of the very different senses which the phrases have borne, are content to admit that their meanings are conventional, but are prepared to contend for the existence of a well-settled and universal convention.

Suppose that at ten o'clock on Monday morning, a person engages to do something in four days. There are four distinct meanings, each of which will bear argument or citation of authority, and each of which may have been in the understanding of the speaker or of the hearer.

First, not counting Monday, on which the engagement is made, Tuesday, Wednesday, Thursday, Friday, may be claimed as appertaining to the four days; in such manner that the pledge cannot be considered as broken, until some moment of Saturday has arrived without its performance. This is one extreme case, and is the debtor's version.

Secondly, counting Monday, the day of the engagement, the four days may be reckoned as Monday, Tuesday, Wednesday, Thursday, and the claimant may consider that at any moment of Thursday proof of performance is due. This is the other extreme case, and is the creditor's version.

Thirdly, different tendencies towards a mixed mode of interpretation may lead to the result that Friday is the day on which the performance may be claimed: and for this many will pronounce, when they consider the question, from mere indecision between the preceding two cases.

Fourthly, those who consider a day as capable of beginning at any moment will say that from ten o'clock on Monday morning to ten o'clock on Friday morning it is four complete days; and that therefore proof of performance may be claimed on Friday, but not before ten o'clock in the morning.

If we had to *make* a meaning for the phrase, we might well fix on the third, which perhaps would be most generally agreed on in our day as the proper interpretation. But we are to consider the meanings which *have* prevailed: this is a very different thing, and will require a little discussion of the stages of the process of counting.

The earliest process of arithmetic is that of counting units, the unit being considered both as the commencement and the ultimate subdivision of the process: in such manner that between ten and eleven, for example, there is nothing imaginable; nothing more between ten and eleven feet than between ten and eleven horses. As to the latter instance, we should still agree with the ancients: we should refuse to admit of any number of horses to ride upon between ten and eleven, being wholly unused to see such chargers as Baron Munchausen's; though we might admit *ten and a half* horses as a possible sale of dog's meat. But with regard to feet we should be inclined to assert that there are lengths between ten feet and eleven feet, and that he restricts modern language, to say the least, who asserts that there is no *number of feet* between ten and eleven, though it may be proper to say that there is no intermediate number of *complete feet*, no *whole number* of feet. But *we* are familiar* with fractions: down to the beginning of the sixteenth century, the Romans, and all the Europeans who used Latin, were so strange to the idea of fractions of numerable units, that the books of arithmetic hardly contain a notion of them. The editions of Boethius (the most common text-book of the learned) published in the fifteenth century, though perfectly free from all allusion to fractions, actually use lengths, by which fractions of a unit can be shown, to indicate the integers of their processes. And this usage is precisely due to the absence of the notion of fractions. A modern teacher uses dots, drawings of pebbles, or horses, or men, by which to inculcate purely *monadic* counting. He avoids length or area, or other simple magnitude, in the first instance, because he does not want to suggest the fraction before its time, which he thinks he should certainly do, if he employed a unit capable of division into parts like itself. But the teacher of the fifteenth century felt

* It is not perhaps allowable to say, as yet, that power of applying notions of arithmetic, except in routine ciphering, is very common. A man of high scientific station, now deceased, who was long a member of the House of Commons, used to say that there never were, at any one time of his continuance in it, more than three men in the House who had a tolerable notion of fractions. The following will show that a palpable absurdity will pass before the eyes of generations of men of letters without notice. In Boswell's Life of Johnson (chapter viii. of the edition with chapters) there is given a conversation between Dr. Adams and Johnson, in which the latter asserts that he could finish his Dictionary in three years. "ADAMS. But the French Academy, which consists of forty members, took forty years to compile their Dictionary.—JOHNSON. Sir, thus it is. This is the proportion. Let me see: forty times forty is sixteen hundred. As three to sixteen hundred, so is the proportion of an Englishman to a Frenchman." No one of the numerous editors of Boswell has made a note upon this, though many things as slight have been commented upon: it was certainly not Johnson's mistake, for he was a clear-headed arithmetician. How many of our readers will stare, and wonder what we are talking about, and what the mistake is?

B 2

no such fear: and there cannot be imagined such a proof of it, as the use of divisible magnitude to signify number in a book so thoroughly monadic as the arithmetic of Boethius. The earliest editions of the *arithmetical* books of Euclid do the same thing.

In our day some notion of fractions is learnt so early and has become so familiar, that the *monad,** or *indivisible unit,* has almost disappeared. There are few elements of computation which we are not accustomed to separate into subordinate elements, which are actually parts of themselves: not merely conventionally, as in the case of the twelve copper pence which are *held* equivalent to the silver shilling (this could be done in the old system), but actually, as in the case of the twelve inches which are the very parts of the foot. We have accordingly forgotten the old maxim that unity has no parts, which was so well fixed in the minds of our forefathers that they likened *unity* in arithmetic to the *point* in geometry. It was in 1585 that Stevinus (one of the most original minds of his day; no less man was wanted) dared to say that it is 0 in arithmetic which answers to the point in geometry, and to pray that the Author of nature would have pity upon the unfortunate eyes of those who could not see it to be so.

We are arriving at the other extreme, or shall do so, if books of arithmetic do not soon begin to inculcate the distinction of monad and magnitude. A man had a letter to send, for which the postage was one penny. Not having a penny stamp by him, he cut a twopenny one in half, and affixed one of the halves to his letter. The post-office clerks, who are monadists, considered this as an unpaid letter, and charged it accordingly; on which the fractionist, considering himself unjustly used, presented more than one memorial to the higher authorities.

It was not impossible, in the old reckoning, to imagine that the monad of the reckoning was only a *part* of its divisible space, the other part being nonexistent: for example, that *time* should be reckoned by *days*, of *twelve hours each*, the intermediate nights being blotted out. The founder of our æra, Dionysius Exiguus, does this: in the last of the rules presently alluded to, he makes his day to be twelve hours, multiplies his number of days by twelve to get the hours,† and adds *three hours* each year to get the intercalary day once in four years: *istæ tres horæ faciunt in IIII annis diem.*

* It may be convenient to revive this old term, in the sense stated. Any magnitude whatever may be considered as a monad, or unit for repetition incapable of division.

† The question will naturally arise, might not Dionysius by possibility be making use of an hour twice as long as the usual one? Fortunately he explains himself on this point, indirectly. He pronounces against a curious notion extant in his time, that the additional day of leap-year was a commemoration of the long day on which Joshua caused the sun to stand still: here *day* means *term of daylight*, but *dies* is still the phrase used. Dionysius had a very odd notion of the meaning of *bissextile*. He thinks that

Something resembling this is the old interpretation of the intercalary day in leap-year. It was not a *new and additional* day, nor counted as such: it would not have been held correct to say that leap-year has 366 days. It is one day repeated: and the two days of *bis*-sextile have but one name. It counted as one day in the calendar: and was allowed for, not by letting it take a letter, and altering the letters of the following days: but by letting the second sextile day keep the letter of the first, *and shifting the dominical letter of the year on its arrival.* Moreover, the double length of the bis-sextile was not allowed to add a day to the moon's age *at the time.* So that the idea mentioned in the note, of the double day representing the *long day* of Joshua, is a misapprehension of easier occurrence than might have been supposed.

A person who is born on the 10th of June, in our day, counts a year as completed so often as a 10th of June arrives. He says, I shall not be of age until the 10th of June; ask him how old he is on the 9th, and he will say, I shall not be of age till to-morrow. If he were born at noon, it is true that he does not complete twenty-one years of days divisible into fractions until noon of the 10th. Nevertheless, in the law, which here preserves the old reckoning, he is of full age on the *ninth:* though he were born a minute before midnight on the 10th, he is of age to execute a settlement at a minute after midnight on the morning of the 9th, forty-eight hours all but two minutes before he has drawn breath for the space of twenty-one years. The law reasons thus;—there are no parts of days; he who is born on the 10th takes the whole of the 10th as part of his life; he is a year old when he has completed 365 days; the 9th of next year is his 365th day; as soon as he has commenced* the 9th, he has lived through the whole of it, for a day has no parts; therefore he has lived a complete year, or is one year old, as soon as the 9th arrives. And the conclusion is unavoidable so soon as it is granted that a day has no parts. The anniversary of birth used to be celebrated as the first day of a new year; it is now considered as the completion of the old one.

the intercalary day was called bissextile, because each year contributed *twice six* points towards it: by a point he means, as he says, a quarter of an hour. Each year then, according to him, contributes *three hours:* so that not only does he himself reckon by days and pass over nights, but he asserts his belief that such had been the custom of his predecessors.

* "'Five years auld exactly this blessed day,' answered the lady, 'so we may look into the English gentleman's paper.'.......'No, my dear, not till to-morrow. The last time I was at Quarter Sessions, the Sheriff told us that......a term day is not begun till it's ended.'—'That sounds like nonsense, my dear.'—'May be so, my dear, but it may be very good law for all that.'"—*Guy Mannering.* Whether the Scotch law differs from the English, or Scott intended the Laird to blunder the point, or blundered it himself, are matters which I must leave to the learned.

We can never without explanation get at the meaning of a person who tells us he was ill for two days. Some will apply the phrase to the last half of one day and the first half of the next; some to two whole days with a fraction before and after; some to an interval of forty-eight hours, made out of one day, and parts of the preceding and following.

But there is another difference between old and new times yet more remarkable, for we have nothing of it now: whereas, in things indivisible, we count with our fathers, and should say, in buying an acre of land, that the result has no parts, and that the purchaser, till he owns all the ground, owns none, the change of possession being instantaneous. This second difference lies in the habit of considering nothing, nought, zero, cipher, or whatever it may be called, to be at the beginning of the scale of numbers. Count four days from Monday: we should now say Tuesday, Wednesday, Thursday, Friday; formerly it would have been Monday, Tuesday, Wednesday, Thursday. Had we asked, what at that rate is the first day from Monday, all would have stared at a phrase they had never heard. Those who were capable of extending language would have said, Why it must be Monday itself: the rest would have said, There can be no first day from Monday, for the day after is Tuesday, which must be the second day; Monday, one; Tuesday, two.

We should say, Monday *does not count*, being the day itself we reckon *from*: in Roman numeration, described by us, it would be, every day *counts*: though a Roman would probably also have said that Monday did *not* count. His scale of numeration beginning at *one*, the mere repetition of *one* would not have been considered as counting, which would begin with the entrance of plurality into the reckoning. We know that it was long usual to deny* that *one* is a number: an assertion derived partly from the idea of plurality being attached to the word number, which would have justified the assertion that *enumeration* begins at two, not at one, and partly from unity being a kind of starting point.

When, at the time of the reformation of the Calendar, the moon of the heavens was full, as we should say, four days before

* This denial lurks in the following old rhyme, which some will remember to have heard, and which Mr. Halliwell has inserted in his collection of nursery rhymes:—

One's none,
Two's some,
Three's a many,
Four's a penny,
Five's a little hundred.

The last line refers to five score, the so-called hundred being more usually six score. The first line, looked at etymologically, is *One is not one,* and the change of thought by which *none*, the denial of *one*, comes to be associated with the denial of *plurality*, is curious.

the ecclesiastical moon, the phrase was *five* days: and when a mode of reckoning this by syllables was invented (see the Companion for 1845, p. 22), *Nova luna hic*, the reckoning started with its first syllable *on* the day of the ecclesiastical moon.

European counting, antecedent to the introduction of the Indian numerals, was entirely fashioned upon the Roman system, in which no symbol for *nothing* exists. The Indian zero, or cipher, in the first instance, was not an express symbol for *nothing*, any more than the blank between two words is an express symbol for no amount of letter-press: it merely served the purpose of the blank type, namely, to keep the rest in their places. The notion of *absence of value*, or *value not yet attained*, as a starting point from which to reckon the introduction of successive amounts of value, was an idea of very slow growth, an ultimate consequence of the suggestion of the symbol 0, but not a part of its first intention. The complete mastery of this notion is among the masonic signs by which one mathematician can detect another in his writings on any subject. But many have it now, to an extent which makes 0, 1, 2, 3, 4, &c., a common series enough; though those who have cast their eyes over books of arithmetic will remember that 1, 2, 3, 4, 5, 6, 7, 8, 9, 0, is a much more usual exposition of the numeral symbols than 0, 1, 2, 3, 4, 5, 6, 7, 8, 9.

Our language now groans under the difficulty of expressing the various ways in which the two terms may be connected with the interval. From Monday to Thursday may mean both inclusive, or both exclusive, or either inclusive and the other exclusive. But this is not the fault of our language, so much as of our imperfect habits: a complete distinction might be made without forcing a single word. For instance—

From Monday to Thursday	Monday, Tuesday, Wednesday, Thursday
Between Monday and Thursday	Tuesday, Wednesday
After Monday to Thursday	Tuesday, Wednesday, Thursday
From Monday before Thursday	Monday, Tuesday, Wednesday.

It is not by any means certain, in our language, whether the word *until* includes what follows or not: "until Thursday," when it refers to an action which does not occupy the whole day, would certainly imply that the day is broken by the action, as in "I remain until Thursday." But take this sentence—"In England, preserve old style until September 14, 1752;" is September 14 the last day of old style or the first day of new? In Roman reckoning it would be both, and this sentence would make us sure that September 15 (that would have been) was the first of the nominal days omitted in the change of style. But, though it be frequently said that old style lasted till September 14, it means that September 2 was the last day of old style, and that, to use the words of the act, "the Natural Day next immediately following the said Second Day of September, shall be called, reckoned, and accounted, to be the Fourteenth Day of September."

There was no confusion as to this matter, of old; for the reckoning always included both its terms or endings, unless otherwise distinctly specified. Whenever it did not do so, the exception required statement. In matters of law, an extension was sometimes admitted by way of privilege, where the usual interpretation would involve penalty or forfeiture: but the distinctness with which this is stated causes the exceptions to confirm the rule. Taking up an old digest of the canon law, we find it *stated* that in the days of a citation the day of service is *not* counted in the term; so that a man cited on Monday to appear within three days need not appear before Thursday. When a benefice lapsed to the bishop by non-presentation on the part of the patron, it is *stated* that the day on which the vacancy occurred was not counted; and the interference of a bishop on the 6th of October, the vacancy having occurred on the 6th of July, was held void. But the case was recent (1703), and the compiler of the digest seems to doubt that the decision was according to the old law.

The French to this day speak of this day week as *huit jours* from to-day (which is therefore included), and of this day fortnight as *quinze jours*. We use *seven* and *fourteen*; but it is not to be inferred that the mode of counting is different. For our old reckoning is by nights, as was that of the ancient Germans; this day week was "this day se'nnight," and this day two weeks is still this day *fortnight* (fourteen nights). Now from Wednes*day* to Wednes*day* there are but seven *nights* intervening, though the inclusion of both Wednesdays may make eight *days*.

In music, the note immediately above another is not merely called the *second* but the second *above*: though, in modern idiom, the ascent being A, B, C, &c., the second above A must be C. And there being seven notes in the scale, the A which comes next above any A is called its eighth, or octave; and the next A the fifteenth, after the manner in which the French reckon days, or that in which the day week of a saint's day is called the *octave* of the Saint. But one of the most decided effects of the old custom of counting both terms as part of the period is the practice of calling the time from the 1st of January to the 1st of January, not a year, but a year and a day. The origin of this phrase we take to be obvious enough, though it may be questioned whether those who have given it have always seen how it arose. Coke expressly lays it down that in the phrase year and day, the day from which reckoning is made must be included; but without any allusion to the mode of entrance of the phrase. Cowell, in his Interpreter, says "*Year and day* . . . is a time thought in construction of our common law, fit in many cases to determine a right in one, and to work an usucapion or prescription in another:" he might have added that it is frequently mentioned in old statutes. And in almost all the cases then cited, it is obviously either an allowance to avoid hardship, or a stretch of the term by the king's prerogative, for the benefit of the crown. In old poetry it is a very common term, and its imitators frequently do not understand it. In Walter Scott's ballad of 'The Noble Moringer,' said by him to be translation from old German, the translation has what we should not believe to be in the original, unless we saw it. The lady has engaged to await her husband's return seven years and a day, according to which, by the old method of counting, she would be at liberty to marry again on any hour of what we should call that day seven years. But the ballad (the translation at least) makes the lady, who is true to the letter of her word, sit waiting till twelve o'clock at night on that day seven years, before she will have the ceremony performed with her new bridegroom. The husband arrives just in time, and the lady says—

> " . . . .Count the term howe'er you will,
> So that you count aright,
> Seven twelvemonths and a day are out
> When bells toll twelve to-night."

We will answer for it, that in the fourteenth century, the lady would not have waited till the odd day was *finished*. In the ballad called the 'Eve of St. John' there is a similar failure of attention to the old custom. The baron of that ballad goes away for "*three* days' space," and on his return the page, who is a spy on the lady, tells him where she has walked for three successive nights; according to which, in the language of the time, the baron was away for *four* days' space.

The necessity of taking in the terminus of reckoning on each side, follows immediately from *one* being the commencement of all counting; those who begin from *nought*, make 0 to represent the initial term, *from* which they reckon. The former reckon *three* from Wednesday to Friday; the latter *two*. The Romans carried the former process to its extreme; or ra-

B 3

ther, never advanced* beyond this rudiment. If the Kalends of February fall on a Friday, the *third* day *before* the Kalends, expressed by the singular phrase *ante diem tertium Kalendas Februarii*, is Wednesday. Any one might suppose that scholars, though aware of this method, had always forgotten to interpret Roman phrases by Roman usage. When Livy speaks of a lunar cycle which begins every *twentieth year*, this plain allusion to the famous cycle of *nineteen years* has never been noticed till our own day (see Dr. Smith's *Dictionary of Greek and Roman Antiquities*, article *Calendar*), and a conjectural emendation of the text had been substituted.

We might have supposed that by referring to the usage of the law, we should be able to settle the fact that the Roman method of reckoning was at one time universally used, and also that later times have either avowedly continued, or avowedly changed, the ancient practice. Instead of this however we find that there is no acknowledgment of the Roman method having once been in use, and that the struggle by which something more modern has been at last instituted, has been made without any distinct knowledge of, or at the least without any distinct reference to, the state of things which once existed.

We feel a right to take for granted, until the contrary is shown, that the original method of the Roman world must have been at one time predominant in systems of law: and we find that the oldest statutes in which reckoning of terms is employed bear us out, unless indeed we are to conclude that legislation had a habit of commencing with the *second* day of a month or year instead of the *first*. Thus the statute 35 Hen. VIII. cap. 12 remits money borrowed by the king *sithence* [since] the first of January: surely this was meant to include all moneys borrowed on any day of that month. In 37 Hen. VIII. cap. 20, *sithen* and *from* are used interchangeably as to one date. In innumerable cases *after* is used with *from and after*, as synonymous, even in the same sentence: and in 1 Edward VI. cap. 1, "after" and "from and after" and "immediately after" the *first day* of May are used synonymously. The number of cases in which "*after the first*" of a month occurs is so great that we cannot imagine how, supposing the *first day* to be excluded, it escaped being a popular maxim that in law a month begins on the second day. In 35 Hen. VIII. cap. 17, there is legislation for "after" the feast of St. Michael, and

* So late as in the seventeenth century, Petavius says that the fourth year *of* the Julian reckoning is the *fifth* year *from* the year of confusion (which preceded that reckoning); we should say that the fourth *of* any enumeration is also the fourth *from* that which precedes the first. None but physicians of our time will understand a joke which, as the newspapers of the day on which we write this inform us, the Romans made on the French before the walls of Rome. The latter made their assaults every other day, on which the former said they had the *tertian* fever.

"before" the feast; and one or other of these certainly legislates for the day of the feast itself. The statute 2 Edward VI. c. 1, prescribing uniformity in the reading of the liturgy *after* the feast of Pentecost, also legislates for *before* that feast, which is included in one or the other phrase: the probability is, that some marked festival would have been chosen as the beginning of uniformity, not as the end of discordance. In 33 Hen. VIII. cap. 12, some crimes committed *since* and *sith* the feast of All Saints are punished, and the statute is "to take effect from" that feast: for other crimes, "from and after" the first day of May. But the following is almost conclusive. In 26 Hen. VIII. c. 3, the bishop must certify to the Exchequer "before the said first day of April, or at any time within four and twenty days next after the said first day of April." It cannot surely be that the first of April was excluded. By the same statute the king is to have first fruits of all clerical persons nominated to benefices "after the first day of January next coming," and all first fruits "from" that day. In a collection of statutes regulating merchandize made in the first of Richard III., c. 8 is for "*after* the feast of St. Michael;" cap. 11 is to "take effect *at* the feast of St. Michael;" and cap. 13 "*from* the feast of St. Michael:" these phrases appear to be synonymous. In 6 Hen. VIII. c. 4, "from the fifteenth day of Easter next coming, or after," is used in one clause synonymously with "at the said fifteenth day of Easter, or after." In this statute, and in those in which the *Quindecim Pasque* is mentioned, the reckoning of weeks is inclusive of both ends. The following instance is more perfect than any. By 21 Hen. VIII. c. 13, spiritual persons must alien certain profits to laymen "*on this side* of the feast of St. Michael," and every lease *to them* of such profits made "*after* the said feast of St. Michael" is void. Will any one say that the *day of the feast* was left open, as by modern interpretation it should have been? We have no doubt, ourselves, that Michaelmas day was included in both, that the penalties would not have been incurred if the alienation had been deferred until that day, as well as that a lease granted on that day would have been void. We have not neglected the possible answer, that Michaelmas day might fall on a Sunday in that year: it fell on a Wednesday. These are a few instances, out of an immense number of the same kind, all tending, even without knowledge of the original mode of reckoning, first to establish the identity of meaning of the phrases "from," "after," "from and after," "next after;" secondly, to show that all of these included the day from which reckoning is made.

The simultaneous use of "after" and "from and after" can

be traced in later times; as in 10 and 11 Will. III. c. 10, which forbids *from and after*, &c.; and punishes those who do it *after*. At this period, it would probably have been settled that the day mentioned is excluded by these words: and the more so, as in cap. 1 of the same session *on or before* is set in opposition to *from and after*. The most modern statutes exclude the day from which the reckoning is made, and thus we often see "from and after the 31st of December." But such explanation is still sometimes thought necessary as is given in 7 and 8 Vict. cap. 76, in which a provision "shall commence and take effect from the 31st day of December, 1844, and shall not extend to any thing done before the first day of January, 1845."

In the time of William III., it was decided that the day from which reckoning was made is included. We are perfectly aware that this decision has since been questioned: to this we pay no more regard, as an antiquarian conclusion, than we do to the decision itself, because neither those who decided nor those who questioned showed themselves aware even of the existence of the Roman method, and therefore neither admitted nor refuted that there must have been a time at which the law must have agreed with the universal practice of the learned, and, for any thing ever shown to the contrary, of the common people also. In Bellasis *v.* Hester (9 William III., Raymond 280,) the point in question was, what was meant by a bill being payable ten days after sight. The parties concerned had neglected to plead the special custom of merchants, which the court therefore refused to consider, and the words were left to take their common legal meaning. In the opinion of all the judges but one, the day of sight was included: that one (Justice Treby) differed on principles of logical interpretation of language. If, he said, the day of sight be included, then the first day after sight is the day of sight itself, which would be absurd. That neither the dissentient judge nor either of the others should have remembered the way in which the Romans reckoned backwards from the Kalends (not of course for any purpose of law, but with reference to the asserted *reductio ad absurdum*), shows how completely the origin of the mode of reckoning which the court pronounced for had been forgotten.

The various classes of decisions which have been made upon this point, as that—when reckoning is made from an act, the day of the act is included, but when from a day, that day is excluded—that the day of an act shall be reckoned or not, according as the party affected is or is not privy to the act—that a day shall be reckoned or not, according as one course or the other will best effect the intention of the party whose intention is to be carried into effect—are all too modern to prove any thing except this, that there has been a struggle between opponent methods.

The old statutes fully satisfy us that, in the middle ages, the time *from* a day and the time *after* a day included that day, and that in the words "*from* and after" we see nothing but the usual iteration of legal phraseology. The common idioms of our language would confirm this, so far as they confirm any thing. We reckon the year *from* the 1st of January, the week *from* Sunday; life dates *from* the day of birth. The same reasoning which has introduced what we may hold to be a more logical use of the words—but which is only so to those whose scale is fashioned upon 0, 1, 2, 3, &c., instead of 1, 2, 3, &c.—has also destroyed other similar uses of language. The term of comparison, however distinct from the things compared with it, was placed among them in speech, just as, in counting, the term of departure was included among the results of departure. Milton gave Lindley Murray and his followers occasion against him when he called Eve the fairest of her daughters, though he wrote recognised English: and such expressions as "this is the most correct of all the others," are not uncommon in old writing.

The following is a striking case in point. The description of Easter day, as given in the old prayer books, is very ancient, and it runs thus:—"*Easter* day . . . is always the first Sunday after the first full moon, which happens next after the One-and-twentieth day of *March.* And if the full moon happens upon a Sunday, *Easter day* is the Sunday after." In this paragraph there are two evidences. Unless "the first Sunday *after* the Full Moon" had been a phrase inclusive of *the Sunday of Full Moon*, the last sentence would have been useless. Again, here is also the full moon which happens *next after* March 21, without any qualification in the case of full moon *on* March 21. And all who know how Easter is reckoned know that this phrase does here include the 21st: if there be a full (calendar) moon *on* the 21st, it is the paschal moon. The legislators of Geo. II., in changing the style, have translated this into modern idiom; their phrase is, "which happens *upon* or next after the 21st day of March." The mistake of reading "full moon" instead of "fourteenth day of the moon," might open an escape from these conclusions, as suggested in the 'Companion' for 1846, p. 5, which would have prevented our bringing them forward, if it had not happened that they are reinforced by another part of the same set of rules: "Ascension day is forty days after Easter." Now among the applications of this rule given by its framers it is found that when Easter day is April 22 or March 26, Ascension day is May 31 or May 4: and it cannot be said that it is forty-days from April 22 to May 31, unless April 22 itself be counted; and the same of March 26 and May 4. The statute 1 Edwd. IV. cap. 2, which begins to take effect "a la quarantisme jour proschein apres le vj[me] jour . . . ." must be supposed to have reckoned this common term, forty days, in the same way.

The reader must not understand us as supporting the position that the day from which reckoning was made was held as belonging *more* to time *after* than to time *before*. According to the principles of ancient counting it would have belonged to *both*, as now to *neither*. We have seen that the unit of reckoning was, from being held indivisible, regarded in the same light as the point, which equally belongs to the line it terminates, and the continuation which it commences.

In mentioning the old statutes, we have hinted our belief that if the common phrase "*after* the first of . . . " did not include *the first*, there would most likely have arisen such a phrase among the people as that in law the second day of a month is the first. We lay more stress upon this than we can venture to propose to any of our readers to do, except to those who are aware how common it is in old English for that which takes the place to take the name. Thus *six score* got the *name* of a *hundred*, because it was common to give 120 to purchasers of 100; and a hundred *and twelve* pounds the name of a hundred weight for a like reason. In assaying metals, the arbitrary piece cut out to try how much *in the pound* was alloy, got the name of *a pound*, and was called the pound *subtile*; and this though the piece cut off were only a few grains. And such uses of language were recognised in their broadest form by statute: it was enacted that the hundred of herrings *shall be* six score.

With so much proof before us that no pains whatever have been taken to preserve the ancient system in ancient history, there is no occasion to shrink from an examination of the views usually entertained of the Julian and Augustan corrections of the calendar: in which it can, we think, be easily made to appear that, for want of permitting Roman words to be significative of Roman meaning, chronologists have arrived at a very unlikely view both of the Julian scheme, and of the Augustan correction.

The error which the priests committed in interpreting the reform made by Sosigenes at the command of Julius Cæsar, consisted in counting every fourth year *by making the year which ends one period begin another*, just as Livy did in describing the Metonic cycle as recommencing every twentieth year: that is, they made the reckoning in strict conformity to the principle that the terminus is included. Thus 1 being leap-year, 4, 7, 10, &c., would be leap-years also: *four* counted from *seven* inclusive brings us to *ten*. As there is no question that this was set right by Augustus, and the superabundant intercalations introduced by the priests allowed for by a sufficient suppression of subsequent ones, it is presumed that we may reckon back as follows. When the Augustan period of correction was passed, it is certain that all the years divisible by four were leap-years. Accordingly A. D. 12 was leap-year, and A. D. 8, and so, *it is said*, would have been A. D. 4, according to the intention of Julius. The next preceding leap-year would have been B. C. 1, the year immediately preceding A. D. 1; the next before that B. C. 5, and so on, each year B. C. being leap-year which divided by four leaves a remainder 1. At this rate B. C. 45 would have been leap-year. Now B. C. 45 was the first year of the Julian reckoning: it is assumed then that Cæsar commenced with a leap-year. The great argument in favour of this is that by the number of leap-years thus introduced we are brought exactly back to what must have been the first of January, B. C. 45. For Cæsar commenced his year with a *new moon*:* and, just taking in such additional days as the preceding system of leap-years gives, we are brought, for this back-reckoned first of January, B. C. 45, to a day at which it was new moon at Rome at 11 o'clock in the evening.

We admit therefore the number of days introduced by the preceding hypothesis to be correct: so that there will be no dispute as to what was the actual day of the back-reckoning on which the first day of the first year of the Julian reform fell. But the supposition that Cæsar made a leap-year at the very commencement, is one of the most forced and unnatural that ever was pressed into the service of an explanation. For the preceding year, B. C. 46, thence called the *year of confusion*, had been made to consist of 445 days! No reason could ever be given why the additional day of February, which was made to allow for the odd six hours of the solar revolution as fast as they amounted to a day, should have been paid in advance: or why, after every thing had been upset by the year of confusion, any want of an additional day could have been felt in the first year, when, if ever, all was straight to begin with.

On looking at the manner in which the Augustan correction is generally stated, it appeared to us easy enough to explain

* Did the Egyptian astronomer know the very day of the full moon, when it happened at 11 o'clock in the evening? Are we to take it for granted that by two hours or more of error he might not throw it into the wrong day? The usual answer is affirmative. To us, however, the accordance with records of the system we put forward rather confirms the astronomer, than the converse.

the manner in which, without sacrificing a single day, the system of leap-years which lasted up to the Gregorian reformation was brought about, namely, that the years which are divisible by four became leap-years. In the following table, the explanation we propose is given on the right, and the most common one on the left. S stands for a sacerdotal leap-year, or one of those which were actually so: J stands for an intended leap-year of the Julian reformation: A stands for a leap-year after the Augustan edict. The years B. C. and A. D. are given, and also those of the Julian reckoning.

Year of		J. Y.	B. C. 46	Confusion.	
J 1	S 1	1	45		
		2	44		
		3	43	J 1	S 1
	S 2	4	42		
J 2		5	41		
		6	40		S 2
	S 3	7	39	J 2	
		8	38		
J 3		9	37		S 3
	S 4	10	36		
		11	35	J 3	
		12	34		S 4
J 4	S 5	13	33		
		14	32		
		15	31	J 4	S 5
	S 6	16	30		
J 5		17	29		
		18	28		S 6
	S 7	19	27	J 5	
		20	26		
J 6		21	25		S 7
	S 8	22	24		
		23	23	J 6	
		24	22		S 8
J 7	S 9	25	21		
		26	20		
		27	19	J 7	S 9
	S 10	28	18		
J 8		29	17		
		30	16		S 10
	S 11	31	15	J 8	
		32	14		
J 9		33	13		S 11
	S 12	34	12		
		35	11	J 9	
		36	10		S 12
J 10	S 13	37	9		
Edict of Augustus		38	8	No leap-year for 12 years.	
		39	7	J 10	
		40	6		
J 11		41	5		

		J. Y.	B. C.		
		42	4		
		43	3	J 11	
		44	2		
J 12		45	1		
			A. D.		
		46	1		
		47	2	J 12	
		48	3		
J 13		49	4		A 1
		50	5		
		51	6	J 13	
		52	7		
J 14	A 1	53	8		A 2

In both systems, the leap-years marked *sacerdotal* are actual; those marked *Julian* are only in the reputed intention of the reformer, except where also marked sacerdotal. Those marked *Augustan* are actual.

The difficulties of the two systems are as follows :—

Common system. First, the year of commencement is made to be leap-year, as already mentioned. Secondly, when Augustus ordained that there should be no leap-year for twelve years, he is made to have ordained that there should be none for *fifteen* years, in our way of reading, or for *sixteen* years in the Roman way. Thirdly, it is assumed that the Julian system began by paying (at the rate of six hours per annum) in advance, while, after the Augustan *vacation*, the payment was made only when due.

Proposed system. When Augustus ordained that there should be no leap-year for twelve years, he is made to have ordained, in our way of speaking, that there should be none for *eleven* years only.

Some persons may think that the final mode of correction is to be interpreted thus; that three sacerdotal (intended) leap-years should be omitted, and that then the reckoning should begin according to the Julian intention. But this would make A. D. 5 to be the first Augustan leap-year, and the common rule for determining leap-year would never have been established.

In asserting the probability of the system we have advanced, it will be observed that we maintain *no leap-year for twelve years* to be a phrase synonymous with *leap-year in the twelfth year*. This is the necessary consequence of a strict, but usual, rendering of the maxim, that the last of the old reckoning is the first of the new, to which Roman enumeration so strictly adhered that there is no first day before the Kalends except the day of the Kalends itself. Putting the difficulties of the two systems against each other, we think it may be safely inferred that the one we propose is very much less than the cumulative amount of the three on the other side. *Twelve* cannot be *twelve* in our sense: shall it be our fifteen or sixteen under no rule at all, or shall it be our eleven under a practice which we know to have been common, and which we see in the divisions of the Roman month?

So much on the question of probability: we shall now look at the words of the historians who describe what actually took place. Of these there are three whose accounts are usually, and justly,* preferred—Censorinus, A. D. 238; Solinus, probably his contemporary; and Macrobius, about A. D. 400. From Censorinus we learn nothing as to the mistake or the correction, only that the intercalary day was to be inserted after each elapsed period of four years, *peracto quadriennii circuitu.* All that has any allusion to the correction, is the information that the month Sextilis received the name of August when Martius Censorinus and C. Asinius Gallio were consuls; and as it is otherwise known that this change of name took place at the Augustan correction of the calendar, and that the above-named were consuls in the year 8 B. C., confirmation is given to the date of this correction. Solinus states that Cæsar added a quarter of a day in the year of confusion, which, as it is impossible to imagine a fraction of a day in any one year, we must take to mean that the year of confusion was considered as furnishing its quotum towards the first bissextile, so that the first bissextile would be the year 3 of the corrected calendar, or B. C. 43. Solinus further states that the priests made the error of adding the bissextile in the fourth year, instead of after the close of the fourth year; and that thus they added twelve days in the lapse of thirty-six years, while only nine ought to have been added, which fault Augustus reformed, and commanded that twelve years should run out without intercalation, *jussit annos XII. sine intercalatione decurrere.* Now observe, first, that in the system we propose, there are *twelve*, and should have been *nine*, sacerdotal leap-years preceding the intervention of Augustus, whereas, taking B. C. 45 as leap-year makes *thirteen* actual and *ten* intended leap-years. Secondly, in our system it takes the priests exactly thirty-six years to make this error; whereas, if B. C. 45 be taken as leap-year, they make the error described by Solinus in thirty-four years, and that which he should have described in thirty-seven years. The two isolated facts stated by this writer—first, that the year of confusion was considered as furnishing its quotum towards

* Some writers are very confused: Pliny, for example, interprets three leap-years omitted by Augustus into three new corrections upon corrections of the whole calendar by Sosigenes himself.

an intercalation; secondly, that the total amount of the sacerdotal error accrued in thirty-six years—support one another.

Macrobius repeats the statement of Solinus as to the thirty-six years, and tells the story of the correction of Augustus in very much the same manner. But he has one sentence more. Not being a Roman, and coming further from the events than his predecessors, it is likely that he should have searched for monuments. He mentions one of a remarkable character, a brass inscription ordained by Augustus for the perpetual preservation of the calendar; and we must presume that in mentioning the arrangement which this inscription perpetuated, he used its words. He tells us* that, after commanding that twelve years should expire without intercalation (*annos XII. sine intercalari die transigi jussit*), he directed that future intercalations should be made *every fifth year*, as Cæsar had ordained. Thus it appears that Augustus, finding the imported phrase of Sosigenes had been mistaken, substituted a more correct one to Roman ears. According to their counting, the selection of 8, 12, 16, &c., after 4 as a commencement, is the selection of every *fifth* number. This proof that the phrase first introduced was changed, in order that the direction might be given in the strictest Roman idiom, will justify us in asserting that every part of the direction, as given by Macrobius from the inscription, is to be as strictly rendered in the same way. Since, then, twelve years are to pass over without leap-year, we interpret it that the twelfth year was the next leap-year. To those who were well accustomed to begin new reckoning from the terminus at which they had arrived in the old one, it would not suggest itself as an impediment that there is logical absurdity in the last of the unintercalated years being the first of the intercalated ones. This brings the first Augustan, and thirteenth actual, leap-year, to A. D. 4, and the fourteenth actual leap-year to A. D. 8: being as if the Julian intention had been that B. C. 45 should have been leap-year. It is essential, as before explained, that the fourteenth actual intercalation should take place in A. D. 8: but the common system can only attain this by demanding that, under an edict of cessation of leap-year for twelve years, there should then be no leap-year *until* four more years had elapsed. This is an inconsistent way out of the difficulty, seeing that the way into it was a demand that intercalations should be considered as payable in advance.

We have not thought it necessary to trace out the origin of

* "Post hoc unum diem secundum ordinationem Cæsaris quinto quoque incipiente anno intercalari jussit [Augustus], et omnem hunc ordinem æreæ tabulæ ad æternam custodiam incisioni mandavit."

the palpably absurd statement which is found in various works in general estimation, namely, that the first of the Augustan leap-years was A. D. 7, after which they proceeded without mistake. How the leap-years afterwards obeyed the rule of falling into dates which are divisible without remainder by 4, is a mystery to those who adopt the statement, and think about it.

When a reckoning is made from 1, the century terminates at 100; but when it is made from 0, through 1, 2, &c., it terminates with 99. About the year 1799, there was discussion whether the eighteenth century terminated at the end of 1799, or at the end of 1800. This was equivalent to a discussion whether the usual reckoning had a year 0, or began with 1. It so happens that the history of our mode of reckoning has been made to have a point of obscurity which may tend to prolong this discussion; and perhaps some may be found to doubt whether this present year 1850, ends the first half of the nineteenth century, or begins the second.

A *century* is any collection of one hundred; its restriction to collection of years is modern. Most readers remember the "century of inventions," and many remember that they thought at first it was the account of some inventive century. Bale's work on English writers is divided into centuries, not of years, but of scholars; and centuries have been published of nativities, and of other things.

A century of years may begin or end with any year, just as a year of days may begin or end with any day; and as the year ending April 7 began at the preceding April 8, so the century ending 1745 began with 1646. But, in like manner as the year of reckoning (as distinguished from a year-space of measurement of time) begins with January 1, so it is presumed that a century is also a unit of reckoning, and has a definite commencement: and that it is so is clear, as to modern times, from the constant phraseology of writers, who talk of the twelfth century, the nineteenth century, &c. But it generally happens that, in speaking of centuries, writers are using a rough denomination: thus no one who finds a paragraph which alludes to the religious troubles of the sixteenth* century, can possibly guess whether that century be meant to begin with 1500 or 1501.

There is no ancient usage as to the beginning of centuries,

* It is to be regretted that we are obliged to talk of centuries under *numeral* figures which contradict the dates. *Fourteen* hundred and twenty is in the *fifteenth* century. We are always obliged to pause a moment before we put a year into its century: and even practised historical writers sometimes make a slip. The second edition of Mr. Macaulay's essays is their third impression; and yet (vol. ii. p. 15) it is said, "We know that, during the fierce contests of the *sixteenth* century, both the hostile parties spoke of the time of *Elizabeth* as a golden age." The italics, of course, are our own.

for the term, as applied to time, is not ancient. Ducange and old Latin dictionaries do not recognise *centuria* as meaning a hundred years. The bull for the reformation of the calendar (1582), when speaking of 1700, 1800, &c., as not being leap-years, calls them *anni centesimi, hundredth years.* But no argument can be derived in favour of an implication that technical centuries *end* with these years; for no such technical term seems to have been then in use.

Again, this very regulation with respect to 1700, &c., affects the calendar rules in such manner, that a rule which lasts from 1700 to 1799 has to be changed for 1800, &c. It is, therefore, matter of necessity that writers on the calendar speak of 1700—1799 as a century. This happens in the tables annexed to the act for the change of style, in which mention is made for instance of "the next century, that is, from the year 1800 till the year 1899 inclusive." Hence many have argued that it is settled by law that the present century begins with 1800. But the body of the act, which is of equal authority, calls 1800 a *hundredth year*, when, if the centuries be settled by the wording of the annexed tables, it should be called a *first year.* But no inference can be drawn; for if Clavius had taken, say 1816, to be one of the Gregorian omissions of a leap-year, then the tables annexed to the act must have spoken of the century beginning with 1816 and ending with 1915, because that particular century-space would have fallen under one rule.

Clavius gives it as the reason why *centesimal years* should be chosen for omission of leap-years, that these are years of great note, being observed by the church as years of jubilee. Had he attached to 1600, 1700, &c., any idea either of commencement or termination of a century, as a unit of reckoning, he would surely have made allusion to it here. What there is shows that, in common usage, the centesimal years were terminations, and not commencements; for a jubilee is a festival of commemoration, not of anticipation. In the year 1800 Mr. Pye, then poet laureate, published his *Carmen Seculare*, with a preliminary dissertation in defence of 1800 being the first year of the new century. Among other arguments, he urges that Prior had done the same in 1700; but he forgets that secular odes have always been retrospective, and properly belong to the last of the old century, not the first of the new. Hear Prior:—

> "Hardly the muse can sit the headstrong horse,
> Nor would she, if she could, check his impetuous force;
> With the glad noise the cliffs and valleys ring,
> While she through earth and air pursues the king."

But Prior's noisy muse was riding on horseback after William

III., not to bring him tidings of future events, but as a convenience for the contemplation of the past.

> "She now beholds him on the Belgic shore,
> Whilst Britain's tears his ready help implore;"

and a great deal more.

We have looked through many of the pieces of this controversy, and have found little or no allusion to how people *did* count; the matter was assumed to demand settlement by the way in which people *ought* to count. Great pains were taken to prove that there must have been a year 0 after the Christian æra; and those who could attribute the habits of a modern mathematician to the old computers—who reckoned I., II., III., IV., &c., and had never dreamed of a zero symbol—made a very plausible figure with those who could not correct them. The astronomers Maskelyne and William Herschel took the side of 1800 as the first year of the century, and of course led many, who did not see that the question is for the antiquarian to decide, not the astronomer, as such. But if astronomers may decide, they have settled the point by what is now universal consent, and not without having had it frequently before them. For they never open the proper page of any common account of the progress of their science without seeing themselves invited to deny, if they think fit, the statement that the planet Ceres was discovered on *the first day of the present century:* it was discovered January 1, 1801. We hold it clear that no usage can exist, except one of very modern times. The present practice of astronomers and chronologers is to make the first year of the reckoning to be the first year of a century, so that A. D. 1—100 is the first century, A. D. 1801—1900 is the nineteenth century.

Remembrances of the monadic system of counting have been before now made to appear in the following statement; that a date, such as 1843, does not mean the whole year 1843, but the indivisible moment at which a certain year begins. If this had been the case, and the term century had been used, then, probably, the moment at which A. D. 100 begins would have been made to terminate the century. That the year ranked as a moment, in reckoning year after year monadically, is true enough; but it had not then a beginning distinct from its end, nor any intermediate parts. It has been urged in support of the above view, that the hours of the clock are reckoned in the same way; thus four o'clock refers to a moment of time, not to an amount of duration. But the phrase contains its own answer, for *four of the clock* merely refers to the place where IV. is written.

An appeal is also made to the intention of Dionysius Exiguus, who introduced the present mode of reckoning in the sixth century. Intentions, unless carried into effect, make no rule in chronology: we do not date from the Christian æra because Dionysius so pleased, but because those who followed him succeeded in establishing a usage; and their usage, not the intention of Dionysius, is the rule. Nevertheless, we mean to enter upon this point, not for its importance, but only to give the reader an idea of the manner in which chronological conclusions have been treated.

We hold chronology to be a subject into which more learned confusion has been introduced than into all others put together. We have given a notable instance of this (see the *Companion* for 1845, page 8), in the fact of so diligent a reader and accurate a scholar as Delambre pronouncing, on the mass of mingled citation before him, that the synodical epistle of the Nicene Council had not been preserved. The mistake originated with the laboured attempt which Clavius and others had made to fasten upon the Council, by subsequent evidence, a proceeding of which the epistle shows no trace. Dionysius Exiguus has been treated in the same manner as the Nicene bishops: every possible kind of assertion as to his system and his meaning, has been fearlessly brought forward and easily granted, upon the testimony of writers who lived many centuries after him.

There is no better proof of want of precision in chronological writers than this, that their most technical term, *æra,* cannot have its meaning settled without dispute from their writings. Is the æra a point of time *from* which reckoning is made, or the whole duration *in* which reckoning is made? When we talk of the year 1849 of the Christian æra, do we understand *of* in the sense of *after*, or in that of *part of?*

It may be matter of opinion what the usage is of the world at large upon this point. To us it seems that people in general would divide time into that which is before, and that which is after, the Christian æra, not into *before* and *during.* Writers who define, generally make the æra a moment of time. Thus we light upon the lexicographers Laurentius and Forcellini, the first of whom calls it a beginning of time, *temporis initium, a quo supputationes astrologi incipiunt;* the second, a definite and noted term from which the following years are numbered, *terminus certus et insignis (ut apud nos Christianos est Nativitas D. N. J. C.) a quo sequentes anni numerantur.* The chronologer Strauchius, who formally defines his terms, makes *æra* and *epoch* of identical meaning, *termini solemnes, a quo tempora putamus.* Æra, says Dr. Hutton, is in chronology the same as epoch. Dr. Carey (1677), whose *Palæologia Chronica* is very learned and clear, strives to use *æra* as the duration beginning from the *epoch;* but he occasionally confuses the two words. John Gregorie, 'De Æris et Epochis,' 1649, uses the words synonymously. Joseph Scaliger uses æra for the duration, on account of finding many cases, out of chronology, which show that the oldest use of the word was in the sense of *number*, so that A.D. 500 might be called the 500th *æra.* Calvisius counts the *æra* from the *epoch*, as we collect, for he does not define. Petavius uses the word doubtfully in many cases; but at times *æra sive epocha* occurs in his writings. Riccioli avoids the term in great part, preferring to use *epoch:* but he often uses it with Scaliger: thus there occurs, "If the æra should exceed 38," &c. But throughout the writers who distinguish *æra* from *epoch*, occur continual instances in which the former word is used in the sense of the latter.

In the first page we opened of the *Art de vérifier les dates* our eye was caught by the assertion that the year 715 of Rome is the 39th before our vulgar *æra* (should be in that work *epoch* or *beginning of æra*), and that the Spanish *æra* precedes (*devance*) the Christian *æra* by 38 years.

It seems that those who define are almost all at variance with many who use. Hence it arises that in a recent technological dictionary the æra is made a fixed point of time at the beginning, while in the middle of the article we read of an æra *commencing from* a certain point. And in Dr. Smith's Dictionary of Antiquities, it is "a point of time from which subsequent or preceding years may be counted." But still, we almost immediately read of an æra which *begins* at a certain year; meaning that the *counting* then begins.

The term itself, as used in chronology, appears to have been introduced by the Spaniards, and appropriated to the æra just mentioned, being the commencement of their reign of Augustus. As *æra* is a very doubtful Latin word (that is, as a singular noun) various methods have been tried to explain it. The translator of Alfraganus derives it from the Arabic, as a corruption of *Tarikh*, which, according to D'Herbelot, is used, among other and non-chronological senses, in that of *epoch:* others speak of an Arabic verb *arah*, to count. Some have suggested an abbreviation of *Annus ERat Augusti*, by picking out the letters here given as capitals; as if two letters would have been selected from the unimportant verb. The following conjecture (which is mentioned without source by

D'Alembert* in the *Encyclopédie Méthodique*, and which we do not find in the old chronologers) is far more respectable, almost even plausible: it derives æra from the initials of *Ab Exordio Regni Augusti.* But there is little occasion to seek for any other origin than the later Latin. Forcellini cites from Salmasius a sentence taken from some old writer on mensuration, in which *æra* means a *datum*, a number to begin from: thus in the question 'given a pentagon of ten-foot side, to find the area,' 10 is the *æra*.

On a review of the whole question, and after consulting many writers not here mentioned, as well those who have used the term, as those who have both used and defined it, we are satisfied that the word *æra* is most generally used as a point of time (or a year of time, if years be used monadically) to reckon *from*. Hence the 100th year of an æra should be understood as the 100th year *after* it: and it would avoid confusion if it were so expressed.

To return to the question of the Christian æra, as introduced by Dionysius Exiguus. This reformation is described by its author, rather scantily, in two† remaining letters: the first addressed to the bishop Petronius; the second to Boniface and Bonus, the *primicerius notariorum*, and the *secundicerius*. The second letter has perfect internal evidence that it was written A. D. 526: the first was probably written the year before.

Dionysius begins his first epistle by referring to numerous requests made to him for an explanation of his paschal system, and to the various unskilful modes in which others had proceeded, in contempt or ignorance of the Nicene rule, which proceeded rather from the light of the Holy Spirit, than from that of secular knowledge. He then proceeds to describe the well-known period, which we know to have been invented by Victorinus, with his own arrangement and use of it, and the rules by which it is applied to any current year. The second epistle has a peculiar object, which will presently appear.

Two questions arise: What did Dionysius mean by the year 1, and what was that year? In what month, and on what day of the month, did his year begin? These questions we shall

* D'Alembert says that *æra* is a term of astronomy used in the same sense as *epoch* in chronology. Is not this a slip of the pen? Transpose the two words in Italics, and the sentence would be read without any remark. Curiously enough, the Alphonsine Tables (cited by Gregorie) have a definition in which any one would suppose the words *æra* and *ævum* had changed places: "Æra Hispanis dicitur tempus limitatum ab ævo aliquo sumens exordium."

† These letters were first published by Petavius, at the end of his *Doctrina Temporum*, afterwards by Bucherius (*Comp. Alm.* 1845, p. 9), and again by J. G. Janus (or Jahn). We cite the epistles of Dionysius from the collected edition of the memoirs of Janus, by C. A. Klotz (Halæ, 1769, 8vo. pp. 211). It was first published, according to Fabricius, in 1718.

endeavour to answer from Dionysius himself: inferences from other writers we shall treat as conjectural.

The principal passage * from the first epistle is as in the note, of which the following is a literal translation. Dionysius is speaking of the paschal cycle of Cyrillus, containing ninety-five years, or five Metonic cycles of nineteen years each.

"This cycle of ninety-five years we set ourselves to abolish by the attention to the subject with which we have gained the mastery over it; bringing forward in our own work the last, or fifth [Metonic] cycle of Cyrillus, because there are six years of it yet to run; and then we assert that we have arranged five other cycles according to the rule of the same prelate, or rather that of the Nicene Council often mentioned. But since Cyrillus began his first [Metonic] cycle from the 153rd year of Diocletian, and finished the last in the 247th year; we, beginning from the 248th year of that tyrant rather than prince, refuse to connect the memory of a blasphemer and persecutor with our cycles, but rather choose to note the dates of our years from the Incarnation of our Lord Jesus Christ...."

To this epistle are appended the last of the five Metonic cycles of Cyrillus, the first five of the twenty-eight Metonic cycles of Dionysius, and a collection of calendar rules, framed by the skill of certain *Egyptians*, and adopted by Dionysius. Though the epistle, which itself is called a *preface*, makes the most express mention of both tables and rules, the two latter were not † printed, either by Petavius, or (according to Fabricius and Jahn) by Bucherius, but only by Jahn himself. The table tells us that A. D. 532 is 248 of Diocletian, and 1 of the cycle of Dionysius: accordingly A. D. 1 would have been 2 of the preceding cycle of Dionysius. And the rule given by Dionysius confirms his table.

According to the received mode of counting, we are to presume that Dionysius meant A. D. 1 of his own æra for the year of the Incarnation. But some time after Dionysius, it is certain that the year commonly received as that of the Incarnation

* "Nonaginta quinque igitur annorum hunc cyclum, studio, quo valuimus, expedire contendimus; ultimum ejusdem B. Cyrilli, id est, quintum cyclum, quia sex adhuc ex eo anni supererant, in nostro hoc opere præferentes; ac deinceps quinque alios juxta normam ejusdem Pontificis, imo potius sæpe dicti Nicæni Concilii, nos ordinasse, profitemur. Quia vero S. Cyrillus primum cyclum ab anno Diocletiani centesimo quinquagesimo tertio cœpit et ultimum in ducentesimo quadragesimo septimo terminavit; nos a ducentesimo quadragesimo octavo anno ejusdem tyranni potius, quam principis, inchoantes, noluimus circulis nostris memoriam impii et persecutoris innectere, sed magis elegimus ab incarnatione Domini nostri Jesu Christi annorum tempora prænotare.."

† That Scaliger had seen the rules at least, is evident from his quoting, as from Dionysius, a rule which is not in the *preface*. That Petavius had not seen either tables or rules, may be presumed (though the contrary has been affirmed) from his not printing either. That Riccioli had not seen them is clear from his making it inferential, from the *words* of Dionysius and Bede, that A. D. 532 was 1 of the cycle of Dionysius, when the table has it expressly.

was not the first year after (or *of*, if the reader please) the Dionysian æra, but the first year before it. Three accounts have been given of this discrepancy. First, it has been supposed that the Dionysian reckoning has been misunderstood, and that the year usually called B. C. 1 is that which Dionysius meant to be A. D. 1: so that this present year would have been called by him 1851. Secondly, it has been thought that he intended to have a *zero-reckoning*, calling 0 the year of the Incarnation, and A. D. 1 the year following. Thirdly, it has been thought that he commenced his year, not with the 1st of January, but with March 25, and that his year 1 begins with the March preceding the January of our year 1.

The first supposition is worthy of no attention, since the appearance of the table which Petavius, &c. knew nothing of. There are 95 years in it with their Easters given, and each described by its *annus domini;* and these Easters agree with those of the rules in the 'Companion' for 1845, page 32.

As to the second supposition, it is for those who affirm Dionysius to have made departure from usual methods to prove it. He explains himself so clearly, and gives the circumstances of the existing state of things, and his own proposed alterations, with so much precision, that it is exceedingly improbable he should have made a departure from usage in his mode of reckoning, without giving the most express warning. In speaking of cases of division in which the remainder is 0, in which the divisor is to be substituted (as in finding the golden number by dividing by 19, in which case the number 19 itself must be taken when the remainder is 0), he takes care, after instances, to enunciate this as a general rule,* *per omnem computum.* If such a writer should, in one distinct case, count after the prevailing method, we are bound to assume that he always did so, in failure of special notice to the contrary. Now it does so happen that there is an instance, but rendered rather doubtful by a misprint in that or another instance, and settled by a third instance. In one of his rules (No. 9) there occur the following phrases, which it will be convenient to number.

1. Count the months from September to March (*a Septembri usque ad Martium*) they make *six;* add two, which makes *eight.*

2. Count the months from September to March, they make *seven;* add two, which makes *nine.*

* It is general, in all chronological computations in which the divisor is a period. In our article on Easter ('Companion' for 1845) we have, in one place, omitted to mention it. In pages 27 and 33 and in division XIV. of the rules, instead of "divide by 7, and keep the remainder," it should be "divide by 7, and keep the remainder, or 7, if there be no remainder."

3. If you count from September to December, always add three in these *four* months.

Either 1 or 2 must contain a misprint, and from the correctness of the sums it is not in the figures; nor can it be in the word *September*, which is the initial month throughout. The advocates of the old method will say that in 1, *March* should be *February*, and then September is reckoned in both cases: of the modern method, that in 2, March should be April, and then September is omitted in both reckonings. And so the question would be left, perfectly balanced, if it were not for 3, in which from September to December is called *four* months: but the intent of the whole passage marked 3 is very obscure. We rely much more on the presumption that ordinary language, used by a writer who is generally perspicuous, is to be interpreted in the manner usual in his time, if no reason can be given to the contrary. Accordingly, we hold that the year 1 of our æra, from which the common reckoning is made, is the year of the nativity according to Dionysius, and also the second year of his paschal cycle. This is the way in which Bede, two centuries after, understood Dionysius; accordingly, those who have thought that our common way of reckoning is not according to the intention of Dionysius, have imputed the alteration to Bede.

The next question is as to the time at which the year of Dionysius commenced. On this point we are to remember that he was an ecclesiastic; that he wrote at Rome for an ecclesiastical purpose, the settlement of Easter; that his paschal *indicia*, such as the golden number, &c. always change on the first of January; and that the ecclesiastical year always began on the first of January. We are not aware that any one of these positions has ever been disputed. The natural inference is, that all the presumptions are in favour of his having made the year of which he wrote begin on the 1st of January. But the *Art de vérifier les dates* assures us that by the common consent of the learned (*tous les savans conviennent*) Dionysius himself established in Italy the practice of beginning the year with the 25th of March, and that he did this at the introduction of his new æra. Since the work we cite* is one which

* We would not by any means disparage the *Art de vérifier les dates*, a work which, in all its peculiar parts, is of the highest merit: but it should be praised with discrimination. It is mainly the work of one man, Maur Francais d'Antine, of the congregation of Benedictines of St. Maur (born 1688, died 1746). It contains an immense collection of dynastic and genealogical chronology, extending down to most of the families of historic note in France, many in Germany, and some in Italy, &c. But we cannot find in the preliminary dissertations and the matters of general chronology any sufficient ground for the eulogies which this work has received, and which are totally unmeaning if they do not amount to a declaration that with this one work alone, the student needs no other. If we wanted the dates connected with a king of France, or an emperor,

deals very much in references and quotations, this mode of shifting such a point on to the shoulders of all the learned in general and none in particular, is far from satisfactory. We choose from among the learned, Petavius, perhaps the most learned of the chronologists, certainly one of those who are most cited. On looking into his work *De Doctrina Temporum* (the edition we use is that of Harduinus, Antwerp, 1703, 3 vols. fol.), we find, in book vi. cap. 10, as the description of one of the paragraphs, *Dionysius a xv. Paschali annos orditur*, Dionysius begins his year from the fifteenth of the Paschal moon. The paragraph itself begins, "*In his* vides Dionysius a decima quinta Paschali annos inchoare," which does not quite bear out the side description of the index maker* or editor; for all we are told is that *in his*, that is, in what has immediately preceded, Dionysius does as stated. Now, first, the 25th of March, and the fifteenth of the paschal moon, are two very distinct things; secondly, we must examine what Dionysius is doing. The extracts discussed by Petavius are from the epistle to Boniface and Bonus above mentioned. Here Dionysius sets forth that he had hoped that all ambiguity and opposition had been removed by his former letter, but that as the parties to whom he wrote had brought out from the archives of the Roman church the writings of Paschasinus,† in which there was mention of *common* and *embolismic* years, and many were anxious to know whether this year agreed with the *paterna regula*, or rule of the Nicene council, he (Dionysius) thought it necessary to show that there was no disagreement. He then proceeds to discuss the year used by Paschasinus, which was the ancient lunar year, founded upon that of the Jews. Petavius seems to have taken Dionysius as describing a year *of his own*, or at least has been so construed, both by followers and opponents. In the first epistle, from which we have quoted above, the only matter in which the beginning of a year is mentioned is a discussion (for the sake of Easter) on the Jewish year, as settled in the books of Exodus and Deuteronomy, which are cited as authorities; so that nothing can be drawn from either of these epistles in support of the notion that their writer began his year in March or April.

even of Japan, or a viscount of Fezenzaguet, or a count of Goritz, we should turn to the *Art* &c.; but for the settlement of all points of general chronology, such, for instance, as those connected with the common æra, we should look elsewhere.

* Descriptions of subject contained in indexes or headings, not made by authors, are not to be relied on. In the *Journal Littéraire de la Haye* for July and August 1713, p. 464, is given a letter of Hudde, which shows that he knew how to find the subtangent when the equation of the curve had no irrational quantities. But the index maker has it referred to thus, "Calcul Differentiel, qui en est l'inventeur."

† Paschasinus was one of the legates whom Leo I. sent to the Council of Chalcedon, A.D. 451. His epistle to Leo on the feast of Easter is extant.

Whence, then, came the assertion that, by the consent of all the learned, Dionysius introduced the method of beginning the year on the 25th of March?

It is perfectly true that, according to the common reckoning of the middle ages, the Annunciation and the Nativity were taken to be events of the year B. C. 1, to those who begin the year with January. It is also true that it became very common to begin the year with March 25, and that the beginning of A. D. 1 was made to be in the March *preceding* the Jan. 1, A. D. 1, from which we reckon. That these things have a connexion with one another we have no doubt; but we suspect the connexion to have originated in a misconception. If the year (Jan. 1—Dec. 31) A. D. 1 were considered by Dionysius as containing the Annunciation and the Nativity, and if those who reckoned from March 25 threw them into *their* A. D. 1, that is, into the year Mar. 25, B. C. 1—March 24, A. D. 1 of Dionysius—the misconception might easily have arisen if those who restored the *reckoning* of Dionysius happened to forget, or did not know, that the placing of the above events had shifted with the reckoning.

There is no occasion to settle this point either one way or the other, for our present purpose, which is to point out that no reasonable ground exists for citing any intention or declaration of Dionysius in favour of any meddling with the received mode of reckoning; and further, to put those who may need it on their guard against the undiscriminating learning of the sixteenth and seventeenth centuries, and the chronological logic of the nineteenth, which does not build on antiquity at all.

A. De Morgan.

University College, *London.*
August 6, 1849.

COMPANION TO THE ALMANAC

FOR

1851.

PART I.

GENERAL INFORMATION ON SUBJECTS OF MATHEMATICS, NATURAL PHILOSOPHY AND HISTORY, CHRONOLOGY, GEOGRAPHY, STATISTICS, &c.

I.—ON SOME POINTS IN THE HISTORY OF ARITHMETIC.

EVERY person who has attempted research in the history of the exact sciences knows by experience that the writers on the subject are open to correction, on almost every minor detail, by those who have the power of examining works to which they had not access. The works of which we speak have long been seldom looked at and never read; in many instances their titles are not preserved in records of books: and the historian, who perhaps would have taken pains to consult them, if he had but known of their existence, may have had no means of arriving even at this first step, unless he happened to pick up copies in his casual visits to the auction or the shop. Such a state of things leads to greater errors than those of omission: for instance, it has a tendency to foster the habit of describing discovery as made *per saltum*, as the work of one man at one definite time; and also to accumulate inventions unduly upon the celebrated names which cannot escape notice. It is natural that any new thought or process should be attributed by the historian to the first *on his list* who has proved his right to it: but to make true history, that list must be complete. Until it can be made so, there will be use in detached examinations, even of points which have been much discussed; to say nothing of those to which no attention has been paid. Such examinations, if they were to wait until the inquirer could present them complete, would never make their appearance; but they may be secure of being read with profit, and even with interest, if every point which does not rest on the authority of the examiner himself be distinctly attached to the source from whence it comes. The present paper is especially on the introduction into arithmetic of *decimal fractions*, and of the word *interest*: what is here given may suggest to those who are in the habit of looking for old books the means of completing what is left undone.

It might be supposed that nothing could be more definite than the mode of introduction of decimal fractions, and of the simple extension of the principle of Indian numeration by which they are expressed. Was it, the reader will ask, any very great effort to imagine the descent from .... thousands, hundreds, tens, units, into tenths,

B

hundredths, &c., so that 111·111 or 111̂ 111, or 111 | 111, should represent a hundred, a ten, a unit, a tenth, a hundredth, and a thousandth, put together? The answer is, that it was too great an effort for one mind, and even for one age, as the following statement will show.

In the year 1525 (according to Heilbronner*) Orontius Fineus (A. B., 1535, 16), in extracting the square root of a number approximately, annexes ciphers in pairs, and, proceeding as we now do, obtains what we should call some of the decimal places. In the case of 10, he extracts the approximate integer root of 10 00 00 00, or 3162. Then, separating 162, which with him is not a fraction, but only a means of procuring fractions, he directs to multiply by the *numerus articulus* time after time, and to separate three figures. If 10 be this articular number, he would produce, as he states, decimal fractions; if 20, vigesimal, &c.; but he prefers the sexagesimal system. Thus $162\times60=9720$; $720\times60=43200$; $200\times60=12000$: whence 3 9′ 43″ 12‴ is his approximate value of $\sqrt{10}$, sexagesimally expressed, according to the usage of the time. Had he left off here, we might have placed him, as to decimals, on a par with those practical men of our own day who multiply by 10 with the multiplication table and carriage; but he concludes his chapter by stating, without a process, the interpretation† of the several places; namely, that in 162, 1 is a tenth, 6 six hundredths, &c. Here then, so far as this one rule is concerned, nothing is wanted to put it on a level with our own time, except *merely* (as we should be apt to say) the agreement to distinguish the unit's place by a mark, as in 3·162. In all probability, the above, so far as it goes, is due to Orontius himself. Tonstall (A. B. 1522, 13), who had looked far and wide into the writings of his time, does not touch the approximate square root at all. Tartaglia (A. B. 1556, 21) gives a full account of it (book ii. fo. 28, 29), and attributes it entirely to Orontius. He gives the fraction as $\frac{162}{1000}$, but passes over the meaning of the *separate figures*. He prefers the old rule, given by his Italian predecessors, and derived, as he supposes, from the Arabs, as being, in his opinion, generally more correct, and particularly when applied to higher roots than the square. But his main reason is one which is more illustrative of Tartaglia's age, than of Tartaglia himself; and which is, above all, illustrative of the reason why the hint given by Orontius bore no fruit. Demonstration is bound by laws of thought,

* Of most of the works herein cited on history the reader will find more detailed descriptions in the *Companion* for 1843, in 'references for the History of the Mathematical Sciences.' On arithmetical works we refer the reader to 'Arithmetical Books from the invention of printing to the present time, being brief notices of a large number of works drawn up from actual inspection,' London, 1847, 8vo., by the author of this article. The references are here made as follows: Orontius Fineus (A. B. 1535, 16) refers to a work of O. F., of which the earliest edition inspected in the above work is of 1535, and is described at page 16. Thus, in the present case, the writer of this article answers for the (probably the third) edition of 1535, but is obliged to rely upon Heilbronner for the date of the first edition, which he has never met with. This will be enough for the general reader, who probably will not care whether the work be an *Arithmetica Practica* or a *De Arithmetica libri tres*: the more special student must be content to follow the reference to its source.

† Posses tamen, inventa radice 3162, accipere 3 pro integris, veluti supra fecimus: sed 1 pro decima unius integra parte, 6 autem pro sex decimis ejusdem partis decimæ, 2 tandem pro duabus decimis unius decimæ alterius decimæ partis integri, denaria numerorum observata ratione. (Page 17.)

and writing and expression by laws of taste and experience; but invention, we should say, is wholly free; a new fact, or a new power, when well verified, have just the same value, come by them how we may. It was not so thought at the time of which we speak, but rather that modes of investigation should be restricted. Tartaglia objects to the rule given by Orontius that this addition of ciphers bears too much the mark of natural sagacity,* and too little that of geometrical procedure; *Questa tal regola* di aggiongere di nulle, *eglie manifesto esser stata trouata piu presto per vn certo natural discorso, ouer giuditio, che pe ragion geometrica, ouero arithmetica* .... In our day, we are learning to bear it in mind that these ciphers are always to be understood, when not expressed, to every possible extent; that our scale of numeration is ...0001·000...., ...0002·000...., ...0003·000...., &c., usually abbreviated into 1, 2, 3, &c. The gradual development of the idea of which Orontius must be held to have given the earliest glimpse, has afforded the greatest help which arithmetic has received in modern times.

If there were any quarter from which Orontius could have taken the hint of this process, it would probably have been the Hindu algebra and arithmetic, or the Arabic, which is, so far as it goes, the copy of the Hindu. But nothing of the kind can be traced in either, and if it could, Tartaglia, whose knowledge of the sources of European arithmetic was greater than that of Orontius, would have been unlikely to have attributed the rule in question to the latter. Orontius must then be considered as having a much higher character for original invention than has been conceded to him. Montucla (vol. ii., p. 574) describes him as a man *assez célèbre* in his day, not useless in the re-establishment of mathematics, who wrote some elementary treatises, and believed he had squared the circle. The blame thrown upon his process by Tartaglia for its over sagacity is a stronger eulogium than the account given of him by his own countryman; and is of itself enough to entitle his writings to such an examination as they have certainly never received in modern times.

This rule, when mentioned by writers on arithmetic, has usually been attributed to the celebrated Peter Ramus. The date must have been misplaced; Wallis, indeed, makes the publication to be of 1560, or earlier, but Heilbronner has nothing of Ramus earlier than 1586; Dr. Peacock makes the first edition to be of 1584, and till lately we had seen nothing earlier than 1592. But we have since found that there are two distinct works of Ramus on arithmetic, and both

* That the limits of expression and of method for the time being are natural limits, is in the creed of some, and in the practice of many: we may all be sure that we hold, in different degrees, the objection of Tartaglia to the intrusion of uncertificated sagacity, and his repugnance to new points of view when started by others (he abounded in them himself). We are accustomed, in particular, to suppose a natural connexion between any idea and the mode in which it first suggested itself. There is a number, the use of which runs through every branch of mathematical science, but which, having first presented itself as the ratio of the circumference of a circle to its diameter, is usually attached to the circle by name and definition. But it equally appears in the formulæ by which the probable fluctuations of a number of hazards from the mean are determined. Hence an approximation could be made to it by observing a large number of hazards and recording the results. Or, as we might express it, though we dared not have done so without the preceding explanation, we might approach to the ratio of the circumference to the diameter with no instruments except a pair of dice, and no operation except throwing them and recording the results.

B 2

of earlier dates. The second edition (we have not seen the first) of the smaller one 'Arithmeticæ libri *tres*,' Paris, 1557, 8vo. (small) is full of enunciations in Greek from Euclid and others; the rule of Orontius is given at p. 119. What appear to be the first and second editions of the second work are 'Arithmeticæ libri *duo:* geometriæ septem et viginti,' Basle, 1569 (and again in 1580) 4to.; and the same rule is given in books viii. and xxiv. (pp. 90 and 161 of the first). Robert Recorde also gave it in the *Whetstone of Witte* (A. B. 1557, 21), but combined with the decimal answer a fraction derived from the remainder, from which we should now obtain more decimal places by the contracted method. Buckley's Arithmetica Memorativa (A. B. —, 20), of the date of which all we can say with certainty is that we have seen a copy of 1570, gives the rule as follows:

> Quadrando numero, senas præfigito cyphras
> Productum quadra, radix per mille secetur.
> Integra dat quotiens, et pars ita recta manebit
> Radici ut veræ, ne pars millesima desit.

But it is worthy of note that Tartaglia, Ramus, Recorde, and Buckley, all give examples with *three* pairs of ciphers, which the last even incorporates in his rule: thus showing the probability of all four being followers of Orontius.

When Wingate (1630), as presently cited, gives his account of how he first came upon decimal fractions, he says, "The truth is, there is no man much verst in *Calculations*, but must needs upon some occasion or other fall upon it: for my part I confesse the first light I received of that way, was out of *Ramus* in the Extraction of the square and cube roots; for by annexing Cyphers unto the square and cube numbers, the broken parts of the roots are converted into *Decimals, ipso facto;*".... To this, however, we may add, that it was long before *ipso facto* decimals were recognised as a system; though the correctness of Wingate's first assertion might be verified by examples. For instance, it was long known that in dividing by 1000... the dividend separates of itself into quotient and remainder, by the very meaning of decimal notation. This hint, we shall immediately see, led Stevinus to a formal system of decimal fractions; but others had probably made such a use of them, *ipso facto*, as was done, for instance, by Masterson (A. B. 1592, 29), whom it is impossible to suppose cognizant of a work published in Belgium (with the ruler of which we were at war) a year or two before he began to write. When we see (p. 125) the following mode of dividing £337652643 by a million, and reducing the result to shillings and pence, we may at first sight think it certain that the author had a complete notion and command of decimal fractions: though nothing is clearer from the work itself than the total absence of any glimpse of other fraction than shilling and penny.

facit	*l.*	337	652643
	s.	13	052860
	d.	—	634320

Stevinus published his Arithmetic containing the treatise *La Disme enseignant facilement expedier par nombres entiers sans rompuz, tous comptes se rencontrant aux affaires des Hommes,* in

1585, in French* (A. B. 1585, 26). It is stated that there had been a previous Dutch edition; but this we cannot trace to any very good authority.† It will be observed that Stevinus does not propose fractions, but substitution of integers for them: the idea of a fraction, distinct from an integer, but *treated by the same rules*, had not yet arisen. It is also to be noticed, that the contrivance is specially for commercial and other practical affairs: this system, power over which has always distinguished the mathematical arithmetician from the commercial one, was invented for the particular use of the latter. The reason was, that Stevinus had written the year before upon compound interest, and the continual necessity of division by 1000... had suggested to him the formation of a general system: while Masterson and his predecessors, as far back as Orontius, had never arrived at more than the *ipso facto* use of decimals, as Wingate called it, in one process each and no more, without any power of extension or assimilation.

Stevinus announces his method in terms at which we should now smile. What, says he, is this proposal? peradventure some admirable invention? no certainly, but so simple a thing that it does not merit the name of invention. He adds, that there is no more self-love in praising it than could be attributed to the discoverer of a new island, when he described to his king its vegetable and mineral products. Previous to any description of it, we must remind the reader, that Stevinus represented an unknown quantity and its successive powers by inclosing in circles what we should now call the exponents: thus x, x^2, x^3, were what, for want of the exact type, we may write as (1), (2), (3). Whether this idea preceded or followed that of decimal fractions in his mind, can hardly be settled now.

But, in describing what we should write as 27·847, he writes it (we cite Girard's edition) as 27 (0) 8 (1) 4 (2) 7 (3): in using it in operation, he puts it thus,

$$\begin{matrix} (0) & (1) & (2) & (3) \\ 27 & 8 & 4 & 7 \end{matrix}$$

There is now nothing more to describe: for, except in this superfluity of notation, the modern system is that of Stevinus. That the utmost simplicity should not have been attained, is no ways remarkable. Discoverers, particularly in matters involving new modes of expression, have often acted towards the new principle as a right-minded

* Stevinus was a genius of the sort which cannot be appreciated in its own day; indeed, it is only in the present century that it has been done full justice to. A few years ago, when the question of raising a statue to him was entertained in Belgium, a member of the legislature, *and of the Academy of Brussels*, raised his voice against the proposal, offering to bet that not one in a thousand had ever heard of Stevinus until his statue was proposed, and avowing that he himself was of the number. M. Quetelet enlightened the Belgians generally on the history of their great countryman (of whom even the year of birth and of death had been forgotten), and the opposing academician received a severe punishment from a foreign minister of the Belgian king, in a tract printed for private circulation, entitled (we suppress an unfortunate name), 'Simon Stevin et M. ———' Nieuport, 1845, 12mo. (pp. 148), and headed by a letter signed with the fictitious name of J. du Fan. Assuredly, a Belgian Pantheon without Stevinus would have been a joke against the nineteenth century through all which are to follow, and against Belgium through all the nineteenth century.

† It is true that Albert Girard, in his collection of Stevinus, describes this *Disme* as having been published in Dutch, in the heading of the tract. We say his *collection*, not his *edition*, for it was published after his death; and the real (anonymous) superintendent of the printing may have introduced this heading.

man acts towards his benefactor: that is, they have forborne to inquire whether they had got the utmost out of it. And we shall see that the passage from the mode of Stevinus to the use of the simple decimal point was not the work of a moment. Invention in that day was more under the fear of opinion than now: others, perhaps, of a later date than Tartaglia, objected to natural sagacity. In the next century, when Wallis, then a young man, had occasion to notice that the square root of 12 is twice that of 3, he hesitated long (A. B. —, xxiii.) before he durst write $2\sqrt{3}$ instead of $\sqrt{12}$, because he "did not know of any to have used" it before him.

We search in vain through the tract of Stevinus for any evidence of his having seen, or seeing having thought it worth notice, that his invention is the completion of the Indian mode of denoting numbers. It is now easy enough to teach a child who understands distinctly what is meant by a tenth or a hundredth part, that the same *device of place* by which we agree to distinguish thousands, hundreds, tens, units, must, on being carried further, present a succession of places devoted to tenths, hundredths, &c.: all that is necessary being some new contrivance to distinguish the unit's place, which is no longer the first on the right. It will be worth while to watch the expressions of Stevinus, in order to see how completely the integer side of a number is to him a parcel tied up by his predecessors, which he does not think of opening for the purpose of comparing his new goods with the old ones.

"The Disme," says Stevinus, "has two parts, definitions, and operation. In the first part, the first definition declares what thing Disme is: By the second, third, and fourth, [are declared] what signify Commencement, Prime, Second, &c., numbers of Disme. In operation are declared by four propositions, Addition, Subtraction, Multiplication, and Division, of numbers of Disme." The first definition is as follows: "Disme is a species of Arithmetic, invented by the progression of tens, consisting in characters of cyphers (*ciffres*) by which any number is described, and by which we dispatch all calculations of human affairs by whole numbers without fractions." But for the last clause (which sums up in a new light) we should here think we had got an opening description of the Indian notation in general, both as to integers and fractions. The explanation appended to this definition begins in like manner, and seems to contain just what we want to find. Stevinus notes that in 1111 each unit is the tenth of the preceding, and that in 2378 each unit of the 8, is the tenth of each unit of the 7. "But because it is convenient that the things of which we treat should have names, and because this method of computation is found by consideration of this tens or *disme* progression, that is, consists entirely of it, as will appear, we properly name this treatise the DISME, *by the which we can operate with whole numbers without fractions.*" Here again, before the clause in italics, we imagine we see a progress towards what we want: but at the moment when we expect to be told that, by extension of notation, integers and fractions are treated *by the same rules*, we are disappointed by finding instead that *there are no fractions*, nor any thing except integers. Community of rules of operation cannot exist in the mind of the writer separate from community of matter. The second definition tells us that every proposed whole number (no matter how many its places) is called a commencement, and its sign is (0); the next that each tenth part of unity is a prime denoted by (1), each tenth part of a prime is a second, denoted by (2), and so on: in such manner that 3(1) 7(2) 5(3) 9(4) is $\frac{3}{10}$ $\frac{7}{100}$ $\frac{5}{1000}$ $\frac{9}{10000}$ or $\frac{3759}{10000}$. But how can this be when there are to be no fractions? While Stevinus thinks of the matter of his computations, he admits fractions: the moment he passes to their form, he converts fractions used as integers into integers. Had Stevinus carried his discovery its full length, he would have entertained the question whether the complete system, on both sides of the decimal separator, should or should not be introduced into elementary numeration. This has been advocated by some: and though it may be doubted whether it could succeed in instruction (in the absence of such previous notion of fractions as a decimal system of weights and measures would give), we doubt, also, whether it has received sufficient discussion. The plan was advocated in Mr. Walker's excellent work (A. B. 1827, 90), but the earliest adoption of it, that we know of, in our language, is in a Dublin* work, which is quite forgotten.

We must leave it to any one who can to show that this admirable tract produced any immediate effect. The civil troubles of the period, and the state of the communication between England and the Netherlands during the war of the Armada, were not favourable to its diffusion. It was translated into English by Norton; and H. Lyte (A. B. 1619, 36) published a professed treatise on decimal arithmetic. We cannot now refer to the second work: the first is 'Disme: the Art of Tenths, or Decimall Arithmetike, . . . invented by the excellent mathematician, Simon Stevin. Published in English with some additions by Robert Norton, Gent.' London, 1608, 4to. The additions are only an elementary introduction to arithmetic. In the body of the work the notation is less cumbrous than that of *Girard's* Stevinus, and probably follows the original more closely; thus, instead of 3(1) 7(2) 5(3) 9(4), we have $3^{(1)}$ $7^{(2)}$ $5^{(3)}$ $9^{(4)}$.

The next work of which we know any thing is that of Witt on Interest (A. B., 1613, xxiv and 33), in which tables are given, purporting, like those of Stevinus, to contain only numerators, with 1000. . . to be supplied when wanted as a denominator, by cutting off as many places as there are ciphers. Witt's practice of multiplying and dividing by 100. . . by altering the place of the decimal separator, and the frequent occurrence of such processes as that of Masterson, quoted above, would secure to him the right of being considered the first user of the single decimal separator, if it were clear that he looked upon his results as representations. But it is to be suspected that he looked upon his decimal fractions as modes of contributing towards, rather than of expressing, results. He had either (which I suspect) not seen the tract of Stevinus, or

* The 'second volume of the Instructions given in the drawing school established by the Dublin Society..1768..under the direction of Joseph Fenn, heretofore Professor of Philosophy in the University of Nants,' Dublin, 1772, 4to. This work is, perhaps, the first which introduces into our language the notation of the Differential Calculus.

else he did not adopt the full system: for there does not occur such a process as multiplication or division of two decimals. The *Rabdologia* of Napier (A. B. 1617, 35), which soon followed, has not the claim which has been made for it: an accidental representation by means of a decimal comma, which occurs once, is nothing but a preparatory step to putting on the whole notation of Stevinus, which is the final mode of representation in each of the two instances which Napier gives.

The first edition of the *Arithmatick* of John Johnson (the *surraighour*), of which the second is described in (A. B. —, 104) was published in 1623. The book on decimal arithmetic is as complete as the tract of Stevinus, and uses his notation, dispensing only with the circles. Thus £3·22916 is written

$$\pounds 3 \mid \overset{1.}{2}\,\overset{2.}{2}\,\overset{3.}{9}\,\overset{4.}{1}\,\overset{5.}{6}$$

When this fulness of description is avoided, as is often done in text-paragraphs, our modern practice is frequently arrived at; thus, 31 | 2500, 34 | 2625, occur without explanation, just as we might write 31·2500, 34·2625. But this is sometimes thought unsafe: thus 358·49411 is described as "358 | 49411 fifths." It is a good illustration of the manner in which those who are on the historians' list come in for more than their share, that as to two persons, Napier and Wingate, to whom the first distinct use of the single decimal point through all kinds of operation has been attributed, Richard Witt has a better claim than Napier, and John Johnson than Wingate. Briggs, in 1624 (A. B. —, xxv), adopted the practice of underlining the fractional portion: thus 5·9231 is written by him 5 $\underline{9231}$. Albert Girard (A. B. 1629, 37) very distinctly used the decimal point (or rather comma) in his brief 'Nouvelle Invention,' &c., but only on one occasion; he finds one root of a cubic equation to be $1\frac{532}{1000}$, and then explains that the three roots will be 1,532 and 347 and —1,879, which he says are expressed in *disme* as far as *tierces*. His second root should have been ,347; and we are left in doubt as to whether it was a printer's omission, or whether, which is not unlikely, Girard thought the comma unnecessary except for the separation of integers.

The first edition of Wingate's arithmetic, which we have not seen, was published in 1529 or 1530, according to different authorities. The second was edited by John Kersey, during the life of the author, London, 1650, small 8vo. Some of the succeeding editions—about a dozen—were so much altered (A. B. 1673, 48, and 1760, 73) that they must not be cited as Wingate's. The second edition, however, had a valuable editor, who has taken such pains to describe his own additions, that a reprint of the first might be made from it. Though the system was hardly settled, its invention was in dispute. "The invention of *decimall arithmetique* writes not many yeares; and since the first invention thereof, time and practice hath added much perfection thereunto: divers challenge the first invention of it, how truely I know not." It is by no means unlikely that more than one, upon such hints as were current, had constructed a system more or less resembling the modern one; and we may be tolerably sure that Wingate had not seen the tract of Stevinus.* Speaking now of his first edition, it appears that it does not treat methodically even of common fractions, and that the short chapter on *reduction* of decimal fractions is merely, so far as subsequent matter is concerned, a preparation for the process of Orontius, which, as before seen, Wingate derived from Ramus. Nevertheless, Wingate actually gives what is, so far as we know, the first clear and complete direction how to construct the modern system, though he hardly makes any use of it. Owing to the scarcity of the first and second editions, the following extracts are desirable (pp. 5, 6, of the second edition):—

"When a single broken number hath for his Denominator a number consisting of an unitie in the first place toward the left hand, and nothing but Cyphers towards the right, it is more particularly called a Decimall......A decimall may be exprest without the denominator by prefixing a point before the numerator: so $\frac{5}{10}$ may be written thus, .5; and $\frac{25}{100}$ thus, .25...... In decimalls, when the numerator consists not of so many places as the denominator hath cyphers, fil up the void places of the numerator with cyphers: So $\frac{5}{100}$, $\frac{50}{1000}$, and $\frac{25}{10000}$, are written thus, .05, .050, .0025...... In decimalls thus exprest, the denominator is discoverable by the places of the numerator: for if the numerator consists of two places, the denominator is an unity with two cyphers......"

It will be observed that we miss here that complete sense of the meaning of the separate places which is shown by Orontius and Stevinus. Of course Wingate was fully cognizant of the meaning of the separate figures in ·1234, but it did not strike him as necessary to make these meanings a part of his explanation. Even in this little matter there are two distinct schools. Some writers define by numerator and denominator, as Wingate does; others extend the Arabic system by the invention of the separate columns of successive fractions.

Gunter (A. B. —, xxv) fell into the decimal point very gradually; but his work was not published till after that of Wingate. We do not trace even the method of Stevinus into note on the continent till the publication of Herigone's course (A. B. 1634, 40). Oughtred (A. B. 1631, 37) and his readers, till the beginning of the eighteenth century, used such notation as 12 | 345 for 12·345. This, the exception in English books, was almost the rule on the continent. In 1690, Dechales, in whose course of mathematics (A. B. 1690, 53) is the most extensive list of arithmetical books which had then been published, gives no account of the tract of Stevinus, and the notation used is as in 12[345.

The preceding hints are by no means a complete history of the subject: but they may be filled up by time and observation of un-

* As late as 1651, one Robert Jager published (London, small 8vo.) his 'Artificial Arithmetick in Decimals,' in which he says that the common way of natural arithmetic being tedious and prolix, God in his mercy directed him to that which he published. It is a system of decimals in which 16 | $\dot{7}\dot{2}\dot{4}\dot{9}$ would be what we write 16·7249. It is by no means impossible that this may have been a real invention, and not an impudent fraud; for the present system was not well established in 1651, and not so well as people would think who judge by the works which have lasted, in 1700.

B 3

known works. The only great single step is that of Stevinus, who is justly called the inventor of decimal fractions: and his method was complete. It by no means diminishes his rights, that a more convenient form of expression was afterwards adopted; but it adds to his fame that there was no genius great enough to introduce this form by one effort of thought so decisive as his.

Perhaps the reader may ask whether, in logarithmic calculation, the common decimal fraction, as now written, was not adopted almost from the first publication of logarithms. We answer that, in the earliest logarithms, no fractions at all were used, the radix being always a high integer. And when the *point* first entered, it was not as a separator of integer from fraction, but as a convenient mode of dividing the integer. Thus in Gunter's logarithms, as late as 1636, the apparent decimal comma is nothing but a separator of the last five figures; and the logarithm of 20 is 1301,02999. Even the very table of Briggs, in the preface of which the decimal separation above noticed is propounded, has no decimal points, though *one* of the commas used for separation falls in the proper place.

It is fortunate that in so easy a subject, and one familiar to so many, the usual course of discovery can be completely illustrated. It would be much too strong a simile to compare the man whose name is in the mouths of all to the engineer who lays the match to a train, and startles the world by an explosion, while no one asks who bored the rock or laid the powder. But though such a comparison would err in degree, it will serve to remind us that, in every great achievement of human intellect of which it falls within the power of history to see the antecedents, we notice a gradual preparation, which is seldom adequately described. The consequence is, a succession of disputes about the authors of discoveries. If a work on history give us to understand that A.B was the originator of a certain power, in terms which imply that he was the first who did *anything* towards it, and if some one afterwards find that C.D did *something*, the assertion of the historian is contradicted; and the contradiction is often carried to the extent of making C.D take the place of A.B. If any one were to conclude that it appears from the present paper that Orontius Fineus was the inventor of decimal fractions, he would only make an inference which has had many parallels. Was not Orontius the first who used a decimal fraction, and invented a rule the results of which were expressed in decimals? Undoubtedly he was, so far as here appears. Nevertheless, it can hardly be necessary to insist upon the differences between him and Stevinus, in which the claim of the latter consists.

We have seen that Wingate does not appear to have known Stevinus, and speaks of various claimants: a later and by no means unlearned writer, Willsford (who published in 1656), observes that the invention of decimal arithmetic is nowhere recorded; that of late years it was put into method, and had its axioms and rules, but that to search for men's names enshrined long since in dust would prove vain.

The question about the introduction of the word *interest* into arithmetic, seems to differ from the former in never having had any attention paid to it. The word is used in several senses: but which is the original and which are the derived ones, has never been asked, so far as we know, by the arithmetician or the canon lawyer; though the first ought to have asked it, and the second could have answered it.

Usury, or the receiving of money as a compensation for money lent or payment deferred, is supposed to have been unequivocally condemned by the Mosaic law; as in Exodus, xxii. 25, "If thou lend money to any of my people that is poor by thee, thou shalt not be to him as an usurer, neither shalt thou lay upon him usury:" and in Leviticus, xxv. 39, "If thy brother be waxen poor...then thou shalt relieve him...take thou no usury of him or increase...thou shalt not lend him thy money upon usury, nor lend him thy victuals for increase:" and in Deuteronomy, xxiv. 19, "Thou shalt not lend upon usury to thy brother; usury of money, usury of victuals...unto a stranger thou mayest lend upon usury." In a state of things under which money could not usually be employed productively, but in which every man lived on his own land, a borrower would always be a distressed man, and the above precepts seem to contemplate that he would be as likely to ask for food as for money. In such a case, usury would be barbarity;* and it is to be remembered that the rates of usury prevailing in ancient times were very high. Aristotle grounds his declaration against usury upon this asserted unproductive character of money, from which, he says, it is against nature to take interest. Accordingly, the distinction between money lent for relief of distress, and money advanced that the borrower might improve it, was not contemplated either in ancient times or in the laws of the middle ages; all usury was strictly forbidden, and the name of it became odious. So far was this carried, that Alexander de Nevo, doctor of law, in his treatise 'Contra Judeos Fenerantes,' printed in 1478, declares that he would not permit even a Jew to lend at usury, though it were to save a Christian from starving. The distinction between money lent to relieve distress, and money lent to be profitably employed, was condemned by the Catholic writers, and supported by the Protestants. Molinæus maintained the difference, and Alsted. (See Scaccia *De Commerciis*, 1648, p. 69; and Alsted, *Encycl.* vol. iii. p. 124.) Accordingly, the earlier books of arithmetic have little or nothing to say upon interest, particularly when written by clergymen. Neither Tonstall nor Clavius (A. D. 1522, 13; 1583, 104) make any reference to it. That it is hardly mentioned by Recorde and other writers of the time of the Reformation depends upon another circumstance, the very great simplicity of the rate of interest in use, usually ten per cent. One instance would be enough for illustration of taking the tenth part of a sum of money; particularly as fractions of years were seldom or never considered. It is more to the purpose that even treatises on book-keeping do not recognise it. In Mellis (A. D. 1588, 27) one account in the

* It has been contended that these laws are a part of the political, and not of the moral, system of the Pentateuch. Against this it may be proposed for consideration that the case contemplated is something like that of lending at twenty or thirty per cent. to a man who must spend the loan in buying food, to remove casual distress, and whose ordinary means are not great enough to reduce the accumulations of such a rate. Nor does the permission to lend at usury to a *stranger* of necessity make the precept political; it may be that the stranger was contemplated as being a *merchant*, an improver of money.

ledger is created by a sum lent in ready money, which sum is repaid without interest.

In almost all the books of the sixteenth century there appears a class of questions which seem to indicate a method employed among merchants of evading direct usury. A man lends to his friend 145*l.* for sixteen months; when the latter is asked to return the favour he can only command 94*l.*; how long ought he to lend this last sum in requital of his own obligation?

The word *interest* was known to the law, in the sense of usury, before it was to be found in arithmetical books. In the statute 37 Hen. VIII. c. 9. it is forbidden to take more than ten per cent. for forbearance of payment "by way or mean of any corrupt bargain, lone, eschange, chevisance, shift, interest of any wares, ...;" but the act is not to extend to "other than in cases of usury, interest, corrupt bargains, ..." This Act was repealed by 5 & 6 Edw. VI. c. 20, which forbids all "usury or increase." It was revived* by 13 Eliz. c. 9, which recites that usury had increased by way of sale of wares and shifts *of* interest. This last phrase occurs twice, once misprinted "*ships* of interest," in the statutes at large; and afterwards the word *shift* is used where we might expect the word *interest*. Whether this *shift of interest* mean anything which can now be explained, we must leave to legal antiquaries; it is enough for our purpose that we show the word *interest*, in connexion with usury, to be older in law than in arithmetic. This is confirmed by R. Witt, before cited, in whose title-page the word does not occur, and in whose preface nothing but "allowance for forbearance." But in his first page he speaks of "1. per 10. per terme, gain, and gain upon gaine . . . . . Or (as commonly men speak) 10. per 100. per ann. interest, and upon interest . . . . ." Having thus introduced his reader to the word, he uses it freely. Forbearance still was the legal word, even beyond the time of James I.

But we must not merely look to England, since the word was established in France as early as here, and afterwards in Italy and Germany. Interest, *interess*, *quod interest*, *interesse*,† was known to the Roman, and thence to the modern civil and canon law, in the older sense which it still retains, as that of a something belonging to. And this use, even in English law, is far older than the meaning

* It seems to have been revived that it might be put to death in another form by additional provisions; but this we have nothing to do with. None of the statutes authorize usury, except as implied in forbidding more than a given rate under penalties. It was considered as against the law of God throughout the seventeenth century; but the necessities of commerce prevailed over all profession. Gerard Malynes, in his *Lex Mercatoria* (first published about 1622), says, "We have usury like a wolf by the ears, dangerous to be kept, and more dangerous to abandon the same."

† This word is declined as a substantive in a charter cited by Ducange *in verb.—carta quam fecit de interessis Episcopi....* Shakspere uses the word *interest* often. But in the Merchant of Venice, it is worthy of note that the Jew, who lives by it, calls it *usance* while Antonio, who despises it, calls it *interest*:—

"He lends out money gratis, and brings down
The rate of usance here with us in Venice.
* * * *
And he rails
Even there where merchants most do congregate
On me, my bargains, and my well-won thrift,
Which he calls interest."

It would seem, then, as if the word was one in common, as well as legal, use, before it became arithmetical.

above discussed. But this original use of the word is not immediately explicable in its application to interest of money as distinguished from principal; for the creditor's interest, in the genuine sense, equally includes the *usury* and the returnable principal. If the word had been derived from the creation of a perpetual annuity, such as our national debt, in which the original creditor looked upon repayment as a distant contingency, and as rather a redemption of the annuity than a return of the principal, then the genuine meaning of the word interest could more appropriately have been applied to the annuity. But how comes a word which included both the loan and the profit of it to stand for the latter only?

When the word *usury* was found too offensive as well as dangerous, it was natural that more gentle terms should be substituted for it. As the word was said in Hebrew to mean that which *bites*, some writers distinguished unlawful usury by the phrase *biting* usury; one of the statutes distinguishes illegal usury by the word *corrupt*. At first, in law, interest and usury seem to have been words of much the same colour; afterwards, legal usury only was interest. The Protestant writers generally did not care to make the distinction at first, and there is not much allusion to it. But sometimes there is such a thing, as in the following extract from the poetical opening to Webster's Tables (A.B. —, 40), probably published in 1605:—

And though in Interest thus thou deal'st, thou not approu'st at all
Of vsurie, which may (for thee) beneath iust censure fall.
Thou not conclud'st such contracts made are lawfull yea or no,
But truly to performe the same (by parties both) dost shew.
In, onely this, thou art a guide, but else, as is most fit,
Thou to the guidance leauest all of grace and holy writ.

In England, we had usage, increase, forbearance. In Italy, even in the fifteenth century, Pacioli (A. B. 1494, 2) found the words *merito* and *meritare* (desert or earning) established. The same word was used by Ghaligai in 1521 (A. B. —, 102), but his chapter on the subject has no head-word at the tops of the pages, as is the general rule of his book: probably he thought there was no occasion to advertise what he was doing. Sfortunati (A.B. —, 16) heads his chapter boldly *delle usure*, which he says people call *meriti*, as if it were a virtue. He then proceeds to give his rules by way of warning against these *meriti* or *dannamenti dell' anima*, and having thus discharged his conscience, he is only the arithmetician thenceforward. By the time of Tartaglia (A. B. 1556, 21) usury had been referred to *compound* interest, and all were not prepared to admit so much (lib. xi., p. 190).

It appears with tolerable clearness, on inquiry into the immediate derivation of *interest*, that the term was skilfully borrowed from one of the permissions of the canon law. Long before *interest* was used in the modern sense, Matthew Paris (cited by Ducange) adverted in one sentence to *usury*, penalty, and *interesse*—*usuras, pœnas, et interesse*—as all connected together. The connexion is thus traced. The principal* circumstances under which receipt of money in

* There is one which we omit, as irrelevant, but mention in a note as curious. Will it be credited that not a few theologians, while declaring against usury because it was forbidden by the Mosaic law in the terms above quoted, maintain that the risk of loss from the poverty of the debtor removes money paid on that account from under the definition of usury?

return for a loan was *not* usury were as follows:—First, where the money so paid was *pœna*, a fine stipulated for in the event of the debt not being paid at a fixed time: but evasion was guarded against by making it essential that neither lender nor borrower should have reason to think it unlikely that the money would then be paid. Secondly, where the money so paid was either *interesse damni emergentis* or *interesse lucri cessantis*, that is, compensation to the lender for some loss accruing, or gain ceasing, to him, in consequence of the loan. Thus, if a person by lending his money was unable to pay a tax, and incurred a penalty, the borrower might pay the penalty, or *damnum emergens*, without imputation of usury upon the lender. In order to secure this result, it was necessary that the loss or cessation of gain should have become certain before the contract or understanding about the *interesse* was made. One case, however, of the *lucrum cessans* was in itself enough to legalise the whole practice of interest. If a man could and would have bought annual rents* or returns of any kind, which, however, he did not buy, or of which he deferred the buying, that he might make a loan, the borrower might pay him those rents without usury. That is to say, anything which a man could and would have made of his money in another quarter he might without usury take of his borrower. Hence the origin of our word *interest*, as the *interesse lucri cessantis*. The obvious difficulty of prohibiting usury under so easy a mode of evading it did not escape notice: it was said that a professed money lender had only to have on his hat and cloak, and to be going into the market, to make it impossible of detection. To this the reply was that the gain the money would make must be a certainty, which could not be said of the sum for which the usurer had the investment yet to seek. But it had been admitted that even an uncertain gain might be estimated by a proper arbiter, and adjudged to be the *lucrum cessans* of the particular case; and common sense would tell any one that profitable employment for money, in the shape of yearly returns, could always be obtained in land or houses. For these and other reasons, a much less willing assent was given by the lawyers to the *lucrum cessans* than to the *damnum emergens:* the former was considered much nearer to usury than the latter; probably none but a canonist ever had any clear notion of the difference.

The end of all these distinctions and the marvellous minuteness and precision of the cases (some of the canonists expressly lay it down that a yearly *loss* is not usury in the loser) made it practically useless to carry a case into the ecclesiastical courts, to which the jurisdiction over usury originally belonged.

A. De Morgan.

University College, London,
Oct. 7, 1850.

* "Ut cum quis paratus habet pecunias ad emendos redditus annuos, qui venales sunt, nec emit, ut indigenti et roganti dictas pecunias mutuet: quo casu sine dubio licet pacisci circa quantitatem, quam haberet ex annuis redditibus." Bassus, *Biblioth. juris canonico-civilis* (in voc. *usura*) vol. iv. p. 429. See further Scaccia *de Commerciis*, p. 166, &c., where the question is argued on both sides at great length, but so as much to enforce the derivation here given of the word in question.

COMPANION TO THE ALMANAC

FOR

1852.

PART I.

GENERAL INFORMATION ON SUBJECTS OF MATHEMATICS, NATURAL PHILOSOPHY AND HISTORY, CHRONOLOGY, GEOGRAPHY, STATISTICS, &c.

I.—A SHORT ACCOUNT OF SOME RECENT DISCOVERIES IN ENGLAND AND GERMANY RELATIVE TO THE CONTROVERSY ON THE INVENTION OF FLUXIONS.

THE celebrated controversy on the invention of fluxions has, any one would suppose, been so fully argued that it would be difficult to make out a reasonable case for introducing the subject again. It is nevertheless true that several disclosures of great importance in the way of evidence have never been made at all until very lately.

This controversy resembles one of those well-worn law cases which must be cited and discussed whenever a certain question arises. Every dispute about priority of mathematical invention* revives it. At the same time, the main and turning points of it can be presented without any such amount of mathematical language as would render an article upon the subject unfit for the majority of readers. We therefore propose to present some of these points, with an account of the recently published materials, and of their bearing on the result.

When, after some petty and indecisive controversy, Leibnitz appealed (1711) to the Royal Society for protection against imputations of plagiarism which had at last assumed a distinct form, the Society, in 1712, appointed the celebrated *partisan*† committee to maintain the side of Newton. The report of this committee, published with epistolary evidence in 1712, under the name of *Commercium Epistolicum*,‡ contains the following sentence, which is the whole of that report, so far as it insinuates that Leibnitz did take, or might have taken, his method from that of Newton;—"And we find

* One most fortunate circumstance about it, as a precedent, is that it fixed the meaning of the word *publication* to the genuine and legal sense. It is the sufficient answer to any one who would restrict this word to its colloquial sense of circulation by means of type.

† We have shown the committee to have had this character in *Phil. Trans.*, part ii. for 1846, and in the life of Newton in *Knight's British Worthies*; and nobody has contested the point. It was, however, universally believed that the intended function of the Committee was judicial, and both Newton and Leibnitz speak of it as if it had been so. But though the Committee itself overstepped its own proper function in the form of its decision, and thereby gave rise to the misconception, we hold the intention of its proposers to have been stated with perfect clearness.

‡ We cannot here detail all the circumstances. The reader may consult the articles *Commercium Epistolicum* and *Fluxions* in the *Penny Cyclopædia*, the life of Newton already cited, Brewster's life of Newton, that in the Library of Useful Knowledge, or Weld's history of the Royal Society.

B

no mention of his (*i.e.* Leibnitz's) having any other *Differential Method* than *Mouton's* before his Letter of 21st of *June* 1677, which was a Year after a Copy of Mr. *Newton's* Letter, of 10th of *December* 1672, had been sent to *Paris* to be communicated to him; and above four Years after Mr. *Collins* began to communicate that Letter to his Correspondents; in which Letter the Method of *Fluxions* was sufficiently describ'd to any intelligent Person."

The committee in their English have "any intelligent person"; in their Latin, subjoined for foreigners, they have "idoneo harum rerum cognitori." Raphson, no stickler for accurate description, as we shall see, could not second this; so he converts the Latin into the original, and gives his own English translation, "to any proper judge of these matters." But even this was too much; so some one else (copied by Hutton in his Dictionary; we do not think Hutton did it himself) has invented a new report, in which we find "a man of his sagacity."

How far this celebrated letter deserves the character here given of it, is one question; whether Leibnitz actually received it, is another. Comparatively little notice was taken of either; so that in many subsequent writings it reminds us of the tree which was cut down that the action for trespass might try the ownership of the estate. It gives, nevertheless, the only possibility, such as it is, which the evidence offers of Leibnitz having seen anything to the point from the pen of Newton.

In order to prove the passage quoted above, it is stated that there existed, among the papers of Collins in the possession of the Royal Society, in the handwriting of Collins, a parcel (*collectio*) of papers containing extracts from Gregory's letters, together with the letter of Newton above-mentioned (but which was not alluded to in the title or docket which Collins placed on the parcel), and that the parcel was marked as to be communicated to Leibnitz, and was accompanied by a copy of a letter to Oldenburg, the party who was to make the communication. Not a word is said on the date at which the parcel was transmitted: so that the committee, in their report, actually added a statement for which there was no pretext of evidence, namely, that Newton's letter was transmitted about a year before June 21, 1677. Further, the evidence does not mention the date at which Collins died (1682), nor how his papers came into the possession of the Society, nor whether there was any guarantee that papers found tied together in 1712 had been so tied up by Collins before 1682, nor whether there was any evidence that Collins had fulfilled his intention of sending the parcel on to Oldenburg, &c. When Leibnitz, who did not remember receiving any such letter, declared that he did not think it necessary to answer anything so weak, his contempt for this unattested statement was of course construed by the other side as being of that kind which parties who cannot answer find it convenient to assume.

The editors, whoever they were, of the reprint* of the *Commercium Epistolicum*, made under the sanction of the Royal Society in

* We say *reprint*, and not *second edition*, because even the old title-page and the old date (1712) were reprinted. Everything was done which could lead the reader to suppose that he had in every respect a repetition of the original work, preceded by a preface of the new editors.

1722, took the liberty of secretly making a few additions* and alterations. Among these, they add the date at which Collins died, and the date of transmission of the parcel: they say it was sent June 26, 1676. How they got this date is not said; but as the next parcel sent by Oldenburg to Leibnitz was stated to have been sent on June 26, it may have happened that the revisors of the second edition borrowed this date for their purpose.

So the matter rested until recently, when the publication of a portion† of Leibnitz's papers took place. And it now appears that if the manuscripts which Leibnitz left behind him, and which found their way into the royal library at Hanover, had been examined, it could have been ascertained *what Leibnitz really did receive from Oldenburg.* It appears that the latter wrote to the former from London, with the date of *July* 26, 1676, not forwarding Collins's parcel, but describing its contents‡ himself. He gives various matters connected with Gregory's researches, and then proceeds to allude to a method in a letter from Newton of December 10, 1672. But though he gives, almost *verbatim*, what we may call the *descriptive paragraph*§ of this letter, he does not even allude to the *example of the method*, in which, according to the report of the committee, the method of fluxions is sufficiently described to any intelligent person. So that, with reference to this asserted *description* of the *method* of fluxions, there is now clear and positive evidence that Leibnitz did not receive it as stated, but received only an account of the *rest* of the letter, which describes the *sort of results* attainable.

Towards the end of 1850 the Master and Fellows of Trinity College, Cambridge, published (from among their manuscripts‖) the

* This fact was discovered by us in 1848; and the additions are exposed in the *Philosophical Magazine* for June 1848. The first edition is now scarce.

† Leibnizens mathematische Schriften, herausgegeben von C. J. Gerhardt. Berlin, 8vo. Erste Abtheilung, Band i., 1849, Band ii., 1850. We have not seen any more, if indeed any more has yet appeared.

‡ Collins had desired, in the title of the parcel, that the contents, after being read by Leibnitz, should be returned to himself. Oldenburg appears to have thought it more prudent to write his own account than to trust the papers to accident by land and sea. (At least, this was our impression before we came to the discovery presently mentioned.)

§ "Defuncto Gregorio," says Oldenburg, "congessit Collinius amplum illud commercium litterarium, quod ipsi inter se coluerant, in quo habetur argumenti hujus de seriebus historia: cui Dn. Newtonus pollicitus est se adjecturum suam methodum inventionis illius, prima quaque occasione commoda edendam; de qua interea temporis hoc scire præter rem non fuerit, quod scilicet Dn. Newtonus cum in literis suis Dcbr. 10. 1672 communicaret *nobis* methodum ducendi tangentes ad curvas geometricas ex æquatione exprimente relationem ordinatarum ad Basin, subjicit hoc esse unum particulare, vel corollarium potius, methodi generalis, quæ extendit se absque molesto calculo, non modo ad ducendas tangentes accomodatas omnibus curvis, sive Geometricas sive Mechanicas, vel quomodocunque spectantes lineas rectas, aliisve lineis curvis; sic etiam ad resolvenda alia abstrusiora problematum genera de curvarum flexu, areis, longitudinibus, centris gravitatis etc. Neque (sic pergit) ut Huddenii methodus de maximis et minimis, proinde que Slusii nova methodus de tangentibus (ut arbitror) restricta est ad æquationes, Surdarum quantitatum immunes. Hanc methodum se intertextuisse, ait Nowtonus [sic], alteri illi, quæ æquationes expedit reducendo eas ad infinitas series; adjicit que, se recordari, aliquando data occasione, se significasse Doctori Barrovio lectiones suas jam jam edituro, instructum se esse tali methodo ducendi tangentes, sed avocamentis quibusdam se præpeditum, quominus eam ipsi describeret."

The word *nobis*, put by us in Italics, should be *ei*; Oldenburg forgot that he was describing, not copying, the account Collins had given him.

‖ Correspondence of Sir Isaac Newton and Professor Cotes....now first published from the originals in the Library of Trinity College, Cambridge, together with an appendix....by J. Edleston, M. A. Fellow of Trinity College, Cambridge. London, 1850, 8vo.

B 2

correspondence of Newton and Cotes, with what is called a synoptical view of Newton's life. This is far below sufficient description; for the synopsis is followed by a body of notes of such research and digestion as make it difficult to give adequate praise to the whole without appearance of exaggeration. We differ much from the editor as to many matters of opinion and statements the character of which is determined by opinion; and we take particular exception to the following account (p. xlvii) of the point before us:—

"Doubts have been expressed whether these papers" (*Comm. Epist.*, p. 47, or 128, 2nd ed.) "were actually sent to Leibniz. We have, however, Collins's own testimony that they were sent as had been desired (*Comm. Epist.*, p. 48, or 129, 2nd ed.), besides Leibniz's and Tschirnhaus's acknowledgments of the receipt of them (*Ib.* pp. 58, 66, or 129, 142). It may also be observed that the papers actually sent (in a letter dated July 26, 1676) to Leibniz by Oldenburg have been recently printed from the originals in the Royal Library at Hanover (Leibn. *Math. Schrift.*, Berlin, 1849), and that in them, as in Collins's draught, which is preserved at the Royal Society ('To Leibnitz, the 14th of June, 1676 About Mr. Gregories remains,' MSS. lxxxi.), we find the contents of Newton's letter of Dec. 10, 1672, except that instead of the example of drawing a tangent to a curve, there is merely allusion made to the method. Collins's larger paper (called "Collectio" and "Historiola" in the *Commercium Epistolicum*), of which the paper just quoted "About Mr. Gregories remains" is an abridgment, and which contains Newton's letter of Dec. 10 without curtailment, is stated in the second edition of the *Commercium* to have been sent to Leibniz, but whether that was the case may be fairly questioned."

There are two things in which we have never failed. We have never examined a point of mathematical history without finding either error or difficulty arising from bad bibliography: and we have never come fresh to this controversy of Newton and Leibnitz, without finding new evidence of the atrocious unfairness of the contemporary partisans of Newton. Nor had we a perception, until we *wrote out* the preceding paragraph, of the full extent of what it proves. It proves that at the time when the Committee of the Royal Society mentioned the "collectio" which contained Newton's letter *uncurtailed* of any part relating to fluxions, and asserted in their final report (without venturing to mention it in its place) that this letter had been forwarded to Leibnitz—they had, and must have seen, among the papers they were appointed* to examine, *Collins's own abridgment* of this *collectio*, headed "To Leibnitz," and containing Newton's letter *curtailed of the very part of which they asserted that it described the method of fluxions sufficiently for any intelligent person.* Of this abridgment they make no mention. We *now* see why the statement that the *collectio* was sent to Leibnitz was not allowed to appear in its place; that is, when the *collectio* was mentioned in the body of the work. Had the blot been hit, they would have pleaded some mistake or forgetfulness, would have produced

* There is not the least reason to suppose that any papers of Collins's ever came into the possession of the Royal Society after the *Comm. Epist.* was published.

the abridgment, and would have taken their stand on the fragment of the letter descriptive of results. We neither believe, nor would have others believe, that in the proceeding just described we are necessarily to impute guilty unfairness to the Committee of 1712, or to some of them: though all the circumstances make it impossible to avoid including this hypothesis among the probable ones. Independently of our knowledge of what *hero-worship* can lead to, even in our own day, we are bound to remember that all the notions as to what is fair and what is unfair in controversy, have undergone much change since the commencement of the last century. And above all, the idea that a party in literary controversy resembles one in a court of law, who may, with certainty of allowance, choose his own evidence, suppress what does not suit, and mystify what does, is now much less in force. In the particular case before us, perhaps something is to be allowed for hurry. The Committee was appointed in parcels on March 6, 20, 27, and April 17; and their report was read on April 24. But the hurry, if any, was their own fault. This striking fact, that the very papers which were examined in 1712 prove that the celebrated letter was not* sent to Leibnitz, but only a description (amounting to extract) of a part of it, and that part not the one which most appears to sustain the report of the Committee, throws into the background the remarks which we intended to make on part of the paragraph above extracted from the synoptical life of Newton. These must now be mixed up with remarks on the whole.

The editor begins by stating that doubts have been thrown on the question whether "*these papers* were *actually* sent to Leibnitz." By these papers the reference tells us we are to understand the *collectio* which has been spoken of. To remove the doubts and prove that "these papers" were actually sent, we are first referred to Collins's own testimony. The reference given would exclude Newton's letter, since nothing is there mentioned as sent to Paris except either Gregory's writings, or what had been done on the *method of series:* the drawing of tangents to curves was a perfectly distinct thing in the language of the day. But this reference leads us to a proof (though one is not needed) that the Committee actually saw the abridgment *which was sent*, and contrived to introduce reference to it in an unintelligible way: so that no one who was ignorant of the existence of the abridgment could infer that anything was sent except the complete *collectio*. The reference is to *Comm. Epist.* (pp. 47, 48, 2nd ed. 129), where we find a letter from Collins to *David* Gregory (the brother of James, whose papers were in question) of August 11, 1676, in which Collins says that he had put together an *historiola*

* It is now clear that the Royal Society owes the world more publication from its archives than has yet taken place: unfortunately, it is not yet alive to the feeling that such disclosures as those of the surreptitious additions to the reprint of the *Comm. Epist.*, and of the suppression now noted, would come most gracefully from itself. It is on record, that in 1716, the Abbé Conti, a friend of both parties, spent some hours in looking over the letter books of the Royal Society, to see if he could find anything omitted in the *Comm. Epist.* which made either for Leibnitz, or against Newton; and that he found nothing. But it now appears either that he did not know what to look for, or that there were papers which did not come in his way. Be it one or the other, the credit of his search is now upset; and Mr. Edleston's discovery proves that another is wanted.

of the writings of his brother and others, in about twelve* sheets, for preservation in the archives of the Society; and that he would find from what followed the letter (ex *sequentibus* comperies) that care had been taken to satisfy the wishes of the French mathematicians. Annexed to the letter is a memorandum to the effect that the *sequentia* had been sent both to the members of the French Academy,† and to David Gregory. Here then are two things; the *historiola* mentioned *in* the letter, and the *sequentia of* the letter: the latter was sent to Paris, and therefore by the *sequentia* we are to understand Collins's abridgment. That is to say, the Committee, which extracted as much from Collins as would prove that something was sent, did not give a word to explain what was sent: and inserted in their report a deliberate statement that the whole of what they chose to call the fluxional part of Newton's letter had been sent.

We are next told that *Leibnitz*‡ acknowledged the receipt of "these papers:" we look at the reference indicated, and we find that Leibnitz does (August 27, 1676) acknowledge letters of July 26, which the editor himself immediately proceeds to inform us, both from the Hanoverian publication and from Collins's draught, did not contain "these papers," but only an abridgment. Finally, the editor concludes that it may be "fairly questioned" whether the transmission ever took place. How can this be? the doubts as to the transmission he has just told us are removed by the testimony of Collins the transmitter, and Leibnitz the receiver. The answer is, that the editor himself immediately proceeds to prove, both from the transmitter and the receiver, that what was transmitted was not the *collectio* of the *Comm. Epist.*, but an abridgment. We cannot but suppose that the editor imagined the existence of the abridgment to be known, and having no idea that he himself was the first to draw it from its retirement, considered the *collectio* and its abridgment as convertible documents, and the information they conveyed as substantially the same. We, however, had never found a trace, in any writing upon the subject, of any mention of the smaller document: and it is clear that the omission of the example of Newton's

* It is now, Mr. Edleston informs us, extant in thirteen sheets: from which it is clear that this *historiola*, as Collins calls it, is what the Committee called the *collectio*; as the editor notes.

† Among these was Leibnitz, who, as we learn from the letter of Collins to Oldenburg, attached to the *collectio*, was one of the French Academy who had desired to have an account of Gregory's writings. In fact, Leibnitz was at Paris when he received Oldenburg's account of Collins's abridgment. The Committee who say that Newton's letter was sent to Paris to be communicated to him, may seem by this phrase to have supposed him to have been at Hanover.

‡ Our extract says, Leibnitz and *Tschirnhaus*. Now though the latter did write from Paris, in September, acknowledging something, yet he does not sufficiently say what, and even the Committee have put a note to his letter, doubting, from its internal evidence, whether he could have seen those extracts from Gregory which were sent to Leibnitz. So that the Committee knew nothing positive as to what was transmitted to Tschirnhaus. Moreover, Tschirnhaus was not Leibnitz. The whole of the passage on which this note is written struck us as so singular, so contrary, in the antagonism of its two portions, to the usual clearness of the whole of which it forms a part, that we could not help suspecting that the editor had been misled by some predecessor. And at last we found out by whom. Keill, in the account of the *Comm. Epist.*, published in English in the *Phil. Trans.* for 1715, and in Latin as a preface to the reprint, has the whole argument, with the affirmation of Collins and the replies of Leibnitz and Tschirnhaus. Keill was more noted, while alive, for getting his friends into embarrassments, than for his discoveries: will he never leave off his old tricks?

method, poor as the pretext against Leibnitz would have been even if it had been there, destroys the pretext* altogether.

We shall join the complete elucidation of the last assertion with the establishment of another statement of Leibnitz, namely, that the Committee of the Royal Society had been guilty of gross suppression of facts unfavourable to themselves, and within their own knowledge. We, who have not right of access to the archives of the Society, can of course only further show this (beyond what is shown by the suppression of the abridgment) by proving suppression of documents which had been already printed; that is, by showing that the Committee either entirely suppressed what they ought to have brought forward, or contented themselves with reference where they ought to have produced extracts. We shall confine ourselves to what is immediately connected with the unlucky fragment of Newton's letter, which was never sent.

First, the Committee refer to the method which Sluse had given for drawing tangents (p. 106, we quote the second edition as more accessible than the first), and which was *printed* in the *Phil. Trans.* as early as 1673. They give Oldenburg's communication to Sluse of Newton's letter, in which Sluse learns that what he had communicated was already known to Newton (p. 106). They also give Newton's admission (p. 107) that Sluse not only had probably an actual priority of discovery, but that, whether or no, he was the first promulgator. All this, so far as it goes, is fair, though it militates strongly against the conclusion of their report with respect to Leibnitz. But it was not fair to suppress all account of the manner in which this celebrated letter of Newton was drawn out. When they state that Collins had been for four years circulating the letter in which the method of fluxions was sufficiently described to any intelligent person, they suppress two facts: first, that the letter itself was in consequence of Newton's learning that Sluse had a method of tangents; secondly, that it revealed no more than Sluse had done. In the third volume (1699) of Wallis's works (in Latin p. 617, in English p. 636) is a fragment of a letter from Collins to Newton, of June 18, 1673, in which he reminds Newton, for what purpose does not appear, of his having communicated the fact of Sluse's discovery, and having received an answer (which was no doubt *the* letter) for the purpose of transmission to Sluse. Again, this method of Sluse is never allowed to appear: reference is made to the *Phil. Trans.*, though many things which had been printed before appear in the *Comm. Epist.* when they serve the right purpose.

To show what we assert we shall compare the two methods.

The paragraph of Newton's letter, from the original in the Macclesfield collection, is as follows (December 10, 1672):—

* If the editor meant that Newton's letter is substantially the same as to the real information it could give, whether with or without the example of the method of tangents, we not only agree with him as to the fact, but should have agreed, if he had asserted that a sheet of blank paper (after what Sluse had already published) would have done just as well. But our reader must remember that it is not the rational interpretation of the letter which is the matter in discussion, but the interpretation of the Royal Society's Committee.

"I am heartily glad at the acceptance, which our rev. friend Dr. Barrow's Lectures find with foreign mathematicians, and it pleased me not a little to understand that they* are fallen into the same method of drawing tangents with me [eandem .... ducendi tangentes methodum]. What I guess their method to be you will apprehend by this example. Suppose C B, applied to A B in any given angle, be terminated at any curve line AC, and calling AB x and BC y, let the relation between x and y be expressed by any equation, as $x^3-2xxy+bxx-bbx+byy-y^3=0$, whereby this curve is determined. To draw the tangent C D, the rule is this. Multiply the terms of the equation by any arithmetical progression according to the dimensions of y, suppose thus $\begin{matrix} x^3 & -2xxy & +bxx & -bbx & +byy & -y^3 \\ 0 & 1 & 0 & 0 & 2 & 3 \end{matrix}$ also according to the dimensions of x, suppose thus $\begin{matrix} x^3 & -2xxy & +bxx & -bbx & +byy & -y^3 \\ 3 & 2 & 2 & 1 & 0 & 0 \end{matrix}$. The first product shall be the numerator, and the last divided by x the denominator of a fraction, which expresseth the length of B D, to whose end D the tangent C D must be drawn. The length of B D therefore is $-2xxy+2byy-3y^3$ divided by $3xx-4xy+2bx-bb$."

Not many days afterwards (January 17, 1673) Sluse wrote an account of the method which he had previously signified to Collins, for the Royal Society, by whom it was printed (*Phil. Trans.* No. 90; also Lowthorp, vol. i. pp. 18—20). The rule is precisely that of Newton, the exponents are multipliers, without any subsequent reduction of the exponents (which prevents both explanations† from describing the method of fluxions to any intelligent person), and instead of dividing by x, Sluse changes one x into BD, and then equates the two results. To have given this would have shown the world that the grand communication which was asserted to have been sent to Leibnitz in June, 1676, might have been seen in print, and learned from Sluse, at any time in several previous years: accordingly, it was buried under a reference. But, worse than this, the Committee had evidence before them that it *had* been so seen *by Leibnitz*, and this evidence they deliberately mutilated.

March 5, 1677, Collins wrote to Newton, giving him certain extracts from a letter of Leibnitz, dated November 18, 1676. This was printed (1699) in the third volume of Wallis. Leibnitz had seen Hudde at Amsterdam, and had found that Hudde was in possession of even more than Sluse; and this he states, referring to the published method of Sluse, as known to himself. He gives also an example, or rather its result, not as showing the method, which was known, but in order further to show how to eliminate one of the

* There is no end of the curiosities of this Committee. After their Latin for the word *they*, they inserted in brackets [Sluse and Gregory], the latter not being a foreigner. If they had given the letter of Collins, just referred to, of June 18, 1673, the reader would have known that Sluse and Ricci are the parties understood.

† If Newton's example had been sent to Leibnitz, and the latter had not known the method already from Sluse, the direction to multiply by the terms of *any* arithmetical progression (a mere slip of the pen on Newton's part, properly preserved by the Latin translator) might have puzzled any *idoneus harum rerum cognitor*.

co-ordinates from the result. The Committee omit this example, without any notice of omission, though they give the passages between which it lies.

We are obliged frequently to recur to the assertion of the Committee that Newton's example, which we have translated, was description enough of the method of fluxions for any intelligent person. That this, which we shall believe to be the most reckless assertion ever made on a mathematical subject, until some one produces its match, was solemnly put forward by the Committee, is not in our day excuse enough for dwelling upon it. But the sufficient excuse is that writers of note, upon the Newtonian side of the question, still quote the assertion with approbation. In Sir David Brewster's life of Newton, for instance, the whole report of the Committee is printed, and a virtual adhesion given to it. On the other hand, the defenders of Leibnitz, most of whom are not English, prefer to establish his rights independently, and evade an encounter which is rendered repulsive by its dealing more with the comparison of old letters than with mathematical explanations.

Some little question has arisen as to the position in which the Royal Society stands in this matter. According to Leibnitz, Chamberlayne wrote to him to the effect that the Royal Society did not wish the report to pass for a decision of its own. Mr. Weld (*Phil. Mag.* 1847; Hist. R. S. vol. i. p. 415) found the minute in question (passed May 20, 1714), in which it is stated that "if any person had any material objection against the *Commercium*, or the Report of the Committee, it might be reconsidered at any time." This Mr. Weld considers as an *adoption* of the Report of the Committee: in which we cannot join, though we admit that it throws the question open, which as long as Chamberlayne's communication stood unanswered, was settled: and enables us to infer adoption from previous acts. In all probability he informed Leibnitz that the report of the Committee was not to pass for a *decision*, meaning the stress to lie there, and stating why: and this would be correct, for a question which may be reconsidered at any time is not decided, except in a technical sense. And very likely he added "of the Society:" for it was the full impression of the time that the Society was one with its Committee. There can be no doubt that the hearty adherence given by the Society to the conclusions, the circulation of the *Comm. Epist.* throughout Europe, the admission of Keill's *recensio* into the Transactions, the sanction of the reprint ten years after, and the obstinate determination, which lasts down to our own time, not to confess one atom of the error nor right one atom of the wrong, amount to an adoption which could not be more than adequately represented by any quantity of minutes.

It seems the fate of this controversy that whatever the English partisans of the eighteenth century supposed to have happened between the two parties really happened the other way, the places of the parties being changed, and to no effect upon the question. Much stress was laid on Collins transmitting from Newton to Leibnitz an example of the method of tangents: it appears that the example was not sent; that the *abridgment* sent did not contain it;

B 3

but it appears that Collins really forwarded a result from *Leibnitz* to *Newton*, which was the only one that passed between them. Not that this gave Newton any information; but neither would Newton's example, if sent, have given any to Leibnitz, after Sluse's publication and Hudde's oral communications.

Again, it was frequently stated that the differential calculus was only the method of fluxions with the notation changed. Now the fact is, that as to every thing elementary that was *published* with demonstration under the *name of fluxions*, up to the year 1704 (when Newton *himself* first published anything under that name) the method of fluxions was nothing but the differential calculus with the notation changed. We know that Newton's letters did not treat of fluxions, nor contain anything from which the writer of a system could draw his materials. No one ventured to print an elementary treatise in England until the seed had grown into a strong plant under the care of Leibnitz, the Bernoullis, &c. When De L'Hopital, in 1696, published at Paris a treatise so systematic, and so much resembling one of modern times, that it might be used even now, he could find nothing English to quote, except a slight treatise of Craig on quadratures, published in 1693. He mentions all that he could of Newton, and even says of the *Principia* that it was full of the *calculus*, which is not true; he should have said it was full* of the *principles* on which the calculus is founded, and of application of them in which the *reader* (whatever might have been the case with the *author*) is directed by thought without calculus. But the distinction is one which was not then appreciated: in fact it needed the calculus, such as it became, to show it. It must be remembered that when De L'Hopital *wrote* (for he could not then have seen the first volume of Wallis), there neither was, nor had been, one word of accusation or of national reflection, to create any bias for or against any one. The first thing of this kind took place in 1695, when Wallis, in the preface to the first volume of his collected works, not only claimed the differential calculus as derived from the method of fluxions, but (in ignorance, as he afterwards knew) grounded the claim upon the two celebrated letters of Newton to Oldenburg, of which little notice is taken here, because not even the Committee of the Royal Society venture a mention of them in their report, as any ground of confirmation against Leibnitz.

The note of alarm thus sounded, our countrymen began to write upon fluxions. Some writings are so advanced that they do not define their terms: from these therefore we cannot tell whether $\dot{x}$ means the velocity with which x changes, or an infinitely small increment of x. Such (at least so we suppose from the enlarged second edition of 1718) was the little tract of Craig, to which De L'Hopital refers, as we have seen: and such were Dr. Cheyne's tract on fluents (1703) and De Moivre's answer to it (1704). Newton himself, in the Principia, was not a fluxionist, but a differentialist. Though imagining quantity

* "C'est encore une justice dûë au sçavant M. Newton, et que M. Leibnis luy a renduë luy-même: Qu'il avoit aussi trouvé quelque chose de semblable au calcul différentiel, comme il paroit par l'excellent Livre intitulé . . . . . . *Principia* . . . . . lequel est presque tout de ce calcul."—*Preface.*

generated by motion or flux (in the celebrated Lemma in which he gives a brief description) he calculates, not by velocities but by *moments*, or "momentaneous increments and decrements," which are infinitely small quantities, for "moments, so soon as they become finite magnitudes, cease to be moments." Of Wallis we shall presently speak. De Moivre (*Phil. Trans.* 1695, No. 216) represents fluxions as momentaneous increments or decrements. And the only elementary writers, Harris* and Hayes,† are strictly writers on the differential calculus, as opposed to fluxions, in every thing but using $\dot{x}$ instead of dx. Harris says, "By the Doctrine of Fluxions we are to understand the Arithmetick of the Infinitely small Increments or Decrements . . . ." These he says Newton properly calls fluxions; and he proceeds to show that his own ideas are not very clear, by asserting that "'Tis much more natural to conceive the Infinitely small Increments or Decrements of the variable and Flowing Quantities, under the notion of Fluxions (that is, according to him, of infinitely small increments or decrements) than under that of Moments or Infinitely small Differences, as Leibnitz . . . . chose rather to take them." And then he proceeds to speak of *velocities*: in fact he jumbles De L'Hopital, whom he did understand, with Wallis, whom he did not. Hayes, a much clearer writer, begins thus: "Magnitude is divisible in *infinitum* . . . . . the infinitely little Increment or Decrement is called the *Fluxion* of that Magnitude . . . . . Now those infinitely little Parts being extended, are again infinitely Divisible; and these infinitely little Parts of an infinitely little Part of a given Quantity, are by Geometers called *Infinitesimæ Infinitesimarum* or *Fluxions* of *Fluxions*." And again (p. 5), "...suppose half the infinitely little increment of X to be $\frac{1}{2}\dot{x}$, and half the Fluxion or infinitely little Increment of Z to be $\frac{1}{2}\dot{z}$." And thus it appears that all explanation that was tendered in print, up to the year 1704, whether by Newton himself, or by any of his followers (except only Wallis as presently mentioned), was Leibnitian in principle. But when Newton, in 1704, published the treatise on the Quadrature of Curves which he had written before Leibnitz communicated the differential calculus to him, he starts with nothing but the notion of quantity increasing or diminishing with velocity, and this velocity or *celerity* is the fluxion. And in the Introduction, written at the time of publication, he says, "I do not consider mathematical quantities as consisting of the smallest possible parts (partes quam minimæ) but as described by continuous

* The first elementary work on fluxions in England is a tract of twenty-two pages in "A New short treatise of Algebra . . . . . Together with a Specimen of the Nature and Algorithm of Fluxions." By John Harris, M.A. London, 1702, octavo (small).

† "A Treatise of Fluxions; or, an Introduction to Mathematical Philosophy. Containing A full Explication of that Method by which the Most Celebrated Geometers of the present Age have made such vast Advances in Mechanical Philosophy. A Work very Useful for those that would know how to apply Mathematicks to Nature. By Charles Hayes, Gent." London, folio, 1704. This work, which has had very little notice (Hayes, born 1678, died 1760, wrote many works, but never set his name to any but this), is a very full treatise, nearly three times as large as that of De L'Hopital, having 315 closely printed folio pages on fluxions, besides an introduction on conic sections.

motion. This is the first public declaration of the meaning of a *fluxion* that was made by the author of the word, *in his own name.*

It may appear strange that we defer till now to mention a very *fluxional* view of fluxions which appeared as early as 1693. But we wish to give prominence to what is really Newton's first publication on the subject, though it has received but little notice until lately. The *second* volume of Wallis's works, containing the Algebra, in which the matter spoken of occurs, was published in 1693, the first in 1695, but false title-pages* make them appear as of 1699. Again, those who look at the preface to the first volume see that Wallis excuses himself from mentioning the differential calculus, because it was nothing but the fluxions which Newton, he says, had communicated to Leibnitz in the celebrated Oldenburg letters, and which he (Wallis) had described, from those letters, nearly word for word, in his *Algebra*. No one of later times would thereupon refer to this Algebra for information; since they would know that nothing upon fluxions could be given word for word, but only letter for letter. For all that is said upon fluxions, in those celebrated epistles, is, as is well known, in two anagrams, one of which is

6 a c c d æ 13 e f f 7 i 3 l 9 n 4 o 4 q r r 4 s 9 t 12 v x,

the information given being that whoever can form a certain sentence properly out of six *a*s, two *c*s, a *d*, &c., will see as much as one sentence can show about Newton's mode of proceeding. No one but Raphson† imagined that any human being derived any information from this; and probably therefore few would be induced by Wallis's preface to consult the work. They would not know (and we shall see that Wallis himself could hardly have anything to make him remember) that Wallis had been in communication with Newton, had obtained not only the key of the anagrams but their meaning, and had added a brief account of fluxions, with an extract from what Newton afterwards published in the treatise of 1704, besides other matter expressly obtained from Newton in explanation of the second anagram. The reader cannot detect the new information, except in that additional part which explains the second anagram: all that can be said of the rest is, that to a reader who compares chapters 91 and 95 there are a couple of sentences which would perhaps puzzle a person who did not know that a new

* The *Comm. Epist.* says that two volumes appeared in 1695; probably the second volume got a new title-page in that year. The third volume was published in 1699, and then the first volume certainly got a title-page of that date. This vile practice of altering title-pages will be put down by the scorn of all honest men, so soon as its tendencies are seen. A person who reads Wallis's collected works under the date of 1699 easily convicts the author, as honest a man as ever lived, of the grossest unfairness, upon his own testimony.

† The sentence was *Data Aequatione quotcunque, fluentes quantitates involvente, fluxiones invenire, et vice versa*, given any equation involving fluent quantities, to find the fluxions, and *vice versa*. Many writers have called this a *cipher*, which it is not: a cipher gives, in some way, the order of the letters as well as substitutes for the letters themselves. Raphson declared that Leibnitz had first deciphered the anagram, and then detected the meaning of the word fluxion, after which he forged a resemblance. But Raphson was the unscrupulous man of the time, if any one could deserve that name. Newton stated distinctly that Leibnitz sent him the details of a Method which was his own in all respects except language. Raphson says (*Hist. of Fluxions*, p. 1) that Leibnitz "writ in answer that he had found out a Method not unlike it, as Sir Isaac himself has hinted, page 253, *Princip*....." The impudence of this paraphrase is one of the minor gems of the controversy: and we could rub it brighter if we had room.

source of information was referred to in these sentences. The reviewer of Wallis in the *Acta Eruditorum*, in complaining of the suppression of the differential calculus, hit the real reason, namely, Wallis's ignorance of a good deal of what had been done abroad: and Wallis, who wrote to Leibnitz the day after he saw this review, acknowledges that he knew nothing of what Leibnitz had written, except two slight and old papers, and had never heard the name of the differential* calculus until the preface was in the press, when a friend mentioned with indignation that Newton's fluxions were current in Belgium under that name. Then, and probably without consulting what he had written, Wallis added the sentence we have mentioned to his preface. In the third volume, Wallis printed all his correspondence with Leibnitz, and all the correspondence with others on the subject which he could collect, and mentions fluxions and the differential calculus as two distinct things in the preface. What we have here to do with, however, is the nature of the publication of fluxions which was made in 1693.

We now come to the independent proofs of the separate invention of Leibnitz, as contained in his recently published papers. Preliminarily, however, to these, we may notice one which was published in 1671, and which shows the way in which the current of his ideas was setting. Dr. Hales, in his *Analysis Fluxionum* (Lond. 1800, 4to.), says that Leibnitz had given no obscure germs of his differential method in his *Theoria Notionum Abstractarum*, dedicated to the French Academy in 1671: and Dr. Hutton (Math. Dict. Art. *Fluxions*) refers to this theory of abstract *notions*. Both are wrong in the name; for the paper which Leibnitz dedicated to the Academy in that year is *Theoria Motus Abstracti* (*Op. Leibn.* vol. ii. part ii. p. 35). This paper is certainly a witness to character; throughout it there occurs a frequent approximation to the idea of infinitely small quantities having ratio to each other, but not to finite quantities. One extract (translated) will serve as a specimen: "A point is not that which has no parts, nor of which part is not considered; but which has no extension, or whose parts are indistant, whose magnitude is inconsiderable, inassignable, less than any which has ratio (except an infinitely small one) to a sensible quantity, less than can be given: and this is the foundation of Cavalieri's method, by which its truth is evidently demonstrated, namely, to suppose certain rudiments, so to speak, or beginnings of lines and figures, less than any assignable." So that, in 1671, it was working in Leibnitz's mind that in the doctrine of infinitely small quantities lay the true foundation of that approach to the differential calculus which Cavalieri presented.

Dr. Gerhardt, the editor of the correspondence already referred to, found among the papers of Leibnitz preserved in the Royal Library at Hanover various original draughts, containing problems in

* Nevertheless, Leibnitz and the differential method are mentioned in the second volume, that is, in the account of fluxions on which we are writing; but (as discovered by Professor Rigaud) Wallis's copy preserved in the Savilian library has manuscript additions which note and explain this forgetfulness. It appears that the whole communication is Newton's, and is inserted in Newton's words: an author can hardly remember another person's writing, to which he gives admission, as if it were his own.

which both the differential and integral calculus are employed, and has published them in a separate tract.* The editor dwells so much on the matter and consequences of the manuscripts, that he forgets to satisfy curiosity as to their form, the circumstances of the discovery, &c.: they ought to be republished with proper fac-similes of the hand-writing. Not that we at all doubt them; for, independently of the full credit due to Dr. Gerhardt, we do not believe that human ingenuity could have forged so genuine a mess of spoiled exercises. We cannot attempt a full account of them; but this is of little consequence, since they will of necessity be fully described in more appropriate quarters, so soon as they are better known to exist.

These papers are seven in number, dated† November 11, 21, 22, 1675, June 26, July, November, 1676, and one without date. They are not descriptions of the principles, but study exercises‡ in the use, of both differential and integral calculus. Except out of the problems themselves we learn nothing of the extent to which the structural operations were in the power of the writer. We find strange mistakes of operation, such as beginners now make: and it is clear that the writer is trying to push his calculus forward into discovery of new results in geometry before he has either sounded its extent or settled its language. In the first of the papers he enters (among other things) upon the examination whether $dx.dy$ is the same with $d\,(xy)$ and $d\left(\frac{x}{y}\right)$ with $\frac{dx}{dy}$: at first he inclines to the affirmative, but in the next page decides in the negative. This will not surprise the mathematician of our day, who remembers that these are the private memoranda of a discoverer in the very process of investigation: but nevertheless he will look to find some particular cause of confusion of ideas at the outset. We suspect it to be as follows. Leibnitz frequently supposes $dx=1$, or $dy=1$: that is, he establishes two kinds of *units*, without any symbolic distinction, the unit of finite, and the unit of infinitely small, quantity. In integration, he halts between the use of $\int y$ and of $\int ydx$, as the expression of an integral. There are also obvious slips of the pen, and operations set down for thought, which lead to nothing.

The first problems treated are in the direct and inverse method of tangents, in which the method of Sluse is referred to by name. The two following extracts, in which the Latin is literally translated, of the date of November 11, 1675, will be as much as we can afford room for. They give two of the earliest problems solved, the first and third.

* 'Die Entdeckung der Differentialrechnung durch Leibniz. Von Dr. C. J. Gerhardt.' Quarto. No date nor place; preface dated "Salzwedel, im Januar 1848."

† The editor tells us that some one had been meddling with the date of the first paper, and had turned the 5 of 1675 into a 3. Leibnitz, speaking from recollection in 1714, says that his discovery was made, as near as he could remember, in 1676.

‡ Professor Rigaud has published, from the Macclesfield collection, a manuscript draught of Newton, of Nov. 13, 1665. But this is formally written out, proposition, resolution, and demonstration. An earlier essay, of May 20, is not given, which is to be regretted. But from the description we see that Newton used the peculiar notation of fluxions in May, and abandoned it in November. His formal proposition uses distinct letters for fluxions of other letters. In Leibnitz, everything in language is progression: no step gained is ever abandoned.

The problem is to find a curve in which the subnormal (w) is reciprocally proportional to the ordinate. Putting z instead of dx, Leibnitz proceeds thus:—"It appears from what I have shown elsewhere, that $\int wz = \frac{y^2}{2}$, or $wz = \frac{y^2}{2d}$." The d in the denominator is the symbol of differentiation of the whole: it frequently happens *in the first papers*. "But from the quadrature of the triangle this is y." We should write ydy, but Leibnitz tacitly makes $dy=1$, and he afterwards says he has here thought of making an abscissa of the ordinate. "Now from the hypothesis $w = \frac{b}{y}$..... whence $\frac{bz}{y} = y$, and $z = \frac{y^2}{b}$. But $\int z = x$. Therefore $x = \int \frac{y^2}{b}$. But $\int \frac{y^2}{b} = \frac{y^3}{3ba}$ by the quadrature of the parabola; therefore $x = \frac{y^3}{3ba}$." This a is not of easy explanation. It is afterwards given to make the subnormal reciprocally proportional to the abscissa. "Here $w = \frac{a^2}{x}$; but $\int w = \frac{y^2}{2}$, whence $y = \sqrt{(2\int w)}$, or $\sqrt{\left(2\int \frac{a^2}{x}\right)}$. Now $\int w$ cannot be found except by the help of the logarithmic curve. Therefore the figure required is that in which the ordinates are in the subduplicate ratio of the logarithms of the abscissæ."

If the Committee of the Royal Society had had these papers before them, they would have justly contended that the calculus of Leibnitz, of which the principles and algorithm were settled, received a great accession of working power when Newton communicated the binomial theorem in the *epistola prior* to Oldenburg; which *epistola prior*, by the rule of contraries already instanced, has been much less insisted on than the *epistola posterior* with its anagrams.

On August 27, 1676, Leibnitz acknowledged the receipt of this communication; and his paper of November, 1676, shows that Newton's algebra had borne its fruit. Previously to this date, we cannot find any fractional power differentiated except the square root. In pure algebraical discovery, Leibnitz does not rank with Newton: and he always acknowledged that in the *method of series* (the phrase by which the algebraical improvements of the day were designated) Newton was before him and beyond him. We have every right to presume, from his conduct, and from the manner in which all subsequent disclosures establish his veracity, that had he lived to publish his own *Commercium Epistolicum*, he would have pointed out the difference between the invention of the differential calculus and the improvement of the algebra which gives it language and guides its mechanism, and would have illustrated from his own papers the power which Newton's improvements in algebra enabled him to add to his existing differential calculus. We believe (with John Bernoulli) that Newton might have made a similar acknowledgment to Leibnitz as to the idea of a fixed and uniform method of denoting operations in the fluxions of which he had already possession.

We have not alluded to the faults on the other side of the controversy, partly because they were much less gross in character, partly because they have been amply insisted on in this country. Nor have we, indeed, in this paper, given anything like a history of the unfair proceedings in this country, but have, for the most part, confined ourselves to points which are particularly affected by recent information. Whether there be anything still to be drawn out must be matter of conjecture, and will be matter of suspicion, until we can be well assured that all the private depositories of information have been exhausted.

A. DE MORGAN.

University College, London,
October 2, 1851.

COMPANION TO THE ALMANAC

FOR

1853.

PART I.

GENERAL INFORMATION ON SUBJECTS OF MATHEMATICS, NATURAL PHILOSOPHY AND HISTORY, CHRONOLOGY, GEOGRAPHY, STATISTICS, &c.

I.—ON THE DIFFICULTY OF CORRECT DESCRIPTION OF BOOKS.

We have often had occasion, in articles contributed to this work, to notice error and difficulty arising out of incorrect or insufficient description of books. The study of *bibliography*, that is, of books as books, in all matters which are requisite to avoid the errors and difficulties just alluded to, has been left to librarians and to bibliomaniacs, as they have been called. Recent events, however, have brought bibliography into collision with the want of it, in a remarkable way.

The year 1850 turned the attention of literary men to the subject, both in England and France: but in very different ways in the two countries. In England, the report of the Royal Commission appointed to examine the state of the British Museum became public, and with it the evidence on which it was founded. This report and evidence contained the details of a severe contest between bibliographers on the one hand, and literary men opposed to bibliography on the other hand, as to the mode in which book-catalogues should be made. The report of the Commission, the comments of the leading reviews, and the subsequent silence of the journals which had for years attacked the librarians of the Museum, gave the victory to the advocates of detail sufficient for accuracy, as one side called it, or of unnecessary minuteness leading to confusion, as the other side called it. And the great extent to which both the antagonist philosophies taught by examples, makes this report, with its evidence, an excellent collection of exercises, and a manual, so far as that term can be applied to a blue-book, of practice for the young bibliographer.

The corresponding display made in France was not altogether so creditable to the literary aspirations of the nation. In the year 1850, appeared the *act of accusation* against M. Libri,* an emi-

* The reader will find some account of the details of this persecution in 'Bentley's Miscellany' for July, 1852, and in the 'Athenæum' for May 27, 1848, and May 12, 1849. How completely the charges are to be attributed, in the first instance, to political and private malice, is now sufficiently known. "He was condemned," says the 'Times,' "for stealing books, many of which are now to be found in the very places from which he was said to have taken them; he was condemned for stealing books which he was proved to have bought of Messrs. Payne and Foss in London; he was condemned for stealing books which it was beyond the power of the French courts to identify, or even to describe correctly." All this we know to be true, with the exception of what is implied in the word *even*: correct description is no such every-day matter.

B

nent mathematician and bibliographer, and a member of the Institute, charged with robbing the public libraries to the value of many thousands of pounds; on which, by default of appearance, he was condemned. The amount of incapacity which either belonged to the framers of this indictment, or was presumed by them to belong to the courts and the literary public before which it was to come, far exceeds all that was exhibited by the ignorers of bibliography in England. None of these last ever thought, or wished to make others think, that the stamp* of a convent *library*, imprinted on the front of an old book, is evidence of an intention, on the part of the stamper, to pass the book off as *printed* in the town where the convent is.

Except, however, to express our belief that these recent events in France and England will be of some effect in widening the circle within which bibliography is studied, we have nothing to do with them here, though we may cite them as among our encouragements for presenting an article on the subject. Our intention is to show, by instances, to how great an extent inaccurate bibliography prevails, both in the descriptions which are given of books, and in those which they give of themselves. We began, in pursuance of this intention, and that we might produce a new case† or two, by taking the first four old books that we happened to lay our hands on, the selection being dictated by the mere accidental proximity of the volumes on our shelves. If no one of these four volumes had given us either error produced, or difficulty likely to produce it in time to come, our associations would have been rudely invaded; for we have been accustomed to consider it almost impossible to take two old books at hazard without encountering one or the other. It happened that *all* the four gave us what we wanted to illustrate.

The first book was a collection of four geometrical and two astronomical treatises by John Werner of Nuremberg, quarto, 1522, beginning "In hoc opere hæc continentur. Libellus Joannis Verneri Nurembergen. Super vigintiduobus elementis conicis. . ."

* The examiners of M. Libri's books found the Aldine Catullus of 1515, Venice, with what they read as "Bibliothecæ S. 10 in Casalibus Placentiæ" either stamped in old type, or in manuscript, (they could not tell which!) on the front leaf. The "S. 10," had they known how to read, would have been "S. Jo." and the whole would have shown that the book once belonged to the Library of the Convent of St. John of the Canals at Piacenza. They impute to M. Libri that *he* stamped these letters, first, to hide the marks of another stamp which they assert to have been erased, next, to pass off the work as *printed* at Piacenza. The terms in which they crow over their unanswerable proof, as they take it to be, that the book had been stolen, will perhaps be cited in bibliographical treatises for centuries to come: " . . . le titre annoncait une édition de Plaisance, et la bibliothèque [de Montpellier] avait perdu une édition de Venise . . . . Pour dissimuler les traces du grattage dont il a été parlé, on avait mis à la place de l'estampille ces mots, ... BIBLIOTHECÆ S. 10. IN CASALIBUS PLACENTIÆ. Manuscrits ou appliqués avec de l'ancienne fonte, ces caracterès jouent l'impression. Mais la fraude ne pense pas à tout: tandis que le titre falsifié annoncait une édition de Plaisance, la dernière page révélait une édition de Venise . . . . . De tels faits ne se discutent pas, ils s'exposent." The supposition that a practised bibliographer, desiring to falsify the place of printing, would forget that it is *almost always* at the end in very old books, is more amusing to those who are looking at the last pages of such books every day, than those who do not look into them can easily imagine. Their proverb ought to have been, *la fraude ne pense à rien*.

† We are not indebted, throughout this paper, to any one instance which was introduced in evidence before the Commissioners, either by ourselves or others.

It is said that this book was so rare in the time of Tycho Brahé, that he could not find it in all Germany, though he secured a copy at last in Italy. The two last treatises being astronomical, we turn to Lalande's 'Bibliographie Astronomique,' and we find at the right year, 1522, that this book consists of the two astronomical treatises, followed by an epistle of Regiomontanus to Cardinal Bessarion on the meteoroscope [instead of preceded by four geometrical treatises of Werner himself]. The authority is Weidler, who, says Lalande, adds two other tracts as contained in this work, of which Scheibel observes that they have never been printed at all. Here is a heap of confusion, in which three noted writers of mathematical history are concerned. Looking at Weidler (at the page cited), we find reason to think the case stands as follows. Weidler, after hinting that Werner printed the works of others as well as his own, gives a list as extant, in which he takes no care to distinguish between what Werner only printed, and what he both wrote and printed. In the middle of this list comes the letter to the cardinal. The last five of the list are five of the treatises which really are in the work before us, the sixth being omitted. Then, says Weidler, these last five works appeared at Nuremberg in 1522. From this it would appear as if Lalande had selected two astronomical works of Werner, the letter of Regiomontanus, and two others which he does not name because Scheibel said they were never printed.

We had turned to Weidler's History, because Lalande cites it (p. 334). We then turned to Weidler's Bibliography, and here we really find that the Nuremberg quarto of 1522 is said to contain five treatises, the three given by Lalande, with two others by Werner, not any of those yet named. And Weidler refers to p. 334 of his own book, in which, as already seen, he gives a very different and more correct account. So that the confusion is as follows. Weidler describes the book in his *History* with nothing but an omission. In his *Bibliography* he gives a totally wrong description, for which he refers to his own more correct *History*. Lalande adopts the account given in the Bibliography, and joins to it the reference to the History, without stating that his reference to the History is only a copy from the Bibliography. No one, without the book before him, could have unravelled this skein of mistakes. We took the work of Lalande because it is decidedly the best piece of scientific bibliography which, at its appearance, had ever been in existence, and therefore gave the best chance of a correct description. But, like other descriptive works which make a commencement of correctness upon books which the authors had examined for themselves, it relies in a great degree upon works prior to the introduction of any effort at minute description.

In the last instance, it happens that the mistake can be traced to its source in a manner which leaves no doubt that it is a mistake. But the unpractised reader must not come to such a conclusion too rapidly. If Lalande had not named his authority, as often happens with him, we should have had three alternatives to consider. 1. A mere mistake. 2. The circumstance of his having happened

B 2

to fall in with a book in which some one had bound together some astronomical tracts of Werner with a copy of Regiomontanus's epistle. 3. The possibility that Werner made two distinct publications at Nuremberg in 1522, one containing his own *six* tracts, the other joining the last *two*, which are astronomical, with the astronomical epistle in question. Either of the first two hypotheses is credible enough. The third looks very unlikely. But it must be remembered that it is utterly impossible to enumerate the number of odd things which occurred in the first century of printing, before authors and publishers had fallen into a common understanding upon their modes of proceeding. Anything imaginable may have taken place in one or more instances; and it happens sometimes that the unlikely thing, stated by a writer who is frequently inaccurate, turns out to be the truth, in spite of the more probable account of a generally more accurate writer. And a strange assertion, which appears to be an obvious distortion of one which is known to be true, may nevertheless be one separate truth, with or without some admixture of the matter of the other. For instance, a poor authority on books, Granger, says that Roger Palmer, afterwards the notorious Earl of Castlemaine, husband to one mistress of Charles II., and ambassador to the Pope of James II., invented and wrote on a "horizontal globe." Now since *John* Palmer, in 1658, did certainly write on the 'Catholique Planisphær,' and since the phrase *horizontal globe* looks very much like an awkward rendering of the word *planisphere*, we at one time took the liberty of thinking that Granger or another had confused the two Palmers; and we were not without our suspicion that the *Catholic* planisphere had perhaps assisted in the transfer of the book to a *Catholic* author. Nevertheless, we afterwards found* that Lord Castlemaine published in 1679, a work on what he called the 'English Globe.' Again, the rule of three, in middle Latin, is *regula detri*, so that, seeing *Detri* mentioned among arithmetical authors, we took it to be pretty certain that, as has sometimes happened, the name of the subject of a book had been substituted for that of the author. Nevertheless, we have since seen in a careful sale catalogue, the description of the work of N. Detri: in which we believe, in spite of the existence of another work by N. Petri.

The second of our four instances is the *Cosmographia* of the celebrated Maurolycus, 4to. Venice, 1543, the year of publication of the great work of Copernicus. At the end it is stated that this work was finished in 1535, and the preface is dated 1540. It is in dialogue; and the teacher says (p. 12) that nothing more need be said about the earth, unless diversity of opinion and human fickleness should so far increase, that there should be ground to suspect some one of believing and maintaining that the earth turns on its axis. I should hardly think, says the pupil, that such a strange opinion would enter the head of any one. Why not, rejoins the teacher, many teach themselves greater absurdities; but be this as

* If Granger had only looked into the 'Catalogue of Royal and Noble Authors' by Horace Walpole, to whom his own work is dedicated, he would have seen an accurate title-page of this work.

it may, to remove all possibility of such opposition, I will demonstrate that the earth cannot move. If all this were first published in 1543, in spite of the date of the preface, we should reasonably presume that the intention of Copernicus had reached the ear of Maurolycus, and had given rise to the introduction, at the last moment, perhaps, of what precedes. For in 1540, Rheticus* published at Dantzig his account of the forthcoming work of Copernicus. In much less than two years, this might be circulated over Europe, for everything found† its way easily to and from Rome, and opinions travelled by epistolary description much more than now. But, if what precedes *were* written before 1540, it shows that, anterior to the publication of Rheticus, there was a feeling that discussion on the earth's motion was at hand. This would be worthy of note, for no one has hitherto shown that, in the case of the earth's motion, there was that previous expectation of change which has marked the approach of most other new doctrines. Had there been, no doubt the work of Copernicus would have given the signal for that sort of opposition which was reserved for Galileo. All this would lead us to suppose that the remarks of Maurolycus were suggested by the special publication of Rheticus, and not by any knowledge, on the part of Maurolycus, of a diffused disposition to think about the actual question‡ of the earth's motion.

But now comes a difficulty. A preface, dated in February 1540, of a work published in 1543, gives some presumption, not a very great one, of a previous edition in 1540 or 1541; rather too much§ to neglect, though far from enough to pronounce upon. Lalande, relying again upon Weidler, affirms that this work of Maurolycus *was* first printed in 1540; and Weidler makes the statement both

* The best chance any reader will have of seeing this remarkable precursor of Copernicus, will be by looking for the second edition (Basle, 1566, folio) of Copernicus himself, to which it is attached. We have never seen either of the two separate and previous editions of the tract of Rheticus; but a letter from Gassarus of Lindau, prefixed to that of 1566, mentions the receipt of the first edition from Dantzig, and is dated 1540. So that neither Lalande nor Weidler is wrong on this point.

† There is reason to suppose that foreign books of second-rate name, travelled from one country to another, during the earlier ages of printing, in larger numbers than now; that is, immediately after publication. At the present time, in the case of a book of no great note, published in France or Germany, hardly more than two or three straggling copies will forthwith find their way to England. But in 1670—80, the bookseller always imported immediately: and the mathematical bookseller complained that he could not sell more than *twenty or thirty*, until the book had gained reputation, in a manner which implied that even this state of things was a falling off.

‡ The reader should be aware that both Rheticus and Copernicus propounded the theory of the earth's motion only as an hypothesis, to *save appearances:* using this phrase in the old sense, though most historians suppose that they also intended the thing signified by its more modern meaning. The phrase to *save appearances* is a cast off phrase of physics; we now say to *explain phenomena.* Thus the supposition that the earth turns on its axis preserves the diurnal appearances of the heavens, and makes them follow: and the old explanation does the same. Copernicus contends for the supposition of the earth's motion as the most simple mode of deducing and calculating the celestial phenomena: leaving the question of its actual truth or falsehood open. The utmost extent to which he commits himself on this point is (lib. I. cap. 8) the affirmation, that on the balance of *à priori* reasons, the motion of the earth, especially the diurnal motion, is more probable than its stability.

§ Castiglione, who published Newton's *Opuscula*, knew that the Optics were published in 1704, and had a copy of 1706. He took for granted (pref. p. vii.) that there could not be two editions so near in time, and therefore announced that by the printer's negligence the edition of 1704 had 1706 on the title-page. The fact is that there was an English edition in 1704, and a Latin one in 1706.

in his History and in his Bibliography. And, what is more, Riccioli (in 1651) makes the same assertion. It matters little or nothing that the work of 1543 is not called a second edition, for it not unfrequently happens that a reprint shows no sign of that character. And though neither the Abbé Scina, in the life of Maurolycus, nor the compiler of the list of works presently mentioned, notes any edition earlier than 1543, yet neither seems to have made much search, and both, to judge by their modes of description, would rest content with the earliest edition they happened to have seen. Thus, though inclined* to believe that the edition of 1543 is really the first, and therefore that the remarks we have quoted are specially directed against Rheticus, we should not be at all surprised if an edition of 1540 or 1541 were to turn up.

It is known that there were among the ancients some who maintained the diurnal motion of the earth, and some who maintained the annual, at least as possible; Ptolemy alludes to them, and gives his reasons against them. Down to the time of Copernicus, we are not told of any (except Cardinal Cusa,† who is not worth alluding to on this point) who really thought anew on the subject, so as to produce fresh arguments either for or against. Nevertheless, it appears, though we cannot find it mentioned by any historian, that Regiomontanus had seriously considered the subject. One of the greatest preservers of his writings was John Schoner, of Carlstadt (1477—1547). In the collection of Schoner's works, first‡ published in 1551, Nuremberg, folio, is an *Opusculum Geographicum*, the first chapter of which is a *disputatio* of Regiomontanus on the subject of the earth's rest or motion. In this short discussion, while deciding the question against the earth's motion, on grounds resembling those of Ptolemy, he cites, as from the ancients, the comparison of the earth to meat roasting on a spit, and of the sun to the fire which cooks it; as also the argument that it is the business of the mutton, which wants heat, to turn round before the fire, and not of the fire to turn round the mutton. To what old writer he refers, we cannot tell, as we cannot find this simile in any of the passages which have been quoted from classic authors. We mention the discussion in which it occurs to point out that it would not be a very easy matter to ascertain whether Copernicus (who died in 1543) could or could not have seen it.

* Maurolycus, in 1553, received a pension expressly to enable him to publish his works: which makes it likely that some of those previously published had been delayed, and the more so as there was remarkable delay even after the receipt of the pension.

† The Cardinal's argument was founded on the non-existence of a centre, deduced from the non-existence of a circumference, to the universe. A book might be written on the manner in which purely subjective notions of the centre and its necessary properties influenced the arguments on this subject, from those of Cusa to those of the Sieur de Beaulieu (1676), who says that the presumption of Copernicus led him to "advance in geometry a proposition as absurd as it is against faith and reason, by making the circumference of a circle fixed and immoveable, and the centre moveable, on which geometrical principle he maintained the stability of the sun, and the motion of the earth."

‡ Weidler, in the History, gives a correct account of this work: in the Bibliography, which refers to the History (p. 337), he makes the mass of its contents to belong to the subsequent edition of 1561, and retains only the three last treatises in that of 1551. Lalande copies him, together with the reference "p. 337," and thus again seems to misstate the matter of his own reference.

According to the preface, the date of composition of this *Opusc. Geogr.* is 1533; from which Lalande says it was printed in 1533; but we can find no notice of any impression previous to that in the collected works of 1551. Weidler says this collection contains some things which had not been previously published: but this can only mean that *he* had not found them.

The third of the books in our list is a quarto printed at Leyden in 1649, the title of which tells us that it is the Geometry of Descartes, first printed in French in 1637, and now rendered into Latin with notes, &c., by Francis Schooten. This is then certainly a second edition, at least. Now we learn in many places that in the second edition of Descartes's Geometry, by Schooten, the additions contain papers by Van Heuraet, Hudde, &c. of the greatest note in the early history of the differential calculus. Not the smallest trace of these things appears in the book before us. Some persons must have been puzzled by this: the truth is, that instead of naming the second edition of Descartes, by Schooten, writers should have named the *second of Schooten's* editions* of Descartes, Amsterdam, 1659, which has on the fly-title, "Renati Descartes Geometria, Editio Secunda"—a wrong description. Thus it appears that the titles of the books themselves may contain the very errors which it is the tendency of bad catalogues to create when they do not exist, and which it is very difficult to avoid, or to correct, even in good ones. This instance is particularly appropriate, for it is of the simplest kind. "It may help to enforce a truism which seems of late years to have been almost entirely lost sight of"—[a gentle mode of expression for vigorously denied and opposed]—"that the making of catalogues correctly, like the making of dictionaries, requires in the 'harmless drudge' who practises it an amount of qualifications which those who despise him are often far from possessing." This quotation comes out of the review of an attempt to catalogue the library of the linguist Mezzofanti, made by a Roman bookseller,† who entered a Cingalese grammar, printed at *Colombo*, under the United States of America, and a Gaelic translation of Thomas à Kempis as a work of Chr[istopher?] Leanmhuinn, the words *Leanmhuin Chriosd* being Gaelic for *De Imitatione Christi*. It is not necessary to choose collections of so recondite a character before the opinion we have quoted can be given: if it were, it should then be added that a great public library like that of the Museum is not

* A difficulty of this kind is far from uncommon. An editor leaves us in doubt as to whether the numbering of the edition refers to impressions, or to the impressions which that particular editor has superintended. It would be well if the word *impression* were used in the general sense and *edition* in the particular. Thus, if A publish four editions of his own work, and if the commentator B then publish three more, there will be seven impressions in editions of four and three; and the sixth impression of A's work will be B's second edition.

† An English auctioneer was brought forward to give evidence upon the catalogue of the British Museum, who declared that cataloguing was not only easy, but very simple indeed, with the assistance of the librarians of the Museum, or of his own clerks. This gentleman was no way to blame, but those who imagined that a sale catalogue would serve the purposes of literature: if the Museum library were to be sold off, his evidence would be valuable; but the librarians of the Museum must not be employed, as he proposed. For these gentlemen have no idea, with a volume of six tracts before them, of entering the title of the first, followed by "and five others:" moreover, they waste time in writing down names of authors in their nominative cases, when the books before them give genitives; and in other ways.

only a larger collection of languages than that of Mezzofanti, but of all other special pursuits as well. And the instance of which we have spoken is a better illustration than any blunder which detects itself by its own absurdity. The English or the Gaelic scholar will not be deceived by the cases we have quoted: the worst that can happen is, that the inquirer who looks under Ceylon for Cingalese misses a book, and it is as if Mezzofanti had not possessed that book. But if, as might possibly happen, Schooten's Descartes of 1649 were entered as what it really is, the second edition; and if that of 1659 made part of a set of Descartes's works, as it often does; and if, as is very common, the *Opera omnia* were insufficiently detailed in the catalogue; and if, as generally has happened, a mathematical historian were somewhat easy on the point of bibliography—the works of Hudde, &c. might disappear from history, as other works have done in a similar way, without disappearing from libraries.

The fourth book in question is another work of Maurolycus, the *Opuscula Mathematica*, Venice, 1575, 4to. On the title-page it appears that this collection was then published for the first time. It consists of a collection of tracts, and of a work on arithmetic with a second title-page of the same place and date as the first. Here, as often happens, is a source of confusion: in binding, these works are separated, each title-page being made the beginning of a separate work. Two things are very common at the date now before us, the binding up of different publications in one, and the distribution of one publication under different title-pages, often without any mark by which to know that all the titles belong to one work. Hence catalogues sometimes represent different publications as one, and sometimes represent one publication under several heads; the binder being the authority in both cases. On looking* more narrowly, to see whether the work itself gives any information on this point, we find, on the verso† of the first title-page, a table of contents, at the end of which is "Quibus omnibus arithmeticorum libri duo demum accesserunt." This is conclusive as to one difficulty, but it introduces another. The word *demum* usually indicates that the edition in question is not the first: *at last*, we are told, the two books of arithmetic are added. Either then there are previous editions without the arithmetic, or at last the arithmetic is added, and a new title-page, probably of later date, printed before all. Nevertheless, we can find no indication of any earlier edition or earlier title-page. None is mentioned: the arithmetic

* The *Anti-bibliographers* contended that any one could make a catalogue who could write a title-page: they did not appear to be aware of the necessity of examining the book. In one book we have before us three treatises of Ozanam, on conic sections, loci, and equations, all Paris, 1687, 4to., all from one publisher, whose residence is described in one way in the first and second, and in another way in the third. Unless they are three separate works, or all one work, either of which is very possible, the presumption furnished by the title-pages is that 1 and 2 were published together, and 3 separately. But an examination of the prefaces shows that 1 was published separately, and afterwards 2 and 3 together.

† The *recto* and *verso* of a leaf are the two pages in the order in which they come. We must use the technical term here, because, if we had only said that on turning over the title the table of contents was seen, it might have been on the recto of the next leaf, and no reprint of the title could have been inferred.

now under consideration has in it a list of works, distinguishing and dating those which were printed, but not containing anything to our present purpose. The maker of a catalogue would be compelled to raise the doubt of a second edition, or of a title-page with a new date, unless he happened to know that Maurolycus died in 1575. We now learn the meaning of *demum:* the work had been waiting for the arithmetic, which the author could not or would not finish, and his death *at last* enabled the publisher to obtain the manuscript, and complete the undertaking. This view of the case is enforced by our finding the arithmetic wholly destitute of preface or introduction, and with some gaps in the manuscript.

These circumstances, apparently so unimportant, help to decide an historical question which is not without interest. We have seen that the *Cosmographia* has a passage which indicates a lurking fear that the doctrine of the earth's motion was likely to be maintained. Though, by 1543, there was plenty of time to become aware of Rheticus's announcement of Copernicus, yet Maurolycus tells us, with the utmost definiteness, at the end, that the work was "finished at Messina, in the straits of Sicily, on Thursday. October 21, indiction ix, in the year of grace 1535, being the day on which the Cæsar, Charles V., returned to Messina from his African expedition." May we, in the face of such an announcement, suppose that this work afterwards underwent augmentation? If not, we have such presumption of the doctrine in question being in agitation, as it might be difficult to find elsewhere. But this presumption is destroyed by the work on arithmetic, which, though certainly unfinished, is terminated by the announcement, that it was finished "at the eighteenth hour of the Sabbath, July 24, when the viceroy Jo. Cerda was expected at Messina, *cum multo pontis et arcus apparatu*, indiction xv, 1557."

These four books, taken down for a first chance, merely to make an opening, have caused great inroad on our space. We shall take a few other illustrations. *Publication* is now commonly confounded with *printing*, though history swarms with instances in which the first was long prior to the second. There are those who would contend for the equivalence of the two words; but perhaps there is no instance more to the point, in proof of the general aptitude to distinguish between the two, than the case of the Academy of Sciences. This body did not begin to print its periodical volumes of transactions, in the manner done by the Royal Society from 1665, until after the *renouvellement* in 1699. It was not until 1729-1733 that the Academy published the collection in eleven volumes (fourteen parts) containing the memoirs from 1666 to 1699, which is now considered as a commencing part of the series. Nevertheless, no one ever referred to the memoirs therein contained, as published at any other date than that at which the subsequent printed volumes showed them to have been communicated to the Academy. The real earlier publications made at the instance of the Academy are the 'Mémoires de Mathématique et de Physique,' in two parts, Paris, 1692, 1693, folio; the 'Divers Ouvrages de Mathématique et de Physique,' Paris, 1693, folio; and

B 3

the 'Regiæ Scientiarum Academiæ Historia,' by J. B. du Hamel, of which the second* and enlarged edition is Paris, 1701, quarto.

The time at which the confusion between publication and printing is most injurious is that at which the printed book was not the exclusive medium, and the manuscript had not altogether disappeared; a period which includes at least a century after the invention of printing. For so long, at least, did writers who had no particular pretension to be antiquaries, cite manuscripts and printed books indiscriminately; and very often without distinction of character: so that subsequent writers, who thought only of printed books, have taken as printed all they could find cited as published. In this way we have the two unprinted works of Werner, as already mentioned, incorporated by Weidler with the printed ones.

We have noticed, as an introduction, and by way of amusement, the manner in which the French *experts*, as they are called, made the bibliographer forge a title at the *beginning* of an old book, by way of altering the edition, forgetting the description at the end. Those who have experience in books, even of a very moderate extent, know that they must always look at the end; because publishers of a former day did sometimes change the venue: not indeed by stamping in the names of convent libraries, but by printing special title-pages. We have before us a folio which, according to the title-page, is Candalla's Euclid, Paris, 1602. Though a tolerably good copy, and in old morocco, with gilt leaves, it was picked up on a mean stall in the open air, at a very low price. The fact is, that in its descent, it did not meet with any real *expert*, who looked at the end, where it appears that it is the Lyons edition of 1578, and that the Parisian title-page is a substitute. It is the only edition of Candalla that contains all the three books which he added. We have not called such a proceeding a trick and a forgery, because it was often something else. A long time elapsed before the characteristics of a book became matter of settled convention. At first there were *no* title-pages; all the description came at the end; and a word or two after the publisher's preface, if any, such as, "Joannis de sacrobusto anglici viri clarissimi Spera mundi feliciter incipit," was the reader's introduction to his subject. Afterwards, very short fly-titles or half-titles, as they are now called, were introduced in a blank leaf. Thus in one book we have, "*Ad inveniendum novam lunam et festa mobilia. Liber perutilis*;" in another we have, "*Questo e ellibro che tracta di mercatantie et usanze de paesi.*" As regular title-pages were introduced, the full descriptions at the end being still generally retained, the publishers seem to have frequently made use of them as a kind of advertisement *prefixed* to the book, of which they were hardly yet considered *as a part:* just as, in our time, we do not consider the lettering at the back as part of the book. Hence, when a stock of any book came into the hands of a bookseller who was not the original publisher, he frequently printed a new title-page to

*This work must be considered as the accredited contemporary corporate early history of the Academy of Sciences; and Brunet (in his earliest edition at least) makes it head an article on this Academy.

attract attention to the place of deposit, the original place, date, &c., being still to be read at the end. But the same practice continued when the *colophon*, or final description, fell into disuse, and the practice then ceased to have any justification, since the title-page had become the principal direct means of identifying the book. And thus it happens that, in all time, difficulties occur with titles. Nor do we see any hope of their final disappearance, as to books yet to be published; unless indeed an increased taste for bibliography should direct opinion against the following practices.

First, new titles are frequently printed, with new dates, sometimes with, and sometimes without, the words *second edition.* Sometimes the words *revised and augmented* are added, without any change whatever in the book. An author may thus lose his priority of discovery, of adaptation, or of introduction. A printer may thus lose his character as an artist; he may be judged in 1852 by his type of 1842; and similarly, the skill and knowledge of the author in 1852, may be set down as being what they were in 1842. It often happens that the author has no knowledge of what has been done: the edition may pass from the hands of the original publisher into those of another, with whom the author has nothing to do. Sometimes the alteration is made by the original publisher.*

Secondly, a title sometimes undergoes alteration which, whether by intention or carelessness, gives an account very different from the truth. We have before us a book, in which the genuine title describes it as containing matter from 1700 to 1846, mostly German; the substituted title describes it as containing all matter up to 1846, in Germany and the adjacent countries.

Thirdly, even in the original title, it is not uncommon to make the date a year later than the actual date of publication. When the book is published in the last months of the year, so that the right date will soon make it appear a year old, the next date is frequently used. Authors should look to this practice, by which their priority may be seriously compromised. Fifty years hence, a discovery, or other matter of merit, under the date 1851, will certainly be held to have preceded the same under the date 1852. But if a publication made in September 1851, be dated 1852, there is time enough for another to republish it under the date 1851, and thus, with or without intention, to secure it in future history. From the preface of the Latin edition of Wallis's Algebra, it appears that this practice of advancing title-pages was common in the year 1685. The truth is, that the year alone is not now definite enough: every title-page should bear the *month* of publication, as well as the *year*. It would also be of much advantage, if there were an understood place, as at the end of the preface, where the author should mark the last date at which any matter was added to the work, not including the verbal alterations which take place in correcting the press.

* We have heard of a case in which a publisher contracted to pay a certain sum to an author, on the appearance of a second edition. Forgetting this contract, and finding the book sell but slowly, he tried to help it forward by the bait of a new title-page, with the words *second edition*. The author immediately claimed his due, which the publisher was obliged to pay. *O si sic omnia!*

Fourthly, it is becoming common to publish books without a date, whenever they are of a species which rapidly grows old, as in the case of atlases, and of popular astronomical books.

All these things are objectionable, and will certainly cause confusion. The accurate date of any book, no matter how obscure in its own day and in that which follows, may become of importance at a still later epoch.

But though title-pages have frequently made erroneous announcements, still more frequently have their contents been misrepresented; in no particular more frequently than as to the name of the author. There is a loose system of description, under which any prominent proper name is taken for that of the author. If the modesty of a commentator should lead him to print his own name in smaller capitals than that of his original, it is very possible that his comment may be entered as the original work. If a friend or patron should contribute a preface, he will perhaps get credit for the whole; thus Billingsley's English Euclid has been entered under the works of John Dee, who wrote the introduction. The inventor of logarithms has before now figured as the author of the *Bloody Almanac*, which to an unattentive title reader is "by the noble Napier." The reason is that John Booker, the real author, announces his work to contain an "Abstract of the prophecies . . . . by the noble Napier." The Latin forms of names do their part: we remember to have noted some confusion between the contemporaries, Francis Patrizi and Francis Barozzi, arising out of their descriptions as *Franciscus Patricius* and *Franciscus Barocius Patricius Venetus*. Must a librarian set down J. Ralphson, F.R.S., the author of a mathematical dictionary in 1702, as a different person from J. Raphson, F.R.S., who wrote at least four other works of a contemporary date? He will be wrong if he do; but nothing except an examination of the lists of the Royal Society will enable him to be certain. Out of such trifles as these spring many mistakes, such as can hardly be avoided, except by knowledge beyond what the books themselves can give. And as to the books themselves, nothing short of a studied examination will show the difference between a perfect and imperfect volume. A folio collection of astrologers (1533) which has at the end of the contents '*Postremo Othonis Brunfelsii* . . . .' has the work of this Otho first instead of last. We have seen many volumes which were really perfect marked *imperfect*, on the assumption that the contents and their tables of contents must tally in order, as in a modern work. But there is a source of confusion about very old works which has not been much noted hitherto, and which promises to give rise to much inquiry. The copies of the *same edition* of the *same work* do not agree with one another. Sometimes there is a discrepancy of this kind. The impression seems to have been printed without any of the large and ornamented capital letters: these were stamped into *a part* of the impression afterwards, leaving the remaining copies with empty spaces for those who preferred to have these letters wholly the work of the illuminator. Sometimes different headings were put in to suit different tastes. For example, we have before us a copy of the first printed edition of the Alphonsine Tables, Venice, 1483. The heading or title is in red ink, as follows ;—'Alfontii regis castelle illustrissimi celestium motuum tabule: . . . . . .' Hain's description (Repert. Bibliogr.) shows that he had inspected a counterpart of this. But Captain Smyth (Cycle, &c., vol. ii. p. 215) has given a facsimile from another copy of this same edition, by which it appears that the heading is in black ink, having a picture of some astronomers looking at an armillary globe, imbedded in the following inscription; 'Tabule Tabularum Celestium Motuum Divi Alfonsi Regis Romanorum et Castelle illustrissim.'

Again, the *Summa de Arithmetica*, &c., of Lucas Pacioli, 1494, has the first pages * differently printed in different copies, ending with different words. This one book begins in three different ways, certainly; perhaps in more.

The reader who wishes to find more extensive accounts of bibliographical difficulties is referred to the report of the Commissioners mentioned at the beginning of this article, and † to its appendix. He may also consult the *Quarterly Review*, No. 143, May, 1843, or the *Dublin Review*, No. 41, September, 1846. We need not add more examples; we content ourselves with the production of enough to show those who have only seen the popular view of the controversy, that there is a case on the other side which it is easy to support by instances. And this case might be much strengthened by having recourse to examples from literature instead of science: the former subject presents a wider field, more causes of accidental confusion, and more cases of intentional obscurity.

Much of the misapprehension which has prevailed on the question of library catalogues in this country, has probably arisen from the anomalous position in which the Museum library has been placed. On the one hand, it is the resort, daily or occasional, not only of those who know what accurate research is, but also of those who are learning it, who arrive thither to make some investigation, and are led on, by the *genius loci* and the temptation of ready means at hand, until they attain a depth far beyond their first intention. It is difficult to overrate what this national library has done, and is doing, for the cause of accuracy. And though a certain writer who describes himself, by implication, as of *delicate intellect*, sneers at the *manufacture of the stuff called useful knowledge*, which is carried on at the Museum, yet all whose understandings deserve a sounder title will see how much better that indispensable manufacture must go on, with such a library at command of the workman. This workman, fifty years ago, could obtain nothing but what his publisher could lend him

* The author of this article showed, in his 'Arithmetical Books,' that there are two commencements of this edition. Prince Boncompagni, to whose researches the early scientific bibliography of Italy is much indebted, and will be more, has since found a third.

† Particularly (No. 12, p. 378) a letter addressed by Mr. Panizzi to Lord Ellesmere, the chairman, at the commencement of the proceedings: this letter ought to be republished in a separate form.

in nine cases out of ten. To all of whom we have hitherto been speaking, a correct description of books is most essential; and by half of them, at least, old books, such as we have been examining, are frequently consulted. On the other hand, the library is frequented by many who only require the books of most easy description, and by many who come but for books of amusement. These classes might be suited by a very easy catalogue, as to most of the books which they want; probably such entries as 'Encyclopædia Britannica' and 'Guy Mannering' would serve their usual purposes. But these classes have not been useless. It may be suspected that the respect with which the House of Commons has treated the Museum library is due to the system under which most voters may obtain admission; and also that, if a library of research had been set apart for men of research, its interests would have been joked, yawned, or sneered out of the House by the unlearned majority. Nevertheless, so soon as literature can run alone, there are many and obvious reasons why a separation should take place between the libraries of the *reader* and of the *investigator*.

The mistakes into which professed bibliographers once fell have been illustrated in this article, but not their application; for which, at length, we had not room. We had, however, no doubt that, before our conclusion arrived, we should casually meet with something new and striking on this point, which might serve as an instance; and we were not disappointed. The 'Bibliotheca Philosophica Struviana ....' Gottingen, 1740, 2 vols. 8vo., by L. M. Kahle, is a professedly bibliographical work; and dates from about the time when Newton's system began to find general favour on the continent. After describing Motte's English version (1730) of the Principia, Kahle adds that one instance will be quite enough to show the bad faith of the version. He then quotes the celebrated scholium in which Newton admits the claim of Leibnitz, and quotes Motte's translation, which is of course of a very different purport; adding that the English translator, in order to deprive Leibnitz of honour, has been impudent enough (*eo usque procedit impudentiæ*) to alter Newton's words. Had the bibliographer remembered, or taken care to ascertain, that Newton himself published three editions, he would have found that Motte was correctly translating from the third of them, and that the substitution was made by Newton himself. At the same time, Kahle's blunder may serve to warn translators that they ought to be very precise in stating the editions on which their versions are made, and the most important, at least, of the variations: together with a sufficient description of the previous editions. And further, foreigners should take notice that English writers are well able to pay in kind any confusion made among the writings of Newton. In proof of this, we have, since the preceding sentences were written, fallen in with a recent work in which Kahle is placed under suspicion of having, under the name of *Kayle*, answered Voltaire by plagiarizing an answer written by *Kahle* seventeen years before Voltaire wrote.

If we ourselves should have fallen into any mistakes, they will serve our purpose, as helping to prove the truth of our title. They will do us a service of the same kind which a lapse of memory of Mr. Macaulay's does for him. In his review (which, like the work itself, is much too short) of the Pilgrim's Progress, speaking of the tediousness of the Fairy Queen, he observes that "very few and very weary are those who are in at the death of the Blatant Beast." The reviewer himself, no doubt one of the few, was also one of the weary; for the blatant beast is *not* killed, and the very last verse extant of the poem shows us that Spenser kept him alive for good reasons of his own.

A. DE MORGAN.

University College, London,
October 4, 1852.

COMPANION TO THE ALMANAC

FOR

1854.

PART I.

GENERAL INFORMATION ON SUBJECTS OF MATHEMATICS, NATURAL PHILOSOPHY AND HISTORY, CHRONOLOGY, GEOGRAPHY, STATISTICS, &c.

I.—ON A DECIMAL COINAGE.

In the Companion for the years 1841 and 1848 we advocated the introduction of the pure decimal system into our coinage. The first article was written before any perceptible feeling, either for or against the change, had been excited; the second followed Dr. Bowring's* motion in the House of Commons, in consequence of which the Government conceded the first step, namely, the introduction of a coin to designate the tenth part of a pound (since called the florin). In the interval between these two articles we found, as we stated, that nearly every man of business with whom we spoke on the subject expressed no fear of the change, but rather a desire for it. The present article follows an investigation of the subject by a Committee of the House of Commons, and the publication of the report of that committee, with the evidence on which it was founded. The desirableness of the change may now be considered as granted: and, unless some unfortunate pressure of other matters should distract attention, it may confidently be expected that the Chancellor of the Exchequer will propose a measure in the course of the next Session, though the magnitude of the operation, in the part which belongs to the Mint, may require that the change shall not be actually made during the year 1854, or even that which follows. As time will be required to circulate information, the necessary deferment of the actual change is not to be regretted; but as this information will never be circulated in earnest until the date of the coming event is precisely known, it is to be hoped that the formation and announcement of a plan, and its discussion by the Legislature, will take place during the next Session, be the date of practical operation what it may.

* Of all the witnesses examined by the Committee, Dr. Bowring is the only one who has had experience of decimal and non-decimal coinage in many and remote countries—Spain, Portugal, Germany, Russia, China, Japan, &c.; and of all the witnesses he is the warmest (though nearly all are decided) in his advocacy of the change.

The main recommendations of the committee are as follows:— 1. That the pound sterling shall continue to be, in name and value as at present, the highest *coin of account.* Some of our readers may have but an indistinct apprehension of this phrase. Our present *coins of account* are pounds, shillings, pence, and farthings: half-sovereigns, crowns, half-crowns, sixpences, fourpences, threepences, though all in circulation, are not used in accounts. The coins of account are the legal denominations: a man could not sue for 7*l.* 10*s.* under the description of six sovereigns, two half-sovereigns, and four half-crowns, though those coins would collectively constitute a legal tender for his claim. 2. That the tenth, hundredth, and thousandth of a pound shall be the other coins of account, under names to be selected, and which in the report are *florin*, *cent*, and *mil.* 3. That the present half-sovereign, shilling (50 mils), and half-shilling (25 mils, no longer *six pence*) shall be retained. 4. That copper coins of 1, 2, and 5 mils, and silver coins of 10 and 20 mils, shall be added, with such others as experience shall show to be desirable.

The proposed coins are then as follows;—

Gold.—Sovereign (1,000 mils), half-sovereign (500 mils).
Silver.—Florin (100 mils), shilling (50 mils), sixpence* (25 mils), two-cent piece (20 mils), cent (10 mils).
Copper.—Five-mil piece, two-mil piece, mil.

The mil is only four per cent. less than a farthing, and will probably be called a farthing by the people: two mils will probably still be called a halfpenny, and four mils a penny. The following table will show how nearly we may think in old money, if we designate the mil as a new farthing, &c.

The *pound* or *sovereign* is 10 florins, 20 shillings, 40 new sixpences, 100 cents, 1,000 mils or new farthings.

The *florin* is 2 shillings, 4 new sixpences, 10 cents, 100 mils or new farthings.

The *cent* (four per cent. less than 2½*d.*) is 10 mils or new farthings.

The great change in the passage of the smaller coins from hand to hand—indeed the only real change, the florin being now established—is 25 instead of 24 lowest coins to the *sixpence*, 50 instead of 48 to the shilling. *Sixpence* is 25 mils or *new farthings* instead of 24 old ones; a shilling is 50 mils, or *new farthings* instead of 48 old ones.

The great change in the money of account is the substitution of florins, cents, and mils, for shillings, pence, and farthings, and the introduction of what appears to those who are used to the latter a magical shortness in reduction. Thus, 6 florins 7 cents 8 mils is 67 cents 8 mils, and also 678 mils: 208 mils is 2 florins 8 mils. The florins, cents, and mils are but the hundreds, tens, and units, in an account kept in mils: the pounds are the thousands.

Accounts may be kept as easily in mils or new farthings as in pounds, &c. The distinction of the *decimal point* will be much

* The name *sixpence*, accented on the first syllable, will not disappear; but *sixpence* will no more be *six pence* than a *cardcase* is a *card case*.

used to mark off the pounds, as in 61·238*l.* in which we see 61*l.* 2*fl.* 3*cts.* 8*mls.* Omit the dot and we have 61,238, the number of mils. Make the florin the principal coin, and the dot separates the florins from the cents and mils: thus, 238·16*fl* is 238 florins 1 cent 6 mils, or 23,816 mils. The point thrown one place to the right converts the reckoning from pounds into florins, from florins into cents, from cents into mils. Thus, 2·178*l.* is 21·78*fl.*, or 217·8*cts.*, 2178·*ml.*, the point being useless in the last case.

To transfer old silver into new, throw it into mils at once, as follows:—Allow 100 for every florin or pair of shillings, 50 for the odd shilling, if any, and one mil for every farthing left, with one mil more for sixpence. Thus 7*s.* 10¼*d.* is 300 and 50 and 41 and 1 mils, or 392 mils, that is, 3 florins 9 cents and 2 mils: and 4*s.* 2¾*d.* is 200 and 11 mils, or 211 mils, or 2 florins 1 cent 1 mil. This method is *universal:* many isolated cases are easier. These reductions are not exact, except at the shillings and sixpences, nor can they be made exact without introducing fractions of a mil; the error never amounts to a mil. Those who wish to have absolute exactness will remember that so many farthings as there are in the excess above the last shilling or sixpence, so many 24ths of a mil are wanting in the new money. Thus in 7*s.* 10¼*d.* there are 4 *d.* or 17 farthings above the last sixpence; accordingly 392 mils is not enough, 17-24ths of a mill are wanting. This correction will never be of any use in ordinary transactions: fractions of a farthing or mil will be talked about while the change is pending, but will subside into their old insignificance before the new coins have circulated for a week.

To transfer new silver into old, allow 2 shillings for every 100 mils, a shilling for 50, if left, and farthing for mil on the remainder, deducting one if the mils left be 25 or upwards. Thus 7*fl.* 8 *ct.* 3*ml.*, or 783 mils, is 15 shillings and 33 mils, say 32 farthings or 8 pence: and 222 mils are 4*s.* 5½*d.* These answers are a little too great, except at exact shillings or sixpences. In all other cases, each farthing above the last shilling or sixpence is too great by its 25th part; thus 15*s.* 8*d.* is 8-25ths of a farthing above 783 mils.

A mil must gain its 24th part to become a farthing.

A farthing must lose its 25th part to become a mil.

In the preceding remarks much of our former articles is briefly recapitulated, with the adoption of the terms florin, cent, and mil. To the names *cent* and *mil* there is one decided objection. It is clear from all that has taken place that the educated part of the public almost desires, and certainly foresees, a complete decimal system of coins, weights, and measures, to which the alteration of the coins, important and useful as it may be, is but a forerunner. Let the new coinage work out its own consequences, and it must lead to the introduction of all the rest of the system. The words *tenth, hundredth, thousandth* are among the most difficult to pronounce in our language: apart from which, it would be otherwise desirable to have separate names to denote these fractions. If we appropriate the words *cent* and *mil* (*mill* would be better, as an English word) to signify coins, we deprive ourselves of the power of using *dime, cent,* and *mil*, to stand for the general words tenth, hundredth, thousandth. When we have the pound sterling decimally divided into, say 10 florins, 100 groats, and 1,000 farthings, and also the pound avoidupois into 10 ounces, 100 drachms, and 1,000 grains (if these names be still used), it will be highly convenient to be able to say that as many grains as are bought for a farthing, so many ounces for a florin, because the grain is the cent of an ounce, and the farthing the cent of a florin. The words *tenth*, *hundredth*, *thousandth*, will never become truly vernacular without some concession to facility of pronunciation which will confound them with *ten*, *hundred*, *thousand:* but, when a complete system is established, both the first and second sets must be used very much oftener than at present. If we now sacrifice the generic words *dime*, *cent*, and *mil*, and let them come to signify merely the parts of a pound sterling, we lose our best method, and our only opportunity, of drawing a very wide and easy vocal distinction between the decimal ascent of tens, hundreds, thousands, &c., and the decimal descent of dimes (tenths), cents (hundredths), mils (thousandths), decimils (ten-thousandths), centimils (hundred-thousandths), millimils (millionths), &c. The Committee gave no decided preference to the terms cent and mil, but seems to have taken them because they had been brought forward.

It is hardly to be expected that the term *mil* can be introduced. If the mil were to circulate with the farthing, as a different coin, undoubtedly the new coin would want a new name, and the name given by the Legislature would obtain immediate adoption. But it is not so: it is proposed that a man shall take off his coat at night with a certain coin in the pocket called a *farthing*, and shall put it on in the morning with that same coin called a *mil.* We doubt if the change will take place. The Legislature can, and will, make a trifling alteration in the value of that coin; it can and will enact that 50 of them shall make a shilling instead of 48. But they will still be *farthings*, because, in popular apprehension, that name identifies certain pieces of copper of certain form and appearance, more than the value for which they pass. They may be *new farthings* for a fortnight; but farthings they will remain.

The new coin of 10 farthings, which it is proposed to call a *cent*, must have a name given to it; if any other name could be found (groat, star, cross, doit,* bit, &c.), the great advantage of a distinct nomenclature for decimals in general might be realized. If we are right in supposing that the mil will almost immediately become a farthing in name, so that of the new terms *cent* only will keep its ground, some confusion will arise from the manner in which that word is already used.

Five for a cent and *five per cent.* sound too like each other: if the word cent be the only new one which gains adoption, it will be wished, on this ground only, that some other had been chosen. And

* The *duyt* was a Dutch coin which once passed in England at various very small values. The word *doit* is used in English without any precise value attached, as very small money: oddly enough, both *doit* and *twopence-halfpenny* (the cent, nearly) stand for undefined trifles of value; we would not give a doit for such a twopenny-halfpenny opponent as one who could deny this.

further, the increasing knowledge of the United States' coinage, in which the cent differs but little from our halfpenny, will sometimes cause confusion. Such objections as the last two would be of little weight against a word as a part of a system; they have force on the supposition that the new word stands alone in its novelty.

That a decimal coinage is desirable—that it can be easily introduced—that it should be introduced speedily, and may be done at one step—are points which all the evidence goes to support, and may be considered as settled. There is another point which may be considered as carried by a very large majority, namely, that the present pound sterling should be a coin, and should be recommended as the highest coin of account. But on this question there was a slight contest between the penny, the shilling, and the pound.

The proposed change throws both the penny and the shilling out of account; no column will be ruled for either: and as to the penny, it becomes but a vernacular name for four of the mils or new farthings. There will be no coin which exactly represents the present penny, four mils being 1-25th less than a penny. The consequence is that postage and receipt stamps, penny tolls, &c., must either bring four per cent. of revenue less than at present, or must be raised to five mil coins, and must* yield 20 per cent. more. In revenue matters, in which the public deals with itself, the question is one of convenience only: in reference to tolls, it is one of justice. It was proposed that in the case of tolls levied by a private company, the additional mil should be granted for such a number of years as would be compensation for the subsequent reduction. It may well be supposed that most toll-owners would prefer the increased popularity and custom which would follow a reduction of four per cent., to the risks of an augmented toll. But however this may be, no reason was shown for any fear of great difficulty in the adjustment of either tolls or stamps. Nevertheless, great stress was laid by one witness upon the supposed impracticability of substituting for the copper coins others of nearly equal value; and a proposition was made to allow the farthing to remain, and to proceed upon it as a basis, inventing coins of 10, 100, and 1000, farthings. The highest coin of account was to be 1,000 farthings, or 1*l.* 0*s.* 10*d.* present money; this "Victoria," to be introduced gradually, was to circulate in conjunction with the present sovereigns, until the latter gradually found their way back to the Bank.

It was taken in more quarters than one for a serious difficulty, that if the copper were reduced four per cent. there would be a loss to the revenue of nearly 100,000*l.*, unless the postage stamps were made five mils, which would probably be an unpopular measure. It seems to have been doubted whether so small a reduction as four per cent. would create a proportionate increase in the number of letters. This is a point on which it is difficult to decide without experience; but, judging from the uniform manner in which facts de-

* Five mils is 96 per cent. of 5 farthings, or 96-20ths of a farthing, or one penny and one-fifth of a penny; and one-fifth is 20 per cent. The discussion which took place on this point in the newspapers arose out of an ambiguity in our language. A man turns 4*l.* into 5*l.*, how much per cent. does he gain?—Answer, 25 on what he began with, 20 on what he left off with.

pending upon human will present in the bulk the regularity of physical phenomena, we are strongly inclined to suspect that, if Government were to carry only 501 letters (supposing coinage to allow it) at the same rate at which it now carries 500, the 501st letter would find its way into the post-office, and perhaps the 502nd, or a fraction of it.

It was denied on all sides that the penny is so essential a matter in the thoughts and associations of the working classes as to make it necessary to retain a coin of its exact value. Perhaps the first thing which would strike a person new to the subject would be that the penny is the monetary corner-stone of the poor man's habits of calculation, and that any change in it would be to him a complete loss of power. So far is this from being the case, that the small tradesmen—meaning dealers in very small amounts—are the witnesses who most unequivocally, and in the fewest words, declare that the substitution of 25 mils for 24 farthings would give no trouble worth mentioning. This is the opinion of a man who takes 1,000 farthing coins over his counter every week.

Another witness styles the labouring classes *keen calculators*, and the description given by several of the way in which sharp habits of calculation are acquired is worth notice. Since coffee, tea, sugar, &c., are sold in quantities which require fractions of a farthing to pay for them exactly, and since the shopkeeper charges the full farthing in the absence of smaller coins, it is the frequent occupation of the customer to examine, in his own mind, whether, by taking the same amount of a higher-priced article, he will not be able to get some or all of the surplus fraction out of the shopkeeper, without risking the next farthing. So that the labouring man thus acquires, in most instances, an expertness in mental reckoning with farthings, which his superiors in knowledge would hardly give him credit for, and do not themselves possess: and he performs arithmetical manœuvres which are, individually, as complex as any single operation which falls in the province of the banker's clerk. It is probable that the main alteration in giving change—25 farthings to the *sixpence* instead of 24—will be more readily apprehended and more easily made a habit by the labouring classes than by any other.

There is much reason to apprehend that the difficulty of the extra farthing or mil, as affecting the working man, will be exaggerated in the approaching discussions: it would be amusing, if, as we believe, it should turn out that the whole of it is to be the portion of the classes which fear it only for the uneducated. Not that there is much to fear for any. If a tradesman were to advertise that he would in future consider a silver sixpence to be six pence and a farthing, and a silver shilling to be twelve pence and a halfpenny, and that he would act upon these rules in *giving* change for sixpences and shillings, a discerning public would rush to his shop with much silver in its hand, without one thought about the new process of calculation: and no one would be puzzled.

Suppose that, on the very day on which this tradesman began his operations, the Government should happen to recal the threepenny and fourpenny pieces, and issue silver bits (whether called cents or not) of 2½*d.* each, or five to the shilling-halfpenny, and copper bits of five farthings each, or five to the sixpence-farthing; then with regard to this one tradesman, and change given by him, the public would have the new coinage.

There is, no doubt, in the proposed change, a violation of principle, an absolute injustice, affecting each individual who has copper money in his pocket or his till, at some midnight to be settled by Parliament. It will be at the rate of 1-25th of the whole, on whatever amount of farthings, halfpence, pence, threepenny and fourpenny pieces he may happen to have in his possession. The number of persons who will be damaged to the amount of a fraction of a farthing is uncountable; but they will get no sympathy. A tradesman who has five pounds' worth of the small money enumerated above will find it suddenly worth only *4l.* 16*s.*, or *4l.* 8*fl.*, as he pleases to reckon it. If Government were to give full value in new (but not small) money, that is, in sovereigns and florins, for all even amounts of small money from 40 shillings upwards, for some days before the change, it is not very likely that the demand would be great. In all probability the inconvenience of beginning the new coinage without any small change in the till would be dearly bought at four per cent. of the money required.

To the above injustice there is some set-off. The broad-rimmed copper penny, which weighed at its issue a full ounce, and which is so easily distinguished from the smaller and lighter coins, it has been proposed to pass as *five* mils or new farthings, that is, as half a cent. If, then, a person having 2*s.* 5*d.* all in copper, should chance to find four broad-rims among them, he may rest in peace, for the change will leave him almost as it found him; his 2*s.* 1*d.* (ordinary), will become 2*s.*, but his 4*d.* (rimmed), will become 20 mils, or nearly 5*d.* What the proportion of rimmed pence is we do not know, but should it be more than 1-7th of the whole (and we do not often see six *pence* without one at least of the rimmed kind), the value of the copper money will, on the whole, be raised, and not lowered; but the same will probably not be true when the threepenny and fourpenny pieces are included.

Next to the question of preserving the farthing unaltered, as a constituent of the new coinage, came that of making the shilling a decimal constituent; that is, of making the highest coin of account to be a ten-shilling piece, under a new name, and dividing the shilling into 100 equal parts, each four per cent. less than half a farthing. This would be the proposed new coinage with every coin of account of half the value now contemplated. This suggestion proceeds upon two suppositions: first, that the shilling is more essential than the pound in our present associations; secondly, that less coins than a farthing are needed.

Since the shilling coin actually remains in the new system, if not as a coin of account, yet as the half of one, it is clear that all its uses and associations as a coin of transfer remain as before. But on no point was there more decided agreement than on this, that with regard to convenience, nothing is more essential than the preservation of the pound or sovereign, unaltered, as the highest money of account. As a matter of arithmetic only, it would be easy enough to change reckoning by pounds and shillings into reckoning by half-sovereigns and shillings: thus 5*l.* 12*s.* is 11*v.* 2*s.*, using *v* to denote half a sovereign. But with numbers of pounds are connected all our associations of value: 500*l.*, as a magnitude, is a notion as well as a reckoning; a noun substantive of many meanings, which are conveyed to the mind without even a distinct presentation of the number as a number. This *five hundred* pounds suggests a certain station, as an income; a certain success, as the price of a copyright; a certain importance of the case, as a barrister's fee; a certain part of the town, as the rent of a furnished house; a certain start in business, as a capital; all without any distinct reference to the number five hundred on every occasion of its occurrence. Let it be necessary to say *one thousand* where we now say *five hundred*, and we introduce a moment of mental calculation into numberless repetitions of the most common phrases—an inconvenience which will last a very long time. In fact, we alter a number of substantives in the worst possible manner; we do not introduce new names, but we make one old word take the place of another. Phrases composed of many words often become compound names, though it may not be the practice to join the constituent words by hyphens; and these compound names are used, like those of a more acknowledged character, with a very faint and almost dormant reference to the meanings of the several constituents. Thus "the fifteen judges" is a phrase often used without the least positive remembrance of fifteen as one more than fourteen, the phrase being thus synonymous with "the whole of the judges." In like manner, "a hundred pounds," "a thousand pounds," &c. have a number of meanings, in which what the sums will do, or what they will suggest, are represented to a mind which takes the representation as it was meant to be given, without picturing to itself the amount of money as an amount. If any one should doubt this, it is only because it cannot be done on purpose, to try the truth of the assertion: we cannot pronounce "one hundred pounds" with an express intention not to remember the separate force of the word *hundred*. But if we alter the highest coin of account, which is always the coin of estimation, we shall be under the necessity of learning, by constant acts (no matter how slight) of mental arithmetic, to attribute different meanings to phrases which have been familiar from childhood. The Astronomer-Royal, in his evidence, said that an alteration of the pound would unhinge every estimate and every contract. No doubt it would; and among the estimates which would be unhinged would be all those which are conveyed in a large class of phrases. This would, in the end, give hundreds of times more trouble than the alteration of all the predictions of probable cost which are technically called estimates, though this alone would be bad enough.

It is long before the confusion arising from wilful alteration of words is made to disappear, when the alteration is decisively marked. A certain managing body, consisting of educated men, was in the habit of adopting or rejecting propositions on the reports of committees, by ballot. Accordingly, "yes" on the ballot-box, meant confirmation of the committee's report; *yes* or *no* to the proposition as the case might be. Thus, "yes means no," was the phrase for voting in confirmation of a recommendation to reject. At last it was resolved to vote on the propositions, and not on the reports: thus, "no" confirmed the report of a committee which recommended rejection. But, though twenty years have elapsed since this simple change was made, it still suggests itself to some one, now and then, to inquire, as the box goes round, whether yes means yes or no.

The various disadvantages of a change in the highest coin of account—the unhinging of settled phrases, contracts, and estimates,—the alteration of all the modes of expressing exchanges—the destruction of a universal commercial term (for the pound sterling is now, all the world over, nearly what the Spanish dollar was, in matters of commerce)—the creation of a great subdivision in the history of monetary affairs, dividing all books on the subject into two classes, written in different languages—might all be compensated, possibly, by the introduction of some greater advantages. The whole question is one of balance; any change has its inconveniences, and its conveniences. There would be an advantage in retaining the farthing unaltered; there would be an advantage in retaining the shilling as a coin of account; there would be an advantage in adopting the French franc, the American dollar, nay, even the Indian rupee. But in no one of these cases is it held that there would be a balance of advantage, when the loss of the pound is in the other scale. The difference is that the rupee has no advocate against the pound, while the unaltered farthing and the shilling of account are not wholly without advocates.

The argument in favour of the *ten-shilling* pound, derived from its mil being about half the present farthing, is founded on the assumption that a smaller coin than the farthing is needed. On this question there is one strong fact on each side. It was proved that, in the absence of a smaller coin, it is common in many places for the shopkeeper to make up the purchase by giving a small portion of snuff, tobacco, or whisky. On the other hand, it is known that half farthings have long existed at the Mint, that an attempt was made to keep them in circulation, and that it failed. It cannot be supposed that the shopkeeper, who is willing to eke out with pinches of tobacco or spoonfuls of whisky, discourages the half-farthing merely for the sake of that portion of the pinch or spoonful which is his profit. Probably he sees that, in selling small quantities, the half-farthing would be often subdivided, so that the trouble of serving the small make-weights would remain, with all the inconvenience of the small coin. Those who are best able to judge are of opinion that no smaller coin than a farthing is needed.

Among the minor points discussed, was the question whether it should be recommended to keep accounts in florins, cents, and mils, three columns, or florins and mils, two columns, with, of course, 100 mils to the florin; that is, whether it should be, for instance, *7fl.* 3*ct.* 2*m.*, or *7fl.* 32*m.* In the United States, accounts are kept in dollars and mils; the dime, or tenth of the dollar, has disappeared from account. This question is not for the Legislature; people may be left to suit their own convenience. It must be remembered, however, that the analogy brought from the United States does not point to the mode of subdividing the florin, but to that of subdividing the pound. If only one subdivision of their highest coin of account be employed by them, we may ask whether, with us, it will not end in using only one subdivision of *our* highest coin? The question then is, whether pounds and mils will not be our mode of *account?* This will certainly be the case if the decimal point be adopted; 15·685*l.*, which in coin will be 15 sovereigns 6 florins 8 cents 5 mils, will enter the books, and appear to the accountant, as 15*l.* and 685 mils. This use of the decimal point will be adopted by the better class of arithmeticians from the beginning: they will think with the wise and talk with the vulgar. One line will be ruled to divide the pounds from the fractions, and the three places following the line will be named florins, &c., but thought of in addition as decimals of a pound. The best plan (the one a mathematical computer follows in any long calculation of tables) would be to *rule every place*, ruling a double line for the separation of pounds: thus 21,482*l.* 7*fl.* 8*ct.* 9*ml.* would appear in the books as—

| | | 2 | 1 | 4 | 8 | 2‖7 | 8 | 9

This would tend to prevent a person who is adding up rapidly from getting into the wrong column, to which he is even now subject.

Some persons, seeing the dollar and its cents in America, the franc and its centimes in France, &c., think that the florin should be the highest coin of account, divided into 100 cents; the pound remaining as the representative of 10 florins. This, again, is not a question for the Legislature to settle; every person may suit his own convenience. Small tradesmen, whose items seldom amount to a pound, may rule their ledgers for florins, and read their pounds, when such things occur, in the tens of florins. All that is required is that every account, or page of account, shall state the unit which is employed; this being done, any one may reckon in pounds, florins, cents, or mils, as he pleases.

One of the most difficult questions on which to come to universal agreement, will be the *point of rejection* in large accounts. At present the Bank of England, and all the private bankers, enter nothing below one penny. Now it is certain that this rule will be useless, as a saving of trouble, and will even give trouble, in the new system. The question then will be, is the lowest entry to be five mils (half a cent), or one cent (2½*d.* nearly). If the half cent be taken, the last column will never show any figure except 0 or 5, and the well-known inconvenience of French accounts will be incurred, in which the *sou* is practically the lowest money of account, represented by 5 centimes. On the other hand, the idea of rejecting everything under a whole cent will startle many. Nevertheless, we suspect that to this it will come, and ought to come, especially at the Bank of England. It would also be desirable that the changes in the funds should be *tenths* instead of *eighths;* we do not mean, however, that this should affect the brokers' commission.

The tradesman and the accountant, on whom the change will be forced many times in each day, will be expert at the new money before the second day has expired. Those who have less to do with calculation will have more trouble with it. The likeness between the new and the old farthing will be their mainstay. The sixpence and the shilling actually remain, as now: but the sixpence takes 25 farthings, the shilling 50: whatever name the Legislature may give to the *new farthing*, they will find it their interest to hold by the word 'farthing.' Any one among the explanations published which confines itself to addition and subtraction, and the simplest multiplication and division, will be enough for their purpose.

Simple tables will be published, in various forms. Some will go all the way up to 20 shillings, in their comparisons of the old and new coinage: others will stop at the florin. One of the second kind, on a small card, giving every farthing of the old money up to a florin, in terms of the new money, with change for a florin opposite to each, will be found useful by many, for a short time. A few minutes' exercise at mental subtraction from 100 (as in 36 and 64 make 100, and the like), will enable many to avoid even this table. But the *carry-one* process must be avoided: take the first figure from *nine*, the second from *ten*, and treat the first figure first. Thus: —24 and 76, the 7 being the complement of 2 to 9; 38 and 62, &c. This process will be of permanent use: the connexion of the old and new money will be transitory. Many will hardly believe that 6 *ct.* 8*ml.*, are reduced to mils by striking out the symbol *ct.*, as in 68 *ml.*: they will suspect that such facility leads to error, and will comfort themselves with '10 times 6 are 60, and 8 are 68.' We have seen men of business refuse to trust an annexed cipher for multiplying by ten, and try it by the multiplication table.

The attention of the public will now be turned, after the change of the money has been made, to the question of decimal weights and measures. This is a much harder subject, and will not perhaps excite great attention until the advantages of decimal money begin to be fully appreciated. But the decimal weights and measures can afford to wait for full discussion, since the change in the coinage is a very large proportion of the whole alteration required. If, as we believe, nineteen out of twenty of all the calculations made relate to money, the decimalization of the weights and measures is but five per cent. of the matter so far as the whole public is concerned. The parties specially concerned with calculation* in weights or measures, engineers, custom-house officers, contractors, &c. &c., must depend upon their own exertions for forcing the next step upon the attention of the Government. If the experience of a decimal coinage should make them active in the matter, they will probably find that the same experience will have disarmed opposition. But the question what the units are to be, will present many points of controversy. The sooner this controversy begins, the sooner will practical attention be drawn to the subject.

A. De Morgan.

University College, October 31, 1853.

* Every tradesman is concerned with weights and measures; but, certain classes excepted, far the most of their *calculations* are in terms of *money*.

COMPANION TO THE ALMANAC

FOR

1855.

PART I.

GENERAL INFORMATION ON SUBJECTS OF MATHEMATICS, NATURAL PHILOSOPHY AND HISTORY, CHRONOLOGY, GEOGRAPHY, STATISTICS, &c.

I. THE PROGRESS OF THE DOCTRINE OF THE EARTH'S MOTION, BETWEEN THE TIMES OF COPERNICUS AND GALILEO; BEING NOTES ON THE ANTEGALILEAN COPERNICANS.

Any reader of the common accounts of astronomical history might suppose that the moment Copernicus sowed the dragon's teeth, a host of armed controversialists arose from out of the ground, and proceeded to mutual slaughter. This arises from the manner in which popular writers are naturally led to pass over an intervening time, and to bring their readers at once to the story of Galileo and the Inquisition. But when it is remembered that this period of excitement begins with the construction of the telescope in 1609, and that the work of Copernicus was published in 1543, it will appear that there is a term of no less than 66 years during which the progress of opinion is to be accounted for. That is, the contemporaries of Galileo looked back upon the announcement of the motion of the earth just as *we* look back upon the commencement of the first French revolution.

During this term of 66 years, the question was not, properly speaking, discussed. Various writers gave opinions, but no book was written against another book. Some leaned towards the actual motion of the earth: some got no further than the admission of that motion as a very simple and efficacious way of deducing the planetary motions. Some thought that the *diurnal* motion only could be maintained: others were equally in favour of the motion round the sun.

The difference between a *physical* and a *mathematical* use of the Copernican or any other theory, is one which is rather puzzling to a reader unaccustomed to such considerations. A *mathematical* Copernican was one who saw that, come how it might, the heavenly appearances are such as *would* take place *if* the earth *did* move about the sun, and also about its own axis: and that, consequently, the supposition of such motions, true or false, would be a convenient and efficacious mode of explaining and predicting celestial phenomena. A *physical* Copernican added to the above the belief that the reason why things appear as they would if the earth had these motions, is that it really *has* them. The first said that the hypothesis *explains* or *demonstrates* phenomena; the second said that the hypothesis is a true statement of the causes which *produce* phenomena.

Every person who knows the heavenly motions, as they appear before our eyes, and has a little knowledge of geometry, *must* be a mathematical Copernican: he cannot fail to see that a Copernican universe would show the same appearances as that in which we live. Accordingly, from the moment when the work of Copernicus appeared, the beauty of the explanation was fully acknowledged, and the author took his place at once among mathematicians of the first order, both for his own novelties, and for his additions to the old system. The highest terms of praise are found in the writings of those who were most opposed to the physical truth of the hypothesis. Those who were inclined to blame the *novelty* of the system—for in every age the production of new opinions meets with reproach—had their mouths stopped by reference to those among the ancients who were known to have believed the actual motion of the earth.

It is said that Leonardo da Vinci held the motion of the earth, as appears by his manuscripts, about 1500: but it does not appear that he constructed any system of explanations. There was a work of Calcagnini actually written and made known (*divolgò*, says Tiraboschi, which we suppose does not imply printing) before that of Copernicus, in which the earth's motion was defended: but as the author was known to have travelled in Poland, where Copernicus had long been teaching his system in every way except through the press, as well as for other reasons, the Italians suppose he must have had knowledge of Copernicus and his opinions. In 1533, J. Albert Widmanstadt, afterwards known as an oriental scholar, explained the views *of Copernicus* before Clement VII., for which explanation that Pope presented him with a Greek manuscript, which was preserved at Monaco in the last century with an inscription by Widmanstadt recording the gift and the reason.

We shall now give some detached accounts of the mode in which the system presented itself to those whom we may term its *Antegalilean* supporters or opponents. The dates attached are those of death, but all the writers belong to the sixteenth century.

Nicholas Copernicus (1543).—The question whether Copernicus himself was a *Copernican* in the modern sense of the word is not easily settled. His phraseology is almost always that of a mathematical Copernican (*Comp. Alm.* 1853, p. 9, note ‡). In a very few places, and cautiously, he leans to the physical truth as probable, and to the diurnal motion as more probable than the orbital. When the Congregation of the Index, in 1620, propounded the alterations under which they could allow his book to be read as a mathematical hypothesis, they found those alterations very few in number: and, though confessedly disposed to cancel the whole of Chapter VIII., as treating of the truth of the motion of the earth, they were nevertheless able to allow it to stand, because the author seemed to be speaking *problematically;* whence they only imposed a few verbal alterations. Riccioli (*Alm. N.* ii. 294) affirms Copernicus to have been a physical* Copernican, and cites the passages which, in his opinion, prove it; but he does not feel able to get farther than the qualified statement that Copernicus maintained the motion of the earth as more probable, or even as demonstrated. The reader who can compare the passages in the page cited with the alterations demanded by the Congregation in p. 496, will be able to judge for himself: see also the life of Copernicus in the *Penny Cyclopædia.*

George Joachim Rheticus (1576).—This celebrated friend of Copernicus, and one of the principal instigators of his publication, was himself the first announcer of the forthcoming system, in his *Narratio*, &c., published in 1540 (*Comp.* 1853, p. 9). It may be collected from this letter that Rheticus was, more than Copernicus, inclined to express his belief in the motion of the earth as an absolute truth. But the passages which prove this are imbedded in accounts of the manner in which the system explains phenomena, and it would take more space than we can give to put in evidence† the distinction between the phrases of Copernicus and those of his herald. The following paragraph, however, will show the manner in which the account struck another person.

Achilles Gassarus.—His letter (*Comp. Alm.* 1853, p. 9, note *), written in 1540, upon receipt of a copy of the *Narratio* from his friend Rheticus accompanied by a very full private letter, *epistola harum rerum refertissima*, was printed at the head of the edition of the *Narratio* referred to in the note just cited. No doubt this private letter gave very explicit statement of the view actually taken by Rheticus. Delambre (*Astr. Mod.* vol. i. p. 138) imagines that it was the work of Copernicus himself which was sent to Gassarus; and concludes, from copies being thus issued three years before the date of the book, that Copernicus delayed its publication even after it was printed. But this is a mistake: we know perfectly well that Copernicus never opened his own book, and received the only copy he ever saw on the very day of his death. Gassarus describes the publication of Rheticus as of novelty enough to stupefy any one, as most contrary to the doctrines of the schools, and as what the monks would certainly call heretical. He instances the motion of the earth as among the wonders, but enters into no detail, apparently because he sends on the work itself to his correspondent. We infer, then, that Rheticus was a *physical* Copernican. This Gassarus was the constructor of a series of almanacs, under the title of *Prognosticon*

* All the writers who make a similar affirmation in the sixteenth century are opponents, except only Thomas Digges, who stoutly and expressly denies (in 1594) that Copernicus meant his assertions "onely as mathematical principles fayned, and not as philosophicall truely averred."

† Some give a very different account. In the *Penny Cyclopædia* (Rheticus), apparently after Zedler, it is stated that this *Narratio* is absolutely written in opposition to Copernicus, "to show that the rotation of the earth about the sun is not a mere probable hypothesis, as Copernicus had thought fit to announce it, but an incontestable truth." And many writers, though not going so far as this, are yet decided in their statements that Rheticus was a physical Copernican. We arrive at the same opinion with more difficulty.

Astrologicum. That for 1546 is dedicated to Rheticus, who, says Gassarus, after the example of his master Copernicus, gave easier introduction and clearer demonstration to the celestial motions. It is probable enough that, next after Rheticus, Gassarus was the first who ventilated the name of Copernicus in print.

Erasmus Rheinhold (1553).—He was the friend of Copernicus, and an abettor of the publication; but his view was very different from that of Rheticus. In his Prutenic (Prussian) tables, he makes use of the *observations* of Copernicus, and he constructs tables of the planets both on the old and on the Copernican data. He enters into no discussion, and (according to Delambre) finds no occasion to speak either of the motion of the earth or of the sun. It must be remarked that several of the early Copernicans, whatever they might think of the system physically or mathematically, found their chief source of admiration in the changes which Copernicus made in the *numerical* data of the planetary, and especially of the lunar, theory. These numerical data might with ease have been transplanted into the Ptolemaic system. Rheinhold, then, so far as he declares himself, heads the school of *numerical* Copernicans.

Peter Ramus (1572).—This celebrated leader of opinion is reckoned rather among the *philosophers* than the *mathematicians*. In the common language of our day, those who investigate *matter* have usurped the name of philosopher, or rather, perhaps, have had it usurped for them; while those who were once called philosophers * are now usually called metaphysicians. This is of no consequence,† except in the confusion it creates: we shall in this article use the term *philosopher* in its old sense. Ramus, a thorough opponent of the old philosophy, began life, when he disputed at Paris for his degree of master of arts, by offering to maintain the contrary of any assertion whatsoever of Aristotle. His writings‡ were censured by the University of Paris in 1543, the year in which Copernicus published his great work. He was a correspondent of Rheticus, and states (*Sch. Math.* book ii.) that he endeavoured by letter to persuade him to free astronomy altogether from hypothesis; and that if Rheticus had not been obliged by circumstances to betake himself to the practice of medicine, the mathematics might have had another Copernicus to celebrate. Ramus, then, was for astronomy without hypotheses, but with a very high opinion of Copernicus and of his system as against previous ones; so far his meaning is not very clear. Mæstlinus took him literally, as advocating merely

* It must, however, be observed, that mind and matter were both subjects of the ancient philosophy, physics being *natural* philosophy. The new school advocated *experimental* philosophy, professing to draw *all* their conclusions from experiment. The two terms were antagonists for some time; but at last they coalesced: and for more than a century writers and teachers have talked of natural *and* experimental philosophy, a combination which still lingers in the prospectuses of schools and lectures.

† It is impossible to keep words to their meanings. The words *physician* and *naturalist* are of the same original meaning, namely, investigators of the external world; but the first now means a person who looks after the health of men, while the second means one who looks after the classification, &c. of brutes and vegetables.

‡ It is a coincidence of date worth noting, with reference to the attack on the old system, that the astronomy of Copernicus and the logic of Ramus were published in the same year.

the determination of numerical data: What, says he, will mere numerical proportions in the human mind move the heavenly bodies? But Ramus had a very different meaning. In the work above cited, published in 1569, he presumes that those who preceded Aristotle, and especially the Chaldæans and Egyptians, were in possession of an astronomy founded on observation and experiment; that Eudoxus invented the *orbs* (the crystal spheres in which the Ptolemaists placed the heavenly bodies), and Aristotle amended them; that these spheres were not taken as fictions, but as real and existing; that the Pythagoreans complicated the system, and made it more ridiculous, by epicycles and excentrics; that Copernicus, an astronomer not merely comparable to the ancients, but one especially to be admired, rejected a whole antiquity of hypotheses, and revived others, not new indeed, but most excellent, which demonstrate astronomy, not from the motion of the stars, but from that of the earth. He afterwards adds that astronomy wants such men as Regiomontanus, Copernicus, and Rheinhold, who would contrive a system, not upon feigned hypotheses, but upon geometry and arithmetic applied to the truth and nature of the stars themselves. Such an astronomy, he says, the Chaldæans once had, and the Egyptians and Greeks (*semper feriatur Leo*) before Aristotle; such an astronomy he is satisfied the Germans* might construct, if they would abandon fabulous hypothesis worked up into something like method and science. He would throw away all the notions of the ancients, and even their observations, and begin with the heavens as if they were only just created, and by aid of new and careful observations, he would appeal to logic, geometry, and arithmetic, to induce and infer a general explanation, if indeed there be anything fixed in celestial phenomena. But if everything really do change from age to age, then there can never be any science of astronomy. If Ramus had only so far modified his assertion, as to allow that changes might possibly follow laws which could be detected, and that the oldest observations might be useful in detecting the very slow changes, there would have been nothing to object to in his lecture. As it is, we see that he is one of the most rational of the Copernicans. Bacon, indeed, calls him a skulking-hole of ignorance, a pernicious bookworm, who grasps nothing but the chaff, and so on. But this is in a paper intended as a kind of brief to counsel against all the philosophers, and the expressions must be looked upon as wilfully exaggerated. But for this, we should have supposed that the English opponent of the overgrown Greek could not bear the idea of a French alliance.

* By this marked reference to the Germans, the distinguished Frenchman probably means that there would be theological difficulties in a Catholic country. Luther set science free from the fear of direct interference. But it must not be supposed that the rule of the church over opinion in philosophy was an invention of the sixteenth century. In the preceding ages, when every man of learning was a priest, it was exceedingly common for those who cultivated philosophy to invent applications to, and illustrations from, theology. To suppose that the church would not watch this philosophy would be exactly the same thing as supposing it would not watch the theology taught by its own priests; and none were more on the watch than the philosophers themselves, each over the rest, a few over themselves. Hence the habit which procured even for the laymen of the sixteenth and seventeenth centuries that blessed order of things which is briefly illustrated in the table of contents of Hallam's *Literature of Europe*, as follows:—"The Jansenists take a distinction, page 271; and are persecuted, page 272."

A 3

Francis Maurolycus (1575).—On the fear shown by Maurolycus (*Comp.* 1853, p. 9) that the question of the earth's motion was likely to be agitated, we may add to our mention of Calcagnini and Widmanstadt the tradition that, so early as 1530, Copernicus was ridiculed on the stage in his native country. Maurolycus disposes of the whole question by saying that Copernicus is more worthy of a whipping * than a refutation. In palliation, it must be remembered that the day of reviews was not yet come, when Maurolycus might have quietly carried his views into practice, under profession of impartial examination. New truths often go through a time of whipping before refutation comes on, and when the refutation is quite complete, they begin to be admitted. Then comes another fermentation, in which either the reputed author is shown to be not the real original, or else he is blamed for not going far enough: this stage over, his statue is erected, he receives divine honours, some of the doings of his successors are attributed to him, and those who would have whipped him, if he had arrived in their day, make him a bulwark against further progress.

Francis Bacon (1626).—Bacon asks whether there be a *system at all*, that is, a spherical universe with some one body immoveable in its centre. He remarks that all, except Copernicus (he means the ancients and Gilbert), of those who have maintained the motion of the earth, sprinkle the stars through the universe like islands in the ocean, and reject the common centre, which Copernicus maintained, placing the sun in it. If there be a system, the sun and the earth contend for the central place. The sun, as of greatest efficacy, and as vivifying and animating the universe, seems very properly placed in the middle; and the more so, as Mercury and Venus, at least, are his satellites. But the Copernican system has the inconveniences of loading the earth with three different motions, of separating the sun from the planets, of introducing a great deal of immobility (all the fixed stars being reputed immoveable), and of connecting the moon with the earth as in an epicycle. If the motion of the earth be allowed, it seems that there should be no system at all. All this is

* We have always supposed, with all who have read the passage, that Maurolycus intended to say that Copernicus ought to have a whipping; but on looking once more, we are inclined to suspect that his meaning was not quite so savage. It should be noted that Maurolycus is apt to express disdain rather than indignation. Of Cardan he says, that he is so ridiculous that he is more worthy of contempt than of reprobation; of Erasmus, that he must not call himself a theologian who plays the parasite in silly colloquies. In the article now in question he has been describing the sphere, and proceeds to say that he does not pretend to supersede other books. He then names some erroneous authors, against whose faults he hopes the reader may be protected by what he has written. He then goes on as follows:—*Toleratur et Nicolaus Copernicus, qui solem fixum et terram in girum circumverti posuit; et scutica potius, aut flagello, quam reprehensione dignus est.* This we conjecture may mean that he would also tolerate N. Copernicus, who makes the sun stand still and the earth whirl round and round [after the manner of a boy's top]; and is more worthy of a whip [to keep his plaything up with] than of a grave rebuke. The usual translation hardly consists with itself; the notion that to whip an astronomer is to tolerate him is utterly rejected at Greenwich and at Somerset House; and a person cannot merit the greater punishment more than the smaller one. He who deserves a whipping, *à fortiori*, deserves a rebuke; but here the *scutica* is spoken of as the lesser, the *reprehensio* as the graver, punishmeut. Whichever way it may be, it is to be noted that Maurolycus did not give the last corrections to this work, as he died while it was being printed, or perhaps beforee th printing commenced.

not meant for conclusion, but for reflexion preliminary to inquiry. Bacon afterwards, in the *Novum Organum*, lays it down as for examination, whether the motions of the earth be real, or only convenient hypotheses for calculation; and this, separately and distinctly, both for the diurnal and the orbital motions. In his ideas of astronomy he perfectly agrees with Ramus, but goes further. It is degraded, he says, by being placed among the mathematical arts, when it should be the noblest part of physics. It ought to show the substances, motions, and influences of the heavenly bodies as they really are; but it gives us only the numbers and motions of the stars. The observations and hypotheses lead only to ingenious representations, but are not the real causes and truth of things. So that astronomy is like the stuffed hide which Prometheus imposed on Jupiter for an ox, with a fair outside, but neither flesh nor entrails. So far as masses and attractions are substances and influences, what Bacon desired is now obtained: he wants something beyond simple geometry. In the *Novum Organum* Bacon expressly lays down the question of the earth's motion as one to be examined. In the *De Augmentis* he says that the absurdity and complication of the Ptolemaic system has driven men to the doctrine of the earth's motion, which is clearly false, *quod nobis constat falsissimum esse*. In another place he lays it down that this doctrine cannot be opposed by astronomy, but can be opposed by the principles of natural philosophy, correctly exhibited. We rather suspect, putting all the passages together, that when Bacon impugns that doctrine as manifestly false which he elsewhere propounds for inquiry, he is taking, for the moment, an advocate's license, in aggravation of the case against the Ptolemaic system. This licence, it seems to us, he often takes. The whole of what he has said on this subject, when put together, does not justify Hume's assertion that he rejected the Copernican system "with the most positive disdain." Like so many others of his day, his view is of one colour or another, according as he is thinking of astronomy or of physics. We take the opinion of the *Novum Organum* to be better cured than that of the *De Augmentis*—because Bacon valued the first more than any other of his writings,—because the first is wholly systematic and argumentative, the second more approaching the discursive and rhetorical,—and because, though the Latin of the second was published three* years after the first, yet the second is known to have been written before the first. We are not among the strongest admirers of Bacon, yet we cannot help thinking that, on this point, he has not been fairly represented.

Joh. Bapt. Benedictus (1590).—In a letter published in 1585, but probably written long before, he gives a cautious reasoning in favour of the earth's motion, without committing himself, as being what a person might say who used his reason without any light from above (*i. e.*, from the Inquisition). He rejects the mechanical difficulty, because the air and water would have the *same natural impetus* of motion as the earth itself. He even gives a hint of the plurality

* The dates of publication are 1620 and 1623. But it is incredible that Bacon should have absolutely pronounced, in 1623, upon a question (and such a question) left open among things to be carefully examined in 1620.

of worlds. Those who follow Aristarchus and Copernicus will never believe that the whole creation is merely meant as an arrangement in favour of *the centre of the lunar epicycle.* Benedictus, or Benedetti, is placed by Libri very high among the Italians of the sixteenth century, though little known to those of our day.

Tycho Brahé (1601).—The system of Tycho Brahé, imagined in or about 1587, combines, as he intended it to do, the mathematical advantages of the system of Copernicus, with absence of any contradiction to the mechanics and physics of his time. In asserting the absolute stability of the earth, and the daily motion of the whole universe round it, he agrees with the Ptolemaists; but he makes all the planets (except the moon) move round the sun, while the sun itself moves round the earth. Thus Tycho was a mathematical Copernican and a physical Ptolemaist. He is the strongest of all the admirers of Copernicus. Of Longomontanus, we need only mention that he was a follower of Tycho, whose writings do not come within our period.

Christopher Rothmann (1596).—He was the friend of Tycho Brahé, and, at one time, his follower. But, as appears by a letter of Tycho written in 1595, he was then a defender both of the annual and diurnal motion of the earth. He was, then, a physical Copernican at the end of his life, a fact not noted by the historians. He is known as a writer only by his published correspondence with Tycho Brahé.

Didacus à Stunica.—He was of Toledo, and published a commentary (afterwards among the prohibited works) on the book of Job, of the date of which we only know that one * edition is said to be of 1584. The writer appears to be a physical Copernican, and his work is the first in which the Bible argument is discussed by a Copernican. It was afterwards prohibited; and what Lalande calls the prohibited *letter* of *Zuniga,* is probably a reprint of so much of Stunica's† comment as related to the subject. This writer may be, probably, the originator of the argument found in his book, that the words of Job—"Who shaketh the earth out of her place, and the pillars thereof tremble,"—prove that the earth moves. The other party replied triumphantly that the very words prove that the earth has a *place,* out of which she can only be moved by the special interference of the Creator.

Francis Patrizi, or *Patricius* (1597).—Here is another philosopher, even whose name has not been noticed by the mathematical historians of this subject. Riccioli has, indeed, got Peter Ramus in his list, and asks whether he be the medical man who is mentioned by Junctinus (a collector of nativities) as having been born in 1532, at 23 minutes after 3 in the afternoon of February 22. He knows

* The index of prohibited books ought to give some information on this point; but the bibliographer has to regret that this catalogue is the worst of all catalogues. We have before us one published at Rome, about 1745. Very few books are dated or described; and we learn that every anonymous *Disputatio de Festo Corporis Christi* must be forbidden; every *Historia de Germanorum origine;* and every book having the title *De disciplina puerorum, recteque formandis eorum studiis et moribus.* Such is the effect of prohibiting anonymous works, under nothing but titles.

† Drinkwater-Bethune inadvertently speaks of Diego Zuniga on the book of *Joshua.*

only the author of the *Scholæ Mathematicæ*, and seems never to have heard that this writer was also the philosophical heretic who shook the schools of Paris. Of Patricius he knows nothing, though he makes one quotation from him. In a former article (*Comp.* 1836), we see that Fienus of Louvain associates one *Bernard* Patricius with Copernicus, and cites no other maintainers of the earth's motion by name. For eighteen years we have looked, from time to time, at biographies and histories, that we might discover this lost Copernican, if possible. We have recently found that Fienus made a mistake in the Christian name, and that *Francis** Patricius is the person intended. This philosopher was a noted opponent of Aristotle, but not in the usual manner: he strove to revive Platonism. Among his modes of doing this, we may mention his collection of points on which he conceived Plato was orthodox, and Aristotle was not, according to the orthodoxy of Rome. His own orthodoxy does not seem to have served his turn, for De Thou tells us that his doctrine was censured by ecclesiastical authority, and that he was obliged to retract. Tiraboschi (who has given a fuller general account) doubts this, on the ground that the work of which we shall speak is studded with the *postillæ* of a certain priest, da Lugo by name, whose business it was to make such remarks as would be sufficient preservatives against heresy. This very circumstance seems to show that the author was a suspected person: and we may hazard a conjecture that he had fallen into disgrace, had made his *pur si muove* recantation all right and regular, and had been allowed to publish, on condition of admitting the inspection and comment of a dry nurse of safer principles than his own. Nevertheless, Patricius was invited to Rome by Clement VIII., in 1592, and, in spite of Cardinal Bellarmine and other powerful Aristotelians, taught the Platonic philosophy there until his death.

The work of which we are to speak is the *Nova de Universis Philosophia*, Venice, 1593,† but having subordinate titles, dated Ferrara, 1591, and evidently ready in that year. Patricius describes the systems of Copernicus and Tycho Brahé, and declares altogether against the orbital motion of the earth; he thinks it impossible. In common with a great number of natural (*and* experimental) philosophers of every age, he sets out with his notion, founded upon the knowledge he possesses, of what is possible and what is impossible in physics: and, in common with the same philosophers, he is very little disposed to allow his reader any appeal from his own preconceptions. When he comes to examine the diurnal motion of the earth, he cites in favour of it Nicetas, Philolaus, Heraclides, Ecphantus, Seleucus, and Aristarchus, before, whose authority that of Aristotle and Ptolemy falls to the ground. He adds Copernicus, whom he calls the greatest astronomer of the age. Having thus cleared the field, he begins to examine the possibilities of the case. Since the stars appear to move in a circle, they must either be

* The person of all others with whom Patrizi was likely to be confounded was Telesio, whose Christian name was *Bernardino*.

† It is often stated that it was published in 1591; and it may be suspected that the title-page of 1593 is a new one. The excessive rarity of the work has prevented it from being cited in modern times. Fabricius makes two editions: but this is a mistake.

carried round with the heavens, or each must have its own separate motion, or the earth must have a rotation. The first supposition he pronounces impossible, on account of the immense rapidity of the motion. The second supposition is impossible as to the more distant stars, possible, *perhaps*, as to the nearer ones, and possible enough as to the nearest. That the earth should revolve is more consonant to reason than either of the other suppositions. He cites cases of apparent motion, such as that of the shores to those in a ship, and pronounces a similar deception to be very likely as to the stars. It is remarkable that Father da Lugo, the dry nurse, says not a word against these speculations; his annotations on the chapter which contains them refer only to the question whether the sun and planets are animated bodies. And we must infer, from the later facts of the life of Patricius, that from 1592 to 1597, the probable diurnal motion of the earth was taught at Rome, under the patronage of the Pope.

Christian Urstisius (1588).—He was a professor at Basle, who wrote on the planetary theory of Purbach, and published a book on arithmetic, which was translated into English. Nothing more is known of him, except the one circumstance which will immortalise his name—he publicly taught the theory of Copernicus, in some lectures which he gave in Italy, and is supposed to have made a convert of Galileo. (Drinkwater-Bethune, *Life of Galileo*, p. 7).

William Gilbert (1603).—This is the celebrated writer on the magnet; and in his work (1600) he declares in the strongest terms for the *diurnal* motion of the earth, as not only probable, but manifestly true. His chief reason is the enormous motion which the heavens would otherwise have. He passes over the orbital and precessional motions, as being beside his purpose to treat of. Bacon describes him as distinctly opposing all except the diurnal motion; and this description, probably the result of private conversation, was somewhat confirmed by the posthumous publication (in 1651) of Gilbert's work, *De mundo nostro sublunari philosophia nova*. Gilbert here finds Tycho Brahé's difficulty of the immense void between the solar system and the fixed stars, and Bacon's difficulty of the *three motions*. But he does not draw a positive conclusion, and he treats with contempt the notion that the earth *must* be the centre.

Edward Wright (1615).—This eminent discoverer in the art of navigation does not give the least appearance of abandoning the old system, either in his work on navigation or in that on the sphere. But in an encomiastic address, as he calls it, written for Gilbert, and prefixed by Gilbert to the work on the magnet, he discusses the diurnal motion, points out in strong terms the improbability of the great motion of the whole sphere of the heavens, decides against the force of the Scriptural argument, does not see why those who advocate so difficult a system as the common one should not give permission to others to adopt something more simple, and ends by declaring the diurnal motion probable enough (*satis probabile esse*), though he does not think that any higher certainty will ever be arrived at. Wright is evidently doing his best; but, looking at the character of the panegyrics which it was common for authors to write in the prefaces of the works of other authors, we are afraid we cannot with certainty infer more than that he was so near the line of demarcation that it did not hurt his conscience to step over it to serve a friend. This address was written before 1600; in 1613 Wright published his work on the sphere, the very title-page of which indicated a Ptolemaist in an unusually marked manner. The book has not even an allusion to the motion of the earth, nor any thing to the point, except the statement that the *primum mobile* is " imagined by the astronomers to show the reason of that dayly motion." The old system is here reduced to an hypothesis.

Christopher Clavius (1612).—This celebrated Jesuit, whose depth of learning was the admiration of his age, calls Copernicus (in 1570) the excellent restorer of astronomy, whom all posterity will gratefully celebrate and admire as a second Ptolemy. But he rejects the actual motion of the earth as absurd, contrary to the senses, and rash (*temeraria*, the word applied to doctrines suspected of heresy before they were actually condemned). He brings forward the arguments against the earth's motion from the ancients and the Scriptures. It is said that, in his collected edition of his own works, prepared in 1611, just after the discovery of Jupiter's satellites, he gave it as his opinion that the old system would no longer do, and that another must be looked for; but we have no reference to the place, and the collection is in five folio volumes. Our authority for the assertion is Foscarini, who made it in 1615.

Francis Vieta (1603).—It would be difficult to find anything to the point in the printed writings of the greatest French mathematician of the sixteenth century, except in one word. He refers to mathematical tables made by certain *rhapsodists*. As he was well versed in Greek, and pedantically fond of introducing words from it, we should have passed this over, if we had not found the word rhapsodist in contemporary writers, used in our modern sense. The makers of tables were Rheticus and Rheinhold, both Lutherans and both Copernicans; for which reason Vieta, when forced to allude to them, only names them by a term of contempt. Those who have examined the manuscript of his *Harmonicon Cœleste* report that he thinks the excellence, if any, of the system of Copernicus, is destroyed by the badness of its geometry. This is the most singular of all the opinions which were uttered on the subject.

Michael Mæstlin, or *Mæstlinus* (1631).—Successively a Professor at Heidelberg and Tubingen, made his first appearance, at what age is not known, as a writer on the new star of 1572. Tycho Brahé (*De Nova Stella*, pp. 543-8) has inserted a writing of his entire, and says of him, that, though he had but a thread to observe with, he had come nearer the mark than several who had used elaborate instruments. Mæstlinus is known as the preceptor and correspondent of Kepler, as one of the opponents of Clavius in the matter of the calendar, as the constructor of ephemerides, and as the author of an epitome of astronomy, which was several times reprinted. The common story of his having lectured on the Copernican doctrine in Italy, and having made a convert of Galileo, is sufficiently disposed of by Drinkwater-Bethune (*Galileo*, p. 7): it seems clear enough that

Vossius confounded Mæstlinus with Urstisius of Basle. Had Copernican opinion been a little more closely watched, it would have been evident that Mæstlinus could not have taught Galileo an extent of Copernicanism which he himself never held. Weidler, and many others after him, place the death of Mæstlinus in 1590; but his correspondence with Kepler extends to 1620, and he edited an enlarged edition of his own epitome in 1624. Our date of his death is given without authority by Drinkwater-Bethune. Riccioli places Mæstlinus among his Copernicans, on the strength of additions made by him to the *Narratio* of Rheticus, in the republication of that Copernican writing, in the *Prodromus* of Kepler in 1596. But any one who will examine the *Appendix M. M. de dimensione orbium et sphærarum cœlestium juxta Tabulas Prutenicas ex sententia N. Copernici*, either in the *Prodromus* itself or in Kepler's correspondence, will see that Mæstlinus professes no more than adherence to the dimensions given by Copernicus. He held by the system as an hypothesis for the calculation of the places of the heavenly bodies. His private opinion (as given in correspondence with Kepler) was that astronomers should let physics alone, as rather disturbing than informing the reader. Throughout all his writings, from that published by Tycho Brahé up to his last edition of his own epitome, there is much on Copernican *numerals*, but little on Copernican *opinions*. And yet he seems to have had a feeling in favour of the *reality* of his notions; and on one occasion, unwittingly, his favourite word *hypothesis* was but a synonyme for an agent. We have already cited his remark on Ramus; now an hypothesis in the human mind will no more move a heavenly body than would a numerical proportion. The most decided expression of opinion which we can find, occurs in one of the additions which he made to his epitome in 1624, posterior, it must be noted, to the time when the Inquisition began to interfere in Italy. Speaking of the enormous revolution of the *primum mobile*, he says, "Among all the reasons which gave Copernicus occasion to think of other hypotheses, and other dispositions of the heavenly bodies, more consonant to reason, *nature*, and observation, this incredible velocity is not the last, if indeed it be not the very first." The word in italics is the strongest proof we can find that the reputed master of Galileo was at all inclined even to hint at what he is said to have taught.

Jordanus Brunus, or *Giordano Bruno* (1600).—He was first a Dominican priest, then a Calvinist; and was roasted alive at Rome, in 1600, for as many heresies of opinion, religious and philosophical, as ever lit one fire. Some defenders of the papal cause have at least worded their accusations so to be understood as imputing to him villanous actions. But it is positively certain that his death was due to opinions alone, and that retractation, even after sentence, would have saved him. There exists a remarkable letter, written from Rome on the very day of the murder, by Scioppius (the celebrated scholar, a waspish convert from Lutheranism, known by his hatred to Protestants and Jesuits) to Rittershusius, a well-known Lutheran writer on civil and canon law, whose works are in the index of prohibited books. This letter has been reprinted by Libri (vol. iv.

p. 407). The writer informs his friend (whom he wished to convince that even a Lutheran would have burnt Bruno) that all Rome would tell him that Bruno died for Lutheranism; but this is because the Italians do not know the difference between one heresy and another, in which simplicity (says* the writer), may God preserve them. He then proceeds to describe to his Protestant friend (to whom he would certainly not have omitted any act which both their churches would have condemned) the mass† of opinions with which Bruno was charged: as that there are innumerable worlds, that souls migrate, that Moses was a magician, that the Scriptures are a dream, that only the Hebrews descended from Adam and Eve, that the devils would be saved, that Christ was a magician and deservedly put to death, &c. In fact, says he, Bruno has advanced all that was ever brought forward by all heathen philosophers, and by all heretics, ancient and modern. A time for retractation was given, both before sentence and after, which should be noted, as well for the wretched palliation which it may afford, as for the additional proof it gives that opinions, and opinions only, brought him to the stake.

The work of Bruno in which his astronomical opinions are contained is *De Monade*, &c., Frankfort, 1591, 8vo. He is the most thorough-going Copernican possible, and throws out almost every opinion, true or false, which has ever been discussed by astronomers, from the theory of innumerable inhabited worlds and systems to that of the planetary nature of comets. Libri (vol. iv.) has reprinted the most striking part of his expressions of Copernican opinion.

Nicolas Raimarus Ursus Dithmarsus (1600?) published works on astronomy in 1588 and 1597, in which he claims the invention of Tycho's system, and charges Tycho with plagiarism. Tycho made the same charge against him; and probably neither charge is true: both were scurrilous. But he makes a great mistake when he imagines that his system agrees with that of Tycho. The difference lies in this, that Tycho combined the reputed advantages of both systems, while Ursus (to choose the name which best represents his style) managed to introduce the alleged disadvantages of both. Tycho, by fixing the earth, avoided the mechanical difficulties which were then unanswerable; while, by making the sun a centre of motion for all the planets, he introduced nearly all the mathematical advantages of the

* That is to say, they knew the difference between a live heretic and a roasted one by actual inspection, but had no idea of the difference between a Lutheran and a Calvinist. The countrymen of Boccacio would have smiled at the idea which the German scholar entertained of them. They said Bruno was burnt for Lutheranism, a name under which they classed all Protestants: and they are better witnesses than Schopp or Scioppius.

† In this medley of charges the Scriptures are a dream, while Adam, Eve, devils, and salvation are truths, and the Saviour a deceiver. We have examined no work of Bruno except the *De Monade, &c.* mentioned in the text. A strong though strange *theism* runs through the whole, and Moses, Christ, the fathers, &c. are cited in a manner which excites no remark either way. Among the versions of the cause of Bruno's death is *atheism*: but this word was very often used to denote rejection of revelation, not merely in the common course of dispute, but by such writers, for instance, as Brucker and Morhof. Thus Morhof says of the *De Monade, &c.*, that it exhibits no manifest signs of atheism. What he means by the word is clear enough, when he thus speaks of a work which acknowledges God in hundreds of places, and rejects opinions as blasphemous in several.

Copernican system. Ursus, by allowing to the earth a diurnal rotation, introduced all the mechanical difficulties; while, by giving orbital motion to the sun, he lost some of the simplicity and truth of the Copernican system without anything in exchange.

Joh. Ant. Maginus (1617).—Maginus was one of those whom we have called numerical Copernicans; he published works in which the *observations* of Copernicus are used in conjunction with those of Tycho Brahé. Of Copernicus, he thinks that there is only Ptolemy with whom he can be compared; but when he speaks of the hypothesis physically, he calls it absurd.

Simon Stevinus (1620).—He describes the mathematical hypotheses of Ptolemy and Copernicus, without a word of comment on the great difference between them: insomuch that, when he ends by desiring every one to think as he pleases, we cannot even gather that he intends this permission to include the opinion of the earth's motion; it may refer merely to the mathematical hypotheses which he has been describing.

John Kepler (1630).—Though Kepler was the friend and fellow-labourer of Galileo, yet he has his *antegalilean* period of reputation, and it is but fit that he should close such a list as the present. He lets us know very distinctly that about 1590 he was a defender of the system of Copernicus on physical or metaphysical grounds; and that not till after he had arrived at this stage did he see the mathematical superiority of that system, which at length he did see, partly by the instructions of Mæstlinus, partly by his own efforts. How he first came by his opinion of the physical truth of Copernicanism, he does not inform us. The work of Bruno appeared in 1591, in Germany, and Kepler shows his acquaintance with that work in after life. He took the degree of Master of Arts in August 1591, and tells us that he defended Copernicus in the physical disputations of the candidates. It is just possible that Kepler may have spoken but vaguely when he says, "about six years ago," and that Bruno may have been his leader. It should also be noted that a work professing Copernicanism was published by Bruno (according to Bayie), so far back as 1584.

We have before spoken of the earliest English Copernicans, Recorde, Field, and Digges (*Comp.* 1837, pp. 35-40; and *Penny Cyclop.*, "Motion of the Earth"). Robert Recorde, in 1556, avowed, in an elementary work, but in a very cautious manner, his leaning towards physical Copernicanism. John Dee, in 1556, in his epistle prefixed to Field's Ephemeris for 1557, said he had hoped that the herculean labours of Copernicus, Rheticus, and Rheinhold, would have been heard of in England. He describes them as restoring the science of astronomy, and supporting their views by the strongest weight of reasons: disgusted by the neglect of these divine lucubrations, he had instigated Field to construct an ephemeris from the data of Copernicus, &c. Field, the pupil of Dee, expresses a similar opinion: both may be suspected of holding the physical system, but their expressions are not conclusive. Thomas Digges, in the preface to the *Alæ &c.* (1573), engages to publish a work in which the hitherto exploded paradox of the motion of the earth shall be supported, at least by probable arguments, and *perhaps* by the strongest demonstration. In 1594, he added to a reprint of his father's work an account and defence of physical Copernicanism, with good answers to some of the mechanical objections. Rejecting Rheticus and Rheinhold, the personal friends of Copernicus, the whole Continent does not produce, by 1573, so large an amount of unbiassed adherence to any view of the Copernican system as we find given in England to its physical truth.

But, both in England and abroad, it would seem that the Copernican, when he ventured his opinion, considered his work as consigned to circulation among astronomical heretics; and that he never gave it in any book which he did not wish to risk. Thus the indignant Copernican, John Dee, would not drop a word on that system in the preface to Billingsley's Euclid (1570), but makes it the office of astronomy to measure the distance of the planets from the centre of the earth. John Blagrave, in 1585, after describing the Ptolemaic system, adds, "This I say after the opinion of old writers, though Copernicus hath ascribed another order." He could go no further; but this reduction of the old system to an opinion looks suspicious. Perhaps Wright's reserve, already mentioned, is to be explained in the same way.

There is one more English anti-Copernican to be mentioned, Thomas Lydiat, who died in 1646. He was a good scholar, and incurred Joseph Scaliger's severest abuse by beating him on points of chronology. In his *Prælectio Astronomica* (1605) he so far departs from received opinion as to maintain that the stars are not fixed in solid orbs, but pendent in æther. He almost ignores Copernicus. In the one mention which he makes of his system, he refers to the argument derived from the immense velocity of the fixed stars, and says, that if the followers of Copernicus had never seen anything swifter than the flight of a bird, the velocity of an arrow or of a cannon-ball would have been equally incredible. This was, perhaps, as good an answer as the celebrated objection deserved. The moderns, who blame their predecessors for not being willing, with Patricius, to declare a certain amount of velocity *impossible*, received with perfect satisfaction the doctrine that light consists of particles of matter flying 200,000 miles in a second. Had they followed their own maxims, this velocity would have made them keep the undulatory theory in view. But though the Copernican system was never allowed to drop out of sight, in spite of the unanswerable objections to it, the undulatory theory was in total neglect for a much longer period than that of which we are writing. Two hundred thousand miles in a second is a much greater velocity than the Ptolemaists wanted for their *primum mobile*, a much greater velocity than that which our school-boys and school-girls are taught to point the finger at them for admitting.

We have now named all the writers of most note, who treated of the subject before the invention of the telescope. The discovery of Jupiter's satellites was the first great blow to the ancient system: it led many at once to the doctrine of *other worlds* besides our own. Bruno and Kepler had maintained this opinion before the telescope was known, and Tycho Brahé had held it; but after the discovery of the satellites, and the resolution of the milky way, it ceased to be so monstrous as it had till then appeared to be. Scioppius makes this opinion the head and front of his charges against Bruno, and ends his attempt to settle terms of combustion with his Lutheran friend by observing that Bruno was gone to tell the inhabitants of the worlds he had invented how blasphemers were treated at Rome. After the establishment of the distribution of stars through space, and of separate planets with satellites, the idea of maintaining that no star except our own had its inhabitants of some kind seemed to most like a declaration from the insects on one leaf that there was no life except their own in the whole forest. Truly, says James Hume (a Scotchman, settled at Paris, well known in the mathematical world) in 1637, if I could only believe with the Copernicans that the earth moves round the sun, I should at once be persuaded that the earth and planets are of one kind, and that all of them, if not the fixed stars also, have air, water, earth, beasts, birds, and fishes, either such as ours or different. He omits *man*, probably on account of the theological question it would have raised; this question is at the bottom of the dispute now in progress.

It must be remembered that the system imagined (and perhaps believed as true in fact) by Copernicus is not altogether that which we now call by his name. It is what Bacon calls a system in which the whole universe is one: everything is *solar system* with Copernicus. The sun is the immoveable centre of the whole: this idea of an immoveable centre was seldom lost sight of, except by such ultra-heretics as Bruno. The fixed stars, instead of being distributed through space, were placed on a sphere at the outer confines of the creation. The laws of planetary motion not being yet discovered, their primitive circular orbits were complicated with the subordinate epicycles, &c. of the old or Ptolemaic system: and Copernicus does not explicitly reject, and in some passages almost seems to favour, the idea of the solid crystal orbs in which the planets were wheeled round, and which Tycho Brahé dismounted by showing that comets pass through them. Copernicus may here be held to have deferred to learned opinion: for those who set out in physics with clear ideas of the possible and impossible had ruled that no heavenly body could move except with its heaven, and as a knot in the board by the motion of the board, to use their own illustration. But, however this may be, the reader who desires to understand this controversy, must remember that those who spoke in favour of the Copernican system (it was otherwise with many of the opponents) understood by it the system taught in the book of Copernicus, upon the supposition that the author meant what he said. Whether he did mean what he said, or whether, on the other hand, he believed more than he said, may be matter for argument: but this question was never raised in the time of which we are speaking. When Kepler and others wanted to show themselves believers in the actual motion of the earth, they declared themselves followers, not of Copernicus, but of Philolaus or Aristarchus. When they speak of Copernicus and his *hypothesis*, they soon let us see that this is from no jealousy of the great mathematician, of admiring whom they are never tired, but from the necessity of being distinctly understood.

Again, most readers will find a difficulty in remembering that the names now unknown to them, and which sound like names of nobody, are to be looked at as if they were names of persons known over the whole literary world. Who is Rheinhold? who is Ursus Dithmarsus? who is Patricius? The questions asked ought to be, Who *was* Rheinhold? &c. The coupling of such names with Copernicus and Galileo may have a strange appearance; but we can cite something so much more strange that we hope our reader will be reconciled to what we have done by comparison. An excellent literary epitomist of the last century (Stolle, 1728) says he remembers to have read that no one could make progress in poetry, oratory, or philosophy, without algebra. This, he says, is a good joke (*egregia concludendi ratio*); as if Virgil, Ovid, Opitius, Gryphius, Hoffmannswaldavius, Lohensteinius, Abschatzius, Neukirchius, Besserus, Brokesius, and others of poetical fame, owed it to algebra. This proves beyond a doubt, he says, that mathematicians* can talk absurdly as well as philosophers.

The reader who is unaccustomed to think of scientific history may transfer more modern views to the credit of older systems, and may not be able to learn that names which are now unknown to general fame are essential to a sufficient view of history; and in both these errors he may receive some encouragement from many who ought to know better. But there is one still greater error which he will actually learn from the writings of the best historians, and from the conversation of those who are best qualified to read the histories: namely, to judge the merit and demerit of a former age by the comparison of their methods with our own, instead of with the methods of those who went before them. No one is so conspicuous a teacher of this folly as Delambre, the greatest of astronomical historians: but the fault is that of his time. We are in the midst (let us hope near the culmination) of a long reaction consequent upon the long period of excessive reverence for antiquity. The nineteenth century will be known in history as the most uplifted of the self-glorifying centuries: and those of a remote time, to whom the difference between the sixteenth and the nineteenth centuries will, as viewed from a distance, not seem quite so great as to us, will be amused by our crowing.

By investing Copernicus with a system which requires Galileo, Kepler, and Newton to explain it, and their pupils to understand it, the modern astronomer refers the want of immediate acceptance of that system to ignorance, prejudice, and over adherence to antiquity. No doubt all those things can be traced; but the ignorance was of a

* This means that the writer lived at a period at which the new philosophy of the material universe was busy in depreciating psychological studies. Time brings about a reaction, and psychology and metaphysics take their *innings*, if we may venture such a word; after which the other party begins again. Common sense, the diffusion of sound thought, and the growth of good manners, act as resistances to the swings of this absurd pendulum, so that each is less than the one before it. In process of time, the philosophies will know each other's value, and it will then be historically expedient to publish "The Nursery of Knowledge, showing how the children quarrelled and scratched about which was the prettiest," in 10 volumes quarto, date and place not yet ascertained.

kind which belonged equally to the partisans and to the opponents, and which fairly imposed on the propounder of the system the onus of meeting arguments, which, in the period we speak of, he did not and could not meet. It must be remembered that, in the sixteenth century, the wit of man could not imagine how, if the earth moved, a stone thrown directly upwards would tumble down upon the spot it was thrown from. Easy experiments verify the law of motion which now explains this; but to be proved by experiment, a law must be conceived and imagined. To be put under discussion, it must be proposed. Now the advocates of the earth's motion never, before the time of Galileo, even conceived this law, never proposed it, and of course never proved it. It might be supposed that they would at least conjecture that progress in the mechanics of motion would, at some future time, reconcile the beautifully simple system of Copernicus with common sense: but no such conjecture was ventured on. And for sufficient reasons. It would have been absurd to expect assent to an astronomical system, because of its clearness and simplicity, on the faith of an assurance that opposite arguments, of an equally clear and simple character, would at some future time be refuted. Any person might prove anything if he could get removal of objections discounted. It is to be remembered that no astronomical system can be a verifiable fact, like the satellites of Jupiter, the circulation of the blood, and the efficacy of vaccination. To this day we do not know the motion of the earth as we know either of the things just mentioned. The evidence is inferential and cumulative: no one ever saw the earth move. Even the brilliant experiment which a few years ago attracted so much attention (as well as a beautiful variation of it just now shown at the British Association) is one, the evidence of which not one man in ten thousand is able to follow. Why was this experiment so eagerly welcomed? Who would run after any new proof, as proof, of the satellites, the circulation of the blood, or the good effect of vaccination?

The matter of reproach is, that the Copernican system progressed very slowly. But the proper question is, Did it progress as fast as any improvement in the details of the existing system would have done? We do not think it did; but we do not think it was much behind its natural pace, according to this test.

The inquirers of the sixteenth century, on the whole, took the actual Copernican system at a fair valuation. They saw its beauty, and they saw its difficulties. The author gained his place immediately: there never was a doubt that Copernicus was one of the greatest of mathematicians. Even his opponents placed him by the side of Ptolemy as a master of hypothesis. The astronomical world was divided, so far as the physical question was concerned, into those who would not reject the evidence of the senses, and the arguments in which they believed, without proof that that evidence was deceptive; and those who attempted to answer that evidence used arguments which would now be held no better than those produced on the other side. Modern historians dwell very little on the *Aristotelian* arguments which were urged on the Copernican side of the question, even by Copernicus himself.

Those who were attached to astrology would necessarily be among the most backward to entertain the idea of the earth's motion. It is probably to this that we owe the not being able to cite an opinion of Cardan either one way or the other. We cannot find that he mentions the name of Copernicus, though he makes some* allusion to the dispute about the central body, quotes the Prutenic tables, and records a conversation on astrology with Rheticus in 1546.

The Copernican controversy of the sixteenth century was passionless; system against system, hypothesis against hypothesis, argument against argument, Philolaus against Ptolemy. It was not until Galileo showed in the heavens a model of the system on a smaller scale, that those who would at any hazard preserve the ancient opinions took real alarm. The exhibition of Jupiter and his satellites was an argument which was not only worth much, but was sure to go for more than it was worth with the whole of the educated community. Hence the difference between the state of the controversy before and after the year 1610. The Cardinals of the Index never prohibited the work of Copernicus until men began to read it through Galileo's glass. Up to that time it was a question for astronomers only; but after that time it was the property of the world at large. The intention of this paper is to illustrate the distinction between the two periods of the controversy.

The abiding interest of the later dispute has partly arisen from the assertion of Protestant writers that the Roman Church, claiming infallibility in matters of faith, decided the doctrine of the earth at rest to be a truth affirmed in the Christian revelation. It has been sufficiently shown that this† is an exaggeration (*Penny Cycl. Suppl.*, "Galileo"), and that the most zealous Roman anti-Copernicans, and even Protestants, were aware, at the time, that no decree of any Inquisition could be held as determining a matter of faith, though it might demand obedience of action and suppression of opinion throughout the jurisdiction of *that particular Inquisition*. To the

* Foscarini cites Cardan as the author of a new system of his own; but there is nothing in the chapter referred to which at all concerns the earth's motion, except notice that this question is to be passed over. Cardan speaks only of sublunary phenomena.

† Galileo, following the example of his opponents, entered upon the question of the interpretation of the Scriptures, and advanced or implied various canons of interpretation from his private judgment. No Protestant thinks it unreasonable to affirm that, so long as a man remains Roman Catholic, he has no business to do any such thing. Men's minds were much stirred on this question, and *some* interference was inevitable, on principles which all the disputants admitted. The Inquisition interfered, and made a blunder. Instead of enjoining silence on all parties, as to the question of modes of interpretation, they relied on the earth's motion being a tenet which would soon go the way of most novelties, and pronounced it false and heretical. This was the proceeding of 1616; that of 1633 was its necessary consequence: it is seldom permitted to governments to escape saying B when they have said A. Most of the defenders of the Papacy have found themselves hampered by some necessity which we cannot understand of defending the conduct of the Inquisition; as to the manner and direction of the interference. If Protestants continue to believe, much to the dissatisfaction of the old church, that this same infallible church made a *demonstrable* mistake, the fault lies very much in the manner in which Roman Catholics have argued their defence. Tiraboschi finds it necessary to omit that the Inquisition declared the doctrine of the earth's motion false (*falsa philosophia*), and only mentions the *heresy*. See his Memoir, *Sulla Condanna del Galileo e del sistema Copernicano*, in which he gracefully compares himself, when finding some fault with Galileo, to Galileo himself finding spots in the sun. The sophism of this paper is, that the condemnation of the earth's motion is quietly assumed to be justified so soon as it is shown that the conduct of Galileo cannot be fully justified.

authorities cited in that article, it may be added that even Riccioli, from whose work all writers draw their account of the trial and sentence of Galileo, himself one of the strongest theological opponents of the earth's motion, most expressly declares (and is allowed by the censorship of the press to declare, as had been many who went before him) against any declaration having been made by the Church itself. As follows:—

"The sacred congregation of Cardinals, taken apart from the Supreme Pontiff, does not make propositions to be of faith, even though it should actually define them to be of faith, or the contrary ones heretical. Wherefore, since no definition upon this matter has as yet issued from the Supreme Pontiff, nor from any council directed and approved by him, it is not yet of faith that the sun moves and the earth stands still, by force of the decree of the Congregation; but at most, and alone, by the force of the Sacred Scripture, to those to whom it is morally evident that God has revealed it. Nevertheless, Catholics are bound, in prudence and obedience, at least so far as not to teach the contrary. But of this subtility* of theology I have treated in my treatise *De Fide.*"

It must also be borne in mind that the Inquisition did not interfere until both sides had got into a warm and angry dispute about certain interpretations of the Scriptures of which both sides admitted every interpretation to be the exclusive property of the Church. From many circumstances we feel a right to suspect that if Galileo and his followers had never made any other answer to the Scriptural argument except declining to assume the power which, by their own concession, belonged only to popes and councils, they never would have been called to account. The general temper of the higher orders of the clergy seems to have been unobjectionable. When Galileo applied to Maraffi, the general of the Dominicans, complaining of an indecent attack made upon him by one of the order, he received a written answer, expressing regret and the vexation of the writer at being liable to be compromised by all the brutalities (*bestialità*) which might and did take place among twenty or thirty thousand monks. Had each offender against common sense and common modesty been a pope, a cardinal, an abbot, or a monk, acting in his single capacity, we should have been pleased to remember how vast and how many have been the obligations which both literature and science have owed to those orders. And we should have recalled the great encouragement which the Copernican system received from popes and cardinals at its first promulgation; nor should we have forgotten that immediately after the first proceedings against Galileo, a pope invited Kepler, the greatest and most noto-

* We have spelt this word after the Latin, that the reader may not fall into Delambre's error, who says, "As if he repented of what he had just written, he adds that this solution is *only a theological subtlety.*" Truly, we should like to see, even in our day, a Roman Catholic priest apply such a phrase, in Delambre's sense, to the difference between the decree of a general Council and of the Inquisition. If Delambre had ever, by any accident, dipped into Occam or Duns Scotus, and found the word *quidditas*, he would have said, "The schoolman admits that his assertion is nothing but a quiddity." A *subtlety*, in the old use of the word, is a distinction which requires thought and explanation: all knowledge swarms with subtleties. Nothing was more common than to say, in answer to the Scriptural arguments, that the Scriptures do not enter into physical and geometrical *subtleties*. Certainly Delambre would not have supposed that his Copernican friends meant to speak scornfully of the difference between the two systems. The word subtlety has led to some curious mistakes. At one time it was partially disused, and, how is not explained, the word *calculation* took its place. Suisset wrote, in the fourteenth century, a book which he headed "Calculator," meaning the *maker of distinctions*. The earlier historians of mathematics got hold of this title (the work itself was scarce), represented him as a great improver of arithmetic, and even as an inventor of algebra. An easy subtlety was much wanted here.

rious of the Copernicans next after Galileo, to be his professor of astronomy at Bologna. But the Inquisition, the common sewer of the *odium theologicum*, has no redeeming point in its general history, though the case before us has this much of alleviation, that (taking for granted that Galileo was *not* put to the torture) there is not enough of atrocity to take off the edge of the ridicule.

What inference is to be drawn from the Church allowing this scandal to remain for two centuries without interference, cannot be settled until the point has been argued by both sides. But the Protestants prefer to stand by their opinion that the Inquisition is the Church; and the Catholics are much too acute to invite the enemy to occupy Sebastopol if the enemy prefer Odessa.

The position of the Roman Inquisition of the seventeenth century is ridiculous enough without any exaggeration. We do not laugh at their declaring the earth's motion *heretical*, but at their declaring it *absurd* and *false*. *E pur si muove* has done its work, and is doing it still; misrepresentation will but undo some of it. To take a hint from Maurolycus, it would be much better to give the Inquisition a few fathoms of stout chain cable to make the earth fast to the spot from which Archimedes proposed to move it, than to compliment the holy office, for party purposes, with the functions of a general council. And it might be worth considering whether the price of the cable should not be raised by a tax upon all who have, in a land of private judgment, endeavoured to prevail upon public opinion to attempt against geology what the cardinals attempted against astronomy.

A. DE MORGAN.

University College, London,
October 16*th*, 1854.

COMPANION TO THE ALMANAC

FOR

1856.

PART I.

GENERAL INFORMATION ON SUBJECTS OF MATHEMATICS, NATURAL PHILOSOPHY AND HISTORY, CHRONOLOGY, GEOGRAPHY, STATISTICS, &c.

I.—NOTES ON THE HISTORY OF THE ENGLISH COINAGE.

The discussions which have taken place on the decimal coinage give a present interest to the history of our coinage in general, and will justify us in putting together a few notes on the subject.

The most complete work on the coinage is the *Annals*, &c., by the Rev. Roger Ruding (London, 1817, 3 vols., 4to, and Suppl. 1819). Next to this comes the *Historical Account*, &c., by Stephen Martin Leake (London, 1726, 1745, 1793, 8vo). Camden's *Remains* (1605 and various other editions, 4to) contain a chapter on *Money* which has been much quoted: the same made be said of Bishop Nicolson's *English Historical Library*. Further references may be found in Ruding, Leake, Nicolson, and the *Penny Cyclopædia*, art. *Coinage*. A good popular book on the coinage is much wanted, which shall properly combine *numismatics*—coins considered as historical monuments—with the monetary history of the nation.

A *Coin* is a certain amount of precious metal, with the state mark upon it as a guarantee for weight and quality. In no coinage is it more necessary to remember this definition than in our own, in which, originally and for a long period, large sums were left to be weighed, and coinage was but a convenience for avoiding the trouble of weighing smaller sums. The Saxons and the early Normans coined only silver, and of this nothing higher than the penny. The Saxons had gold *bizants* or *bezants*, but these were coined at Constantinople (Byzantium). In the early Norman times, Italian, Flemish, and Jewish money-lenders brought foreign gold into the country. But it must be remembered that up to the time of Edward III. (excepting only the *gold penny* of Henry III., meant for twenty silver pence, and probably not circulated) there was no coinage of gold by an English sovereign. Pence, halfpence, and farthings, in silver, formed the national currency for money *counted* and not *weighed.* When Edward I. coined *fourpences*, they were called *groats*, or *great* coins: but this coin did not become generally current till the time of Edward III. The Saxons had copper coins of eight to the penny; but no copper was coined after the Conquest, as a national coin, until the time of James I. Copper was considered a base metal; a kind of token: a copper halfpenny struck, but not circulated, in the reign of Elizabeth bears, not the word *halfpenny*, but the words *pledge of a halfpenny:* even King James's farthings were considered as tokens.

A look at the time of Richard I., in the novel of *Ivanhoe*, will illustrate what we have said. The knights ransom their armour in *zecchins:* the *zecchin* was an Italian gold coin, introduced by the Italian money-lenders. The readers of the *Arabian* Nights* know it as the *sequin*. The Jew and the Abbot ransom themselves from Robin Hood in *crowns*, French or Flemish (for there was no such English coin), and probably gold. The baron who proposes to torture the Jew out of a thousand pounds of silver, produces his scales, and demands Tower weight. He offers to take a mark of gold for each six pounds of silver, the regular terms of the day: this again was weight, for the *mark* was never a coin, at any time, but only two-thirds of a pound. So far we have no fault to find. But when Prince John offers the archer twenty nobles, we may ask where the Prince was to find either the name or the coin. When he proposes to fill the horn with "silver pennies," we can imagine the yeoman wondering what choice of pence his Highness could possibly have, except silver ones. When the Saxon peasant grumbles at the Jews for not flinging him "a mancus or two," he was more unreasonable than Sir Walter meant him to be: for the old Saxon mancus, mancusa, or manca, never was a coin, but only a money of account in the books, and a number of pence in payment; to say. nothing of the bearer of a letter complaining that the receiver did not pay him the price of a small flock of sheep for his trouble.

Many readers will be surprised to hear that the early Normans coined nothing higher than a penny: but they ought to be more surprised that the same kings coined nothing lower than a farthing. The price of a sheep was fourpence: at least this was the price at which the king's purveyors compounded for a sheep, when they demanded one; perhaps we may more safely put the market-price at sixpence. How should we get on in our day, if we had no coin smaller than would buy the twenty-fourth part of a sheep fit for the Queen's table? The probable explanation is that the lower orders had very little to do with money; they were serfs who were fed and clothed by their masters.

The *mark*, as we have said, never was a coin: and yet no name is more common in English monetary language. The prince's ransom, and the forester's bet upon his skill in archery, are equally in marks. A mark of silver might be counted in pence, or in foreign crowns, but the name was essentially descriptive of a weight.

* The coin of the original Arabic is the *dinar*, of nearly the same value as the Italian zecchino. This word *zecchino*, or *sequin*, is apparently not Italian, but Oriental: the original of it is a word meaning generally coin as distinguished from bullion, which we may spell *sicca*, since we know it under that spelling in the sicca rupee, &c.

Thus it is no surprise to read that the duke who made a captive of Richard I. received the ransom in marks, of which he offered a portion to the Cistercians to *make*—not to *buy*—censers for their chapel services: an offer which the Cistercians refused, in contempt for the man who had taken so unfair an advantage of a brother Crusader.

It seems to have been an early principle that the great "valuers" of money should never be coins. These valuers were the shilling, a weight of the twentieth of a pound; the mark, a weight of two-thirds of a pound; and the pound, which was at last our pound troy, but which at first was, perhaps, the Tower pound, three-quarters of an ounce less than the pound troy. This troy pound, as every one knows, is of twelve ounces, each ounce having twenty pence, each penny, latterly at least, twenty-four grains. We say *pence*, not *penny-weights*, as now, for the penny *was* a weight; and we find such expressions as eleven ounces twopence farthing, meaning eleven ounces two pennyweights and a quarter of our modern phraseology. The word *pound* has never been the name of a coin. The 20*s.* pieces of James and Charles were *laureats* and *caroluses:* that of George III. is a *sovereign.* To this day we think of the coin as a *sovereign*, of the debt it wipes out as a *pound.* In like manner the shilling was never anything but the twentieth of a pound weight of silver until Henry VII. coined a few shillings, and Henry VIII. circulated this coin extensively. But the name did not come in at once with the coin. That which is now the shilling was the *groat of twelve pence*, and the *testoon*, when first it appeared as a coin.

When the gold coinage was introduced, there was wisdom in not attempting to coin an equivalent to twenty shillings. Both gold and silver were in the standard currency, and both were legal tender to any amount. While the relative values of the two metals were changing, it would have been impossible to preserve any gold coin in a state of equivalence to twenty shillings of silver. Our present silver coins* are only tokens, and pass for more than their value, as compared with gold; care is taken that it shall never be worth while to melt the silver coin into bullion. A coin passing for something very near 20*s.* the pound of account, would have been a serious inconvenience, especially if it had fluctuated, being alternately

* A shilling only differs from a promise to pay a shilling in being of a more costly material than promises are usually written on. This costliness is the means of preventing forgery: the promise of government to receive the token at a shilling is signified on what is so nearly worth a shilling, that imitation could not be done at a sufficient profit. Thus the government will receive back its token, no matter how much worn or battered it may be, provided only so much of it hang together as to make it certain that no other part of it is circulating as another token. But gold circulates in another character, on its value as a commodity. Hence the necessity of government receiving it at value on its return to the Mint, and making a deduction for loss of weight. Many persons do not understand the distinction, and imagine that their loss upon the light gold is a hardship, because there is no loss on the light silver. Hardship or no hardship, the two cases are perfectly different. To make deduction for a light shilling would be much the same thing in the Mint, as it would be in the Bank to make deduction from a 5*l.* note on account of stains, or crumpling, or tearing. Our silver is not *money.* Nobody is bound to take 45*s.* of it in payment of 2*l.* 5*s.* It cannot be legally tendered in sums of more than 40*s.*; and its meaning is, that the Mint will pay a gold sovereign for every 20*s.* of it, just as the Bank will pay five sovereigns for every one of its written promises called five-pound notes.

above and below the pound of account. No wise legislature will ever knowingly introduce into monetary arithmetic the necessity of adjustments which are sometimes additive and sometimes subtractive. Astronomers, many at least, have learnt the convenience of keeping the observatory clocks slow enough to insure the corrections necessary for true time being *always additive*. If, as has been proposed by some, a tenpenny coin were introduced, for the purpose of assimilation to the French franc, now 9½*d.* and a fraction, such a fluctuation in the exchanges as should throw the franc alternately above and below the 10*d.*, would be the great arithmetical nuisance of its century, to all who should be concerned.

It appears from what has preceded, that up to the time of Edward III. the English currency was silver, all the gold in use *as coin* being foreign: that large sums of money were weighed, not counted, the pound of silver being the chief unit, though hardly better known than the mark, or two-thirds of a pound: that the coinage was entirely of pence, halfpence, and farthings; so that coinage was no more than a convenience for preventing the constant weighing of small sums. It is further said that the cross, which was always marked upon the reverse of the penny, was used to facilitate the fracture of the piece into four farthings; but it is a matter of discussion whether this was really done to any extent. The occurrence of the cross on the small money gave rise to the description of extreme poverty conveyed in "He has not a cross to bless himself with."

From a very early period, the English coinage was called *sterling*, a name by which it has been known throughout the world. It is also obvious that by this name came to be signified the superior soundness and good faith of the English coinage; from which, as an adjective, it has come to stand for that which can be fully depended on. The account of the origin of this word which the antiquaries prefer, is as follows:—That in or before the time of Richard I., money coined in the east of Germany began to be in request in England for its purity, and was called *Easterling* money, those Germans being called Easterlings; that shortly afterwards some of their best coiners were sent for into England, and that from that time the money was called Easterling and sterling. Another account derives the word sterling, esterling, or *starling* (as according to Spelman it was sometimes written), from the little stars which were frequently on the English penny, and which were almost universal in the pence of the time at which the English coinage began to be of high reputation. In our minds there is no doubt that the second is the true derivation, for which opinion we proceed to give reasons.

In the first place, the word *sterling* or *esterling* meant *a penny*, not coin in general, but the 240th part of a pound. By statutes of Henry III. and Edward I., the English penny, which is called a sterling (*qui vocatur sterlingus*), shall weigh &c. &c. By an ordinance of Henry II., his French subjects must pay two *pence* each, his English subjects one *sterling* each. By an ordinance of Henry III., every sterling which is under weight is to be melted.

Richard II. gives the King of Scotland, not *tria millia librarum sterlingarum*, 3,000 pounds sterling, but *tria millia librarum sterlingorum*, 3,000 pounds *of* sterlings. This same idiom occurs frequently, namely, a case inflexion which proves that *sterlingus* is a substantive, and not an adjective expressing quality. Nor was it an English substantive only. A Cologne ordinance prescribes that 20 sterlings shall weigh an ounce. David I. of Scotland, who reigned before coiners were ever said to have come from the east into England, ordains that the sterling shall weigh 32 grains of wheat. And David II. ordains that "our money, that is to say sterlings (*moneta nostra, videlicet sterlingi*), shall not be carried out of the kingdom." We might multiply testimonies of this kind: those we have cited are all in Ducange (at the word *Esterlingus*), with proper references, except the 3,000 *millia*, &c., which is from a charter in Hoveden. So late as Henry V., the word sterling meant a penny; for in the last year of Henry IV., there is a proclamation against Venetian pence, of which it is said that three or four are hardly equal to one sterling (Ruding). We may also note, that in becoming the name of a coin, the word sterling also became the name of a weight. An old explanation of the purity of the silver of the time of Edward I. (Camden), states that eleven ounces two easterlings and one ferling [farthing] should be silver, and the rest alloy. The next paragraph refers to this as eleven ounces and two pence ferling.

But if the word *sterling* mean the *little-star* coin, how comes it to be *esterling* in at least half the mentions of the word, and, specially, in all the French records without fail? How come the French to be more addicted than the English to that part of the word which, if the syllable mean *east*, does not belong to their own language? In this way, we think: the old French for star is *Estelle*, a form which they still keep as a female name, and from which they get the modern *étoile*. The *r* in *Esterlin* comes from the English, we suppose, but they do not always use it. In an old French ordinance it is said that *Chaque Estellin doit pezer* 3. *oboles Tournois* (Ducange). And the same writer cites an old French romance, in words which we leave the reader to decipher:—

Més je ne suie mie venus
En cest pais ô tant escus,
Et pour ses *estellins* recevoir,
Més pour la terre tout avoir.

From the time of the Saxons it had been customary to put on the reverse of the penny small birds, fleur-de-lys, studs, annulets, stars, &c. The same usage also prevailed elsewhere. The penny with simple round studs or buttons, which might be looked upon as stars, became exceedingly common about the time of Henry II. and Richard I. The great reformer of the coinage was Henry II.: "this king," says Leake, "seems to have been the first, from the Conquest, that made any considerable regulations for money affairs." One of his pence has stars before the bust of the sovereign. We cannot pretend to determine what particular stars, and whether on

A 3

head or *tail*, they were which brought the term *sterling* into use: nor even whether, as we think likely, the term is older English than has been supposed, as it certainly was older Scotch. What seems to us certain is this, that the word is derived from *star*, and not from *eastern*. This has been the opinion of a minority among the writers, Polydore Vergil, for instance, in older times, and Bishop Nicolson in more modern times.

It is certain that the Germans were called Easterlings and Osterlings. It is certain also that one German, Otho, surnamed *Cuneator* or *The Coiner*, was largely employed in England. We have no evidence of any other person from the east being so employed. Bishop Nicolson affirms that the Prussians and other neighbours of the Baltic who are the nations described as Easterlings, had not themselves any silver coin till the 13th century, a period too late for the hypothesis; for which assertion he cites what seems good authority. It is plain enough that the confusion between the two meanings of *esterling* is of an early date; an old monk of St. Edmondsbury (Ruding) defends the common idea by reminding his readers that, in later times, certain gold coins, though English, were called *Florentines* (shortened into *florins*), because the coiner employed was an Italian from Florence. This old monk, Walter de Pinchbeck, may have been the originator of the theory, and the Florentine derivation may have been the means of suggestion.

In all matters of probable reasoning, but especially in etymology, it is to be remembered that a fair percentage of results are such as become, in course of time, highly improbable. The thinker who, as he ought to do, always adopts the most probable conclusion, must make up his mind to be wrong in a certain minority of his cases. A test for determining the circumstances under which the improbable is to be preferred to the probable would be worth, to the human race, the next ten great results of science or observation, even though some things as great as vaccination, railroads, and telegraphs, were to be among them. We can imagine that, a hundred years hence, the *buss*, as it will then be universally called, which runs along the streets, will be undoubtedly set down as the old English word *buss* (from the same root as *box*), meaning any large vehicle, as in the *boat* (a kindred word) called the herring-*buss*. Who would hear without laughter the proposal to deduce the word from the last syllable of *omnibus*, a term derived from carriages open to all, introduced in 1829 from France? What! people would say, do you mean that public carriages were not open *to all* before these long boxes came into use? Here truth would have to walk off under a cloud of improbability, as she is often obliged to do. The derivation of *esterling*, which we here reject, is not rejected because it is improbable—for we consider it probable enough *à priori*—but because we think we see direct evidence in favour of another derivation.

The English coinage, subsequent to the Norman invasion, may be divided into four stages. In the first, lasting till Edward III., and of which we have already spoken, large sums were paid by weight, and small coins served for small transactions. In this stage silver was the only national coinage, of this no coin was higher than the penny, and all the gold in use, as coin, was from abroad. Gold went by weight, at nine times the value of silver. In the second stage, lasting until the time of Henry VIII., a gold national coinage was gradually introduced, and going down, as we shall see, to rather small sums; the groat of four pence was introduced, and remained throughout the period the highest silver coin. In the third stage, lasting till the time when Newton ruled the Mint, gold coins were greatly multiplied, and higher silver was coined. Gold of much antiquity, and very much clipped, was current at the end of this period, rather by weight than by nominal value. In the fourth period, lasting till our own time, the variety of coins has been considerably diminished, and much approach has been made towards the simplicity of a good system. The establishment of one standard, gold, first by the practical effect of Mint regulations, and afterwards by law, has abolished the greatest source of the confusion of old times, and has turned the idea of money which does not pass for its nominal value into a tradition, or else into a notion attached to paper promises not convertible into gold on demand.

The attempt at a double coinage, with two distinct standards, which lasted till nearly our own time, resulted in a never-ending variation of the value of gold as compared with silver. To this must be joined the gradual depreciation of the silver coin. The pennies of William I. to Henry III. give 1*l.* of money in a pound of silver; the shillings of our day, with the same alloy, give 6*l.* 6s. to the pound of silver; and to this we have come gradually. As to the gold, even after allowing for the changes requisite to meet the depreciation of the silver, there is a large variation, due partly to fluctuations of relative value, partly to the exigencies of the times. The florins of Edward III. gave 15*l.* of money to the pound of gold; the nobles of Edward IV. gave 22*l.* 10s. Henry VIII. coined, at different periods, sovereigns of 27*l.*, 28*l.* 16s., and 30*l.*, to the pound of gold; Edward VI. of 34*l.*, 28*l.* 16s., 36*l.*, 33*l.* The gold coins were in circulation at odd farthings a piece; and these values in a state of fluctuation. The consequence must have been, a perpetual adjustment of transactions to the state of the money. On the day we write this, we see in *Notes and Queries* an extract from church-wardens' accounts of 1551, in which they twice credit themselves "in allowance at the fall of the money." In many ways it appears that money was a *commodity*, to be valued like other commodities, and not a measure of value, of which the receiver only inquires whether it be *good*, that is, really issued by the government.

The work on Arithmetic of Robert Recorde, first published in 1540, about the end of our second period, gives the following as the state of the actual coinage. It is taken from the edition of 1561, the earliest we ever met with, and John Dee, the editor, does not think it necessary to make any remark. The spelling is here modernized.

Gold Coins.

	£.	s.	d.		s.	d.
Sovereign	1	2	6	Half Angel	3	9
Half Sovereign . .	0	11	3	George Noble	6	8
Royal	0	11	3	Half George Noble . . .	3	4
Half Royal	0	5	7½	Quarter Noble	1	8
Quarter Royal. . .	0	2	9¾	Crown	5	0
Old Noble (Henry) .	0	10	0	Half Crown	2	6
Half Old Noble . .	0	5	0	Another Crown. . . .	4	6
Angel	0	7	6			

The only difference between the two crowns lay in the first having a rose over the crown, while the second had four fleur-de-lys round it.

Silver Coins.

	d.	
Groat	4	Penny.
Harp Groat	3	Halfpenny.
Penny of 2 pence .	2	Farthing.
Dandiprat	1½	

The farthing was only distinguishable from the halfpenny, to many persons, by a cross on one side and a portcullis on the other.

In his edition of 1573, John Dee repeats this table as that of 1540, and says that the coins of 1570 are very different, meaning that various additions and subtractions had been made. He announces a new table at the end of the book, but he either forgot it, or, which is more likely, he found it a harder task than he had supposed to collect and value all the coins in use. Even Recorde himself only professes to give the coins which were most common. The remaining arithmetical authors of the sixteenth century do not attempt a list of coins.

In the editions following 1630, the editors have given a list of the most usual gold coins, and of the silver coins, agreeably to the proclamation of November 23, 1611. The gold is as in the following table, and the silver is described as consisting of the Edward crown, half-crown, shilling, half-shilling, and three-pence; the Philip and Mary shilling and half-shilling; the Mary groat and two-pence; the Elizabeth shilling, nine-pence, six-pence, four-pence, three-pence, two-pence, penny, three-farthings, and halfpenny.

Usual Gold Coins, 1630.

	£.	s.	d.		£.	s.	d.
Great Sovereign . .	1	13	0	2 parts (thirds) of do.	0	4	7
Double do. (Henry) .	1	2	0	George Noble . . .	0	9	9½
Do. do. (Eliz.) .	1	2	0	Half Noble . . .	0	4	11¼
Royal	0	16	6	First Crown (Henry) .	0	6	11½
Half Royal	0	8	3	Base Crown Henry) .	0	5	6
Old Noble	0	14	8	Sovereign (Henry, best)	0	11	8¾
Half Noble	0	7	4	Sovereign (Henry) .	0	11	0
Angel	0	11	0	Edward Sovereign .	0	11	0
Half Angel	0	5	6	Elizabeth Sovereign .	0	11	0
Salute	0	6	11½	Elizabeth Crown . .	0	5	6

	£.	s.	d.		£.	s.	d.
Half Crown . . .	0	2	9	Cross Dagger . . .	0	11	0
Unit	1	2	0	Half Dagger . . .	0	5	6
Double Crown . .	0	11	0	Rose Royal . . .	1	13	0
Britain Crown. . .	0	5	6	Spur Royal . . .	0	16	0
Thistle Crown. . .	0	4	$4\frac{3}{4}$	Angel	0	11	0
Half Crown . . .	0	2	9	Half Angel . . .	0	5	6

It will easily appear that the counting of money, even in 1540, must have been beyond the power of most, by *our* usual process of taking coin after coin in the hand. In 1540, a parcel of half and quarter royals, half-sovereigns, &c., with some of each of the two kinds of crowns, would have been a payment requiring writing to make up the total. Whether we take the first period, the second, or the third, we see the reason of the old law which requires the tenant to tender his rent to the landlord at such time before sunset as would leave time to count the money by daylight.

It is evident that by this time our coinage had assumed the character which it now bears; namely, that the gold part of it very much exceeded the silver in value. We must remark the smallness at which the gold coin had arrived: and the value was still further lowered before the end of the century, by wear and by clipping. Small gold coin is an expensive article. It presents a larger surface in proportion to its bulk than the larger coins, and has a much greater wear and tear. Our half-sovereign is a costly convenience. No gold coin has ever given long satisfaction, since the *mechanics* of coinage has been understood, which is either very much greater or very much less than the sovereign which we have now.

The confusion of the gold coinage was increased by wear, by fraud, by new coins, and by the lowering of the silver, until it reached its highest point in the last quarter of the seventeenth century. The silver coinage had by that time arrived at a deplorable pitch of badness. The great reformation projected by Charles Montague, afterwards Lord Halifax, and carried into effect under his administration by Isaac Newton, was completed in 1699. The history of this magnificent operation has not been properly written by the historians of the coinage, whose attention has been too much engrossed by questions of *head* and *tail*. Even the records of the Mint are far from sufficient on this important point. We know that the main body of the imperfect gold was called in, though some of the old caroluses and jacobuses were in circulation for nearly fifty years afterwards; we know also that the silver was thoroughly and radically reformed. Had Newton possessed less fame in other things, his biographers would have given us a little more account of an operation which would have been a lasting reputation to any one else; for private biography frequently amplifies public records. But it has been Newton's fate to have his career in the public service not only thrown into the shade, but its origin attributed to any cause but the right one. With some, it was attributed to Montague's regard for Newton's beautiful niece, Catherine Barton, whom it is hardly possible Montague could have set eyes on when Newton received his first appointment. With others, it is a reward for scientific services, and an acknowledgment of genius almost divine. There has always been a school of British *philosophers* who ecstasize when they think of the chief of their body gaining an office of 600*l.* a-year, accomplishing a harder task than had ever been laid upon any holder of the same office, rising in consequence to a higher post of 1500*l.*, and finally gaining a knighthood! With all this science had nothing to do, further than common arithmetic and common chemistry. Montague had sat with Newton in the Convention Parliament, and had had ample means of testing his friend's qualities as a man of business. The operation he projected on the coinage was one of unheard-of difficulty, and many prophecied that it would fail altogether, more, that it would ruin commerce. He therefore secured the strongest scientific assistance he could find, and enlisted Locke, Halley, and Newton in the cause. He could not have stood the risk of any incompetent supporters; for his administrative power depended neither on wealth nor influence, and, though of noble family, he was long called and considered the *parvenu* of political power. And he tried Newton's price like a prudent minister. The hook was first baited with the third-rate office of comptroller, and when it was found this would not do, with the second-rate office of warden.

The coinage had reached the state of complication now to be described by the time of the Restoration. At this period it was all augmented by proclamation. The following table was collected by William Jeake in 1674, in his "Λογιστκηλογία, or Arithmetick Surveighed and Reviewed," which was published by his son in 1696, the year of Newton's appointment. It was copied into Harris's "Lexicon Technicium" in 1704, which shows that the necessity for such a table was not then quite obsolete.

This table shows the values of the coins in 1640, and also after the proclamation. Even Jeake does not profess to give all the gold coins, but only most of them. Perhaps the remainder consisted mostly of those foreign gold coins which were settled by another proclamation of the same year.

Gold Coins.

	1640			1660		
	£.	*s.*	*d.*	£.	*s.*	*d.*
Old Double Rose Noble	1	16	4	1	18	8
Double do. H8, E6, PM. El.	1	16	0	1	18	4
Great Sovereign J. Double Rose Noble J.	1	13	0	1	15	3
Double Rose Royal or Real	1	10	0	1	12	0
Double Old Sovereign	1	6	8	1	8	5
Best Double Sovereign H. Double Sovereign E6, El.	1	3	10	1	5	5
Double Sovereign (Jacobus)	1	2	0	1	3	10
Laureat or 20*s.* piece J. 20*s.* Piece of C1	1	0	0	1	1	4
Old Rose Noble	0	18	2	0	19	4
Spur Royal H8, E6, PM. El.	0	18	0	0	19	2
Spur Royal J.	0	16	6	0	17	7
Double Noble El. Old Noble H.	0	16	0	0	17	1

	1610 £. s. d.	1660 £. s. d.
Rose Royal	0 15 0	0 16 0
Old Sovereign	0 13 4	0 14 2
Best Sovereign H. / Sovereign E 6, El.	0 11 11	0 12 8
Old Angel Noble H.	0 12 1	0 12 10
Last Angel Noble H 8, E 6, PM. El. / First Angel J.	0 11 11	0 12 8
Sovereign J. (Double Britain Crown)	0 11 0	0 11 9
George Noble	0 10 10	0 11 6
Last Angel J.	0 11 0	0 11 9
Half Laureat J.	0 10 0	0 10 8
10*s.* Piece C 1	0 10 0	0 10 8
Angel C 1	0 10 0	0 10 8
Half-Spur Royal	0 9 0	0 9 7
First Crown H.	0 8 0	0 8 5
Single Noble El. / Half Old Noble	0 8 0	0 8 6
Salute	0 7 11	0 8 5
Base Crown H., Rose Crown / Crown E 6, El	0 5 11	0 6 4
Half Angel Noble	0 6 0	0 6 5
Half Last Angel H. / Half Angel E 6, PM. El. / Half First Angel J.	0 5 11	0 6 4
Britain Crown J.	0 5 6	0 5 10
Half George Noble	0 5 5	0 5 9
Half Last Angel J.	0 5 6	0 5 10
New Crown J. / Crown C 1	0 5 0	0 5 4
Two parts [thirds] of Salute	0 5 3	0 5 7
Half First Crown H.	0 4 0	0 4 2
Half Salute	0 3 11	0 4 2
Half Rose Crown / Half Crown E 6, El.	0 2 11	0 3 2
Quarter Old Angel Noble	0 3 0	0 3 2
Quarter Last Angel H. / Quarter Angel E 6, PM. El. / Quarter First Angel J.	0 2 11	0 3 2
Half Britain Crown J. / Quarter Last Angel J.	0 2 9	0 2 11

Silver Coins.

	s. d.
Crown E 6, El. J., C 1, C 2	5 0
Half Crown do. do.	2 6
Shilling E6, PM., El., J., C 1, C 2	1 0
Sixpence do. do.	0 6
Groat, Old H 8, last H 8, M. El. C 1	0 4
Threepence El. C 1	0 3
Twopence H 8, El., J., C 1, C 2	0 2
Three Halfpence El.	0 1½
Penny H 8, E 6, M., El., J., C 1, C 2	0 1
Three Farthings El.	0 0¾
Halfpenny El., J., C 1, C 2	0 0½

We have given these lists at length because we cannot find in the annalists of the coinage any tables of coins current at one time. It would be difficult to collect such tables from contemporary historical documents: and probably it did not strike the writers to whom we have alluded that books of arithmetic are the proper quarters in which to search. There are curious indications of the absence of communion between the numismatists and the arithmeticians. Camden states that Henry VIII. is said to have coined a piece called a *dandiprat*, but he does not know its value: Leake, and Ruding (we believe, but we do not speak so confidently about Ruding's three quartos as about Leake's single octavo) are content to give the same information, without any knowledge of the value of the coin. Our reader has seen it, as three half-pence, in Recorde's list: he will also notice that this coin disappears in the list of 1630, probably omitted by mistake, and reappears again, and as a coin of Elizabeth, in the list of 1674.

The reader will naturally ask what the new coinage was, which was started by Newton, and which must be presumed to be the basis of our present coinage. On this history says nothing, and the records of the Mint emphatically say nothing, for they are complete enough on every other point. The truth is to be inferred from lists of coins. There was no new coinage, that is, there was no new invention of pieces. As to the current work of the Mint, a sufficient reform had long taken place. From the time of Charles II. the issue of gold pieces had consisted of five-pound pieces, double guineas, guineas, and half guineas: that of silver pieces had consisted of crowns, half-crowns, shillings, sixpences, groats, threepences, twopences, and pence. This issue was continued; and the only piece which Newton added was the quarter guinea of 1718, which was soon found inconveniently small, though it remained till the last quarter of the century. A gold coin passing at an odd number of pence over shillings must have been a great inconvenience.

We suppose that the five-pound pieces afterwards passed at five guineas. Various foreign gold coins, principally Portuguese, called *Portugal pieces*, came into use in the last century, so that the list of gold pieces in and about 1770 will remind us of the time of Robert Recorde. It is as follows, the less usual coins being in italics:—

	s.	d.		£.	s.	d.
Sixteenth of a Moidore	1	8¼	Guinea	1	1	0
Eighth of a Moidore	3	4½	Moidore	1	7	0
Eighth of a Johannes	4	6	Johannes	1	16	0
Quarter Guinea	5	3	*Two-Guinea Piece*	2	2	0
Quarter Moidore	6	9	*2½-Moidore Piece*	3	7	6
Quarter Johannes	9	0	*Double Johannes*	3	12	0
Half Guinea	10	6	*Five-Guinea Piece*	5	5	0
Half Moidore	13	6	*Five-Moidore Piece*	6	15	0
Half Johannes	18	0				

The recoinage of silver completed in 1699 amounted to nearly seven millions of money. It was probably not a very good specimen of the art, and it was done under a high pressure of necessity. The coin was soon worn on the faces, and bent. John Philips wrote his poem, *The Splendid Shilling*, in 1703, while the new coinage was yet sound and bright. Branston's imitation, *The Crooked Sixpence*, was written some years before 1744, and the title indicates that a fair proportion of the sixpences had become bent. At the beginning of the reign of George III. the shillings and sixpences had become, in everything but their extreme thinness, those little white plates which so many can remember to have circulated in 1815. But at that period the silver coin had become even more of a token than it was in 1699: and the extension and uniformity of the gold coinage, combined with the use of paper-money, and the addition of the mechanism of the Bank of England to that of the Mint, had placed the monetary system of the country on a footing of convenience wholly unknown at the Revolution. Still, the recoinage of 1816 was a large measure, requiring a great deal of administrative forethought.

This recoinage of 1816 brought our system, except only as disturbed by circulation of foreign pieces, to its present state of simplicity, by the substitution of the sovereign for the guinea. It is very much to be doubted whether either the Government or the nation would listen to any scheme of decimal coinage which requires so great an operation as the remodelling of all the silver at once. Fortunately, though not so generally understood as it ought to be, the scheme which divides the pound into 1000 new farthings, or mils, requires no immediate recall of any portion whatsoever of the coinage. If, by Act of Parliament (which would probably be thought necessary, though the Royal prerogative extends so far by itself) the half-shilling were made to contain twenty-five* farthings instead of twenty-four, the whole of the system would be established, and the commercial classes, and all who keep books, might be safely trusted to find their way to reckoning in pounds, tenths of pounds or florins, hundredths of pounds or cents of ten farthings each, and thousandths of pounds or new farthings, or mils, if that name be preferred. The calling in of the crowns, half-crowns, fourpenny and threepenny pieces need not take place: it would be enough that those which come in should not be allowed to go out again. The half-crowns, in particular, come into the Bank in enormous numbers while the harvest is growing, and go out again as the time† of reaping approaches: they might be arrested, and florins and sixpences might take their places. If a little exertion were made to help in the coins which are

* This article being entirely on coinage, we say nothing about the injustice of damaging those who have six pence of copper money to the extent of a farthing once in their lives. Many persons have a horror of this spoliation. We heard a minister say as follows to a deputation:—"When a poor man has six pence in his pocket, he will have to say to himself, 'Now I have to go and work before I can make this into a silver sixpence.'" We doubt whether there be in the whole of Great Britain a labouring man who ever imagined to himself a farthing's worth of work, or took the trouble to form a notion of the time it would last.

† The reason is that a reaper's wages are almost always half a crown a-day. The coin is then very convenient for payment, but it must have a tendency to cause a very great want of smaller change. A paymaster, with a box of florins and another of sixpences, might pay half crowns nearly as easily as if he had single pieces; and the superior convenience of the mode of payment would be manifest the moment the money had changed hands.

not to go out again, it would be so much the better: but this is not a matter of primary necessity. The issue of silver cents (of ten farthings or mils each) might take place at leisure: and the same with such other coins as should be judged desirable.

The two schemes which have found advocates, in opposition to the retention of the present pound and florin—for which an enormous preponderance of opinion has been declared—are based upon the farthing and the penny. The effects which these systems would have upon the coinage are as follows:—

The farthing system would require coins of 10, 100, and 1000 farthings; a cent of 2½*d.* present money, a florin of 2*s.* 1*d.* present money, and a pound of 1*l.* 0*s.* 10*d.* present money. Either a complete alteration of all the gold and silver coinage must take place, or the *decimal coinage* cannot exist, and *decimal computation* will be but an option, such as people have now if they choose. For there is nothing to hinder any one from using what mode of keeping books he pleases, provided only that he take the trouble to translate his receipts into his own system for himself, and his payments into the common system for his neighbour. But decimal reckoning will never prevail against non-decimal currency.

The penny system would require a small coin of one-tenth of a penny; though its advocates seem to incline towards preserving the farthing for the poorer classes, and making the decimalization ascend upwards from the penny. This would require a coin of 10*d.*, say a franc, and one of 100*d.* or 8*s.* 4*d.*, say an imperial. The inconvenience of such a coin as the imperial, decidedly as much too large for silver as it is too small for gold, strikes every one who approaches this subject. This penny system, like the farthing one, requires the complete alteration of all our coinage, both gold and silver, except the half-crown, which becomes a three-franc piece, and the crown, the reproach of our present coinage, which becomes a six-franc piece.

We have seen, in ancient times, such a confusion as that of two small gold crowns, of the same size and different values, distinguishable only by the difference between a rose and a fleur-de-lys. We have seen this same kind of confusion carried to a much greater extent, insomuch that there existed at one time, of English coins only, more than fifty in circulation, many of them undistinguishable in size and of different values. We do not read that this complexity gave rise to any complaint; it had come on gradually, people were used to it, and probably thought that the time and trouble which their money cost them were necessarily incidental to money affairs. If any thinking man had proposed simplicity for the sake of saving time and trouble, who can doubt that he would have been a person not used to the details of business, not a practical man, and so forth. Any banker would readily admit that the monetary system of 1674 would, if it existed in our day, require three times as many clerks as are now wanted. And yet, in such accounts as we have seen of the debate on the coinage, when Montague (afterwards Lord Halifax) proposed his measure, there is not a word about the multifarious character of the coins, and the difficulties thence arising. If all the packet of coinage which Jeake has recorded, had been sound and unclipped, we have no reason to suppose that a voice would have been raised against it.

The state of things now existing is very different. The introduction of the florin, by the side of the half-crown, has been the subject of much remark, as to the liability of confusion, though there is some distinction in every point to which people must look. The introduction of the threepenny piece by the side of the fourpenny, involving the trouble of looking at the edge of the coin, has been considered a grievance by the bankers, and justly, in comparison with the distinctions between other coins. It is hardly to be supposed that much of this backward progress would be tolerated: and yet we are not without propositions the adoption of which would make the whole of our coinage, for a good many years at least, an imitation of that which formerly existed.

It has been proposed that, the sovereign remaining as now, a gold piece of the value of tenpence more should circulate with it. This would be a luxury in a dark London morning; for 1*l.* and 1*l.* 0*s.* 10*d.* in gold would be undistinguishable by size, and the bankers' clerk would need to read every piece of gold he receives. Unless, indeed, the new coin were made thinner and broader, which would very much increase its wear. Again, it has been proposed that the coins of the penny system, the franc (10*d.*) and the gold imperial (8*s.* 4*d.*) should circulate with our present coinage. The difficulties of distinction would here be thrown upon the franc and the shilling, as well as upon the imperial and the half-sovereign. But there would be a greater inconvenience even than this. Suppose a person were to hand over in payment 2 sovereigns, 3 half-sovereigns, 5 imperials, 6 shillings, and 7 francs. The receiver would have to cast this up in his own way. There is an option in this coinage: a person may either use the present system, or the decimal system of the penny. One party might be expecting 6*l.* 3*s.* 6*d.*, the other might bring it in as 14 imperials, 8 francs, and 2 pence. Against such a mixture it seems to us needless to argue.

The proposers of the farthing and penny systems seem to have started with the impression that their opponents, the advocates of the scheme approved by the House of Commons, the Bank of England, &c., intended an alteration of all the silver coinage. Hence they were under no difficulty in proposing an alteration of their own. When they came to understand how easily the existing silver coinage may remain under the system voted by the House, all desirable alterations being made at perfect leisure, and in convenient time, they were under the necessity of contriving the adaptation of a similar convenience to their own systems. Hence arose their plan of two different and concurrent coinages; a plan which we are satisfied will, independently of other disadvantages, drive their proposals out of the field the moment they come to be examined in detail by the side of the plan which divides the pound into 1000 new farthings.

Since the time when gold and silver passed by weight as well as by coinage, the two, and only two, actions upon the whole of the coinage at once, have been adopted under the spur of necessity.

In the time of William III., the gold had ceased to represent anything like its value, from clipping and wear. It is said that five pounds of coined money was intrinsically scarce worth forty shillings. As to the silver, it was common to sell bullion for 6s. 3*d.* an ounce of the common coin, and for 5s. 2*d.* of good and new coin; so that even the silver coinage passed at varieties of value. And yet this evil had been endured for many years, and no minister had dared to face it. In the reign of George III., it may almost be supposed that the government, not daring to attempt a complete recoinage, would not venture to put out any new shillings, &c., which might bring the old ones into discredit. Small silver was struck in 1763, 1764, 1775, 1778, and 1787; but not enough to be put into general circulation. The consequence was, that *honest* illegal coining made a very good livelihood: little silver plates, exactly resembling government shillings and sixpences, and quite as valuable, were supplied by "private enterprise." It is stated that the value of the counterfeit shillings was 8½*d.*; the genuine shillings, as worn, could have been worth but little more. The continuer of Leake says, in 1793, that "few, very few indeed, of the shillings and sixpences now in use, appear ever to have been legally coined." The coiners evaded the law by putting a faint head of William III. on one side, and leaving the other quite plain; representing the coins, when detected, as buttons not yet shanked; and some actually escaped on this plea. From 1751 to 1793, at least, there was no coinage of crowns or half-crowns. Nothing was done to mend all this till the peace of 1815.

It may fairly be inferred that any establishment of a decimal coinage which contemplates a grand measure of recoinage, to be executed at one time, has no chance of success whatever. All the plans *now* recognise this: all the proposers declare that the existing coinage is to continue, until gradually absorbed by the Mint. But in all the plans, except that which adopts the pound and divides it into 1000 parts, this declaration is, in everything except intention, a subterfuge and an evasion. A system which invites us to begin reckoning in sums of ¼*d.*, 2¼*d.*, 2s. 1*d.*, and 1*l.* 0s. 10*d.*, while our present coins exist, and while we are watching the gradual introduction of coins representing these sums, concurrently with the coins which now exist, will, we are satisfied, invite to nothing but laughter, when it becomes generally understood. The same may be said of the system which proposes 10*d.* and 8s. 4*d.* on the same terms. And any system which only gives the option of decimal reckoning, and leaves the power of continuance in our present reckoning, will assuredly never succeed in introducing decimal reckoning at all.

We are now speaking of *coinage*, and therefore have only to compare with others our system in this one respect, referring to our number for 1854 for its description, &c. The farthing, halfpenny, and penny, might remain in circulation, at 4 per cent. under their present value, or at 25 farthings to the half-shilling. The principal new coin required would be the *cent.*, ten new farthings, the hundredth of a pound. But this coin need not be fully introduced at the moment of the change; it would be enough that reasonable diligence, such as would not seriously inconvenience the Mint, should be used to issue it, and to call in the silver threepences and fourpences, which would, in the meantime, pass for 12 and 16 new farthings. The recall of the half-crowns, and the circulation of florins, might proceed with deliberate steadiness.

Among the ancient difficulties of coinage was that of producing a perfectly round figure. The piece was beaten out to the proper size before it was punched. Hence the outer ring of the design could not come exactly upon the edge of the coin, or at equal distances from it throughout, so that there was an external space of which no one could say how large it ought to be, or how large it had been. The consequence was that clipping the coin was a trade, which many severe laws were passed to prevent. A piece of money was considered of good reputation, if it were not clipped *within the ring*; and various laws designate this expressly as the limit up to which the coin was legal tender. The coining mill, invented in France about 1553, was used in England in 1561, but only for a short time. Some partial revival of its use occurred, but it did not become a permanent instrument till 1662. The milled money was the first in which equal thickness was secured by coining from plates drawn out between rollers; and also the first in which letters were put on the edge, or in which what we still call a *milled edge* was produced. This put a stop to outrageous clipping, and transferred that name, which still remained in Acts of Parliament, to the less violent process of filing; while at the same time it introduced the process of *sweating*. It is said that, about 1770, filing of English gold went on in open day in Parisian workshops, at a gain of two shillings on the guinea. At the same time, both silver and copper were coined too high; so that much of the new coin found its way to the melting-pot before the shine had left its face.

The combined action of the Mint and the Bank has supplied a heart to the circulation of coin; and better knowledge of the conditions of health, in conjunction with the stimulating power of regulated paper currency, has prevented the alternate overactions and underactions of this organ. We have but one more great change to look for, the establishment of a coinage which will naturally lead to a decimal mode of reckoning: and it is fortunate that this can be done without any greater change than trivial and gradual additions to and withdrawals from the silver, a change of only four per cent. in the copper, and no change of any kind in the gold.

University College, London, A. DE MORGAN.
October 20, 1855.

COMPANION TO THE ALMANAC

FOR

1857.

PART I.

GENERAL INFORMATION ON SUBJECTS OF MATHEMATICS, NATURAL PHILOSOPHY AND HISTORY, CHRONOLOGY, GEOGRAPHY, STATISTICS, &c.

I.—NOTES ON THE STATE OF THE DECIMAL COINAGE QUESTION.

WE commence this article by a few additions* to our last year's notes on the history of the coinage.

We were not correct in implying that gold and silver were both legal tender, without interruption, up to 1816. From the first coinage of gold up to 1664, the relative values of gold and silver coins were settled by proclamation; and these proclamations had the force of law in making gold a legal tender. But from 1664 to 1717, the value of the gold was not thus settled, and silver only was legal tender. There is a letter from Dryden to Tonson the publisher, written in or about 1684, in which he says, "I expect forty pounds in good silver, not such as I had formerly. I am not obliged to take gold, neither will I, nor stay for it above four-and-twenty hours after it is due." In 1717 the guinea was declared a legal tender for 21 shillings. This was a little over its relative value, so that, though silver to any amount was still a legal tender, it became customary to pay in gold. It was not until 1816 that silver ceased to be a legal tender for more than forty shillings.

We have said that by 1600, or thereabouts, the gold coinage appears to have exceeded the silver in value. This is not an easy statement either to verify or to refute, as there are no accounts worth credit of the amounts in circulation. The most credible estimates state that, at the Revolution of 1688, the silver in circulation was nearly double of the gold in value. But it is known that, about 1611, an immense influx of silver caused the exportation of the gold to such an extent that for years there was little use of gold. In consequence of this, the gold coins were raised ten per cent. by proclamation, and this brought back the gold so rapidly, and caused so large an exportation

* The following misprints should be corrected. The writer on coinage (p. 5) is *Rogers* Ruding (not *Roger*). The charter (p. 9) is of Richard I., not Richard II. The pound of silver is coined (p. 11) into 66*s*., not 6*l*. 6*s*. Our present coinage is not decimal.

of silver, that the silver, it is said, became as scarce as the gold had been. Such accounts are probably exaggerations; but they seem to us to justify a suspicion that when the balance was finally restored, the gold and silver in circulation were not very different in value. The civil war must have been favourable to the increase of the relative amount of the silver in circulation; not merely because gold is more conveniently hoarded and carried out of the country, but also because of the quantities of household plate which were given up for coinage by both sides. Some estimates cited by Leake (apparently from anonymous writers) make the silver, just before the great recoinage, to be about eleven millions, and the gold no less than eighteen millions and a half.

Since our last article appeared, the Royal Commission for examining into and reporting upon the question of the alteration of the coinage has commenced its inquiries. This Commission consists of Lord Monteagle, Lord Overstone, and Mr. Hubbard. It was appointed in the autumn of 1854, soon after the vote of the House of Commons in favour of the *pound and mil* system which had been recommended by the Committee of the House in the preceding year. In the inquiry before the House of Commons, all the witnesses examined (with one exception) saw no difficulty in the simple decimalisation of the existing pound sterling. They were as follows: —Mr. T. Hankey, Mr. Laurie (who afterwards changed his opinion), Sir C. Pasley, Mr. Airy (the Astronomer-Royal), Sir John Herschel (then Master of the Mint), Mr. De Morgan, Mr. Strugnell (grocer), Mr. Bevan (banker), Mr. Lindsey (grocer), Mr. Meeking (draper), Mr. Arbuthnot (of the Treasury), Mr. Wm. Brown (M.P. and merchant), the Duke of Leinster, Mr. Miller (Bank of England), Mr. H. Taylor (clerk to Whitbread and Co.), Mr. Rowland Hill, Mr. Bennoch (commission warehouseman), Mr. Beard (architect), Mr. Bazley (President of the Manchester Chamber of Commerce), Mr. Dowie (of the Liverpool Chamber of Commerce). Mr. Kirkham (clerk in Liverpool), Mr. Gregory (engineer), Mr. Franklin (accountant), Sir John Bowring. Mr. Headlam, M.P., saw great difficulty in retaining the pound, and proposed that the farthing should remain unaltered. Such a preponderance of witnesses, of all kinds* of occupation, of whom the great majority are strongly in favour of retaining the pound, while the remainder make no difficulty and propose no other plan, laid the Committee open to the charge of selecting their witnesses: though it is known that the same Committee, when it found the evidence going all one way, invited the attendance of persons known or supposed to hold other opinions; but the invitation was not accepted. The Royal Commission naturally made it the first business to inquire into the opinions of those who, in the interval, had written against the proposal of the Committee of the House. It is understood that, during the last spring, six gentlemen who had written in favour of the preservation of the

* It is not uncommon to affirm that the pound and mil system is supported chiefly by *mathematicians from Cambridge*. It is a characteristic of our country that *any* name which indicates knowledge and study can be effectively used as a term of contempt in political discussions.

penny, underwent examination; and it is understood further that one of these gentlemen stated that he had altered his opinion, that he had become in favour of letting things remain as they are, and that he believed very few persons cared about the matter.

Since the Report of the Committee of the House of Commons two associations have been formed, and, so far as we know, not more than two. The first is the *Decimal Association*, united in favour of the House of Commons' plan, or the pound and mil scheme. This body owes its existence mainly to the exertions of Mr. William Brown, M.P. for South Lancashire, the mover of the parliamentary committee. Mr. Brown, as a merchant trading with America, has long been struck by the very great advantage which a decimal coinage gives to the United States. The Decimal Association commenced its proceedings in June, 1854, then having on its list about 230 members of the Lower House, some peers, many influential London merchants, and many mayors and other municipal officers of commercial towns. It has exerted itself in the circulation of pamphlets on the subject, and it contributed greatly to the success of Mr. Brown's resolutions in the debate of 1855.

From the very first the association declared itself in favour of retaining the pound. The following is one of the resolutions carried at its first meeting, June 12, 1854:—"That the magnitude of our mercantile and money transactions, the vast amount of our revenue and banking accounts, and the fixed ideas of income and position which attach to the pound, render it expedient to adopt it as the unit coin of account." We quote these words because, even in the House of Commons, it has been stated that the association was framed, and its funds subscribed, only for the promotion of *decimal coinage* generally; and that the members of its council, in giving an undivided support to the pound and mil scheme, had run counter to the intentions of their constituents.

The other association, calling itself the *International Association*, is the British branch of a society formed at Paris in 1855, during the *Exposition Universelle*. It has for its object to obtain the establishment of common weights, measures, and coinage, throughout Europe at least, and, if possible, throughout the whole world. At the first formation of this society it was thought that the money question offers difficulties which are not incident to the questions of weights and measures: accordingly, in the resolution which forms the basis of the association, it is laid down that *absolute* uniformity in weights and measures shall be aimed at, and all *possible* uniformity in coins.

A great number of proposals exist on the way of making *a decimal coinage*, but only *one* on the way of making *our coinage decimal*; this *one* is the pound and mil system, which we advocate. One of the new systems reckons upwards from the farthing, another from the halfpenny, another from the penny. Other new systems severally propose as the principal coin of account, the half-crown, the four shillings, the crown, the eight shillings, the ten shillings, the twenty-pence, the guinea. Of these, the system which begins at the farthing has received some slight support, and, when the Committee of the House reported, seemed more likely than any other to be prominently brought forward. The ten-shilling coin has had several advocates, but it has not obtained sustained notice. The only proposal which maintains itself within view is that which is now usually called the *tenpenny system*, in which the penny is retained, and the chief coins of account are the penny, a coin of the value of ten pence, and a coin of the value of a hundred pence. This proposal is not supported by any association known to the public, and owes its maintenance entirely to the exertions of a few able writers.

In the *Journal of the Society of Arts* (vol. iii., No. 142, August 10, 1855), is a large and useful list of publications on the subject. The following is a summary. The writings which advocate the pound or a unit decimally related to it are marked A; the penny, halfpenny, or farthing, B; other proposals, C:—

Number of publications advocating	A.	B.	C.
From 1784 to 1852	13	7	4
1853	11	3	0
1854	21	7	2
Half of 1855	18	10	4
From 1784 to 1855	63	27	10

The International Association has published nothing, as yet, concerning the plan of weights, measures, and coinage which they propose. It would obviously be futile to procure a congress of nations to settle the point, until a well-formed national opinion prevails on the principles, at least, of the measure. It is believed, however, from the opinions expressed by some of those who have made the greatest exertions in favour of the scheme, that the French system, called the metrical system, would be the one proposed, so far as it depends on the association. Now, without expressing any opinion on the ultimate practicability of an international system, it is obvious that the plan, be it what it may, cannot be brought into action for a great many years. No one government has yet been induced to communicate with any other on the subject; no one nation has expressed, as a nation, any wish to co-operate. The scheme is of such magnitude, and the time required is so great, that the trifling alteration which the pound and mil plan proposes in our existing coinage does not constitute an opposition. Were we persuaded that the international plan was at hand, and to arrive at the end of twenty years, which is much more than we can think possible, we should still advocate the decimalisation of our own pound, in the belief that it would pay its expenses over and over again within that time. Our readers will remember that *our coinage is decimalised as soon as the coins below sixpence are lowered by proclamation four per cent.* In that case 25 farthings go to the half shilling, and 1,000 farthings to the pound. It may be convenient and desirable to introduce the cent of 10 new farthings, and to issue no more half-crowns, but these are the contingencies of the measure, not its essential parts. Again, it is to be observed that decimalisation, in itself, is a step towards a certain international scheme. Any such scheme would certainly be decimal, and a nation which, in the period of agitation, taught itself how to use decimal money, would have advanced a considerable step. Such a nation would, when the final change came, have to change one decimal scheme into another, altering only the absolute and not the relative values of its coins. Such alteration would be easier than the change of both absolute and relative values at once. If the international plan had met with something like acceptance, and if it had been seen that twenty-five years must produce a plan which would then be set in action, we are satisfied that even though the pound and mil plan had never till then been heard of, it would have been proposed, as an *interim* measure, to decimalise our present coinage, so soon as it should have been seen that such decimalisation required nothing more than the lowering of all the coins under sixpence by four per cent. Should the International Association become a large one, a party will see that this is desirable, and will make the rest see it too. The same party would see it all the more when they reflected that, even according to the views of their original promoters, men sanguine enough to hope for international weights and measures, there is great reason to doubt the possibility of international coinage. The decimalisation of our own coinage is a bird which may be in the hand as soon as we please: the international scheme is a thing of which it is by no means certain that it is a bird at all; and which is not even in the bush, but at the top of a very high tree.

If all the nations in Europe had a common coinage to-morrow, it would no doubt be a convenience to travellers: not all the convenience they expect, unless all the coins were really struck at one mint. Local suspicions and dislikes would produce difficulties, unless all governments united in making the coins of all the countries a legal tender, both between foreigner and native, and also between native and native. This would be very difficult to make a permanent rule. The exigencies of governments have in all time past produced deteriorations of the coinage, and there is no reason to suppose it will be otherwise in time to come. As soon as the deteriorations commenced, the law of legal tender would become a dead letter, and the *agio* which would exist between the various couples of deteriorated coins would be just the old difficulty over again. That is, the inexperienced traveller would never know what he was to have for his money. We are satisfied that many internationalities must accompany or precede international money of permanent practicability; among others a common standard, and an international law for the preservation of the purity of the coinage, any infringement of which should be, if persisted in, a *casus belli.* Further, the commercial difficulties of calculation would not be materially lessened by any circumstance of international coinage, except its *decimality:* and we need not say that we look forward to as practicable, and advocate as most desirable, a complete coincidence of all coinages in this respect. If England and France had one coinage, exchange would not cease. The rate of exchange, that is, the value of the coins of one country measured in those of another, depends partly upon the current necessity of paying debts from one

A 3

country to the other, which regulates the demand for the bills of exchange by which those debts are to be paid. This would make a difference between the coins, though they were of the same weight and fineness. Without entering into the complexities of exchange, any reader will see that a franc in London, which has to be sent to Paris, is not worth so much as a franc in Paris, by the price of carriage and insurance. The majority of persons not engaged in business which brings them into contact with foreign bills of exchange do not understand this point, which may be sufficiently explained without technicalities, as follows:—

French and English money being exactly the same, say francs, suppose we owe a thousand francs to M. Durand, at Paris. If we remit a thousand francs in silver, we must pay carriage, and either stand risk, or pay an insurance. Now suppose that M. Lenoir at Paris owes Mr. Smith in London a thousand francs, and has remitted an acknowledgment in the form of a bill of exchange. If we pay Mr. Smith a thousand francs, and buy Lenoir's bill of him, we need only send this bill to Durand at Paris, who can there claim a thousand francs from Lenoir. But Smith, by the competition of the market, if there exist many transactions in which money must be sent to Paris, will not let us have Lenoir's bill at 1,000 francs. Since it saves carriage and insurance of coin, he will find those who are willing to give him a part, at least, of the expense saved. If, then, by the circumstances of the market, we have to give 1,017 francs 35 centimes for Lenoir's bill, instead of setting down the 17 francs 35 centimes as expense of carriage, &c., the language of commerce would impute an imaginary difference to the coinages, which we have supposed to be identical in weight and fineness; it would be said that 1,000 francs Paris currency are worth 1,017 francs 35 centimes London currency. And, so long as any rate of exchange existed, English money would have to be converted into French, and French into English, by an arithmetical operation, just as now.

If all the coinages were decimal, tables of logarithms would be used with great effect, especially in the more complicated questions in which the rates of three or more countries are involved. But for this purpose it would be of little or no consequence whether the coinage were the same or not; that is, so long as they all agreed in being decimal. If the coins were all of the same weight, the *dictionary* part of the business would be a little easier; most of the different logarithms wanted in a question would lie near the same place in the book.

We are satisfied by the above reasons, and others, that an international coinage would not produce anything near to the conveniences which are expected from it; that it would be a measure requiring a great many years to carry into effect; that by far its greatest advantage would be the universal decimalisation of coinage which it would produce; that this advantage is independent of the absolute values of the coins in different coinages; that every nation which decimalises its own coinage makes a great step; and that, in our own case, even supposing the ultimate adoption of the international plan to be contemplated, the decimalisation of our existing pound is well worth while, as a convenience during the growth of feeling in favour of the international scheme, and as one means of directing a sound formation of opinion upon it.

It seems almost to be taken for granted by the promoters of the international plan, that the French system should be adopted. This system, so far as concerns the actual coinage, would have serious inconveniences. A silver standard we hold to be inadmissible; and if our largest coin must be gold, a gold coin of 10 francs would be far too liable to wear, and a gold coin of 100 francs would be impracticably great. The awkward expedient of such a primary coin as one of 20 francs, which cannot be admitted into pure decimal accounts, would certainly not find favour in this country, unless the advantages of the other parts of the scheme could be put forward in a manner of which we have no expectation whatever.

With regard to the establishment of common weights and measures, though it is very evident that important advantages would result from it, and that the measure itself is within possibility, yet there are circumstances about the existing movement which would prevent our joining it at present. We feel, in common with many others, a suspicion that the International Association is practically pledged to the French or metrical system: and we are by no means prepared to desire this metrical system without further inquiry. We suspect that the support which the Association has received in France is due to the French supporters trusting this practical pledge as much, at least, as we believe in it; and we think, that if a European congress were proposed to debate the *periculum et commodum* of the whole subject, France would not be prepared to acquiesce in any conclusion which would abolish her present system. Moreover, as in the case of the money, so in that of weights and measures, we hold that decimalisation is the most important of the advantages which would be gained, independently of the actual weights and measures. Setting the advantages of the decimal division apart, it is to be inquired into, how much of the whole advantage is thus set apart, and how much remains to be attributed to the simple uniformity of absolute values. We are inclined to suspect that the true answer would be enough to cool the zeal of those who support the whole change, and to induce them, looking at what is practicable within a lifetime, to turn their energies towards the complete introduction of a decimal system into our own country. This distinction is never made in the writings which support the international view. It may not have been clearly seen, but from the time it is put forward, fairness requires, and no doubt will obtain, the recognition of this great division of the question. There are two points under consideration: the absolute weight, measure, or coin; and the manner in which it is divided into smaller ones. As to the latter point, the world is well agreed; measuring ought to be brought to agree with counting—*by tens.* If all the world proceeded by tens, a great point of uniformity would be gained. If, in addition, all the world counted *the same things* by tens, complete uniformity would be gained. We neither use the same measures as others, nor do we proceed by tens. Of all that we should gain, if we did both, how much would arise from our counting by tens, and how much from our using the very same units of counting which others use. It is incumbent upon those who advocate the complete international system to show us that so much of the advantage depends upon the *second* point (which we cannot effect without great inconvenience), as may furnish a very decided balance to that inconvenience. The first point may be looked upon as conceded.

We now come to the consideration of the tenpenny system. As there is no association in favour of this plan, we are obliged to take the account of it from individual writers, of whom there are several who agree with one another in the chief points. With these writers, the main reasons for throwing away the pound from accounts, and the shilling from the easy connexion with accounts which the pound and mil system gives it, are first, the assumed importance of the penny; secondly, the robbery, as it is called, of lowering the existing copper money four per cent. by an Act of Parliament. In order to avoid this, the following plan is proposed:—Accounts are to be kept in *tenpences* and pence; or, supposing we call the tenpence a franc, 2*l.* 17*s.* 11*d.* would be represented by 69 francs 5 pence, or 69·5 francs. The pound and shilling, thus thrown out of all easy relation to account, are to remain as coins until they are gradually absorbed by the Mint, which is to issue francs, and, according to some of the writers, a gold coin of *twenty* francs. And, as silver is not to be a legal tender beyond 40*s.*, a person having to pay any total of account, as 387 francs 7 pence, must find out how often 24 is contained in the francs, to know what sovereigns he is to pay; he is then to make up the remaining francs out of his stock of mixed shillings and francs as he best can. Some of the advocates of the tenpenny system recommend that money should be kept *sorted*, shillings in one purse or partition, francs in another. A man unused to mental arithmetic is to have his wages put into his hand in a mixture of shillings and tenpences, with odd pence, if need be: for we do not at all believe in the will or the power of the shopkeeper or the payer of wages to keep a separate assortment of francs and shillings; nor do we think such a contrivance would be of much use.

For ourselves, we do not believe in the possibility of any system coming under serious discussion, in which *payment* is anything but *counting down*. We do not believe that, even for a great advantage, the people of this country would consent to make the passage of money from hand to hand an affair of mental arithmetic. But when it became generally known that, to procure the disadvantage of never receiving and paying without calculation, all our fixed ideas of money were to be abandoned, we believe the community would soon satisfy any commission which should recommend the scheme, or any House of Commons which should decide in its favour, that it is much easier to alter our copper by four per cent., than to oust the pound from our accounts, and the shilling from our thoughts.

The merchants and bankers of the City of London, in one of the most influentially-signed petitions which ever emanated from the City, presented in 1855, use language on this subject which is hardly that of petitioners. They say, that "the pound constitutes an English national fixed idea of value and position, and is associated with every existing contract, and every comparison of past revenue, expenditure, and price, and *must be retained*." They say also, that every other method, except that from the pound downward, is altogether impracticable. We are so perfectly convinced that their opinion is correct, and we feel so sure that the general assent of those who have attended to the subject goes with it, that we do not apprehend any consequence from the vigour and ability with which a few writers urge the tenpenny scheme, except a handle given to Ministers to stave off the trouble and responsibility of actually bringing into operation the only plan which has power of success.

The progress of this question is one of those remarkable signs which show the tendency of our present politics. If the classes from which legislators and ministers are usually chosen intended, and were justified in intending, to train up the country into a pure democracy, in which the *comitia* of the people should be the real executive government, and Parliament should exist only for the registration of their conclusions, they would deserve admiration * for the steady and cautious manner in which they are and have been proceeding. Instead of having and acting upon a fixed opinion as to what is wise and right, or even professing to have it, both the Cabinet and the House of Commons give the country to understand that their business is only to apply their best endeavour to carry into effect such measures as are dictated by a sufficient pressure from without. It is held to be a good answer to a demand for the consideration of an important question, that there is no outcry for it in the country. Even the House of Commons cannot put the government in motion without this outcry at its back. A few days after the House had voted a strong resolution in favour of the pound and mil scheme, urging its immediate adoption, the Chancellor of the Exchequer informed a deputation that the government intended to—do nothing. Second thoughts, aided, we believe, by the appearance of an intention on the part of some members of the majority to appeal to the House again, produced the Commission of which we have spoken. And yet the subject had been considered by two scientific commissions, and by a Committee of the House of Commons, who had all given an unhesitating opinion in favour † of the immediate decimalisation of the pound. The reason is plain enough. This

* In every point but one. They seem to have a leaning towards an *uneducated* democracy.

† The following is an extract from the speech of Mr. William Brown, in moving the resolutions passed by the House of Commons in 1855:—" The petitions which have been presented are:—Corporations, 31; Inhabitants of Towns, 26; Chambers of Commerce, 16; Scientific and Literary Institutions, 11; Mechanics' Institutions, 15; Schoolmasters, 3; Actuaries, 1; Merchants, Clerks, and Operatives, 22; Sandhurst College, 1; Merchants, &c., 5; Benefit Societies, 1; Insurance Companies, 1; total petitions, 133; signatures, 15,379; but 56, which were signed officially, represented some very large bodies. There are two or three that require especial notice. That from the City of London is signed by 59 bankers, stock-brokers, and money-dealers; 38 merchants, foreign and general; 28 merchants, tea, sugar, spice, coffee, grocery, tobacco, &c.; 22 merchants, wine and spirit; 39 warehousemen, silk, linen, cotton, woollen, lace, ribbons, hosiery, straw-plait, &c.; 13 shipowners, brokers, wharfingers, &c.; 5 miscellaneous; 45 assurance offices; 944 signatures. It commences with the governor and deputy-governor of the Bank of England, and other banks; the chairman and deputy-chairman of the Stock Exchange, and 'many leading houses of great importance (none of their clerks were asked to sign), who if trouble was apprehended in the change, would not petition for it.'

question never did, and never will, cause a political excitement. No electors will make their votes depend upon it—no members will make their general support of government depend upon it—it will never raise a mob in Hyde Park. It is a good which no man has a bad interest in: it has therefore nothing but the slow march of mind to trust to. In such a state of things, a few writers may retard the consummation most to be desired by promulgation of a plan which, to the general run of men of business, of politics, and of science, appears not only utterly inferior to that which finds general support, but wild and impracticable in itself. We can illustrate the length to which mere opposition will go, in the following way. If two persons, proposing the *very same* coins, should call them,—the one bears, lions, and tigers,—the other prawns, oysters, and lobsters,—no man of common apprehension, far less a tried scholar and logician, would pronounce that he had two *plans* of coinage before him. He would say that the two persons advocated the same plan, but differed in their nomenclature. But the Chancellor of the Exchequer, arguing for delay in the debate of 1855, and arguing from the variety of the "plans" proposed by different persons, sets out, under four different sets of names, as four * different plans, the farthing system, which proposes coins of ¼*d.*, 2½*d.*, 2*s.* 1*d.*, and 1*l.* 0*s.* 10*d.* Thus in one system a farthing is called a *cash*, in another a *mil*, and so on; it being the same farthing throughout. A Minister who could thus play with statistics, would be very apt to make capital—the capital where interest is delay—out of the existence of five writers urging an opponent scheme with some system and more ingenuity. This is, with all personal respect for those writers, our sole reason for serious argument against the tenpenny scheme. But for this reason, our only argument against it would be the endeavour to make better known the pound and mil plan. We feel perfectly easy in our reliance on the common sense of the country, that it will not hear of the expulsion of the pound sterling from accounts, while the sovereign is to be retained as means of payment, after division by 24; that it will not hear of the mixed circulation of shillings and tenpences; that it will stick to its old and successful plan of reforming that which is, instead of substituting that which has never been, especially in matters connected with our oldest habits of estimation, usages of action, and associations of thought. But when the tenpenny plan, wild as it is, may become the drag-chain by which a reluctant government may impede the plan which *can* succeed, it assumes an importance which it never would otherwise possess; unless, indeed, a large association of men engaged in business, and desirous, for their own sakes, of the division by 24, and the mixture of silver coins of nearly equal value, were to raise its head among us.

It requires but little attention to realize some of the inconveniences which would attend a mixed circulation of francs and shillings: but it is likely that there are other inconveniences which would not be discovered, or at least not appreciated, until the change actually came into practice.

* One of these he attributed (by mistake, soon acknowledged,) to Sir John Herschel, a decided advocate of the pound and mil scheme. It is desirable to state this, as some persons may never have heard Sir J. Herschel quoted in any other way.

As we now are, the French franc is occasionally passed off for the shilling, sometimes by fraud, sometimes by mistake. A mixed circulation of francs and tenpences would raise the question—"Is this a franc or a shilling?" every time either coin was paid or received. Unless indeed the franc were made sensibly thicker than the shilling, so as to be smaller to the touch; in which case it would not ring, and the most important test of goodness would be sacrificed. And the question would be constantly raised in the dark, or by the light of the street lamps, when it would be difficult to answer with any practicable investigation. We do not remember any allusion to this difficulty in the writers who advocate the tenpenny plan. If anybody could show us that this plan was the best in all other respects, we should be loath to adopt it without making the recal of the shillings a condition. The inconveniences of the joint circulation would make it worth while to face the expense and trouble of an immediate abolition of the shilling. But this abolition is a measure before which the advocates of the tenpence recoil. They know that the mere proposal would be destructive of every chance of gaining further hearing.

The number of shillings in existence is reckoned as not much under 120 millions. To recal them at once is held by all parties to be impracticable. To abolish them gradually, by issuing francs as fast as coin is wanted, and melting down the shillings as fast as they come in, would probably require many years, ten at least, more probably twenty. Indeed, it is difficult to say how long it would be before the voluntary return of the shillings would have brought them all home. Though the silver was recalled in 1699, by authority, there are persons alive who remember that some coins of Charles II. were in circulation just before the recal of 1816. It is not, however, the mere recal which would be the difficulty, were it absolutely necessary. A small premium on the restoration of shillings to the Mint, given to the postmasters throughout the country, with transmission post free, and a supply of francs duly furnished, would bring in nineteen out of twenty shillings in a couple of years. The difficulty is the supply of francs. The shillings being, say 110 millions, 132 millions of francs would be wanted. It would not be impossible to coin 100 thousand francs a-day: but if this were done for 300 days in each year, it would take more than four years to furnish the requisite supply. Substitute the practicable for the not impossible, and allow time for 84 millions of fivepenny bits to replace the sixpences, and it will appear that ten years of high-pressure work at the Mint would be necessary. If, however, the shillings and sixpences could be replaced by an equivalent amount in tenpenny and fivepenny bits, in the course of a single night, and by the wand of a magician, the country would find itself in a poor state in the morning. Every idea of value would be upset: all notions of cheapness and dearness would require translation. A man who had made up his mind over night that he would go as far as 1*l.* 17*s.* for a purchase, must take pen and paper, if not more ready than usual, to find out how many francs he may venture upon. And this would last for years with many, for months with all. And for what? To avoid an alteration of the copper coin, which would amount to a little short of one farthing in sixpence.

What the uneducated man is to do in the case of the actual tenpenny proposal, in which shillings and francs circulate together, we cannot even conjecture. On the only public occasion in which an advocate of that system met the advocates of the pound and mil system in debate on this point, he proposed that even the bankers and shopkeepers should keep their coins sorted. But the poor man has no stock worth separating, nor indeed would the separation avail him much; for he must use his shillings at last, and use them in tenpenny subdivisions. The most important person, in considering the *difficulties* of the change, is the person who can just count money and put sums together into a whole, and no more. The class to which this person belongs is the largest in the community, and the least competent to face an arithmetical difficulty. Their share of the pound and mil scheme is, that they must remember that the half shilling has an additional farthing: their contribution to the simplicity of the change is the lowering by four per cent. of the copper which happens to be in their pockets at the moment of the change. One of them might be addressed by an advocate of the tenpenny system as follows, not without irony, but with perfect truth:—

"My friend, there are a set of people who, for their own commercial convenience, are meditating a robbery of your pocket. They want to declare the copper coin less in value by four per cent.; they want to make 25 farthings go to the half shilling, now six pence. This will probably cost you a fraction of a farthing, perhaps a whole farthing, or even more. If you happen to have by you six copper pence on the day when they visit your pocket, you will find that you want a lowered farthing, 24 parts out of 25 of the old farthing, to make up the silver bit which those coppers would have brought you five minutes before their robbery was committed. It is true that for this depredation they will introduce your children to a much better chance of rising in the world of business. It is true that they will save your children's time, their only property, and set many an hour free for better improvement than mastering rules of arithmetic which seem made on purpose to be perplexing. They say also, and we cannot deny it, that you will, in most cases, get as much for four lowered farthings as you did for four of the old value. It can hardly be that the competition of the shops will allow prices to be raised merely because 100 farthings now mean only what 96 farthings did mean. But still you are robbed. Now, we have got a plan which will give all the advantages and save you from losing a fraction of a farthing, though we confess that, to make your penny a fixture, we shall have to make your shilling a puzzle. You shall have coins of tenpence each mixed with the shillings; your wages shall be handed to you in mixed tenpenny and twelvepenny bits. You get seventeen shillings a-week, perhaps. Now, one week your master will hand you 13 shillings, 4 tenpennies, and a sixpence, and two pence, which, of course, you see in a moment is 17 shillings. Next week you may have 11 shillings, 7 tenpennies, and 2 pence, which you also see must be 17 shillings, without any trouble. In this way you will go on, probably all your life, giving and receiving mixed coins. But then you have saved your farthing, or its fraction. Your master will keep his accounts in tenpennies. Your wages are 20 tenpennies and 4 pence. If he should happen to dislike odd payments, and should think 20 tenpennies enough for you, you lose 4 pence a-week, or else you strike. The people who want to rob you say you had better bear the loss of one out of 25 upon the coppers you happen to have by you on some one day than take the chance of such a contest. It is for you to choose; but we assure you that the penny, not the shilling, is the coin on which your money dealings depend. You can better afford the shilling to be made a confusion to you all your life long, than bear a change in the penny, even though slight and once for all !"

Such would be the account given of the tenpenny plan, if it came into serious discussion at a mechanics' institute. The advocates of the pound and mil would get a bowl of black and white and red and blue counters, to represent shillings and francs and sixpences and fivepences, and, ruthlessly refusing to let the tenpenny men *sort them*, would challenge those men to come forward to the table and pay miscellaneous sums out of the chance-handful. Then, if the challenge were accepted—which, perhaps, it would not be—would all present see the public advocates of the tenpenny—able speakers and clear thinkers, as would have appeared—hammering and reckoning, and puzzled by hang-fire calculations, *instead of counting*, before they could tell out 17*s.* or 8*s.* 6*d.*, or 15 francs, or 10 francs 7 pence, &c. And then all would see them picking out the shillings, and avoiding the francs, or *vice versâ*. And then would the real consequence of the issue of francs be foreshadowed.

For it is clear enough that when these troublesome tenpences first made their appearance, one or two at a time, there would be a dislike to them, on account of the break they would cause in the counting of the shillings. The shopkeeper must take them ; but he would find that he cannot give them—he must please his customers. They would come back to the Mint with a mark of rejection upon them. The nation can refuse a coin, and has done so. If the government persisted in the issue, the opposition would take a form which no government could resist.

Again, the tenpenny system would not *necessitate* decimal reckoning. It might be made the law that accounts should be kept in tenpennies. Accordingly, the pleaders and solicitors would convert litigated sums into tenpennies, which had never seen themselves in that dress before. Or, supposing that no debt could be sued which had not been first demanded in tenpennies, every account rendered would be expressed, in all its items, by tenpennies to satisfy the law ; and, in another set of columns, by pounds, shillings, and pence, to satisfy the debtor. After a while, this useless form would cease, the law would evaporate into desuetude, and the francs would be melted into shillings. There is no such love of decimal calculation in the half-instructed mass as would enable a decimal system, without either pound or shilling in it, to fight its way against the familiar ideas,

The like difficulties do not stand in the way of the pound and mil system. Here the paying coins are all those to which we have been accustomed, with no change except that the copper is lowered, so that 25 farthings go into the half shilling. The new cent of ten farthings or mils, will probably be a convenience; if not, the public may reject it, without in the least hurting the rest of the system. It is not essential to decimal *calculation* that ten mils should travel in one parcel. And beyond avoiding the cent, if his customers dislike it, there is nothing which the shopkeeper can be asked to do.

Again, the pound being decimally divided, decimal calculation may be left to work its own way: it will not be essential, though it may be judged desirable, to make florins, cents, and mils the only legal subdivisions of the pound. Turn the pound into 1000 farthings, with 100 to the florin, and decimal calculation becomes easier than any other. Let any one who chooses reckon in pounds, shillings, pence, and farthings, at 4 farthings to the penny, 12½*d.* to the shilling, and 20*s.* to the pound. He will soon learn his mistake. Turn the pound into 1000 farthings, and leave the florin, and no Act of Parliament could prevent decimal calculation.

It is unnecessary to repeat what we said in the *Companion* for 1855, on the method by which the farthings of the existing money are replaced by their lowered selves in the proposed money. It seems to be admitted that people may manage the transfer, subject only to loss or gain of a fraction of a farthing. The ground of opposition on this point is a determined endeavour to persuade the public that people will continue to think, to estimate, and to buy and sell, in old farthings, after the new ones have taken their places. What, asked a member in the debate of 1855, is a poor applewoman to do, when the penny (meaning the present penny) is ·004166666 *ad infinitum* of a pound? The answer is, that the applewoman will know no more about the penny being ·0041666 *ad infinitum* of a pound, than she will know about the new penny being

$$\frac{8}{\pi}\int_0^{\infty}\frac{\sin xv^2}{v}\,dv$$

per cent. below the old one. She will know, without translating her thoughts into definite integrals, or decimals either, that the penny is a little lowered, so that to sell at four a penny, new money, what she used to sell at four a penny, old money, will leave her a new penny short on every hundred apples she sells. As the penny and the apple adjust their mutual dealings every year without trouble, they will manage to do the same, even though the penny should undergo a permanent change, once for all, not amounting to the fifth part of the common fluctuation of apples from harvest to harvest. Some existing contracts may require that, so long as they last, the items shall be computed in old money, and the results translated into new money. If every contract existing at the moment of the change were to make such calculation desirable during its continuance, it would be but a transient and partial prolongation of the old system in certain processes subsidiary to accounts. Nor would so much trouble be caused by it as by the innumerable converting of pounds into tenpences, which the tenpenny system would require at the time of its adoption.

The prospects of the question seem to us to be, that a decimal system, and that system founded on the pound, will surely be adopted at last. But how long it may be before the measure is carried, depends partly on the inertia of the government, partly on the inertia of the educated world. That any government we are likely to have will put it off as an evil day, is all but certain. So long, at least, as government is overworked, and perhaps longer, the question which excites no political feeling, and receives no political support, must stand by till its betters are served; and its betters grow up session by session. But what can be expected more than that the government should reflect the people, and truly represent their modes of thought and action? Who is ignorant that the community is a perfect community, striving hard, or at least crying out, with discouraging success, for a legislature and a cabinet worthy of itself? Why then may not the government possess one of the perfections of the people? The apathy of the Ministry is exceeded by the apathy of those who, with knowledge to understand what is wanted, and perception of the greatness of the benefit to be received, cannot bring themselves to take any steps in its favour. In June, 1855, the question was within a hundred petitions of succeeding. As it was, the victory obtained forced nothing but a royal commission. Nevertheless, this is a step, and we suspect it will turn out to be a greater step than was intended.

University College,
October 21, 1856.

A. De Morgan.

www.ingramcontent.com/pod-product-compliance
Ingram Content Group UK Ltd.
Pitfield, Milton Keynes, MK11 3LW, UK
UKHW022029190726
13853UKWH00005B/2172